# LAWS AND MODELS

## SCIENCE, ENGINEERING, AND TECHNOLOGY

# LAWS AND MODELS

## SCIENCE, ENGINEERING, AND TECHNOLOGY

Carl W. Hall

CRC Press
Boca Raton London New York Washington, D.C.

**Library of Congress Cataloging-in-Publication Data**

Hall, Carl W.
Laws and models : engineering, and technology / Carl W. Hall.
p. cm.
Includes bibliographical references.
ISBN 0-8493-2018-6 (alk. paper)
1. Science Handbooks, manuals, etc. 2. Technology Handbooks, manuals, etc. I. Title.
Q40.H35 1999
500—dc21 99-29327
CIP



Direct all inquiries to CRC Press LLC, 2000 N.W. Corporate Blvd., Boca Raton, Florida 33431.


International Standard Book Number 0-8493-2018-6
Library of Congress Card Number 99-29327
Printed in the United States of America 1 2 3 4 5 6 7 8 9 0
Printed on acid-free paper

# PREFACE

As distinguished from the laws of society, physical laws have been developed to describe or represent the response of a subject (person, elemental particle, earth, moon, stars) in a certain natural (chemical, biological, physical) or a controlled environment, and expressed in such a way that they can be used to predict the response. The law that is general or covers a wide variety of situations is useful in describing the parameters of less general and more specific conditions. Laws are identified according to the phenomena and/or the person(s) who first stated them, or for whom the relationships were named. Where possible, this book identifies the statement of the law with the name(s) of the person(s) expressing the law, or to person(s) whom colleagues wish to honor. It is not a simple matter to identify the person after whom a law is named. Which Bernoulli? Which Curie? Was it Wentzel or Wenzel? Was it Fisher or Fischer? And when two or more names are used to identify a law—was one person a student of the other? Often, little biographical information can be found on these early investigators after whom laws are named, unless they continued in the field. In the early years, there was a penchant for secrecy of one's findings, which might become known only after several years, because they appear in a foreign-language journal or are included the scientific notes of a deceased person. Sometimes, the earlier persons identifying a phenomenon or relationship are recognized by having a law named after them, and sometimes not. It is not unusual for a law to be named after the person who publicized it rather than the individual who expressed it earlier.

My collection of laws began 40 years ago. Laws were commonly espoused in mathematics, physics, and chemistry books, and I found them intriguing and useful. I began to see laws as primarily developed by scientists as a method of describing the results of their work. Many of these laws, such as those of Kepler, Newton, Bernoulli, etc., are fundamental to scientists and engineers. However, many relationships of the natural world with which the engineer worked did not fit the accepted definition of laws identified by scientists at the time. There began to appear in the literature very useful relationships in the physical and chemical worlds, and belatedly in the biological world, described as *models*, identified by some writers as *laws*, that were not as general as the previous concept of laws. A special effort has been made in this book to include laws and models pertaining to medical, physiological, and psychological laws in fields that are becoming important in engineering. In the literature, one can find numerous examples of laws described as axioms, canons, effects, principles, rules, theorems, or theories, where these can stand alone. It is interesting to note that a law is often described in terms of a principle, a theorem, or an effect. Those not incorporated under laws have been identified as models. One can think of the world being represented by laws as described by scientists and models as used by others, particularly engineers and technologists. It is common that a particular relationship is called a law by one author and a rule, or some other term, by another author. I have included these various designations (axiom, effect, postulate, principle, rule, theorem, etc.), in addition to laws as represented in the literature. No attempt has been made to adjudicate which designation is most appropriate. Dimensionless groups, such as named after Reynolds or Mach, and called *numbers,* form a very important category of models, of which at least 100 have been included. These numbers form an important category of relating variables. In this collection, I have identified as *models* those scientific relationships not given the designation of laws. The laws and models are listed in strict alphabetical order, omitting apostrophes, brackets, commas, and dashes.

Empirical laws are frequently used to describe relationships, in contrast to theoretical laws. Empirical laws, such as the laws of Kepler, are frequently explained by theoretical laws but are only approximately true or accurate and refer to directly observable objects and properties. Theoretical laws, such as the Avogadro law, at least until the present, have not been open to direct observation and measurement.

A distinction needs to be made between universal and statistical laws. A universal law asserts a regularity that holds without exception, such as the law of conservation of charge (the amount of electric charge in an isolated system is always preserved). Statistical laws state that something happens in a certain percentage of cases—such as tossing a coin for heads or tails. Many of the more recent laws, such as in quantum mechanics, are based on statistical relationships. Statistical relationships are often difficult to express without mathematical relationships that appear complicated. For this compilation, these are written in general terms to convey the essence of the law rather than with detailed mathematical expressions, which can be found in the sources cited.

Models are often used to represent variables in a relationship that otherwise cannot be manipulated, or can be manipulated only with great difficulty or major expense. Thus, fields of endeavor dealing with the atmosphere or geological formations or large complicated systems can be modeled. The result could be that a model, perhaps due to its applicability, at a later date earns enough acceptance of people in the field to be called a law. No attempt has been made to distinguish among law, rule, theorem, model, etc., except as used or cited by people in the field. Brief biographical information of the persons after whom the law or model is named is provided, and sources of information are provided for the scholar who wishes to pursue the subject. Cross-references help guide the reader to related laws and models. An attempt has been made to minimize extremely complicated mathematical equations so that the reader can gain the sense of the law or model without knowing all the details. Extensive sources are listed to assist the interested reader in learning more about a law or model. In some cases, an entire book can be written on a specific law or model, so to expect the subject to be completely covered here would be unrealistic.

With respect to mechanics or style, some decisions were made to streamline the manuscript, and to make the presentations user friendly. The apostrophe was omitted from most of the laws. It is a usual practice to omit 's from names of laws named after two or more people, so why not do the same for one person? This practice also made alphabetical ordering more sensible and saved space. Diligence was exerted in attempting to find biographical data about each person after whom or for whom a law or model was named. Perhaps a former graduate student or laboratory associate was involved with the work but did not continue in the field and was lost in the biographical data. Some of these may be recorded in foreign-language biographies, and even though searched, several names could not be located. Some laws are named after two or more people, unknown to each other, who separately found the same relationship—the first may have originally expounded the law or model, and the second found the same relationship or additional information to verify the original theory.

One is impressed by the breadth of activities of early scientists, such as von Baer; the family connections, such as Bernoulli and the Curies; and the long time some early scientists took to publish their work and the secrecy with which some people worked, thus denying the original discoverers their rightful credit.

# LAWS AND MODELS: AN INTRODUCTION

**George A. Hazelrigg**

From the dawn of recorded history, men and women have sought to manipulate their environment in pursuit of a more comfortable, fulfilling, and safer life. The deliberate manipulation or utilization of nature through the use of energy sources and natural resources, with intended benefit, is engineering. The ancient stone nyrage of Sardina, which remain standing today, attest that such feats of engineering date back over 4000 years. With time, our feats have grown in both number and complexity, and today the process of engineering is conducted by some 2 million engineers in the U.S.A. alone.

All feats of engineering, whether stone dwellings or space stations, require a particular sequence of events. First, the engineer must understand the needs and wants of the society or subgroup of society that is to be served. Second, the engineer must formulate concepts of potential designs that might serve the designated needs and wants. Third, the engineer must analyze the concepts to determine their functionality. Fourth, the engineer must optimize selected candidate designs and choose a single preferred design. And fifth, the engineer must design a production system to realize the selected design.

More recently, engineering does not stop even here but continues through the life cycle of a design, even including disposal at the end of the product's life. Throughout this process, the goal of the engineer is to find a design, a production system, and a context for use of designed products, processes, and services that can effectively and efficiently realize the goals imposed by the underlying needs and wants of society. And to do this requires that the engineer be capable of predicting the future—the future wants and needs of society, the future context within which products will be used, and the performance of the as-yet-undesigned products themselves. The engineer must, within reasonable bounds, be able to determine how every proposed design would perform were it selected, and from this make rational choices among design alternatives. This is quite a challenge and leads us into the forest of modeling.

In performing the design process, all available knowledge is drawn upon so as to evaluate the various parameters of a functioning process or design—as might be represented by mathematical, physical, biological, and chemical relationships that are often represented by laws or models. At the same time, additional knowledge is sought, as needed or as perceived as needed, to satisfy the design process.

## A SYSTEM OF LOGIC

Beginning as early as the ancient Greeks, man sought to create a system of logic within which better understanding and predictions could be facilitated. This system has become known as *mathematics*. But, even to this day, we do not have a clear understanding of what mathematics is and why it is such a powerful tool. Barrow (1991) proposes four views of mathematics that are quite insightful.

The first view he calls *formalism.* This view defines mathematics to be "nothing more or less than the set of all possible deductions from all possible sets of consistent axioms using all possible rules of inference." In this formalism, truth is defined as a statement that is consistent with a set of axioms and the web of logical connections they imply. The ability of mathematics to describe or predict nature is purely coincidental, and we get little insight as to why mathematics works.

The second view is *inventionism.* This view "regards mathematics as a purely human invention..., it is a product of the human mind. We invent it, we use it, but we do not discover it." This view leads, indeed, to a concept of why mathematics works. In the evolution of the species, those species whose brains could distinguish between past and future, and between cause and effect, and that could make predictions in concert with nature, were more likely to survive and reinforce the gene pool. Thus, in this view, natural selection led to a set of species for which logic corresponds well with nature. And man is the culmination of this evolution of logic.

The third view is the *Platonic* view. In this view, mathematics exists and man discovers it. Indeed, the Platonic view holds that mathematics existed before the universe was created and will exist after it ends. The Platonic view holds mathematics in much the same regard as God. The physical world *is* inherently mathematical, and thus mathematical theories describe natural phenomena. A logical consequence of this view is that all physical processes can be modeled mathematically, and hence the universe could be modeled on a computer. Furthermore, it would be impossible to distinguish between the simulation and the physical universe.

The fourth view is called *constructivism.* This view was created at the end of the nineteenth century as a result of the logical paradoxes of modern set theory and the strange Cantorian properties of infinite sets. In this view, mathematics is defined to "include only those statements that could be deduced in a finite sequence of step-by-step constructions starting from the natural numbers, which were assumed as God-given and fundamental." This view of mathematics removes such familiar procedures as argument by contradiction, and thus reduces the content of mathematics considerably.

## PREDICTIONS

Today, in all logical activities, we predict the future using models, either explicitly or implicitly. And all predictive models have their basis in mathematics. Indeed, it is necessary that such be the case in order that resulting predictions be logically consistent with our knowledge. So, we must ask, what is a model? For one thing, a model is an abstraction of reality. Nature is simply too complex to understand in toto, and so we abstract from nature's reality those elements that are important to any given circumstance. A model then provides a logical connection between the elements that we include in our abstraction. Thus, a model is also a relationship between elements that clearly implies correlation, but it may also infer cause and effect. It is only through models, and especially inferences of cause and effect, that we gain an understanding of nature. And we validate our understanding by using models of past phenomena to predict future events.

In engineering, we use models to combine disparate elements of knowledge and data to make accurate predictions of future events. For example, an engineer might want to know whether a bridge of a particular design will withstand a given load. It is, of course, possible to guess at the answer directly. But models give us answers in which we have greater confidence. They do this in a two-step process. First, a model enables us to disaggregate a guess from a single quantity—for example, true or false, the bridge will stand—into a set of parameters, each of which we must also guess, but for which we can make more precise estimates. Then, in a second step, the model provides the framework within which we can unify the data logically and consistently to obtain the desired result with, hopefully, less uncertainty than we would obtain from a direct guess. For example, if we want to know the acceleration of a particle, the model $a = F/m$ enables us to guess at $F$ and $m$ rather than $a$. If it is possible to obtain accurate estimates of these parameters, we can use the model to unify them into the desired result, namely an accurate estimate of $a$. In this view, the value of a model lies in its ability to reduce uncertainty in the estimation of a variable.

This concept is illustrated in the figure below. One's ability to estimate the parameter of interest can be represented by a probability density function (pdf), or simply probability function, on the original guess. The model disaggregates this guess into a set of parameters, each of which in turn must be estimated. The idea is that it should be possible to estimate each of these parameters with much less variance, that is, with much less peaked pdfs. Then the model enables these estimates to be combined with a resulting estimate of the parameter of interest that has a more peaked pdf.

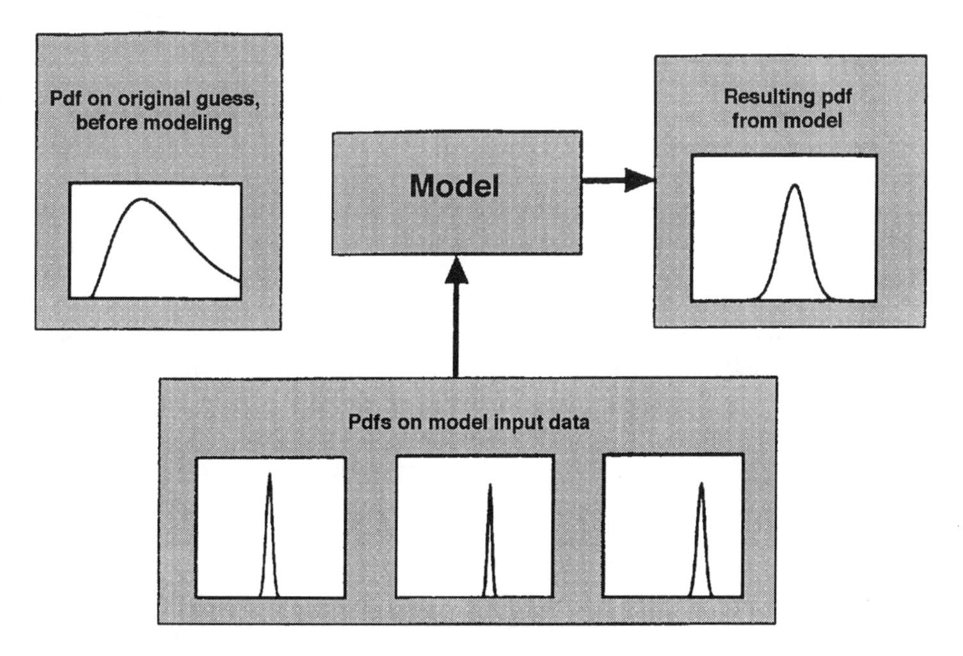


A conceptual view of a model. Reprinted from Hazelrigg, G.A., *Systems Engineering: An Approach to Information-Based Design,* Prentice Hall, 1996, used with permission.

## FUNDAMENTAL ASSUMPTIONS IN MODELING

So far, we have identified that models provide understanding and that they enable improved estimation. But what do models themselves depend on? It is important to recognize that inherent in all models of physical phenomena is the assumption that the fundamental rules of nature, which we call laws, are fixed across all space and time. It is our belief that, whenever we obtain results that differ from those predicted by a model, the reason is (1) the model is incorrect, (2) we have not included a sufficient breadth of elements in the model, (3) in conducting the experiment, we did not isolate the system sufficiently from exogenous corrupting factors, or (4) we erred in collecting the data or in computing the results. We rarely challenge the underlying assumption of invariance of natural laws. Indeed, all physics is based on the assumption of invariance, and the validation of all physical theories uses this assumption; without it, rational prediction would be impossible. But we continually challenge the validity of statements that portend to represent the natural laws, and we validate physical theories by repeated experimentation in temporally and spatially disparate cases.

The need to validate theories and their resulting models by spatially and temporally disparate experiments imposes stringent conditions on new theories. In short, it says that they must be connected to extant theories—they must explain physical phenomena within the context of previous explanations. The conduct of a valid experiment demands that enough be known about the environment that the experimenter be capable of isolating the experiment from corrupting factors. For example, consider the demonstration that gravity acts on rocks and feathers in the same way to produce acceleration. It was long felt that gravity acted differently on different substances. But the understanding that air creates drag, which retards the acceleration of the feather more than of the rock, led to the conduct of an experiment in a vacuum. The corrupting factor was removed, and the experiment produced results that confirmed the hypothesis. This understanding of corrupting factors is adequate only when experiments are conducted on the margins of extant knowledge. As a result, theories tend to build upon one another.

## UNITY ACROSS DISPARATE EVENTS

One goal of science, and indeed that of models, is to unify seemingly disparate events. This is a noble goal, as it extends our predictive capabilities. Note, for example, how models derived from Newton's Laws unify events that span the universe in both scale and distance. They provide that the motion of a feather floating on a pond and the motion of the planets around the sun and, indeed, the motion of the galaxies are all governed by the same concept, namely that the time rate of change of momentum of a body is proportional to the net force exerted upon it. This notion of unification is extremely powerful. Not only does it broaden our predictive capability well beyond the bounds of experience, it also provides a venue for validation of the theory itself.

Models derive from two fundamentally different processes: first, from observation, but also they derive from pure thought—mathematics. Our life process exposes us to many phenomena: sunrise, sunset, warmth, cold, rain, sun, wind, lightning, and life itself. The mere act of observing these natural phenomena in a methodical and careful manner provides us with data from which we can formulate models. But our observations themselves are couched in terms that can be thought of as models, some of which are so fundamental and well accepted that they are recognizable as laws. For example, we speak of *warm* only in comparison to *cold* or *hot*. We imply a model of temperature. As we move to more sophisticated measurements, models become indispensable. The measurement of a voltage implies the existence of a model that defines voltage. At some level, all measurements—that is, all gathering of numerical data—imply the existence of models, and the data themselves have absolutely no meaning in the absence of their associated models. Thus, measurement or data-taking and model building are intrinsically linked and utterly inseparable.

Yet, data themselves are not the basis from which some of the most important laws of nature have been formulated. Consider Einstein's formulation of the theory of relativity. A postulated model of the nineteenth century was that the solar system was moving through an ether. This model held that the speed of light should differ according to the direction in which it travels. Within the context of this model, Michelson and Morley conducted their famous measurements of the speed of light and determined that their hypothesis was incorrect. Indeed, it appeared, within their ability to measure, that the speed of light is the same, independent of direction. Einstein internalized this notion, accepting it as fact. But it is very important to recognize that it was not established fact. He then went on to ask, what are the logical conclusions resulting from the assumption that the speed of light is independent of direction? The theory of relativity is consistent with this assumption and the axioms of mathematics. Thus, the theory emerged from pure, that is, mathematical thought.

## THEORY AND DATA, DATA AND THEORY

A theory that derives from pure thought cannot stand, however, without the test of physical data. And so such data must be obtained through the theory itself. A new theory makes predictions that can be measured. Recognizing this, we can develop experiments to generate the desired data. But there is often a complication. New theories tend to emerge on the margin, near the limits of our ability to make measurements. This is because large discrepancies from extant models are easily detected, and the faulty models can often be discarded and replaced with more appropriate models. Thus, at least at the most fundamental levels, that is, at the level of the fundamental laws of nature, the confirmation of new models demands measurement at the limits of resolution with existing instruments. And so the advance of science with the development of new theories is often retarded by the technology of scientific instruments.

But this is not always the case. Often, new theories emerge around new instruments, and the technology of the instruments leads the science. A good example of this is the telescope, which changed man's interpretation of the universe, or the microscope, which led to the microbial theory of disease, both by direct observation. New instruments often enable new measurements, which lead to unexpected results. An example of this is the invention by Volta of the voltaic cell. This cell enabled the separation of the elements and rapidly led to the discovery of the elements. From this discovery came a whole new collection of models governing chemistry.

In all cases, however, we see that the process of discovery is a process that proceeds on the margins of extant knowledge. We build models on data and collect data on models. We build instruments using models and interpret the readings that instruments give based on models. And the most fundamental of these models we call *laws*.

And such is the subject of this book. Over the past 3,000 years, development and application of physical laws has been a mainstay of the sciences, particularly the physical sciences, and engineering. Beginning with Pythagoras' Law of Strings, we have amassed a rich collection of laws that govern and predict nature in almost all currently observable forms. This book presents that collection in a unique format. Each law and model is given, along with date credited, where available; keywords; authors or persons recognized, with a brief biographical sketch; sources of information; and cross-references.

# AUTHOR

**Carl W. Hall,** Ph. D., P.E., is a consultant, having retired in 1990 as deputy assistant director, engineering, of the National Science Foundation. After three years of infantry service in the U.S. Army, including the awards of CIB and the bronze star, in the ETO during WWII, he resumed his academic studies. He earned his bachelor of science (summa cum laude) degree in agricultural engineering at Ohio State University in 1948; his master of mechanical engineering at the University of Delaware in 1950; and his doctorate at Michigan State University in 1952. He taught at those institutions and at Washington State University, where he was dean of engineering and professor of mechanical engineering, from 1970 to 1982. He participated in the SMG program at Harvard University.

Dr. Hall has held numerous consulting and professional assignments with industry, foundations, universities, the United Nations, and the World Bank. He has served as a national officer and is a fellow of ABET, ASAE, ASME, and chaired the EAC (Engineering Accreditation Commission). He has also been active in NSPE, ASEE, AIBS, is a fellow of AAAS and AIMBE, and has won numerous national and international awards. He is a member of the National Academy of Engineering.

He is author or editor/coeditor of 28 books and has written more than 400 articles and papers. His most recent book is *The Age of Synthesis.* His specialty areas are heat and mass transfer as related to drying and energy.

# ACKNOWLEDGMENTS

Thanks are expressed to the following for their advice and assistance in the preparation of the book: G. I. Barenblatt, Arthur Bergles, Clayton Crowe, Claudia Genuardi, Mildred Hall, George Hazelrigg, Karla Kalasz (illustrations), A.S. Mujumdar, Edward A. Murphy III, and Bob Ross, and many of the public and institutional libraries in the U.S., especially the Arlington Public Library, George Washington University Library, and the Library of Congress.

**Carl W. Hall**

# Disclaimer

Information in this book has been obtained from numerous sources considered to be reliable. However, the editor and publisher cannot guarantee the accuracy or completeness of information. Due to the nature of the book, users of the information may need to go to one or more of the sources of information cited in order to appropriately use it, depending on the application under consideration. The editor and publisher cannot be responsible for any errors, omissions, or damages arising out of the use of this information. The book is provided as a means of supplying information without rendering engineering or other professional services. An appropriate professional should use the information for the design, construction, maintenance, and management of elements or systems.

# Possible First Recorded Physical Law

The first physical law expressed in mathematical terms could well be the Pythagorean Law of Strings:

Pythagoras, c. 582–c. 500 B.C., Greek philosopher and mathematician, stated that harmonic sounds were given by strings whose lengths were in simple numerical ratios, such as 2:1, 3:2, 4:3, etc.

**Gamow, G., 1961, 1988**
**Porter, R., 1994**

# Definitions and Descriptions

*Laws* and *models* can be expressed in a number of ways, as illustrated by the following identifiers. An *effect* or *theorem* identified by one author could be called a *law* by another author. A well known example is the Bernoulli theorem, called the *law of large numbers.* Usage often determines the identification of the statement. For example, Newland Relationship became known as a law; Delta function is not a function; Hamilton Principle is not a principle; and many laws are axioms, such as the Laws of Conservation of Energy. Those relationships not called *laws* are categorized as *models* in this book. Some examples of models are presented here.

**APHORISM**—A concise statement of a principle, truth, or opinion; a maxim; an adage.

**ASSERTION**—A belief or an idea; subject to proof or verification. Mayer Assertion (regarding energy).

**AXIOM, MAXIM**—Self-evident or accepted as self-evident; Newton's first two laws originated as axioms; axiom of choice in statistics; a basic proposition in a formal system presented without proof. Wolff described axioms as indemonstrable, theoretical, and universal, thus belonging to the synthetic. See POSTULATE. Design A.; Grinnell A.; Grice M.; Peano's A.; Playfair A.; Rheology A.

**CANON**—General rule or generally accepted basic principle. Molar Extinction C.; Morgan (Lloyd) C.

**CONCEPT**—An idea. Quark C.

**CONJECTURE**—To infer from incomplete information or evidence. Kepler C.

**CONVENTION**—A belief widely accepted by long usage.

**COROLLARIES**—Propositions that follow with little proof from another already proven.

**CRITERION**—A principle against which a proposition can be treated or tested. Chauvenet C.; Lawson C. (for fusion); Routh C.

**CURVE**—A line that represents an equation or relationship of variables. Linear Growth C.; Lorenz C.; Pareto C.; Wunderlich C.

**DIMENSIONLESS NUMBER or GROUP**—A ratio of two or more numbers with units (dimensions), taken with respect to physically significant units, that provide a meaningful number without units.

**DOCTRINE, DOGMA**—Idea put forth as true, not necessarily supported by evidence.

**DUALISM**—A view that, in some areas of concern, there are two fundamental entities, such as mind and body, good and evil, true and false, and plus and minus. Analysis and synthesis can be considered a dualistic approach.

**EFFECT**—A response to a stimulus. Abney E.; Aschoff E.; Auger E.; Barkhausen E.; Bauschinger E.; Becquerel E.; Bohr E.; Compton E.; DeHaas-Van Alphen E.; Debye E.; Doppler E.; Edison E.; Ettingshausen E.; Fenn E.; Forbush E.; Gudden-Pohl E.; Gunn E. (electron); Hall E.; Joule-Thomson E.; Kerr E.; Kelvin E.; Leduc E.; Magnus E.; Majorama E.; Mass-Law E.; Meissner E.; Mossbauer E.; Paschen-Black E.; Peltier E.; Photoelectric E.; Piezoelectric E.; Piobert E.; Purdy E.; Purkinje E.; Seebeck E.; Sewall Wright E.; Skin E.; Stark E; Thomson E.; Voigt E.; Zeeman E.

**EQUATION**—A statement of equality between two expressions. Airy E.; Arrhenius E.; Clapeyron E.(rate reaction); Bragg E.; Deutsch-Anderson E.; Dittus-Boelter E.; Gibbs-Helmholtz E.; Gibbs-Thomson E.; Hamilton E. (position of a body); Lange-

Hicks E.; Lorenz Force E.; Navier-Stokes E.; Nernst E.; Nutting-Scott Blair E.; Quadratic E.; Stark-Einstein E.; Van der Pol E.; Van der Waals E.

**FORMULA, FORMULATION**—A general principle stated in mathematical terminology. Balmer Series F.; Cauchy F.; Clausius F.; Leibnitz F.; Nyquist F.; Planck F.; Shannon F.

**FUNCTION**—A mathematical relationship. Dirac Delta F.; Gibb F.; Goedel's F. (a parallel to theorem).

**HYPOTHESIS**—A statement of conjecture subject to proof. A hypothesis is that inertial forces arise from the gravitational interaction between the accelerated body and mass of the distant stars. Avogadro H.; Gaia H.; Helmholtz-Lamb H.; Starling H.

**INEQUALITY**—A statement of inequality between expressions. Camp-Meidell I. is called a law by E. B. Wilson.

**LAW**—A general principle that represents all cases where properly applied. DeMorgan L.; Einstein L.; First L. of Thermodynamics, et al; Ohm L.

**MAXIM**—See AXIOM

**METHOD**—A procedure or order to be followed. Runge-Kutta M.

**MODEL**—Tentative representation for testing a phenomenon (structure, equation, relationship, etc.). Burger M.; Charney M. (of Navier-Stokes Equation); Covering-Law M.; Dirac M.; Linear M.; Maxwell M.; Newton-Kelvin M.; Sommerfeld M. (metal); Turing M.; Viscoelastic M.

**NUMBER**—See DIMENSIONLESS NUMBER or GROUP.

**PARADOX**—A true statement that appears to be contrary to other statements or opinions. Russell P.

**PARAMETER**—An arbitrary constant or equation or other expression that can be used for comparison. Coriolis P.; Hydrodynamic P.; Magnetic force P.

**PHENOMENON**—An observable fact or event. Bezold-Brucke Ph.; Gunn Ph.; Loeb Ph.; Magnetoelastic Ph.; Relaxation Ph.; X-ray Ph.

**POSTULATE**—An observation or projection of a relationship based on insight or intuition that becomes a hypothesis if proved to be true; often used synonymously with axiom; but Wolff described postulates as demonstrable, practical and particular, thus belonging to the analytic. Avogadro P.; Hill P.; Koch P. (bacteriology).

**PRINCIPLE**—A basic truth, law, or assumption. Babinet P.; Bernoulli P.; Berthelot-Thomsen P.; Blackman P.; Boltzmann Ionic P.; D'Alembert P.; Exclusion P. (Pauli); Franck-Condon P.; Gilbert P.; Hardy-Weinberg P.; Homeopathy P.; Huygen P.; Lagrange P.; Least Action P.; Le Chatelier P.; Limiting Factor P.; Lister P.; Maupertius P.; Quantum P.; Relativity P.; Saint-Venant P.; Uncertainty P. (Heisenberg); Van't Hoff P.; Venturi P.

**PROBLEM**—A question proposed for examination or solution. First Digit P. (Benford).

**PROOF**—Expressions dealing with mathematical symbolic logic. Goedel P.

**PROPOSITION**—A statement proposed to represent a relationship. See RELATIONSHIP; STATEMENT.

**RELATION(SHIP)**—A thought advanced as true. Bohr R.; Gibb R.; Koschmieder R.; Newland R. (that became known as the Law of Octaves); Ostwald-Freundlich R.

**RULE**—Standard procedure for solving a class of problems. Abegg R.; Antonow R.; Area, R. for; Bergmann R.; Blanc R.; Blondel R.; Constant Proportions, R. of; Ecogeographic R.; Fleming R.; Geiger and Nuttall R.; Gibb Phase R.; Grimm-Sommerfeld R.; Guldberg R.; Fajans R.; Jordan R.; Hahnemann R.; Helmholtz-Thomson R.; Hesse R.; Hildebrand R.; Kramer R.; Lewis-Randall R.; Signs, R. of (algebraic); Matthiessen R.; Maxwell R.; Richards R.: Simpson R.; Species R.; Thomson-Nernst R.; Trapezoidal R.; Weddle R.; Young R.

**STATEMENT**—May be a definition or relationship. Mach S.

**TAUTOLOGY**—A compound proposition that is true regardless of values assigned, such as "A or not A."

**THEOREM**—An idea demonstrably true or proven proposition; a well formed formula in a given logic system. Bayes T.; Chebyshev T., called the Law of Large Numbers T.; Earnshaw T.; Goedel (Godel) T.; Lami T.; Last T. (Fermat); Pythagoras T.; Roberts-Chebyshev T.

**THEORY**—System of circumstances leading to explanation of a phenomenon. Arrhenius-Ostwald T.; Bar T.; Bayes-Laplace T.; Bernoulli Kinetic T.; Buckingham T.; Gravitation, T. of; Helmholtz Accommodation T. (eye); Helmholtz T.(color); Lewis T.; Recapitulation T.; Ptolemy T.; Quantum T.; Relativity T. (Einstein); Usanovich T.; Young-Helmholtz T.

**THESIS**—A proposition forming ideas into expressible form to be communicated to others, usually written but can be spoken or subject of a composition. Church T.

# List of Illustrations


Energy consumed by mammals and birds....9
All or none....10
Simpson rule....15
Trapezoidal rule....16
Change of applied magnetic field....26
Loading and unloading to plastic strain....27
Triple point for water....45
Force between particles....49
Bulk flow....58
Carnot cycle....64
Impulsive motion....64
Color wheel....77
Conjugacy....82
Control volume....87
Cosine....96
Einthoven (EKG, ECG)....135
Elasticity....137
Exponential Failure....155
Fermat....162
Diffraction....172
Friction forces....173
Gerber....183
Movement of glacier ice....186
Gompertz....188
Hall effect....200
Hertz contact....214
Hooke vs. Newton....218
Increasing returns....228
Isotherms....234
Wall pressure....236
Lanchester....261
Lens law....268
Machines....284
Reciprocal deflections....296
Indentation....302
Mohr circle....305
Gaussian distribution and normal distribution....325
Force AC is equivalent to AB + BC....340
Pareto....341
Poisson ratio....356
Power function: hyperbola....360
Power function: parabola....360
Logarithmic power curves....361
Pythagorean....371
Quadrants....374
Reflection of light at a plane surface....388

Reynolds number....395
Stiffness panels....429
Stress law of fracture....434
Survival....438
Tangents....442
Thermocouple....445
Toothed gearing....452
Transonic theory....454
Change in temperature when crushed ice is heated....455
Wall and velocity defect....467
Viscoelastic material models....470
Wagner....476
Wave motion....480
Yoda power law....493
Young-Helmholtz....494
Young modulus....495

# Abbreviations

| | |
|---|---|
| AAAS | American Association for Advancement of Science |
| ACS | American Chemical Society |
| AGI | American Geological Institute |
| AIChE | American Institute of Chemical Engineers |
| AIME | American Institute of Mining, Metallurgical and Petroleum Engineers |
| AIP | American Institute of Physics |
| AMA | actual mechanical advantage |
| API | American Petroleum Institute |
| ASCE | American Society of Civil Engineers |
| ASME | American Society of Mechanical Engineers |
| BET | Brunauer-Emmet-Teller |
| DOD | Department of Defense (U.S.) |
| DOE | Department of Energy (U.S.) |
| e.g. | for example |
| eng., engr. | engineer, engineering |
| erf | error function |
| esp. | especially |
| est. | estimate(d) |
| etc. | and so forth |
| exp(x) | (x) is the exponent of e |
| ff. | and following |
| i.e. | that is |
| ISO | International Standards Organization |
| LH | latent heat |
| ln | logarithm to the base e |
| log | logarithm to the base 10 |
| MTBF | mean time before failure (same as) |
| MTTF | mean time to failure |
| n.a. | not available |
| n.d. | no date; or undated |
| NAE | National Academy of Engineering (U.S.) |
| NAS | National Academy of Sciences (U.S.) |
| NASA | National Aeronautics and Space Administration (U.S.) |
| NBS | National Bureau of Standards (presently NIST) |
| NIST | National Institute of Standards and Technology (U.S.) |
| NIH | National Institutes of Health (U.S.) |
| NSF | National Science Foundation (U.S.) |
| NTP | normal temperature and pressure (25$^{\circ}$ C at 1 atm.) |
| NUC | National Union Catalogue |
| NYPL | New York Public Library |
| TMA | theoretical mechanical advantage |

# Physical Constants (in SI units)

acceleration constant or Newton constant of gravitation, G

$6.67259 \times 10^{-11}$ m$^3$kg$^{-1}$s$^{-2}$

acceleration in free fall due to gravity (standard value), g

9.80655 ms$^{-2}$

Arrhenius—see universal gas constant, R

atomic mass, u (unified) (see Weizacker mass law)

$1.6605402 \times 10^{-27}$ kg

Avogadro constant, $N_A$

$6.0221367 \times 10^{23}$ mol$^{-1}$

Bohr magneton, eh/2m$_e$, $\mu_B$ or $\beta$

$9.274025 \times 10^{-24}$ JT$^{-1}$ (T = telsa)

Bohr radius, $a_o$

$0.529177249 \times 10^{-10}$ m

Boltzmann constant, k (R/N$_A$), or B

$1.38066 \times 10^{-23}$ JK$^{-1}$

classical electron radius, $r_e$

$2.81794092 \times 10^{-15}$ m

Compton wavelength, $\lambda_C$

$2.42631 \times 10^{-12}$ m

Curie constant, C

$N\mu^2/3k$

where N = number of parametric ions per unit volume,
$\mu$ = magnetic moment of each ion, k = Boltzmann constant

e = 2.71828…, the base of the natural logarithm system, a transcendental irrational number

electron radius, $r_e$

$2.81794 \times 10^{-15}$ m

electron rest mass, $m_e$

$9.1093897 \times 10^{-31}$ kg

electron volt, eV

$1.60218 \times 10^{-19}$ J

elementary charge, e

$1.60218 \times 10^{-19}$ C (C = coulomb)

Faraday constant, F

$9.6485309 \times 10^4$ Cmol$^{-1}$ (C = coulomb)

gas constant—*see* molar gas constant

gravitational constant, $G_N$

$6.67259 \times 10^{-11}$ m$^3$kg$^{-1}$s$^{-2}$

Lorentz constant, L (some references use Lo)

$2.45 \times 10^{-8}$ watt-ohmK$^{-2}$

Loschmidt number, $N_L$

$2.686763 \times 10^{25} m^{-3}$ mole/L

magnetic constant, $\mu_o$

$4\pi \times 10^{-7} Hm^{-1}$

mass of alpha particle, m

$6.56 \times 10^{-27}$ kg

molar gas constant, R

$8.314510 Jmol^{-1}K^{-1}$

molar volume (ideal gas), RT/ρ

$22.41410 Lmol^{-1}$ or $22.414 m^3mol^{-1}$ or $2.241383 \times 10^{-3} m^3mol^{-1}$

muon mass, $m_\mu$

$1.8835327 \times 10^{-23}$ kg

neutron rest mass, $m_r$

$1.674929 \times 10^{-27}$ kg

Newtonian constant of gravitation, G

$6.67259 \times 10^{-11} m^3kg^{-1}s^{-2}$ or $Nm^2kg^{-2}$

Number of molecules per cc of gas at NTP, N

$2.705 \times 10^{19}/cm^3$ (or /L)

pi, π, ratio of circumference of circle to its diameter

3.14159... (a transcendental irrational number)

Planck constant, h

$6.6260755 \times 10^{-34}$ Js

proton rest mass, $m_p$

$1.6726231 \times 10^{-27}$kg

R—*see* universal gas constant

radius of an electron, a

$2.8 \times 10^{-16}$ m

Rydberg constant, $R_\infty$

$10.9737315 \times 10^6 m^{-1}$

standard atmosphere, Po

101325 Pa

Stefan-Boltzmann constant, σ

$5.67051 \times 10^{-8} Wm^{-2}K^{-4}$

universal gas constant, R, or molar gas constant

$8.314510 Jmol^{-1}K^{-1}$

velocity of light in a vacuum, c

$2.997925 \times 10^8 ms^{-1}$

Wien displacement law constant, b

$2.897756 \times 10^{-3}$ mK

zero, Celsius scale, $T_o$

273.15 K

**Sources**

Kaye, G. W. C. and Laby, T. H. 1995; Lide, David R. 1994 (75th ed.). *CRC Handbook of Chemistry and Physics.* CRC Press, Boca Raton, FL.

Besancon, Robert M. 1974 (2nd ed.); *The Encyclopedia of Physics.* Van Nostrand Reinhold Co., New York, NY, 1067 p.

Lerner, R. G. and Trigg, G. L. 1991. *Encyclopedia of Physics.* Addison-Wesley Publishing Co., Reading, MA, 1408 p.

# Contents

Laws and Models: An Introduction ... vii
*George A. Hazelrigg*

Possible First Recorded Physical Law ... xix

Definitions and Descriptions ... xxi

List of Illustrations ... xxv

Abbreviations ... xxvii

Physical Constants (in SI Units) ... xxix

Laws and Models ... 1

Sources and References ... 499

## ABBOTT-AUBERT EFFECT

Negative work suppresses or reverses biochemical reactions; thus work done upon contracting muscles by stretching them does not appear as heat. The Abbott-Aubert effect is opposite the Fenn effect.

**Keywords:** biochemical, contracting muscles, heat, negative work

AUBERT, Hermann, 1826-1892, German physiologist

**Sources:** Morton, L. T. and Moore, C. 1992; Parker, S. P. 1987.

*See also* FENN

## ABEGG LAW OR RULE (1904)

The numerical sum of the maximum positive valence and the maximum negative valence of an element has a tendency to equal eight. This relationship is most apparent for the elements of the 4th, 5th, 6th, and 7th groups of the periodic table.

**Keywords:** atom, chemistry, element, periodic, physics, valence

ABEGG, Richard Wilhelm Heinrich, 1869-1910, German chemist, born in Poland

**Sources:** Fisher, D. J. 1988. 1991; Hodgman, C. D. 1952; Morris, C. G. 1992.

## ABNEY LAW OR EFFECT (1877)

The hue of a light with a wavelength less than 5700 Angstroms (Å) shifts to the red end of the spectrum, and of a light with a wavelength greater than 5700 Å shifts to the blue end of the spectrum, when a spectral color is flooded with (by adding) a white light.

The above may be represented by the fact that changes in the hue of light may result from changes in saturation. For example, it appears that yellow-reds become more red with increasing saturation, and blue-greens become more blue.

**Keywords:** color, electromagnetic, hue, illumination, lighting, spectrum

ABNEY, Sir William de Wiveleslie, 1843-1920, English chemist and physicist

**Sources:** Considine, D. M. 1976; McAinsh, T. F. 1986; Physics Today 19(3):34. 1966.

*See also* BEZOLD-BRUCKE; GRASSMANN; PURDY

## ABNORMAL DIFFUSION OR ABNORMAL DISPERSION—SEE KUNDT

## ABRAMS LAW (1919)

The strength of concrete depends on the water/cement ratio. The best ratio is about 0.25, but more water is needed in practice to wet the sand and stone. This law is a special case of the general law expressed by Feret in France in 1896.

**Keywords:** cement, concrete, water

ABRAMS, Duff Andrew, 1880-197x, American consulting civil engineer

**Sources:** Lewis, R. J. Jr. 1993; Meyers, R. A. 1992; Scott, J. S. 1993; Who Was Who. 1996.

*See also* FERET; NUTTING-SCOTT BLAIR

### ABSORBANCE, LAW OF; ABSORPTION, EXPONENTIAL LAW OF

By combining the laws of Bouguer and Beer the absorbance of light:

$$A = -\log T = \log \frac{I_o}{I} = abc$$

and the absorptivity is A/bc
where A = absorbance
T = transmittance
$I_0$ and I = intensities of light, incident and transmitted
a = absorptivity; b = thickness; c = concentration

**Keywords:** absorptivity, light, transmittance
**Source:** Considine, D. M. 1976.
*See also* BEER; BOUGUER; LAMBERT-BEER

### ABSORPTION OF GASES BY LIQUIDS, LAWS OF

The following laws hold for pressures only a few atmospheres above atmospheric pressure.

1. In the same gas, the same liquid, and at the same temperature, the weight of gas absorbed is proportional to the pressure, or, at all pressures, the volume dissolved is the same. This statement is known as the Henry law.
2. The quantity of gas absorbed decreases with an increase of temperature.
3. The quantity of gas which a liquid can dissolve is independent of the nature and the quantity of other gases that may be held in solution.

**Keywords:** absorption, dissolve, gas, physical chemistry, pressure
**Sources:** Clifford, A. F. 1964; Isaacs, A. 1996; Thewlis, J. 1961-1964.
*See also* HENRY; PARTITION

### ABSORPTION—SEE BOOLEAN; HENRY [FOR GASES (ABSORPTIVE)]

### ACCELERATION (BIOLOGY), LAW OF

The development of an organ in a body accelerates in proportion to the importance of that organ.

**Keywords:** embryology, medical, physiology
**Sources:** Bates, R. and Jackson, J. 1980; Kenneth, J. H. 1963.

### ACCELERATION (CONSTANT)—SEE NEWTON SECOND LAW

### ACCELERATION (GRAVITATIONAL)—SEE NEWTON

### ACCELERATION NUMBER, $K_O$ OR $N_{KO}$

A dimensionless group used to represent accelerated flow dependent only on the physical properties:

$$K_o = \frac{E^3}{\rho g^2 \mu^2}$$

where $K_0$ = acceleration number
E = modulus of elasticity

$\rho$ = density
g = acceleration due to gravity
$\mu$ = dynamic viscosity

Also, the Acceleration number equals the Reynolds number × the Froude number squared, divided by the Homochronous number cubed.

**Keywords:** dimensionless, flow, physical properties
**Sources:** Bolz, R. E. and Tuve, G. L. 1970; Parker, S. P. 1992; Potter, J. H. 1967.
*See also* FROUDE; HOMOCHRONOUS; REYNOLDS

## ACCORDANT JUNCTIONS, LAW OF—SEE PLAYFAIR LAW

## ACOUSTICAL—SEE OHM ACOUSTICAL

## ACTION AND REACTION, LAW OF; ACTION-REACTION LAW

When forces are applied to a body constrained by supports, forces are exerted on the supports such that the action and reaction are equal, opposite, and colinear.

**Keywords:** forces, mechanics, physics
**Sources:** Meyers, R. A. 1992; Parker, S. P. 1983.
*See also* NEWTON

## ACTION OF ACID ON SUGAR, LAW OF (1850)

In an acid aqueous solution of sucrose, the sucrose is hydrolyzed to an equimolar mixture of glucose and fructose as described by L. Wilhelmy.

**Keywords:** fructose, glucose, sucrose
WILHELMY, Ludwig F., 1812-1864, German chemist
**Source:** Grayson, M. 1978.

## ADAPTATION LAWS (ECOLOGY)

### Allen Law or Rule

Parts protruding from the body of mammals, such as ears and tails, tend to be shorter in colder climates.

### Bergmann Rule

Individuals tend to be smaller in warmer climates.

### Gloger Rule

Cold, dry climates encourage light coloration in animals; warm and moist climates encourage darker colors.

**Keywords:** cold climates, color, dry, limbs, moist, warmer, wet
ALLEN, Joel A., 1838-1921, German zoologist
BERGMANN, Max, Carl George Lucas Christian, 1814-1865, German biologist
GLOGER, Constantine Wilhelm Lambert, 1803-1863, German zoologist
**Source:** Considine, D. M. 1976.
*See also* ALLEN; BERGMANN; GLOGER

## ADDITION OF VELOCITIES, LAW OF

If a system S′ moves with the speed of v relative to S, and a particle moves with the velocity of w relative to system S′, the velocity of the particle relative to S is the sum of the two velocities, if they are moving in the same direction.

**Keywords:** direction, particle, relative, velocity

**Source:** Brown, S. B. and Brown, L. B. 1972.

*See also* VECTOR ADDITION

## ADDITIVE IDENTITY LAW

Zero added to any real number gives the identical number, such as (a + 0 = a).

**Keywords:** added, algebra, number, zero
**Sources:** Dorf, R. C. 1995; James, R. C. and James, G. 1968; Sneddon, I. N. 1976.
*See also* ALGEBRA

## ADDITIVE INVERSE LAW

For any real number (+n) there is another number (–n) such that their sum is zero. (+n) is considered the additive inverse of (–n), and vice versa.

**Keywords:** additive, algebra, number, real, sum, zero
**Sources:** James, R. C. and James, G. 1968; Sneddon, I. N. 1976.
*See also* ALGEBRA

## ADDITIVE LAW OF PROBABILITY

The probability of either A or B is the sum of the probability of A and the probability of B in which the events must be mutually exclusive:

$$P(A \text{ or } B) = P_A + P_B$$

in which $P_A$ means the probability of A.

**Keywords:** probability, statistics
**Source:** Doty, L. A. 1989.
*See also* COMBINATION; COMPLEMENTARY; CONDITIONING; MULTIPLICATIVE

## ADDITIVE PRESSURE LAW—SEE DALTON

## ADDITIVE VOLUMES, LAW OF

The volume occupied by a mixture of gases is equal to the sum of the volumes of the components of the mixture when the temperature and pressure are equal.

**Keywords:** gases, pressure, volume, temperature
**Source:** Parker, S. P. 1989.

## ADIABATIC LAW

The general relationship which states that for expansion of gases without loss or gain of heat (as distinguished from isothermal):

$$P\rho^{-\gamma}$$

where P = pressure

$\rho$ = density
$\gamma$ = ratio of specific heats at $c_p/c_v$ (pressure/volume)

**Keywords:** expansion, gas, temperature
**Sources:** Lapedes, D. N. 1976; Mandel, S. 1972.
*See also* EHRENFEST

### ADJOINING PHASE REGIONS, LAW OF (MATERIALS)

A region of P phases in a multicomponent materials system must always be bounded by regions of P± (plus and minus) phases; for binary systems, a single phase can be bounded only by two-phase fields and by no other single phase field. This law pertains to phase diagrams in heterogeneous materials.

**Keywords:** binary, heterogeneous, materials, multicomponent, phase
**Source:** Bever, M. B. 1986.

### ADRIAN LAW; ADRIAN-BRONK, LAW OF—SEE ALL-OR-NONE

### ADSORPTION LAW—SEE FREUNDLICH

### ADSORPTION ISOTHERM MODELS—SEE BRUNAUER-EMMETT-TELLER (1938); FREUNDLICH (1909); HARKINS-JURA (1946); LANGMUIR (1916)

### ADVANTAGE, LAW OF; OR PRINCIPLE OF ADVANTAGE

When two or more incompatible and inconsistent responses occur to the same situation, one has an advantage, is more reliable, and occurs more frequently than the other.

**Keywords:** incompatible, reliable, responses
**Sources:** Goldenson, R. M. 1984; Wolman, B. B. 1989.

### AFFINITY LAWS

For specifying pumps for a process or pumping operation, engineers may reduce an impeller diameter to obtain a desired reduction in head and flow. To determine the amount of reduction needed, the affinity laws may be used that compare two different impeller diameters in the same pump at the same speed, in which:

1. The capacity, Q, is directly proportional to the ratio of impeller diameter, that is $Q_2 = Q_1(D_2/D_1)$, where D = impeller diameter.
2. The head, H, is directly proportional to the square of the impeller diameter ratio, $H_2 = H_1(D_2/D_1)^2$, where D = impeller diameter.

These laws must be used with caution. Generally, the lower the impeller specific speed and the larger the reduction in size of the impeller, the larger the discrepancy between the Affinity law calculations and the actual performance.

**Keywords:** capacity, diameter, head, impeller, pump
**Source:** Chemical Engineering 102(1):88, January 1995.

### AIRLIGHT LAW—SEE KOSCHMIEDER

**AIRY EQUATION**

An equation for multiple beam interference for light transmitted through a plane parallel plate. The intensity, I, of the light transmitted is:

$$I = \frac{I_o T^2}{(1-R)^2} \frac{\left(\frac{1 + 4R\sin^2\delta}{2}\right)}{(1-R)^2}$$

where $I_o$ = intensity of incident beam
R = reflectivity of the surfaces of the plate
T = transmissivity
$\delta$ = phase difference between directly transmitted light and light that is once reflected from the two internal surfaces of the plates

**Keywords:** beam, light, plates, reflectivity, transmissivity
AIRY, Sir George Biddell, 1801-1892, English astronomer and mathematician
**Source:** Ballentyne, D. W. G. and Lovett, D. R. 1980.
*See also* BEER; LAMBERT; REFLECTION

**ALBEDO LAW**

A perfect diffuse reflection obeys Lambert cosine law in which the surface brightness of the reflecting surface is proportional to the cosine of the angle of incidence and is independent of the angle of reflection.

**Keywords:** brightness, cosine, diffuse, incidence, reflection
**Sources:** Gray, H. J. and Isaacs, A. 1975; Michels, W. C. 1961.
*See also* LAMBERT

**ALBITE LAW—SEE TWIN LAW**

**ALEXANDER LAW**

Involuntary rapid movement of the eyeball is enhanced if the gaze is turned. A nystagmus (a rapid involuntary oscillation of the eyeballs), produced either by rotation or thermally, can be accentuated by moving the eyes in the direction of the jerky component of the nystagmus, the rapid return movement of the eyes in a direction opposite to the initial, slow movement.

**Keywords:** eye, movement
ALEXANDER, Gustav, 1873-1932, Austrian doctor
**Sources:** Landau, S. I. 1986; Zusne, L. 1984.

**ALFVÉN NUMBER (Al) OR ALFVÉN SPEED OR ALFVÉN GROUP**

A dimensionless group that represents the propagation of acoustic waves in presence of a magnetic field:

$$Al = v\, L\, (\rho\, \mu)^{1/2}\, B^{-1/2}$$

where v = velocity of flow
L = length
$\rho$ = density
$\mu$ = permeability
B = magnetic flux density

H. O. G. Alfvén introduced the term *magnetohydrodynamics*.
**Keywords:** acoustic, dimensionless, magnetic
**ALFVÉN**, Hannes Olof Gosta, 1908-1995, Swedish physicist; Nobel prize, 1970, physics (shared)
**Sources:** Besancon, R. M. 1974; Bolz, R. E. and Tuve, G. L. 1970; Jerrard, H. G. and McNeill, D. B. 1992; Millar, D. et al. 1996; Parker, S. P. 1992; Porter, R. 1994; Potter, J. H. 1967.

## ALGEBRA, LAWS OF—SEE ASSOCIATE; BOOLEAN; COMMUTATIVE; DISTRIBUTIVE; EXPONENTS; MULTIPLICATION

## ALGEBRA OF PROPOSITIONS, LAWS OF; ALGEBRA, FUNDAMENTAL LAWS OF

1. Associative Laws:

$$(p \vee q) \vee r \equiv p \vee (q \vee r)$$

$$(p \wedge q) \wedge r \equiv p \wedge (q \wedge r)$$

2. Commutative Laws:

$$p \vee q \equiv q \vee p$$

$$p \wedge q \equiv q \wedge p$$

3. Complement Laws:

$$p \vee \sim p \equiv t \qquad p \wedge \sim p \equiv f$$

$$\sim\sim p \equiv p \qquad \sim t = f, \sim f \equiv t$$

4. De Morgan Laws:

$$\sim(p \vee q) \equiv \sim p \wedge \sim q$$

$$\sim(p \wedge q) \equiv \sim p \vee \sim q$$

5. Distributive Laws:

$$p \vee (q \wedge r) \equiv (p \vee q) \wedge (p \vee r)$$

$$p \wedge (q \vee r) \equiv (p \wedge q) \vee (p \wedge r)$$

6. Idempotent Laws:

$$p \vee p \equiv p$$

$$p \wedge p \equiv p$$

7. Identity Laws:

$$p \vee f \equiv p \qquad p \wedge t \equiv p$$

$$p \vee t \equiv t \qquad p \wedge f \equiv f$$

Symbols: $\wedge$, and; $\vee$, or; $\rightarrow$, implies; $\supset$, includes; $\sim$, negative; $\equiv$, biconditional

**Keywords:** logic, mathematics, symbols
**Sources:** Brown, S. B. and L. B. 1972; Dorf, R. C. 1995; Flew, A. 1964. 1984; Lipschutz, S. and Schiller, J. 1995; Sneddon, I. N. 1976.
*See also* ADDITION; EXPONENTS; MULTIPLICATION

## ALLARD LAW (OPTICS)

The variation of the flux density of illuminance (I) at a distance (x) from a point source of artificial light in an atmospheric extinction coefficient (σ) is represented by:

$$I = Lx^{-2}e^{-\sigma x}$$

This relationship is more properly correct for monochromatic radiation.

**Keywords:** atmosphere, illumination, light, radiation
ALLARD, Georges, twentieth century (wrote books in French, 1922–1948), French physicist
**Sources:** Morris, C. G. 1992; NUC; Parker, S. P. 1997.

## ALLEE LAW

A particular habitat has an optimal population level for a given species.

**Keywords:** habitat, population, species
ALLEE, Warder Clyde, 1885-1955, American zoologist and ecologist
**Sources:** Gray, P. 1967; Morris, C. G. 1992.

## ALLEN LAW (1877) OR RULE (MAMMALS)

Mammals (and birds) have shorter limbs or extremities, such as ears and tails, in colder regions, resulting in a reduced loss of heat.

**Keywords:** limbs, mammals, temperature of environment
ALLEN, Joel Asaph, 1838-1921, American naturalist and zoologist
**Sources:** Bothamley, J. 1993; Graham, E. C. 1967; Gray, P. 1967.
*See also* ADAPTATION; BERGMANN; GLOGER

## ALLEN LAW (1900) (FLUID FLOW)

There is an extensive region where resistance to fluid motion is a function of both density and viscosity of the medium, and for the following relationship, the exponent n is two-thirds (2/3):

$$R = kd^n v^n \mu^{2-n} \rho^{n-1}$$

where R = resistance
k = constant for particular fluid
d = density
v = velocity
μ = viscosity
ρ = density

**Keywords:** density, flow, fluid, resistance, viscosity
ALLEN, Herbert Stanley, b. 1873, English physicist
**Sources:** Dallavalle, J. M. 1948; NUC.
*See also* DARCY; REYNOLDS

## ALLEN PARADOXIC (PARADOXICAL) LAW (DIABETES)

In normal healthy individuals, the more sugar that is consumed, the more sugar is utilized, whereas in diabetic individuals, the reverse is true.

**Keywords:** diabetes, medicine, physiology, sugar
ALLEN, Frederick Madison, 1879-1964, American physician
**Sources:** Brown, S. B. and L. B. 1972; Landau, S. I. 1986; Stedman, T. L. 1976.

## ALLOMETRIC SCALING LAWS IN BIOLOGY

The dependence of a biological variable Y on a body of mass M is characteristically represented by an allometric scaling law:

$$Y = Y_0 M^b$$

where b = the scaling exponent
$Y_0$ = a constant representative of the organism

Metabolic rates of entire organism scales are $M^{3/4}$, known as the 3/4-power law; life span and embryonic growth scales are $M^{1/4}$, known as the 1/4-power law; and rates of cellular metabolism and heart rates are $M^{-1/4}$, known as the minus 1/4-power law. The larger the animal, the lower the pulse rate (Fig. A.1).

**Keywords:** biology, growth, metabolic, population
**Sources:** BioScience 48(11):888, November 1998; Science 276(5309):122-126, 4 April 1997; Science 278(5337):371-373, 17, October 1997.
*See also* KLEIBER; POWER LAW

## ALL-OR-NONE LAW OF EXCITATION (1871)(1917); ALL-OR-NOTHING LAW; ADRIAN PRINCIPLE; ADRIAN-BRONK LAW

The impulse of a sensory nerve fiber is independent of the way it is stimulated. A single nerve impulse produces one effect on an end organ, such that the stimulus activity does not affect the intensity of the nerve impulse. The weakest stimulus produces the same response. The relationship was first established for all-or-none law for contraction of the heart muscle by H. Bowditch in 1871. The principle was applied to skeletal muscle by F. Pratt in 1917 (Fig. A.2).

**Keywords:** medicine, nerves, physiology, stimulus

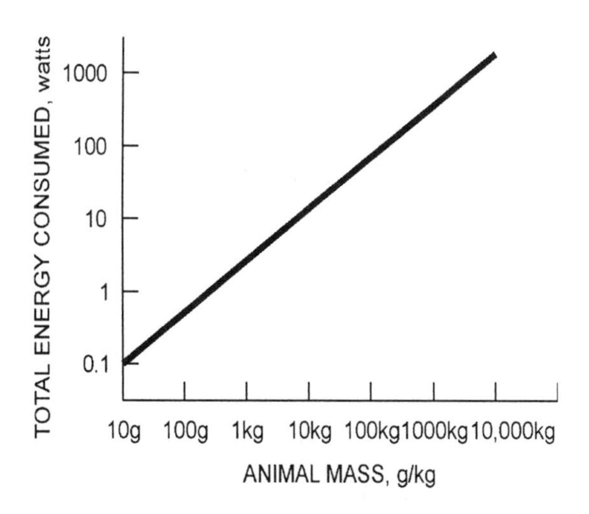


**Figure A.1** Energy consumed by mammals and birds.

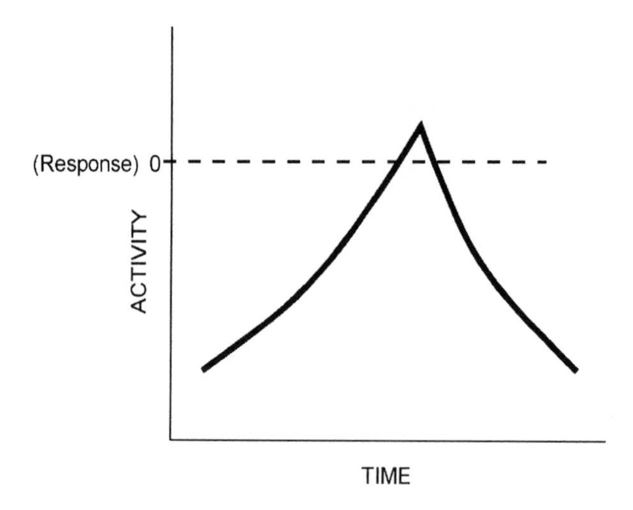


**Figure A.2** All or none.

ADRIAN, Edgar Douglas, 1889-1977, English neurobiologist; Nobel prize, 1932, physiology and medicine (shared)
BOWDITCH, Henry Pickering, 1840-1911, American physiologist
BRONK, Detlev Wulf, 1897-1975, American physiologist
PRATT, Frederick Haven, 1873-1958, American physiologist
**Sources:** Bothamley, J. 1993; Friel, J. P. 1985; Good, C. V. 1973; Landau, S. I. 1986; Lapedes, D. N. 1974; Rashevsky, N. 1960; Schmidt, J. 1959.
*See also* BOWDITCH

### ALTERNATION LAW OF MULTIPLICITIES

Neutral atoms of successive elements in the periodic table have spectral terms that alternate between even and odd multiplicities.

**Keywords:** atoms, chemistry, periodic table, physics, spectral
**Sources:** Honig, J. H. 1953; Thewlis, J. 1961-1964.

### AMAGAT LAW OF ADDITIVE VOLUMES (1877) OR AMAGAT-LEDUC LAW OR RULE

The volume of a mixture of (perfect) gases is equal to the sum of the volumes of the component gases, each taken at the total pressure and temperature of the mixture. Thus:

$$V = V_1 + V_2 + V_3$$

where V, $V_1$, $V_2$, $V_3$ = respective volumes

This law is related to the Dalton law but considers the additional effects of the volume instead of pressure.

**Keywords:** chemistry, gases, physics, volumes
AMAGAT, Emile Hilaire, 1841-1915, French physicist
LEDUC, Stephane, A. N., 1853-1939, French physicist
**Sources:** Fisher, D. J. 1988. 1991; Freiberger, W. F. 1960; Holmes, F. L. 1980; Morris, C. G. 1992; Perry, R. and Green, D. W. 1984.
*See also* DALTON

### AMBARD LAW

With a constant urinary urea concentration, the output of urea varies as the square of the blood urea; with a constant blood urea concentration, the output of urea varies inversely as the square root of the urinary concentration. This law is now considered as obsolete.

**Keywords:** blood, medicine, physiology, urea, urinary
AMBARD, Leon, 1876-1962, German physiologist
**Source:** Stedman, T. L. 1976.

### AMDAHL LAW

The time required to do a computation will always be determined by the slowest part of the computation. This is also expressed as follows: "If S percent of a program is inherently sequential, the maximum attainable speed up is 1/S. So, for example, if 25 percent of the time is spent executing inherently sequential operating systems, then even if all user programs have unlimited parallelism, and there are an unlimited number of processors, the overall system speedup will at most be four."

**Keywords:** computation, computers, parallelism, time
AMDAHL, Gene M., twentieth century (b.1922), American electrical engineer
**Source:** Science v. 261:859,13 Aug. 1993.

### AMMON LAW

The ratio of the maximum breadth to maximum length of the head multiplied by 100, known as the cephalic index, and stature vary inversely.

**Keywords:** anthropology, cephalic index, head, stature
AMMON, Frederich August von, 1799-1861, German pathologist
**Sources:** Lapedes, D. N. 1974. 1976.

### AMONTONS LAW OF PROPORTIONALITY (FRICTION) (1699); OFTEN CALLED COULOMB LAW OF FRICTION

There is a proportionality between the friction between bodies and the mutual pressure of moving bodies in contact, and is not related to the surface area of contact. This relationship is not true at the microscopic or atomic level. Amontons invented the hygrometer in 1687.

**Keywords:** friction, mechanics, pressure, proportionality
AMONTONS, Guillaume, 1663-1705, French physicist
**Sources:** Holmes, F. L. 1980; Millar, D. et al. 1996; Thewlis, J. 1961-1964; Webster's Biographical Dictionary. 1959.
*See also* COULOMB LAW OF FRICTION

### AMPERE LAW (1827); AMPERE-LAPLACE LAW; LAPLACE LAW

A law of electromagnetism in which the force, F, on a conductor in a magnetic field is proportional to the intensity or strength of the magnetic field, H. The magnetic intensity of a conductor is proportional to the current, i, and inversely proportional to the distance, n, squared from the conductor. Thus, H is proportional to $i/n^2$.

**Keywords:** electricity, force, magnetic, physics
AMPERE, Andre Marie, 1775-1836, French mathematician and physicist
LAPLACE, Pierre-Simon de, 1749-1827, French mathematician
**Sources:** Considine, D. M. 1976; Freiberger, W. F. 1960; Haus, H. and Melcher, J. R.; Isaacs, A. 1996; Parker, S. P. 1989; Whitaker, J. C. 1996.
*See also* ELECTROMAGNETIC; ELECTROMAGNETISM; LAPLACE

### AMPERE RULE

If the direction of the electric current through a conductor is away from the observer, the magnetic field associated with the current is clockwise to the conductor, and vice versa.

**Keywords:** current, direction, flow, magnetic field
AMPERE, Andre Marie, 1775-1836, French mathematician and physicist
**Source:** Isaacs, A. 1996.

### ANALOGY, LAW OF; LAW OF ASSIMILATION

Learners tend to reply in a similar way as they replied before to a similar situation. It is a duplication of the association by similarity. The law is credited to E. Thorndike.

**Keywords:** assimilation; learners; reply (response)
THORNDIKE, Edward Lee, 1874-1949, American educator
**Sources:** Goldenson, R. M. 1984; Wolman, B. B. 1989.
*See also* THORNDIKE

### ANDRADE CREEP LAW; ANDRADE LAW OF TRANSIENT CREEP

The creep strain, $\gamma$, of a material, often applicable to many metals and plastics, is proportional to the one-third power of time, t:

$$\gamma \propto t^{1/3} \qquad \gamma = ct^{1/3} \text{ where c is a constant}$$

The creep law is also expressed as the strain rate, $\dot{\gamma}$, which is proportional to $1/t^{2/3}$ of time, and holds generally at high temperatures and stresses.

$$\dot{\gamma} \propto t^{-2/3} \qquad \dot{\gamma} = At^{-2/3} \text{ where A is a constant}$$

**Keywords:** materials, mechanics, metals, plastics, rate, strain
ANDRADE, Edward Neville da Costa, 1887-1971, English physicist
**Sources:** Eirich, F. R. 1969; Thewlis, J. 1961-1964.

### ANGLES, LAW OF

The law expresses the relationship between the sine, cosine, tangent, and cotangent of the angles of a triangle:

$$\sin (A + B) = \sin C$$

$$\cos (A + B) = -\cos C$$

$$\tan (A + B) = -\tan C$$

$$\cot (A + B) = -\cot C$$

where A, B, and C are internal angles of the triangle
**Keywords:** angles, cosine, sine, tangent, triangle
**Source:** Tuna, J. J. and Walsh, R. A. 1998.
*See also* COSINE; SINE; TANGENT

### ANGSTROM LAW (1861)

The wavelengths of light absorbed by a material are the same as the wavelengths given off by the material when luminous. Angstrom measured the wavelength of light in 1868, which was measured more accurately by H. A. Rowland in 1883.

**Keywords:** absorption, light, luminous, radiation, wavelength
ANGSTROM, Anders Jonas, 1814-1874, Swedish astronomer and physicist
ROWLAND, Henry Augustus, 1848-1901, American physicist
**Sources:** Carter, E. F. 1976; Stedman, T. L. 1976; Webster's Biographical Dictionary. 1959.

## ANGULAR MOMENTUM, LAW OF; OR MOMENT OF MOMENTUM—SEE ANGULAR MOTION, CONSERVATION OF

## ANGULAR MOTION, LAWS OF

The laws relating to angular rotation of a body are analogous to those describing linear motion.

**First Law**
A rotating body will continue to turn about a fixed axis with undiminished angular speed unless acted upon by an unbalanced torque.

**Second Law**
The angular acceleration produced by an unbalanced torque on a body will be proportional to the torque, in the same direction as the torque, and will be inversely proportional to the moment of inertia of the body about its axis of rotation.

**Third Law**
When one body exerts a torque upon a second body, the second body exerts an equal torque in the opposite direction upon the first and about the same axis of rotation.

**Keywords:** acceleration, mechanics, physics, rotation, torque
**Sources:** Isaacs, A. 1996; Nourse, A. E. 1964; Thewlis, J. 1961-1964.
*See also* NEWTON

## ANGULAR TRANSMISSION—SEE SNELL LAW

## ANTICIPATION, LAW OF—SEE MOTT

## ANOMALOUS NUMBERS—SEE BENFORD

## ANTONOW (ANTONOV) (ANTONOFF) LAW OR RULE (1907)

The interfacial tension of two liquids in equilibrium is equal to the difference between the surface tensions of these liquids when exposed to the atmosphere. It is one of the limits of the Gibbs adsorption equation.

**Keywords:** equilibrium, interfacial, liquids, surface, tension
ANTONOW, George W., nineteenth-twentieth century, Soviet physical chemist
**Sources:** Bothamley, J. 1993; Chemical Abstracts 53:20993g. Nov. 1959; Considine, D. M. 1976; Fisher, D. J. 1988. 1991; LeGrand, D. G. 1980. Polymer Engr. & Sci. 20(17):1164, Nov.; Morris, C. G. 1992.
*See also* GIBBS

## APEX LAW OR RULE; LAW OF EXTRALATERAL RIGHTS

In the United States statues, a law used in litigation based on a tip, point, summit, or highest point of land, mineral, or other object that may refer to a property right. More specifically, as described in Black's Law Dictionary, in mining in the United States, mineral laws give to the mining claim the rights of the veins that extend into the public domain, the apex of which lie within the surface exterior boundaries.

**Keywords:** boundaries, law, mining, property
**Source:** Bates, R. L. and Jackson, J. A. 1980.

### APPEARANCE OF UNSTABLE FORMS, LAW OF

The unstable forms of monotropic substances are obtained from a liquid or vapor state before the stable form appears.

**Keywords:** liquid, monotropic, vapor
**Source:** Clifford, A. F. 1964.

### ARAGO LAW—SEE FRESNEL-ARAGO

### ARAN LAW

Fractures at the base of the skull result from injuries to the vault with the fractures radiating along the line of the shortest circle.

**Keywords:** fracture, medicine, skull
ARAN, Francois Amilcar, 1817-1861, French physician
**Sources:** Friel, J. P. 1974; Landau, S. I. 1986.

### ARBER LAW

"Any structure disappearing from the phylogenetic line in the course of evolution is never again shown by descendants of that line."

**Keywords:** descendants, disappearance, evolution
ARBER, Agnes Robertson, 1879-1960, English botanist
**Sources:** Gray, P. 1967; Holmes, F. L. 1980. 1990; Morris, C. G. 1992.
*See also* DOLLO

### ARCHIE LAW

Represents flow or formation of reconstructed porous media, as applied in mining and geological fields:

$$F \propto \varepsilon^{-m}$$

where F = formation factor, or dimensionless resistance to flow
m = cementation factor, usually from 1.3 to 2.5 for geological materials
ε = porosity

**Keywords:** cementation, formation, geological, reconstructed
ARCHIE, Gustave Erdman, twentieth century (b.1907), American mining engineer
**Source:** Adler, P. M. 1992.

### ARCHIMEDES LAW (c. 240 B.C.); ARCHIMEDEAN PRINCIPLE; ARCHIMEDES BUOYANCY FORCE

A body floating or submerged in a fluid is buoyed up by a force equal to the weight of the displaced fluid.

**Keywords:** buoyancy, fluid mechanics, physics
ARCHIMEDES, Syracusani, c. 287-212 B.C., Greek mathematician and engineer
**Sources:** Brown, S. B. and L. B. 1972; Daintith, J. 1981; Fairbridge, R. W. 1967; Isaacs, A. 1997.

*See also* ARCHIMEDES NUMBER; FLOTATION; HYDROSTATICS

## ARCHIMEDES NUMBER, AR OR $N_{AR}$; ALSO KNOWN AS BUOYANCY NUMBER

A dimensionless group that contrasts gravitational forces and viscous forces, used to relate motion of fluids and particles due to density differences, such as for fluidization:

$$Ar = (gL^3\rho/\mu^2)(\rho - \rho_p)$$

where g = acceleration due to gravity
L = characteristic length dimension
$\rho$ = density of fluid
$\rho_p$ = density of particle
$\mu$ = dynamic viscosity

**Keywords:** density, fluids, gravitational, viscous
ARCHIMEDES, Syracusani, c. 287-212 B.C., Greek mathematician and engineer
**Sources:** Bolz, R. E. and Tuve, G. L. 1970; Parker, S. P. 1992; Potter, J. H. 1967; West, B. H. et al. 1982.
*See also* ARCHIMEDES LAW; BUOYANCY NUMBER

## AREAS, RULES FOR ESTIMATING SURFACE

### Durand Rule

$$A = h\,(4/10\ y_0 + 11/10\ y_1 + y_2 + y_3 + \ldots + y_{n-2} + 11/10\ y_{n-1} + 4/10\ y_n)$$

### Simpson Rule (Fig. A.3)

$$(n \text{ is even})\ A = 1/3h\,(y_0 + 4y_1 + 2y_2 + 4y_3 + 2y_4 + \ldots + 2y_{n-2} + 4y_{n-1} + y_n)$$

### Trapezoidal Rule (Fig. A.4)

$$A = h(1/2\ y_0 + y_1 + y_2 + \ldots + y_{n-1} + 1/2\ y_n)$$

### Weddle Rule

$$(n = 6)\ A = 3/10\ h\,(y_0 + 5y_1 + y_2 + 6y_3 + y_4 + 5y_5 + y_6)$$

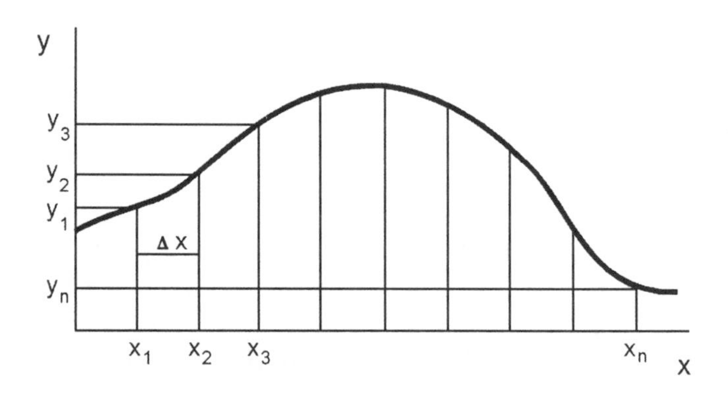


**Figure A.3** Simpson Rule.

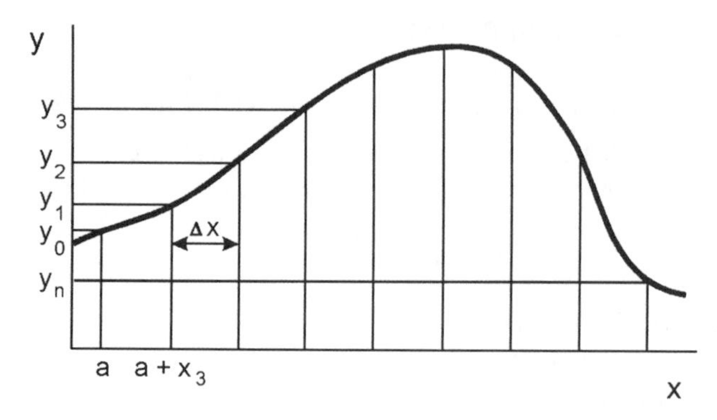


**Figure A.4** Trapezoidal Rule.

where A = the area

y = the vertical distance from a horizontal axis, uniformly placed along the x-axis

**Keywords:** area, Durand, Simpson, trapezoidal, Weddle

DURAND, William Frederick, 1859-1958, American mechanical engineer

SIMPSON, Thomas, 1710-1761, English mathematician

WEDDLE, Thomas, 1817-1853, English mathematician

**Sources:** Baumeister, T. 1967; NAS Biographical Memoirs. 1976; Research and Education Assoc. 1994.

**AREAS, LAW OF—SEE KEPLER SECOND LAW**

**ARISTOTLE (c. 345 B.C.)—SEE ASSOCIATION, LAWS OF; CONTIGUITY LAW**

**ARITHMETIC PROGRESSION, LAW OF; ARITHMETIC SEQUENCE, LAW OF**

A series of numbers, each term of which, not including the first term, is derived from the preceding term by the addition of a constant number, represented by:

$$(a), (a + b), (a + 2b), (a + 3b), \ldots$$

The sum of an arithmetical progression is:

$$S = 1/2(a + L)\, n$$

where S = sum of the terms

a = first term

L = last term

n = number of terms

**Keywords:** mathematics, number

**Sources:** Isaacs, A. 1996; James, R. C. and James, G. 1968; Speck, G. E. 1965.

*See also* GEOMETRIC

**ARITHMETIC SEQUENCES—SEE ARITHMETIC SEQUENCES, LAW OF**

**ARITHMETIC SERIES—SEE ARITHMETICAL PROGRESSION, LAW OF**

## ARNDT LAW; ARNDT-SCHULZ LAW

The amount of physiological activity is related to the stimulus, such that a weak stimulus increases activity and a very strong stimulus inhibits or abolishes the activity. This law is obsolete.

**Keywords:** physiology, stimulus, strong, weak
ARNDT, Rudolf Gottfried, 1835-1900, German psychiatrist
SCHULZ, Hugo, 1853-1932, German pharmacologist
**Sources:** Dorland, W. A.; Landau, S. I. 1986; Stedman, T. L. 1976.

## ARRHENIUS GROUP—SEE ARRHENIUS TEMPERATURE

## ARRHENIUS LAW (1887)

Only solutions of high osmotic pressure exhibit electrical activity and are electrically conductive.

**Keywords:** chemistry, electrical, osmotic pressure
ARRHENIUS, Svante August, 1859-1927, Swedish chemist; Nobel prize, 1903, chemistry
**Source:** Science 273(5281):1512. 13 Sept. 1996.

## ARRHENIUS LAW FOR MIXTURES

The viscosity of a fluid is related to the concentration of solids by:

$$\mu = \mu_o e^{kc}$$

where $\mu$ = suspension viscosity
$\mu_o$ = solute viscosity
c = solid concentration
k = a constant

The above is sometimes referred to as the Arrhenius law which, when modified, becomes the law for mixtures, and is modified for mixtures of sols and suspensions of two or more oils, as:

$$\mu = \mu_1 e^{[p \ln(\mu_2/\mu_1)]/100}$$

where p = the volume, percent of the heavier liquid

**Keywords:** concentration, fluid, mixtures, viscosity
ARRHENIUS, Svante August, 1859-1927, Swedish chemist; Nobel prize, 1903, chemistry
**Sources:** Eirich, F. R. 1969; Hui, Y. H. 1992.
*See also* IDEAL MIXTURES

## ARRHENIUS-OSTWALD THEORY OF ACIDS AND BASES

With the establishment of the concept of ionization of compounds in water, acids were defined as substances which ionized in aqueous solutions to give hydrogen ions [H+], and bases give hydroxide ions [OH–]. The ionization of acid in water can be written as $HA = H^+ + A^-$ and equilibrium equation for acid is $[H^+][A^-]/HA = K_{HA}$.

**Keywords:** acids, aqueous solutions, bases, hydrogen, hydroxide, ionization
ARRHENIUS, Svante August, 1859-1927, Swedish chemist; Nobel prize, 1903, chemistry
OSTWALD, Wilhelm, 1853-1932, German physical chemist
**Source:** Parker, S. P. 1989.

## ARRHENIUS RULE

The rate of a chemical reaction is approximately doubled with each 10° C increase in temperature.

**Keywords:** double, chemical, reaction, temperature
ARRHENIUS, Svante August, 1859-1927, Swedish chemist; Nobel prize, 1903, chemistry
**Source:** Fisher, D. J. 1988. 1991.

## ARRHENIUS TEMPERATURE DEPENDENCY LAW OR ARRHENIUS EQUATION

The rate constant of a reaction is related to the temperature of the reaction as follows:

$$k = k_0 e^{-E/RT} \text{ or } \ln k = \ln k_0 - E/RT$$

where $k$ = rate constant of a reaction at temperature, T
$k_0$ = constant called frequency factor for a particular reaction
$E$ = activation energy
$R$ = universal gas constant
$T$ = absolute temperature

The natural log of k is plotted versus the reciprocal of the absolute temperature. E/RT is the Arrhenius group, used in reaction rates, and is the activation energy divided by the potential energy of a fluid.

**Keywords:** activation energy, rate constant, temperature, universal gas constant
ARRHENIUS, Svante August, 1859-1927, Swedish chemist; Nobel prize, 1903, chemistry
**Sources:** Bolz, R. E. and Tuve, G. L. 1970; Hui, Y. H. 1992; Land, N. S. 1972; Perry, R. H. 1967; Parker, S. P. 1992; Potter, J. H. 1967; Reiner, M. 1962.

## ARTICULATION, LAWS OF

In dentistry, a set of rules to be followed in arranging teeth to produce a balanced articulation by proper meeting of the natural and artificial teeth in opposing jaws.

**Keywords:** arrangement, teeth
**Sources:** Friel, J. P. 1974; Landau, S. I. 1986.
*See also* HANAU

## ASCHOFF EFFECT OR RULE

In continuous light with increasing intensity of illumination, light-active animals increase their spontaneous frequency, while dark-active animals decrease it.

This rule is not applicable to plants.

**Keywords:** activity, animals, illumination, light, response
ASCHOFF, Karl Albert Ludwig, 1866-1942, German pathologist
**Source:** Thomas, C. L. 1989.

## ASSIMILATION, LAW OF—SEE ANALOGY, LAW OF

## ASSOCIATION, LAW OF (GEOLOGY)

The occurrence of certain varieties of asbestos in certain geological formations.

**Keywords:** asbestos, formation, geological
**Sources:** Sinclair, W. E. 1959; Thrush, P. W. 1968.

### ASSOCIATION, LAWS OF

Association of ideas to account for the functional relationships between ideas; the law of contiguity (association) has been used by experimental psychologists in modern studies on conditioning. Also known as the principles of Aristotle which were formulated by him. The law was supported and strengthened by D. Hume and is sometimes referred to as the law of contiguity.

**Keywords:** functional, ideas, relationships
ARISTOTLE, 384–322 B.C., Greek philosopher
HUME, David, 1829-1912, British philosopher
**Sources:** Angeles, P. A. 1981; Encyclopaedia Britannica. 1961.
*See also* CAUSE AND EFFECT; CONTIGUITY

### ASSOCIATIVE LAW OF ADDITION

If a = b and ab are real numbers, the way the numbers are grouped does not affect the sum so that these can be treated as:

$$a + (b + c) = (a + b) + c = a + b + c$$

and for vectors:

$$(\alpha B)a = \alpha(Ba)$$

and for complex numbers:

$$[(a + bi) + (c + di)] + e + fi = (a + bi) + [(c + di) + (e + fi)]$$

**Keywords:** algebra, mathematics, numbers, vectors
**Sources:** Good, C.V. 1973; James, R. C. and James, G. 1968; Karush, W. 1989; Newman, J. R. 1956.

### ASSOCIATIVE LAW OF MULTIPLICATION

For real numbers to be multiplied, the way factors are grouped does not affect the product:

$$a(bc) = (ab)c$$

**Keywords:** algebra, grouping, mathematics, real numbers
**Sources:** Good, C. V. 1973; James, R. C. and James, G. 1968; Karush, W. 1989; Newman, J. R. 1956.

### ASSOCIATIVE SHIFTING, LAW OF

Stimuli associated with the original stimulus-response situation could come, in time, to participate in the initiation of the responses to a learning situation, even in the absence of the original stimulus.

**Keywords:** learning, stimulus, response
**Sources:** Corsini, R. J. 1984. 1987. 1994.
*See also* LEARNING; THORNDIKE

### ATMOSPHERES, LAW OF

The distribution of molecules in an ideal atmosphere is subject only to gravity and thermal vibration.

**Keywords:** gravity, physical chemistry, physics, thermal, vibration
**Sources:** Considine, D. M. 1976; Huschke, R. 1959.

## ATOMIC MASS LAW—SEE WEIZSACKER MASS LAW

## ATOMIC NUMBERS, LAW OF (1913)

By working with X-rays and consecutive elements in the periodic table, H. G.-J. Moseley showed that the wavelength of the X-rays decreased while the frequency increased as the atomic weight increased. He postulated that the charge of the nucleus increased +1, +2, +3, etc., with the atomic number. The atomic number was more fundamental than the atomic weight.

**Keywords:** atomic weight, charge, elements, wavelength, X-rays
MOSELEY, Henry Gwyn-Jeffreys, 1887-1915, English physicist
**Source:** Asimov, I. 1966.
*See also* DALTON; MENDELEEV; MOSELEY; OCTAVE

## AUBERT PHENOMENON

The illusion of the tilting of a vertical light streak to the side opposite to the direction of the head when the head is slowly tilted in a dark room.

**Keywords:** illusion, light, head, medical, tilted
AUBERT, Hermann, 1826-1892, German physiologist and psychologist
**Sources:** Friel, J. P. 1974. 1985; Zusne, L. 1984.

## AUERBACH LAW—SEE URBAN CONCENTRATION, LAW OF

## AUGER EFFECT (1925) OR AUTO-IONIZATION LAW

An atom that contains an inner shell vacancy becomes deexcited through a cascade of transitions that are due to two kinds of competing processes: x-ray emission and radiationless, or Auger transitions. An electron makes a discrete transition from a less bound shell to the vacant, but more strongly bound, electron shell. The theory of radiationless transitions was formulated by G. Wentzel in 1927.

**Keywords:** atom, deexcitation, electron, emission
AUGER, Pierre Victor, 1899-1994, French physicist
WENTZEL, Gregor, 1898-1978, German American physicist
**Sources:** Landau, S. I. 1986; Lerner, R. G. and Trigg, G. L. 1991; Nature 135:341, 1935; Parker, S. P. 1989; Pelletier, P. A. 1994.
*See also* WENTZEL

## AUSTIN-COHEN LAW

The effective field strength from an aerial transmitter can be determined depending on the height, current, and distance between the transmitter and receiver, wavelength, and terrain:

$$E = 377\, I\, h/\lambda\, L\, e^{-\alpha r}$$

where E = effective field strength
h = aerial height
I = uniform current carried
L = distance between transmitter and receiver
λ = wavelength
α = constant, depending on the terrain

**Keywords:** aerial, electrical, field strength, physics, receiver, transmitter

AUSTIN, Louis Winslow, 1867-1932, American physicist
**Source:** Ballentyne, D. W. G. and Walker, L. E. Q. 1959.
*See also* RADIO WAVES

## AUTO-IONIZATION LAW—SEE AUGER EFFECT

## AVALANCHE, LAW OF (PSYCHOLOGY)

Neural impulses spread from a stimulus receptor to a number of other neurons that result in an effect disproportionate to the initial stimulus, as in an epileptic seizure.

**Keywords:** neural, receptor, stimulus
**Source:** Goldenson, R. M. 1984.
*See also* RAMON AND CAJAL

## AVERAGE LOCALIZATION, LAW OF

Visceral pain is localized in the least mobile viscera.

**Keywords:** medicine, pain, physiology, visceral
**Sources:** Friel, J. P. 1974; Landau, S. I. 1986.

## AVERAGES, LAW OF

The distribution of sample means possess less variability as shown by a smaller standard deviation of the sample means is:

$$\sigma_x/N^{1/2}$$

where $\sigma_x$ = standard deviation
N = size of sample

**Keywords:** distribution, sample, standard deviation, variability
**Source:** Van Nostrand. 1947.

## AVERAGES, LAW OF, FOR PLANT VARIETIES

New varieties (of plants) tend to increase the capacity for continued existence, which by the same general law become predominant, replacing previous dominant varieties.

**Keywords:** dominance, plants, varieties
Based on DARWIN, Charles, 1809-1882, English biologist
**Source:** Brown, S. B. and Brown, L. B. 1973.
*See also* DARWIN; MENDEL

## AVOGADRO LAW (1811) OR HYPOTHESIS; POSTULATE OF EQUAL VOLUMES, LAW OF

Equal volumes of different gases contain equal numbers of molecules at the same temperature and pressure. The Avogadro constant or number, which is $6.022 \times 10^{23}$ molecules in one gram-molecular weight or one mole of a substance, was formulated several years later. The number of molecules per unit volume is known as the Loschmidt number (Loschmidt calculated the Avogadro number) and is $2.687 \times 10^{22}$ molecules/L. From this law, one can determine the molecular weight of any gas, as the density compared with hydrogen equals one-half of the molecular weight. Avogadro coined the use of the word *molecule*.

**Keywords:** gases, molecules, physics, volume

AVOGADRO, Amedeo Count of Quaregna e di Cerretto, 1776-1856, Italian physicist and chemist
LOSCHMIDT, Joseph, 1812-1895, Austrian physicist
**Sources:** Asimov, I. 1966; Considine, D. M. 1976; Hodgman, C. D. 1952; Krauskopf, K. B. 1959; Layton, E. T. and Lienhard, J. H. 1988; Mandel, S. 1972; NYPL Desk Reference. 1989; Parker, S. P. 1989; Speck, G. E. 1965; Travers, B. 1996.
*See also* COMBUSTION; VAPOR PRESSURE

## AXES, LAW OF (CRYSTALLOGRAPHY)

Opposite ends of any of the axes of a crystal are cut by the same number of similar faces and are arranged similarly. The elements of crystal systems are based on the angle and amount of angle of the three axes, such as (1) cubic or regular in which the three axes are at right angles and all equal, (2) tetragonal in which the three axes are at right angles of which two are equal, etc.

**Keywords:** crystal, ends, faces, materials
**Source:** Considine, D. M. 1976.

# B

## BABINET PRINCIPLE OR LAW (1838); LAW OF DIFFRACTION BY DROP

Two diffractive screens, one of which is the negative of the other, produce the same diffraction patterns, or those patterns produced by complementary screens. In dichroic crystals, the faster ray is less absorbed.

**Keywords:** diffraction, drop, negative, patterns
BABINET, Jacques, 1794-1872, French physicist
**Sources:** Fisher, D. J. 1988. 1991; Parker, S. P. 1989.
*See also* FRESNEL

## BABINSKI LAW

A normal subject (human) inclines to the side of the positive pole according to the law of voltaic vertigo and a pathologic subject falls to the side to which he tends to incline spontaneously. There is no action if the labyrinth is destroyed.

**Keywords:** medicine, physiology, posture, vertigo
BABINSKI, Joseph Francois Felix, 1857-1932, French neurologist
**Sources:** Friel, J. P. 1974; Landau, S. 1986; Reber, A. S. 1985; Stedman, T. L. 1976.

## BABO LAW (1847)

The vapor pressure of a solvent is lowered in proportion to the amount of substance dissolved when a nonvolatile solid is added to a liquid.

**Keywords:** chemistry, liquid, physical chemistry, vapor pressure
BABO, Lambert Heinrich Clemens von, 1818-1899, German chemist
**Sources:** Denbigh, K. 1961; Hodgman, C. D. 1952: Isaacs, A. 1996; Uvarov, E. B. 1964.
*See also* RAOULT

## BACH LAW (1888)

The yield value and tensile strength of a material are dependent on the speed of testing, which with steel the tensile strength was less with a slower speed, because of the irregular vibrations of the atoms of the material. Bach law is a power law modification of Hooke law. Scott Blair has $s = Kd^n$, where s is the unit stress, K is a constant, d is the unit of deformation, and n = an approximate constant.

**Keywords:** deformation, speed, stress, testing, yield
BACH, Julius Carl von, 1847-1931, German engineer
**Sources:** Ballentyne, D. W. G. and Walker, L. E. Q. 1959; Houwink, R. and DeDecker, H. K. 1971; Scott Blair, G. W. 1949.

## BACTERIAL GROWTH, LAW OF; OR LAW OF ORGANIC GROWTH

The increase of bacteria growing freely in the presence of an unlimited food supply is proportional to the number of bacteria present, represented by the equation:

$$dN/dt = kN$$

where N = number present at t
t = time
k = constant

**Keywords:** bacteria, growth, microbiology, pathology
**Source:** Brown, S. B. and Brown, L. B. 1972.

### BAER LAW (BIOLOGY); VON BAER LAW; LAW OF VON BAER (1837)

The more general features that are common to all members of a group of animals are developed in the embryo earlier than the special features that distinguish the various members of the group. The development of an organism proceeds from the generalized (homogeneous) to the specialized (heterogeneous) condition, so that the earliest embryonic stages of related organisms are identical and distinguishing features develop later. This law was the predecessor of the theory of recapitulation.

**Keywords:** animals, development, embryo, features
BAER, Karl Ernst Ritter von, 1792-1876, Estonian naturalist, geologist, and embryologist
**Sources:** Critchley, M. 1978; Friel, J. P. 1974; Landau, S. I. 1986.
*See also* EMBRYOGENESIS; HAECKEL; RECAPITULATION

### BAER LAW (GEOLOGY); VON BAER LAW

The principle according to which the rotation of the Earth causes asymmetrical, lateral erosion of stream beds.

**Keywords:** erosion, rotation, stream beds
BAER, Karl Ernst Ritter von, 1792-1876, Estonian naturalist, geologist, and embryologist
**Source:** Gary, M. et al. 1973.
*See also* STREAM BANK

### BAGNOLD NUMBER, Ba OR $N_{Ba}$

A dimensionless group that represents the drag force related to the gravitational force for saltation:

$$N_{Ba} = 3\ c_o\ \rho_g\ V^2/4\ d\ g\ \rho_p$$

where $c_o$ = a constant
$\rho_g$ = gas density
$\rho_p$ = particle density
V = fluid velocity
d = diameter
g = gravitation

**Keywords:** drag, gravitational, saltation
BAGNOLD, Ralph A., 1896-1990, British physicist
**Sources:** Bolz, R. E. and Tuve, G. L. 1970; Parker, S. P. 1992; Pelletier, R. A. 1994.

### BAIRSTON NUMBER—SEE MACH

### BALFOUR LAW

The speed with which the ovum forms segments is approximately proportional to the concentration of the protoplasm in the ovum. The size of the segments is inversely proportional to the concentration of protoplasm in the ovum.

**Keywords:** concentration, ovum, protoplasm, speed
BALFOUR, Francis Maitland, 1851-1882, Scottish biologist
**Source:** Lapedes, D. N. 1974.

## BALLOT, LAW OF—SEE BUYS BALLOT

## BALMER SERIES FORMULA (1885)

A series in the visible bright line spectrum emitted by hydrogen is represented by the formula:

$$1/\lambda = R(1/2^2 - 1/n^2)$$

where $\lambda$ = wavelength of the line; $1/\lambda$ = wave number
R = Rydberg constant
n = 3, 4, 5,...; as n increases, the lines become closer

**Keywords:** bright line, hydrogen, spectrum, wavelength
BALMER, Johann Jakob, 1825-1898, Swiss mathematician and physicist
**Sources:** Speck, G. E. 1965; Thewlis, J. 1961-1964; Uvarov, E. B. et al. 1964.
*See also* PHYSICAL CONSTANTS-RYDBERG; RYDBERG-SCHUSTER

## BANCROFT LAW (ECOLOGY)

A generalization that organisms and communities tend to come into a state of dynamic equilibrium with their environment.

**Keywords:** communities, ecology, equilibrium, organisms
**Sources:** Hanson, H. C. 1962; Mettler, C. C. 1947.

## BAND SYSTEMS, LAW OF—SEE DESLANDRES

## BANSEN NUMBER, Ba OR $N_{Ba}$

A dimensionless group relating the heat transferred by radiation to the thermal capacity of the fluid:

$$Ba = h_r A_w / V_m c$$

where $h_r$ = radiant heat transfer coefficient
$A_w$ = wall area of the exchange
$V_m$ = mass flow rate
c = specific heat

**Keywords:** heat, radiation, transfer
**Sources:** Bolz, R. E. and Tuve, G. L. 1970; Parker, S. P. 1992; Potter, J. H. 1967.
*See also* STANTON

## BARBA LAW (1880)

Similar test specimen of metal with plastic deformation deform in a similar manner. For a cylindrical specimen the same value is obtained for the elongation if the gage length to diameter of the specimen is maintained as a constant.

**Keywords:** gage, material, metal, metallurgy, plastic
BARBA, Alvero Alonso, 1569–c. 1640, Peruvian metallurgist
**Source:** Thewlis, J. 1961-1964.

## BARDEEN, COOPER, AND SCHRIEFFER THEORY—SEE BCS

## BARFURTH LAW

The axis of the tissues in a regenerating structure of tissue is begins perpendicular to the cut.

**Keywords:** axis, cut, perpendicular, regenerating, tissue
BARFURTH, Dietrich, 1849-1927, German anatomist
**Sources:** Friel, J. P. 1974; Gray, P. 1967; Landau, S. I. 1986.

## BARIC WIND LAW OR BASIC WIND LAW—SEE BUYS BALLOT; STORMS

**Sources:** Fairbridge, R. W. 1967; Huschke, R. E. 1959.

## BARKHAUSEN EFFECT OR LAW (1919)

The succession of abrupt changes or minute jumps in magnetization occurring when the magnetizing force acting on a piece of iron or other magnetic material is varied due to discontinuities in size or orientation of magnetic domains (Fig. B.1).

**Keywords:** change, iron, magnetization
BARKHAUSEN, Heinrich Georg, 1881-1956, German electrical engineer
**Sources:** Considine, D. M. 1976; Isaacs, A. 1996; Parker, S. P. 1987. 1989.

## BAR THEORY OR LAW (1877)

Thick deposits of salt, gypsum, and other evaporated material form in oceans or seas or lakes as a result of a lagoon separated from the ocean by a bar in an arid climate. As water is lost by evaporation, additional water of normal salinity flows in from the ocean. The salinity constantly increases as some water is evaporating, reaching a point where salts are deposited. The theory was advanced by C. Ochsenius in 1877.

**Keywords:** evaporates, evaporation, salinity, salt deposits, water
OCHSENIUS, Carl, 1830-1906, German geologist
**Source:** Bates, R. L. and Jackson, J. A. 1984.

## BARUCH LAW

When the temperature of the water in a bath is above or below that of the body skin temperature, the effect is stimulating; when the temperatures are same, the effect is sedative.

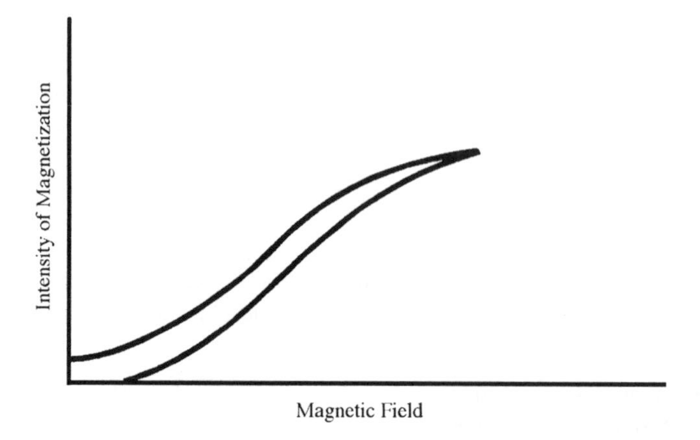


**Figure B.1** Change of applied magnetic field.

**Keywords:** bath, body, medicine, physiology, temperature
BARUCH, Simon, 1840-1921, American physician
**Sources:** Friel, J. P. 1974; Landau, S. I. 1986.

## BASIC WIND LAW—SEE BUYS BALLOT LAW; STORMS

**Source:** Fairbridge, R. W. 1967.

## BASIN AREAS, LAW OF (1956)

The direct geometric relation between stream order and the mean basin area of each order in a given drainage basis expressed as a linear regression of the logarithm of mean basin area to the stream order, the positive regression coefficient being the logarithm of the basin-area ratio that was originally stated by S. Schumm.

**Keywords:** basin area, drainage, stream
SCHUMM, Stanley A., twentieth century (b. 1927), American geologist
**Sources:** Bates, R. L. and Jackson, J. A. 1980. 1984; Gary, M. et al. 1972.
*See also* BAER

## BASTIAN-BRUNS LAW; BASTIAN LAW; BASTIAN-BRUNS SIGN

The tendon reflexes of the lower extremities are abolished if there is a complete traverse lesion in the spinal cord cephalad (toward the head) to the lumbar (near the loin) enlargement.

**Keywords:** alexia, medicine, physiology, reflex, spinal cord, tendon
BASTIAN, Henry Charlton, 1837-1915, English neurologist
BRUNS, Ludwig von, 1858-1916, German neurologist
**Sources:** Friel, J. P. 1974; Landau, S. I. 1986.

## BAUMÈS LAW—SEE COLLES

## BAUSCHINGER EFFECT

If a stable metal has been unloaded and then the stress increased to a level that will cause plastic strain again, plastic flow occurs more easily than if unloading had not occurred, with the difference being particularly large when the direction of strain is reversed (Fig. B.2).

**Keywords:** loading, specimen, reloading, stress, yield
BAUSCHINGER, Johann, 1833-1893, Austrian engineer
**Sources:** Flugge, W. 1962; Goldsmith, W. 1960; Reiner, M. 1960.

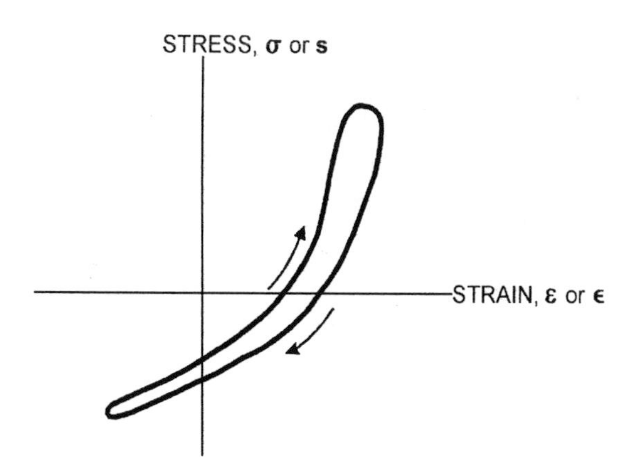


**Figure B.2** Loading and unloading to plastic strain.

## BAUSCHINGER LAWS (1890)

Several relationships involving the evaluation of materials, particularly tension stress, elastic limit, yield point, and modulus of elasticity, some of which have been changed or modified as improved means of measurement have been developed,

(a) A tension stress above the yield point increases the yield point to the applied stress
(b) A tension stress above the yield point reduces the elastic limit
(c) A tension stress which lies above the elastic limit, below the yield point, increases the elastic limit
(d) As a rule, a stress above the yield point lowers the modulus of elasticity
(e) Severe jars lower the elastic limit which has previously been raised by overstress
(f) Heating followed by cooling lowers the elastic limit and yield point which was previously increased by overstress
(g) Rapid cooling causes more effective lowering the elastic limit and yield point than slow cooling

His work, along with others, led to the practical development of reinforced concrete in the twentieth century.

**Keywords:** elastic limit, modulus of elasticity, stress, tension, yield point
BAUSCHINGER, Johann, 1833-1893, Austrian engineer
**Source:** Goldsmith, W. 1960.

## BAVENO LAW—SEE TWIN

## BAYES LAW OR BAYES THEOREM (1763)

This law states exactly how the probability of a certain "cause" changes as different events actually occur. The theorem was published posthumously in 1764. The difficulties in applying the law or theorem depend upon the fact that "a priori probabilities" are not known and are assumed to be equal in the absence of other knowledge.

**Keywords:** cause, events, probabilities
BAYES, Thomas, 1702-1761, British mathematician
**Sources:** Considine, D. M. 1976; Freiberger, W. F. 1960; James, R. C. and James, G. 1968; Sneddon, I. N. 1976.
*See also* LAPLACE LAW OF SUCCESSION

## BAYLISS AND STARLING LAW OF INTESTINE (c. 1913)

A generalization that a bolus moves abnormally in the small bowel because there is contraction above and relaxation below the distended ring, once accepted it is now believed that this relationship is far too complex to express in this simple law.

**Keywords:** bowel, contraction, movement, relaxation
BAYLISS, Sir william Maddock, 1860-1924, English physiologist
STARLING, Ernest Henry, 1866-1927, English physiologist
**Sources:** Friel, J. P. 1974; Garrison, F. H. 1929; Talbott, J. H. 1970; Thomas, C. L. 1989.
*See also* INTESTINE

## BCS THEORY (1957) (SUPERCONDUCTIVITY)

Named after J. Bardeen, L. Cooper, and J. Schrieffer, the theory relates to the superconductivity of some metals and alloys for which the electrical resistance goes to zero when cooled below a certain transition temperature. This relationship was first identified in 1911 and represented by the London theory, named after F. London and H. London.

**Keywords:** alloys, electric resistance, metals, temperature, transition

BARDEEN, John, 1908-1991, American physicist and electrical engineer; Nobel prize, 1956, physics (shared), and 1972, physics (shared)

COOPER, Leon Neil, twentieth century (b. 1930), American physicist; Nobel prize, 1972, physics (shared)

SCHRIEFFER, John Robert, twentieth century (b. 1931), American physicist; Nobel prize, 1972 (shared)

LONDON, Fritz, 1900-1954, German American physicist

LONDON, Heinz, 1907-1970, German English physicist

**Sources:** Besancon, R. M. 1974; Isaacs, A. 1996; Parker, S. P. 1989; Physics Today 16(1):19-28, 1963; Physical Review, 1957.

*See also* BOSE-EINSTEIN

### BECQUEREL EFFECT (1896)

The phenomenon of a current flowing between two unequally illuminated electrodes of a certain type when they are immersed in an electrolyte. Becquerel first reported rays emitted spontaneously by solid substances and is given the credit for discovering the phenomenon of radioactivity.

**Keywords:** current, electrodes, illuminated, electrolyte

BECQUEREL, Antoine Henri, 1852-1908, French physicist; Nobel prize, 1903, physics (shared)

**Sources:** Considine, D. M. 1995; Parker, S. P. 1992.

*See also* CURIE

### BEER LAW (1852)

The transmittance of an absorbing medium at a single wavelength of light varies with the negative (minus) logarithm of colorant concentration. Also stated as, the absorption of light by a solution changes exponentially with the concentration.

**Keywords:** absorption, illumination, light, transmittance

BEER, Wilhelm, 1797-1850, German astronomer

**Sources:** Friel, J. P. 1974; Hodgman, C. D. 1952; Lapedes, D. N. 1974; Layton, E. T. and Lienhard, J. H. 1988; Stedman, T. L. 1976; Thewlis, J. 1961-1964.

*See also* BEER-LAMBERT LAW; BOUGUER

### BEER-LAMBERT ABSORPTION LAW (1852)

The absorbance of a substance is directly proportional to the concentration of the material in the substance which causes the absorption. Or, the amount of radiation transmitted by a material decreases exponentially with increasing concentration for a given length (distance). Thus:

$$A = \log_{10} 1/T_r$$

where $A$ = absorbance
$T_r$ = transmittance

This law allows the qualitative determination of chemicals in solution by measuring the light absorbed in going through the solution. The law is sometime written as:

$$A = \varepsilon\, b\, c$$

where $\varepsilon$ = molar absorptivity
b = length passed through by light ray
c = chemical concentration

There was considerable controversy over the claim as to who developed this law as W. Beer, C. Bernard, P. Bouguer, and J. Lambert worked in the area. The relationship is now often indexed under the Beer-Lambert law.

**Keywords:** absorbance, physical chemistry, physics, radiation, transmittance
BEER, Wilhelm, 1797-1850, German astronomer
LAMBERT, Johann Heinrich, 1728-1777, German physicist
BERNARD, Claude, 1813-1878, French physiologist
BOUGUER, Pierre, 1698-1758, French mathematician
**Sources:** Denny, R. C. 1982; Landau, S. I. 1986.

## BEHAVIOR, LAW OF—SEE PAVLOV; SKINNER; THORNDIKE

## BEHRING LAW

The blood and serum of an immunized person, when given to another person, will render another person immune.

**Keywords:** blood, immunization, medicine, physiology
BEHRING, Emil Adolf von, 1854-1917, German bacteriologist
**Sources:** Clifford, A. F. 1964; Dorland, T. L. 1976; Friel, J. P. 1974; Honig, J. M. 1953; Landau, 1986.

## BELL-MAGENDIE LAW (1822); BELL LAW (1811); LAW OF MAGENDIE

The nerve cell fibers entering the spinal cord are provided two distinct types of functions: those at the back (posterior) of the cord are routes for nerve impulses (sensors) from sense receptors to the brain, and those at the front (anterior) are routes for nerve impulses (motor) from the brain to the muscles and organs. Also, in a reflex the nerve impulse can be transmitted in one direction only.

**Keywords:** cells, medicine, nerves, physiology, spinal cord
BELL, Sir Charles, 1774-1842, Scottish physiologist
MAGENDIE, Francois, 1783-1855, French physiologist
**Sources:** Dorland, T. L. 1976; Friel, J. P. 1974; Lapedes, D. N. 1974; Reber, A. S. 1985; Stedman, T. L. 1976; Thomas, C. L. 1989; Wolman, B. B. 1989; Zusne, L. 1987.

## BELONGINGNESS, LAW OF (1931)

In learning, subjects integrated into a pattern are better remembered than those which merely occur in close proximity. Some subjects or kinds of material are more readily learnable than others as some things seem to go together more naturally than others. The law is a result of the work of A. Thorndike.

**Keywords:** learning, materials, memory, patterns, subjects
THORNDIKE, Ashley Horace, 1871-1933, American educator
**Sources:** Corsini, R. J. 1984. 1987; Good, C. V. 1973; Harre, R. and Lamb, R. 1983.
*See also* LEARNING; THORNDIKE

## BENFORD LAW OF ANOMALOUS NUMBERS (1938)

The frequency of first digits of a random decimal begins with digit p is:

$$\log_{10} (p + 1) - \log p$$

This relationship shows that thirty percent of the random decimal numbers begin with 1 and five percent with 9. Although this law was named after F. Benford, the phenomenon was formulated by S. Newcomb in 1881.

**Keywords:** digits, mathematics, numbers, probability, statistics
BENFORD, Frank, twentieth century, American physicist/mathematician
NEWCOMB, Simon, 1835-1909, Canadian American astronomer/mathematician
**Sources:** Fisher, D. J. 1988. 1991; Kotz, S. and Johnson, N. L. 1982; Rosenkrantz, R. D. 1977.
*See also* FIRST-DIGIT PROBLEM

## BERARD—SEE DELAROCHE AND BERARD

## BERGMANN RULE OR LAW (ECOLOGY)

In a polytropic wide-ranging "species of warm-blooded animals the average size of members of each geographic race varies with the mean environmental temperature." Body size, especially weight and bulk, tend to be greater for better heat conservation in colder areas. There is a correlation, inversely of body size with the mean animal temperature.

**Keywords:** animals, environment, size, species
BERGMANN, Carl George Lucas Christian, 1814-1865, German biologist
**Sources:** Considine, D. M. 1974. 1995; Gray, P. 1967; Lapedes, D. N. 1974 Parker, S. 1957.
See also ADAPTATION; ALLEN; GLOGER

## BERGONIÉ-TRIBONDEAU LAW; BERGONIÉ AND TRIBONDEAU LAW

The sensitivity of cells to radiation varies (1) directly with the reproductive capacity of the cells, and (2) inversely with the degree of differentiation of the cells.

**Keywords:** cells, physiology, radiation, reproduction
BERGONIÉ, Jean Alban, 1857-1925, French physician and radiologist
TRIBONDEAU, Louis, 1872-1918, French physician
**Sources:** Friel, J. P. 1974; Landau, S. I. 1986.

## BERNARD LAW—SEE BEER-LAMBERT; BOUGUER

## BERNOULLI-EULER LAW, THEORY OF FLEXURE (1744)

In simple form, the curvature of the central fiber of a homogeneous bar or beam is proportional to the bending moment, named after D. Bernoulli and L. Euler. In mathematical form, where the cross-section remains constant, the moment of inertia:

$$M = E\ I/R$$

where M = moment of inertia
E = Young modulus
I = second moment of inertia around axis of beam
R = radius of curvature

The bending moment M is related to the radius of curvature of the central line of the beam. Theory is used to predict the deflection of a loaded beam from a knowledge of elastic constants, Young modulus to Poisson ratio.

This law was partially discounted by Saint Venant in 1856. J. Bernoulli earlier developed the equation for elastic curve of a beam. Galileo proposed the new sciences of elasticity and strength of materials in 1638. C. Coulomb presented a complete theory of a cantilever beam and determined its neutral section in 1773.

**Keywords:** bending, center fiber, cross-section, moment
BERNOULLI, Daniel, 1700-1782, Swiss mathematician
BERNOULLI, Jakob, 1654-1705, Swiss mathematician
COULOMB, Charles Augustin de, 1736-1806, French physicist
EULER, Leonhard, 1707-1783, Swiss mathematician
**Sources:** Ballentyne, D. W. G. and Lovett, D. R. 1971; Considine, D. M. 1974. 1995; Thewlis, J. 1961-1964.
*See also* HOOKE; POISSON; SAINT VENANT; YOUNG MODULUS

### BERNOULLI KINETIC THEORY OR LAW OF GASES (1738)

The kinetic energy of an atom of its mass multiplied by the square of its velocity. This law was built on the previous work of the Greek philosopher Democritus.

**Keywords:** atom, energy, mass, velocity
BERNOULLI, Daniel, 1700-1782, Swiss mathematician
DEMOCRITUS, c. 470-380 B.C., Greek philosopher
**Source:** Besancon, R. M. 1980.

### BERNOULLI LAW OF LARGE NUMBERS—SEE LARGE NUMBERS (WEAK LAW)

### BERNOULLI LAW OF LIQUID FLOW (1738); BERNOULLI THEOREM

A relationship that expresses the conservation momentum in fluid flow. If no work is done on or by a flowing incompressible liquid the total head remains unchanged.

**Keywords:** fluid flow, liquid, mechanics, work
BERNOULLI, Daniel, 1700-1782, Swiss mathematician
**Sources:** Considine, D. M. 1976; Layton, E. T. and Lienhard, J. H. 1988; Scott, J. S. 1993.

### BERNOULLI PRINCIPLE (1738); BERNOULLI LAW; BERNOULLI LAW OF HYDRODYNAMIC PRESSURE (FLUIDS)

As the velocity of fluid flow increases its pressure decreased. The Bernoulli principle is expressed as:

$$K = p + 1/2\, \rho\, v^2 + h\, \rho\, g$$

where K = constant
- p = pressure of the fluid
- ρ = density of the fluid
- v = velocity
- h = difference in elevation
- g = acceleration due to gravity

Approximately 50 yr. earlier, E. Mariotte measured the pressure of water exiting a pipe but not through a pipe.

**Keywords:** Fluid flow, pressure, velocity
BERNOULLI, Daniel, 1700-1782, Swiss mathematician (son of Johann Bernoulli)

MARIOTTE, Edme, 1620-1684, French physicist
**Sources:** Asimov, I. 1972; Dorland, T. L. 1976; Guillen, M. 1995: Isaacs, A. 1996.
*See also* MARIOTTE

**BERTHELOT EQUATION**

An equation of state relating pressure, volume, and temperature of a gas with the gas constant, R. The equation is derived from the Clausius equation:

$$pV = R\,T\,(1 + 9\,pT_c/128\,p_c\,T\,[1 - 6\,T_c^2/T^2])$$

where p = pressure
V = volume
T = absolute temperature
$T_c$ = critical temperature
$p_c$ = critical pressure

M. Berthelot developed the field of thermochemistry.*

**Keywords:** constant, gas, pressure, temperature, volume
BERTHELOT, Marcellin Pierre Eugene, 1827-1907, French physical chemist
**Sources:** *Chemical Heritage 16(2), Fall 1998; Considine, D. M. 1976.
*See also* CLAUSIS

**BERTHELOT LAW; BERTHELOT PRINCIPLE OF MAXIMUM WORK (1867); BERTHELOT-THOMSEN PRINCIPLE**

Berthelot was one of the earliest persons who in 1862 proposed equations of rate constants, k, represented by:

$$k = Ae^{DT} \text{ or } \ln k = \ln A + DT$$

where A and D = empirical quantities
T = absolute temperature

Of the various possible low temperature non-endothermic reactions which can proceed without the aid of external energy, the process takes place accomplished by the greatest evolution of heat. The solvents used have been acids of the same nature of the solutions; thus, liquid metals should be used for alloys, and molten oxides for multiple oxides. Several rules for selecting and combining solvents and solutes have been developed to improve the accuracy of calorimetry.

**Keywords:** calorimetry, chemistry, heat, non-endothermic, physical chemistry
BERTHELOT, Marcellin Pierre Eugene, 1827-1907, French physical chemist
THOMSEN, Hans Peter Jorgen Julius, 1826- 1909, Danish chemist
**Sources:** Ballentyne, D. W. G. and Lovett, D. R. 1972; Denbigh, K. 1961; Hodgman, D. C. 1952.

**BERTHELOT-NERNST LAW**

The ratio of the activity of a chemical substance in two phases is constant at a given temperature.

**Keywords:** activity, chemical, constant, temperature
BERTHELOT, Marcellin Pierre Eugene, 1827-1907, French chemist
NERNST, Hermann Walther, 1864-1941, German physical chemist; Nobel prize, 1920, chemistry
**Sources:** Chemical Heritage 16(2), Fall 1998; Honig, J. M 1953.

## BERTHOLLET LAW (1801); CHEMICAL AFFINITY LAWS

Salts in solution react with each other so as to form a less soluble salt if possible. Stated another way, two salts in solution by double composition, if they can produce a less soluble salt than the two salts, will produce a less soluble salt. C. Berthollet showed that acids do not need to contain oxygen.

**Keywords:** decomposition, reaction, salt, soluble, solution
BERTHOLLET, Claude Louis, 1748-1822, French chemist
**Sources:** Chemical Heritage 16(1):25, 1998; Dorland, T. L. 1976; Honig, J. M. 1953; Stedman, T. L. 1976.
*See also* PROUST

## BERTRAND PRINCIPLE OF SIMILITUDE

If two systems differ only in geometrical magnitude, the masses of corresponding parts being proportional to each other, they will be mechanically similar if the forces acting on and the velocities of respective parts bear the relation:

$$\text{force} \propto \frac{\text{mass} \times (\text{velocity})^2}{\text{linear dimension}}$$

The above relationship describes the conditions under which two systems which are geometrically similar may also be mechanically similar, and is applicable for testing small model devices.

**Keywords:** geometric, mechanically, similarity, systems
BERTRAND, Joseph Louis Francois, 1822-1900, French mathematician
**Source:** Northrup, E. F. 1917.

## BERZELIUS AND MITSCHERLICH, LAW OF ISOMORPHISM

Substances which form crystals of similar shape, known as isomorphous substances, have similar chemical properties, can usually be represented by analogous formulas, and the valence of the element can be derived from the formula of an appropriate compound. Berzelius discovered silicon; he is known for arranging elements in order of decreasing electronegativity; he was the first to use terms protein and catalytic force.

**Keywords:** crystalline, materials, shape, substances
BERZELIUS, Jons Jakob, 1779-1848, Swedish chemist
MITSCHERLICH, Eilhardt, 1794-1863, German chemist
**Sources:** Chemical Heritage 16(1):25, 1998; Hampel, C. A. and Hawley, G. G. 1973.

## BERZELIUS LAW; KNOWN ALSO AS DUALISTIC THEORY (CHEMISTRY) (1811)

*See* DUALISTIC THEORY; CONSTANT COMPOSITION (1817); ISOMORPHISM

## BET MODEL—SEE BRUNAUER-EMMETT-TELLER

## BETTI LAW OR THEOREM

When two systems of forces act on an elastic body, the work of the forces A on the displacements B is equal to the work of the forces B on the displacements A. The Maxwell reciprocity law is a special case of this law or theorem.

**Keywords:** displacements, elastic, forces
BETTI, Enrico, 1823-1892, Italian mathematician

**Sources:** Flugge, W. 1962; Gillispie, C. C. 1991.
*See also* MAXWELL (RECIPROCAL); PLASTIC FLOW

### BETZ LAW (1927)

Assuming a wind machine of 100 percent mechanical efficiency, the power that can be extracted is 59.3 percent of the wind's energy:

$$\text{Power} = 0.593\ \rho\ V_0^3/2$$

where $\rho$ = air density
$V_0$ = air velocity
0.593 = constant called the Betz coefficient

**Keywords:** air, coefficient, efficiency, power, velocity
BETZ, Albert, twentieth century (b. 1885), German physicist/engineer (fluid dynamicist)
**Sources:** Golding, E. W. 1955; Hunt, V. D. 1979; NUC; Schlicting, H. 1968.

### BEUDANT LAW (1819)

A principle of combining mineral substances whereby essentially similar compounds dissolved in the same solution will precipitate together, forming a crystal whose properties are common. The interfacial angles of this new crystal will have a value intermediate between the angles of the original compounds and proportional to the quantity of each.

**Keywords:** combining, compounds, crystals, interfacial, minerals, quantity
**Source:** Gillispie, C. C. 1981.
BEUDANT, Francois-Sulpice, 1787-1850, French mineralogist and geologist
*See also* MITSCHERLICH

### BEZOLD-BRUCKE PHENOMENON

The hue of a light is basically related to frequency, but changes in hue of most stimuli also result from changes in intensity. With increases in intensity, only three hues in the spectrum remain invariant, and these are at 475, 505, and 570 millimicrons.

**Keywords:** hues, intensity, light, spectrum, stimuli
BEZOLD, Albert von, 1836-1868, Bavarian physiologist
BRUCKE, Ernst Wilhelm von, 1819-1892, German physicist
**Source:** Physics Today 19(3):34, March 1966.
*See also* ABNEY

### BICHAT, LAW OF

Two major body systems are in inverse relationship, called the *vegetative* and *animal*, with the vegetative system providing for assimilation and augmentation of mass and the animal system providing for the transformation of energy.

**Keywords:** animal, body, systems, vegetative
BICHAT, Marie Francois, 1771-1802, French anatomist
**Sources:** Goldenson, R. M. 1984; Wolman, B. B. 1989.

### BIG BANG MODEL (1927; 1946*)

A theory of how the early universe developed, called the Big Bang model. Atomic nuclei that formed during the first few seconds and minutes of the formation of the universe, made mostly of helium and hydrogen, expanded throughout the universe over 10 to 20 billion years.

The theory explains the expansion of the universe, background radiation, and the abundance of helium in the universe. The theory was first developed by A. G. E. Lemaitre and revised by *G. Gamow.

**Keywords:** expansion, helium, hydrogen, radiation, universe
LEMAITRE, Abbe Georges Edouard, 1894-1966, Belgian astronomer
GAMOW, George, 1904-1968, Russian American physicist
**Sources:** *Asimov, I. 1976; Isaacs, A. 1996; Parker, S. P. 1989; Scientific American 275(6):68-73, December 1996.
*See also* HUBBLE

## BINGHAM NUMBER, Bi OR $N_{Bi}$ OR Bm OR $N_{Bm}$

A dimensionless group applicable to the flow of Bingham plastics is the yield stress divided by the viscous stress:

$$\text{Bi or Bm} = \tau_y L/\mu_p V$$

where $\tau_y$ = yield stress
L = channel width
$\mu_p$ = viscosity of the material
V = velocity of flow

**Keywords:** flow, plastics, viscous, yield
BINGHAM, Eugene Cook, 1878-1945, American chemist
**Sources:** Bolz, R. E. and Tuve, G. L. 1970; Land, N. S., 1972; Parker, S. P. 1992; Perry, R. H. 1967; Potter, J. H. 1967.

## BINGHAM SHEER STRESS LAW

A liquid that obeys the sheer stress law of Bingham, which describes the plastic behavior of many suspensions, such as cement slurry:

$$\tau = \tau_o + n_{pl}\tau$$

where $\tau$ = sheer stress
$\tau_o$ = flow limit
$n_{pl}$ = Bingham plasticity
$\tau$ = deformation rate

**Keywords:** deformation, liquid, plastic, sheer, slurry, suspensions
BINGHAM, Eugene Cook, 1878-1945, American chemist and rheologist
**Sources:** Eagleson, M. 1994; Reiner, M. 1960.

## BIN LOADING—SEE JANSSEN; KETCHUM; RANKINE

## BIOCLIMATIC LAW

The rate at which phenomenological events advance from the lower to higher latitudes and altitudes in spring and retreat in autumn. In North America, events generally occur at the average rate of four days to each degree latitude, 5 degrees of longitude, and 400 ft altitude. This is an arbitrary and formalistic scheme that can act as a standard to which actually occurring conditions can be compared.

**Keywords:** altitudes, climatology, latitudes, meteorology, phenomenological, seasons
**Source:** Huschke, R. E. 1959.
*See also* HOPKINS

### BIOGENESIS, LAW OF REPRODUCTION

All life comes from preceding life.

**Keywords:** evolution, life, physiology
**Sources:** Stedman, T. L. 1976; Thomas, C. L. 1989.
*See also* BIOGENETIC; HAECKEL; RECAPITULATION

### BIOGENETIC LAW; HAECKEL LAW

An organism passes through developmental stages resembling various stages in the phylogeny of its group; ontogeny recapitulates phylogeny.

**Keywords:** developmental, ontogeny, organism, phylogeny, stages
HAECKEL, Ernst Heinrich, 1834-1919, German biologist
**Sources:** Gray, P. 1967; Lapedes, D. N. 1974; Stedman, T. L. 1976; Thomas, C. L. 1989.
*See also* HAECKEL; RECAPITULATION

### BIOLOGIC CONTINUITY, LAW OF

Life on Earth has been continuous without interruption since inception, and there has been no gross change in the physical environment that would kill the temperature-sensitive organisms that exit with a metabolic range of about 0 to 35° C.

**Keywords:** change, environment, evolution, life, organisms, physiology
**Sources:** Fairbridge, R. W. 1967; Oliver, J. E. and Fairbridge, R. W. 1987; Thomas, C. L. 1989.
*See also* BIOGENESIS; BIOGENETIC

### BIOT (COMBUSTION) LAW OR NUMBER OR BIOT NUMBER, $\beta$ OR $N_\beta$

The rate of heat transfer from the center of a burning solid propellant is represented by a number identified as beta:

$$\beta = h\, R_e/k$$

where $\beta$ = Biot number
$h$ = heat transfer coefficient
$R_e$ = external radius of burning grain of propellant
$k$ = constant

**Keywords:** combustion, heat transfer, solid propellant
BIOT, Jean Baptiste, 1774-1862, French physicist
**Sources:** Considine, D. M. 1976; Land, N. S. 1972.
*See also* AVOGADRO; COMBINING WEIGHTS; DALTON

### BIOTIC LAW (1953)

A general classification of physical laws not necessarily relating to life, but as a causal chain connecting determining and determined events of a system. Both mnemic (precedes the event) and teleological (follows the event) laws are included. These laws are expressed in terms of the chronological relationship of cause and event.

**Keywords:** cause, effect, life, physiological, psychological
**Sources:** Braithwaite, R. B. 1953; Lapedes, D. N. 1974.
*See also* MNEMIC; TELEOLOGICAL

### BIOT LAW (1804)

Relates the intensity of solar radiation to the thickness of the atmosphere by the relationship:

$$I' = I\, e^{-at}$$

where $I'$ = intensity of radiation transmitted through thickness, t
$I$ = intensity of radiation of the incident beam of light
$t$ = thickness
$a$ = coefficient of absorption

This is an improvement over the earlier formula of Delaroche.

**Keywords:** atmosphere, intensity, thickness, radiation, solar
BIOT, Jean Baptiste, 1774-1862, French physicist
**Source:** Holmes, F. L. 1990.
*See also* DELAROCHE; CONDUCTION

### BIOT LAW OF ROTARY DISPERSION OF LIGHT (1818); BIOT LAW (OPTIC) (1812, 1818)

The rotation of polarized light produced by a plate of quartz decreases progressively with change of color from violet to red which is represented by the equation:

$$\alpha = k/\lambda^2$$

where $\alpha$ = rotation of light
$k$ = constant of material
$\lambda$ = wavelength

In general, the law can be expressed as optically active media produce a rotation which is proportional to the length of the path, to the concentration, and approximately to the inverse square of the light.

**Keywords:** color, light, optics, physics, polarized, rotation
BIOT, Jean Baptiste, 1774-1862, French physicist
**Sources:** Gillispie, C. C. 1981; Holmes, F. L. 1990.

### BIOT NUMBER, Bi OR $N_{Bi}$

A dimensionless group that relates the internal resistance to heat or mass transfer to the resistance of that flow at the surface of the solid.

For heat transfer:

$$Bi_h = hL/k$$

where $h$ = film coefficient
$L$ = length representing the half thickness of the plate or radius
$k$ = thermal conductivity of the solid

For mass transfer at the interface and mass transfer rate in interior of solid wall:

$$Bi_m = k_m L/D$$

where $k_m$ = mass transfer coefficient
$L$ = thickness of solid wall
$D$ = diffusivity at the interface

**Keywords:** conductivity, heat, mass, resistance
BIOT, Jean Baptiste, 1774-1862, French physicist
**Sources:** Bolz, R. E. and Tuve, G. L. 1970; Land, N. S. 1972; Parker, S. P. 1992; Perry, R. H. 1967; Potter, J. H. 1967.
*See also* BIOT; NUSSELT

## BIOT-SAVART LAW; BIOT AND SAVART LAW (1820); AMPERE LAW; ALSO KNOWN AS LAPLACE LAW

A law in electromagnetism in which the magnetic field in the vicinity of an infinitely long, straight conductor is proportional to the strength of a steady current in the conductor, and inversely proportional to the distance from the conductor:

$$\beta \propto I/s = \mu_0 \, I/2\,\pi\, s$$

or magnetic intensity

$$H = 2\ I/s$$

where $\beta$ = magnetic flux density (magnetic induction)
$\mu_0$ = permeability of free space
H = magnetic intensity
I = current
s = distance from the conductor

**Keywords:** conductor, current, electrical, magnetic
BIOT, Jean Baptiste, 1774-1862, French physicist
SAVART, Felix, 1791-1841, French physicist and physician
**Sources:** Daintith, J. 1981; Encyclopaedia Britannica. 1961; Gartenhaus, S. 1966; Parker, S. P. 1992.)
*See also* AMPERE

## BIRCH LAW

The law is based on the principle that the velocity of each material is linear with density. Based on geophysical research the following law was developed:

$$V_p = a(AW_m) + b\rho$$

where $V_p$ = compressional wave velocity of seismic waves
$\rho$ = density
$a(AW_m)$ = parameter depending upon the mean atomic weight
b = constant

The law is a linearization of the power law over a certain range of density.
This law is used as a tool for design and construction of earth models.

**Keywords:** compressional wave, density, earth, geophysical wave, velocity
BIRCH, Francis, 1903-1992, American geologist
**Sources:** Science 177:261-262, 21, July 1972; Science 257:66-67, 3 July 1992.
*See also* RICHTER

## BIVALENCE, PRINCIPLE OF OR LAW OF

Every statement is either true or false, that is, every statement has a truth-value, and there are only two truth values. It does not consider the "excluded middle." The principle or law of bivalence is a semantic principle or one governing the interpretation of the language to which it is applied.

**Keywords:** false, statement, true, truth-value
**Source:** Flew, A. 1984.
*See also* EXCLUDED MIDDLE

### BIVARIATE LAW; TWO-DIMENSIONAL NORMAL LAW

The joint distribution of two statistically related variables, such as range and deflection, may be related by a generalization of the normal or Gaussian law to two random variables.

**Keywords:** deflection, Gaussian, normal, statistics, range, variables
**Sources:** Freiberger, W. 1960; James, R. C. and James, G. 1968; Sneddon, I. N. 1976.
*See also* NORMAL

### BLACKBODY LAW (1900)

If energy is radiated at each wavelength of light from an ideal radiating body (one that is capable of emitting light at every frequency) at an absolute temperature, T, that body must also be an ideal absorber, known as a blackbody. The law was developed by M. Planck in 1900 based on L. Boltzmann's studies. In 1860 G. Kirchhoff showed that an ideal emitter must also be an ideal absorber.

**Keywords:** absorber, emitter, energy, radiation, temperature
BOLTZMANN, Ludwig Edward, 1844-1906, Austrian physicist
KIRCHHOFF, Gustav Robert, 1824-1887, German physicist
PLANCK, Max Karl Ernst Ludwig, 1858-1947, German physicist
**Source:** Bynum, W. F. et al. 1988.
*See also* PLANCK; STEFAN; STEFAN-BOLTZMANN; WIEN DISPLACEMENT

### BLACK LAW FOR TEMPERATURE OF MIXTURES (1754)

The law describes the equilibrium temperature of a mixture of two liquids. The final temperature of a mixture can be calculated from the average of the initial temperatures with the heat capacities as weights:

$$c_1m_1t_1 = c_2m_2t_2 - (c_1m_1 + c_2m_2)\, f$$

where $t_1$ and $t_2$ = initial temperatures of substances 1 and 2
$m_1$, $m_2$ = masses of the two liquids
$c_1$, $c_2$ = specific heats of the two liquids
$f$ = final temperature

J. Black developed the concept of latent heat of evaporation in 1754 and latent heat of melting ice in 1774.

**Keywords:** equilibrium, heat capacities, liquids, masses, temperature
BLACK, Joseph, 1728-1799, Scottish chemist
**Sources:** Layton, E. T. and Lienhart, J. H. 1988; Schmidt, J. E. 1959; Science 222(4627): 971-973, 2 Dec 1983.

### BLACKMAN PRINCIPLE OF LIMITING FACTOR—SEE LIEBIG

### BLAGDEN LAW (1788)

The reduction of the freezing point of a solution is proportional to the amount of dissolved matter in the solution for moderate concentrations. R. Watson first discovered this law in 1771.

**Keywords:** chemistry, concentrations, freezing point, physical chemistry, solution
BLADGEN, Charles, 1748-1820, British physical chemist
WATSON, Richard, 1737-1816, English chemist
**Sources:** Bynum, W. F. et al. 1981; Considine, D. M. 1976; Dorland, T. L. 1976; James, A. M. 1976.
*See also* (DE) COPPET; HESS; KOPP; RAOULT

## BLAKE NUMBER, B OR $N_B$

A dimensionless group that represents the inertial force divided by the viscous force used for representing flow of fluids in particle beds:

$$B = \rho V/\mu\ (1 - e)S$$

where $\rho$ = mass density
$V$ = velocity of flow
$\mu$ = absolute viscosity
$e$ = void ratio (volume of voids/volume of solids)
$S$ = area-volume ratio of particles

This is a modified Reynolds number.

**Keywords:** flow, fluids, inertia, viscous
BLAKE, Henry William, 1865-1929, American chemical engineer
**Sources:** Bolz, R. E. and Tuve, G. L. 1970; Land, N. S. 1972; Morris, C. G. 1992; Parker, S. P. 1992; Potter, J. H. 1967.
*See also* REYNOLDS

## BLANC LAW FOR MOBILITY OF FREE ELECTRONS IN GASES

The mobility of electrons and ions in a mixture of gases a, b, c,... is given by:

$$1/u = 1/u_a + 1/u_b + 1/u_c + \ldots$$

where $u_a$ represents the mobility in pure gas, a, at partial pressure, $p_a$, etc., for electrons and ions that are sensibly independent of the field strength.

**Keywords:** chemistry, gases, ions, physical chemistry, physics
**Source:** Menzel, D. H. 1960 (does not refer to Blanc).

## BLANC RULE OR BLANC REACTION

Cyclization of dicarboxylic acids on heating with acetic anhydride to give either cyclic anhydrides or ketones depending on the respective positions of the carboxyl groups: 1,4- and 1,5-diacids give anhydrides, while diacids in which the carboxyl groups are in 1,6- or further removed positions give ketones.

Also expressed as a synthetic form of barium sulfate, a white precipitate which is formed with sodium sulfate, used as a pigment extended.

**Keywords:** acids, anhydrides, cyclization, diacids, ketones
BLANC FIXE, from French, for white precipitate formed
**Sources:** Lewis, Richard J. 1993; Morris, C. G. 1992; Potter, J. H. 1957; Sax, N. I. and Lewis, R. J. 1987.

## BLOCH LAW (LIGHT FLASHES) (1885)—SEE BUNSEN-ROSCOE

The critical duration of flashes of light to provide a clear image is about 30 milliseconds.

**Keywords:** flashes, light, image
BLOCH, A. M., nineteenth century, German scientist*
**Sources:** Landau, S. I. 1986; *Zusne, L. 1987.
*See also* BUNSEN-ROSCOE; EMMERT; FERRY-PORTER; TALBOT

### BLOCH THEOREM

The wave function of an electron moving in a periodic potential, as in a crystal lattice, has the form of a plane wave, modulated by a function that has the same periodicity of the original plane wave function.

**Keywords:** crystal, electron, function, potential, wave
BLOCH, Felix, 1905-1983, Swiss American physicist; Nobel prize, 1952, physics
**Sources:** Asimov, I. 1972; Besancon, R. M. 1974. 1985; Parker, S. P. 1989.

### BLOCH $T^{3/2}$ LAW OF MAGNETISM

At very low temperatures, the spontaneous magnetism, $M_s$, may be expressed as a function of temperature by:

$$M_{s,T} = M_{s,o} (1 - CT^{3/2})$$

where C = a constant equal to, for example:

$$C = 0.0587/2s\ (k/2sJ)^{3/2}$$

for a body centered structure of spin s. This law is in quite good agreement with observations in the very low-temperature region but, at a somewhat higher temperature, a $T^2$ term replaces the $T^{3/2}$ term.

**Keywords:** magnetism, spin, temperature
BLOCH, Felix, 1905-1988, Swiss American physicist; Nobel prize, 1952, physics
**Source:** Ballantyne, D. W. G. and Lovett, D. R. 1980.

### BLONDEL RULE

A guideline for perspective that a well balanced stairway occurs when the G + 2H = 2 ft, where G is the tread and H is the height of the step.

**Keywords:** architecture, balance, perspective, stairway, tread
BLONDEL, Nicolas-Francois, 1618-1686, French military engineer
**Source:** Holmes, F. L. 1980. 1990.

### BLONDEL-REY LAW (1911); ALSO CALLED TALBOT LAW

The threshold below which a steady point source of light in invisible, and the apparent brightness above the limit, both depend upon the total luminous flux entering the eye. The law is applicable for flashing light below 5 Hz. The law is expressed as:

$$\beta = \beta_o\ [t/(a + t)]$$

where $\beta$ = near the threshold for white light
$\beta_o$ = brilliance of a point source
t = flash duration in seconds
a = constant, approximately 0.2 seconds

**Keywords:** illumination, lighting, physics, vision
BLONDEL, Andre Eugene, 1863-1938, French physicist and engineer

REY, Jean Alexandre, 1861-1935, Swiss engineer
**Sources:** Bothamley, J. 1993; Morris, C. G. 1992; Thewlis, J. 1961-1964.
*See also* BLOCH; BUNSEN-ROSCOE; TALBOT

### BOBILLIER LAW

A law of design applied to four bar mechanisms of general rigid plane motion, associates point paths and their centers of curvature. By use of the law, one can quickly map center or path points on a desired ray when one other center or path point pair are known on a different ray.

**Keywords:** design, four bar, mechanism, motion
BOBILLIER, Etienne, 1798-1840, French geometer and mechanics
**Source:** Parker, S. P. 1994.
*See also* ROBERTS

### BODE LAW (1772); ALSO CALLED TITIUS-BODE LAW (1792)

An empirical relationship relating the mean distance in astronomical units (A. U.) of various planets from the sun. The relationships are obtained by adding to 4 the following figures: 0, 3, (3) (2) = 6; (6) (2) = 12; (12) (2) = 24; etc. Thus, the astronomical units of distance from the sun are:

| | Bode Distance |
|---|---|
| Mercury | 4 |
| Venus | 7 |
| Earth | 10 |
| Mars | 16 |
| | 28 |
| Jupiter | 52 |
| Saturn | 100 |
| Uranus | 196 |
| Neptune | 388 |
| Pluto | 772 |

Although named after J. Bode, who published the data, the law was previously discovered by J. D. Titius. A formula that approximates the distance of the planets from the sun in astronomical units, A. U. = 4.0 + (0.3N), for which N = 0, 1, 2, 4, 8 for each successive planet.

**Keywords:** astronomy, mathematics, planets, physics
BODE, Johann, Elert, 1747-1826, German astronomer
TITIUS, Johann Daniel, 1729-1796, Polish astronomer
**Sources:** Encyclopaedia Britannica; 1961; Mitton, J. 1993.

### BODENSTEIN NUMBER, Bo OR $N_{Bo}$

A dimensionless group representing diffusion in reactors:

$$Bo = VL/D_a$$

where V = velocity
L = reactor length
$D_a$ = axial diffusivity

**Keywords:** diffusion, reactors, velocity
BODENSTEIN, Max, 1871-1942, German chemist
**Sources:** Bolz, R. E. and Tuve, G. L. 1970; Parker, S. P. 1992; Potter, J. H. 1967.

### BODY SIZE LAW FOR METABOLISM; BRODY LAW; BRODY AND PROCTER LAW (1932)

In general, there is a linear relationship between the logarithm of metabolic rate and body weight. The metabolism rate of a mammal is proportional to the three-fourths power of the weight. Although there are modifications, the basic expression of the relationship, which is empirical, is:

$$M = a\,W^{p}$$

$$M = 70 \times W^{3/4}$$

where M = in kcal per day metabolism rate
W = weight of the mammal in kilograms
a, p = constants

**Keywords:** mammal, metabolism, weight
BRODY, Samuel, 1890-1956, American physiologist
PROCTER, R. C., twentieth century, American physiologist
**Source:** Kleiber, M. 1961.
*See also* KLEIBER

### BOHR EFFECT (PHYSIOLOGY)

The effect of body size on the capacity of the blood to transport oxygen to the tissues, not by a change in blood volume or hemoglobin content, but oxygen-hemoglobin dissociation. This dissociation is known as the Bohr effect. An increase in carbon dioxide concentration in the blood makes the blood more acidic and decreases the efficiency of uptake of oxygen by hemoglobin.

**Keywords:** blood, body, hemoglobin, oxygen, size
BOHR, Christian, 1855-1911, Danish physiologist
**Sources:** Friel, J. P. 1974; Isaacs, A. 1996.

### BOHR RELATION OR BOHR THEORY (PHYSICS) (1913)

If an atom is struck by a plane electromagnetic wave whose frequency, $\upsilon$, is related to the difference in energy of two levels, $E_1$ and $E_2$, the Bohr relation:

$$\upsilon = (E_1 - E_2)/h$$

and if there is an allowed dipole transition between the two levels, then an atom in the lower state can be excited into the higher state with the absorption of a photon, or an atom in the upper state can be stimulated to make a transition to the lower state with the emission of a photon of the same frequency, polarization, and direction as the incident electromagnetic wave.

**Keywords:** atom, dipole, electromagnetic, energy level, frequency, photon, state
BOHR, Niels Henrik David, 1885-1962, Danish physicist; Nobel prize, 1922, physics
**Sources:** Isaacs, A. 1996; Meyers, R. A. 1992.

### BOILING NUMBER—SEE NUSSELT BUBBLE

### BOILING POINT LAW, GENERAL

The decrease in vapor pressure of a nonvolatile solvent at the boiling point is proportional to the increase in mole fraction of solute as related to the moles of solute present (Fig. B.3). Thus:

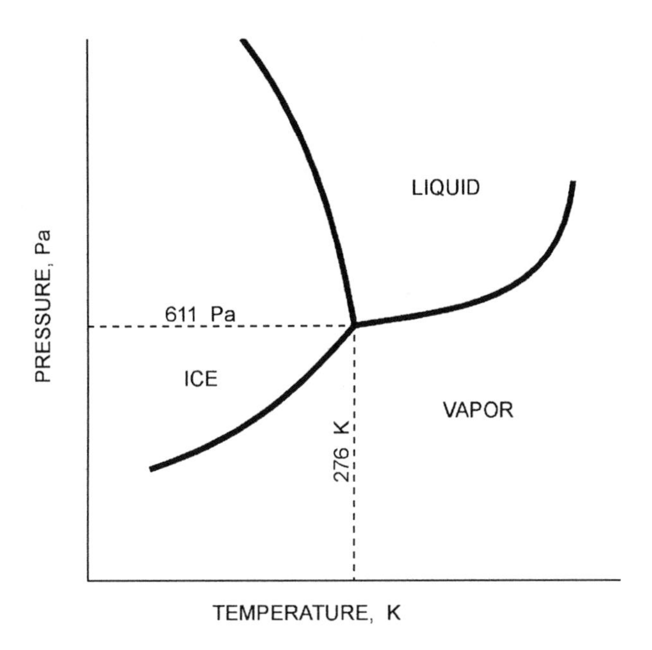


**Figure B.3** Triple point for water.

$$-dp = -p\ dx/x$$

where p = vapor pressure
x = mole of solute

**Keywords:** physical chemistry, solution, thermodynamics, vapor pressure
**Sources:** Considine, D. M. 1976; Parker, S. P. 1987.
*See also* RAMSAY AND YOUNG; RAOULT; VAN'T HOFF

## BOLTZMANN—SEE STEFAN-BOLTZMANN

## BOLTZMANN DISTRIBUTION LAW (1871)

Although we cannot specify the state of a particle molecule at a specific time, at a certain temperature, we can describe the energy state of a fraction of the molecule at a defined state (i and k) as represented by:

$$N_i/N_k = \exp\{(-E_i - E_k)/kT\}$$

where N = number of molecules
E = energy
T = temperature, absolute
k = Boltzmann constant

**Keywords:** constant, energy, molecule, state
BOLTZMANN, Ludwig Edward, 1844-1906, Austrian physicist
**Sources:** Besancon, R. M. 1974; Isaacs, A. 1996.
*See also* BOSE-EINSTEIN; FERMI-DIRAC; MAXWELL

## BOLTZMANN IONIC PRINCIPLE OR LAW

The ionic distribution in a gas is a function of the ratio of the electrical potential energy to the thermal energy of each ion.

**Keywords:** distribution, electrical potential, gas, ions, thermal
BOLTZMANN, Ludwig Edward, 1844-1906, Austrian physicist
**Source:** Honig, J. M. 1953 (partial).

## BOLTZMANN NUMBER, Bo OR $N_{Bo}$

A dimensionless group that is a parameter of thermal radiation exchange that relates the enthalpy of gases and heat flow emitted at the surface:

$$Bo = vc\gamma/\eta T^3$$

where v = flow velocity
c = specific heat
$\gamma$ = specific weight
$\eta$ = Stefan-Boltzmann constant
T = absolute temperature

This is also defined as the Thring radiation group.

**Keywords:** enthalpy, gases, heat, radiation, thermal
BOLTZMANN, Ludwig Edward, 1844-1906, Austrian physicist
**Sources:** Bolz, R. E. and Tuve, G. L. 1970; Land, N.S. 1972; Parker, S. P 1992; Potter, J. H. 1967.
*See also* THRING

## BOND COMMINUTION THEORY OR LAW; THIRD THEORY OF COMMINUTION (1952)

An empirical relationship based on a large number of experimental results, for representing the energy required to reduce a material to a smaller size:

$$W = W_i \ (10/P^{1/2} - 10/F^{1/2})$$

where W = required work input
P and F = (respectively) 80% passing sizes of the product and feed particles
$W_i$ = index, that represents the energy required to reduce a large particle to 80% passing 100 μm (in kWh/t). The energy represented by this relationship is between the values for energy using the Kick and Rittinger laws for crushing.

**Keywords:** energy, engineering, grinding, size reduction
BOND, Fred C., twentieth century (d. 1899), American mining and metallurgical engineer
**Sources:** AIME (mining) Trans. 193. 1952; Kirk-Othmer. 1985. 1998; Morris, C. G. 1992.
*See also* KICK; RITTINGER

## BOND NUMBER, Bo OR $N_{Bo}$ (1928)

A dimensionless group that relates the gravity force divided by the surface tension force in atomization, motion of bubbles and drops, and capillary flow:

$$Bo = \rho \ L^2 \ g/\sigma$$

where $\rho$ = mass density

L = characteristic length
g = gravitational acceleration
$\sigma$ = surface tension

This number is equal to Weber number/Froude number, and to the Eötvös number.

**Keywords:** atomization, gravity, surface tension
BOND, Wilfrid Noel, twentieth century, English physicist
**Sources:** Bolz, R. E. and Tuve, G. L. 1970; Land, N. S. 1972; NUC; Parker, S. P. 1992.
*See also* EÖTVÖS; FROUDE; OHNESORGE; WEBER

## BONE FORMATION, LAW OF

Every change in the form and function of a bone results in formation of rod-like formation and orientation (trabecular) and external form of the bone.

**Keywords:** bone, external, form, function
**Source:** Glasser, O. 1944.

## BOOLEAN ALGEBRA LAWS (1854); OR BOOLE LAWS OF THOUGHT

Boolean algebra consists of a set of objects X, Y, Z, ... so that, for any two elements X and Y, a product of XY, a sum X + Y, and a component X-bar are uniquely defined to satisfy the following basic rules.

**Idempotence Laws**
Multiplying a Boolean quantity by itself or adding a Boolean quantity to itself leaves the quantity unchanged:

$$X + X = X$$

$$X \cdot X = X$$

**Commutative Laws**
The order of addition does not affect the sums. The order of multiplication does not affect the product:

$$X + Y = Y + X$$

$$XY = YX$$

**Associative Laws**
The meaning of the product XYZ or the sum of X + Y + Z is unambiguous.

**Distributive Laws**
A Boolean algebra is distributive for both multiplication and addition:

$$X(Y + Z) = XY + XZ$$

$$Y + YZ = (X + Y)\ (X + Z)$$

**Zero and Unity Laws**

$$0 + X = X \qquad 1 + X = 1$$
$$0 - X = 0 \qquad 1 - X = X$$

**Absorption, Laws of**

$$X + XY = X \qquad X + \bar{X}Y = X + Y$$

$$X + (X+Y) = X \qquad X(\bar{X}+Y) = XY$$

**Involution, Law of**

The value of a quantity twice negated is unchanged:

$$\bar{\bar{X}} = X$$

**Complementarity, Laws of**

The sum of $X + X$ forms a united class and the product $X \cdot \bar{X}$ defines an empty or null class (0):

$$X + \bar{X} = 1 \qquad X \cdot \bar{X} = 0$$

**Keywords:** absorption, algebra, associative, commutative, complementarity, distributive, idempotence, involution, unity, zero.
BOOLE, George, 1815-1864, English mathematician
**Sources:** James, R. C. and James, G. 1968; Sneddon, I. N. 1976.
*See also* THOUGHT

## BORDET LAW

Blood corpuscles added to a hemolytic medium in mass are more rapidly dissolved than when added in fraction.

**Keywords:** blood, hemolysis, medicine, physiology
BORDET, Jules Jean Baptiste Vincent, 1870-1961, Belgian bacteriologist; Nobel prize, 1919, physiology/medicine
**Source:** Friel, J. P. 1974.

## BOREL—SEE LARGE NUMBERS

BOREL, Felix Edouard Emile, 1871-1956, French mathematician

## BORN THEORY OR LAW OF MELTING (1939) (CRYSTAL)

The rigidity modulus of a crystal decreases with temperature.

**Keywords:** crystal, modulus, rigidity
BORN, Max, 1882-1970, German physicist
**Sources:** Bothamley, J. 1993; Thewlis, J. 1961-1964.

## BOSANQUET LAW

In a magnetic circuit the magnetic flux, maxwells, is equal to the magnetomotive force, gilberts, divided by the magnetic reluctance, oersteds. The Bosanquet law is the magnetic analogy with the Ohm law. Flux in a magnetic circuit is $F/R = 0$.

**Keywords:** electricity, gilberts, magnetism, maxwells, physics, reluctance
BOSANQUET, Robert Holford, 1841-1912, English physicist
**Sources:** Considine, D. M. 1974. 1995.

## BOSCOVICH LAW OF FORCES BETWEEN PARTICLES (1745); UNIVERSAL FORCE LAW (1758)

The forces between particles (Fig. B.4) are repulsive at very small distances and become indefinitely greater and greater as the distances are diminished, in such a manner that they are capable of destroying any velocity with which one point may approach another, before the distance between them vanishes. When the distance between them is increased, they are

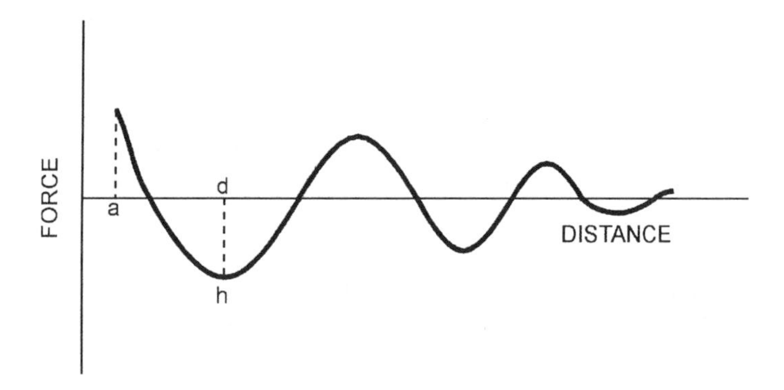


**Figure B.4** Force between particles.

diminished in such a way that at a certain distance, which is extremely small, the force becomes nothing. Then, as the distance is still further increased, the forces are changed to attractive forces, which at first increase, vanish, and become repulsive forces. For quarks, the force is greater for a larger distance between them.

**Keywords:** attractive, distance, forces, particles, repulsive

BOSCOVICH, Rudjer Josip, 1711-1787, Croatian (Yugoslavian) mathematician

**Sources:** Barrow, J. D. 1991; Jammer, M. 1957; Lederman, L. 1993.

*See also* QUARKS

### BOSE-EINSTEIN DISTRIBUTION LAW (1925)

Gives the number of bosons, such as photons, $n_l$, in a particular energy state, $E_l{:}n_l = 1/[e^{E(i)/kT-1}]$. Bose-Einstein statistics are used in developing this law, in which the Pauli exclusion principle is not obeyed, so that any number of identical bosons can be in the same state. The exchange of two bosons of the same type affects neither the probability of distribution nor the size of the wave function. The atoms lose their individual identities as zero deg. abs. (0 K) is reached, forming Bose-Einstein condensate, a phenomenon of quantum mechanics. Individual atoms in a gas at 0 K settle into a single quantum state.

**Keywords:** bosons, energy, exchange, exclusion, probability

BOSE, Satyendra Nath, 1894-1974, Indian physicist [after whom boson named]

EINSTEIN, Albert, 1879-1955, German Swiss American physicist; Nobel prize, 1921, physics

**Sources:** Chem & Eng News 75(5):15, 1997; Gray, H. J. and Isaacs, A. 1975; Science News 148, p. 373, Dec. 2, 1995; Scientific American 273(8), 18 August 1995.

*See also* PLANCK; QUANTUM

### BOUDIN LAW

There is an antagonism between malaria and tuberculosis.

**Keywords:** malaria, medicine, tuberculosis

BOUDIN, Jean Christian Marc Francois Joseph, 1806-1867, French physician

**Sources:** Friel, J. P. 1974; Landau, S. I. 1986.

### BOUGUER ANOMALY LAW

A value for gravity after corrections are made for latitude, elevation, and terrain. The anomaly is obtained by subtracting the attraction of the topography above sea level from the free air anomaly, which is strongly correlated with topography on a local scale. The Bouguer anomalies are normally used in geophysical prospecting.

**Keywords:** elevation, gravity, latitude, terrain
BOUGUER, Pierre, 1698-1758, French physicist and mathematician
**Sources:** Bates, R. L. and Jackson, J. A. 1983; Lerner, R. G. and Trigg, E. L. 1981.

### BOUGUER LAW (1729); BOUGUER AND LAMBERT LAW; BEER LAW; LAMBERT LAW OF ABSORPTION (1768); BEER-LAMBERT-BOUGUER LAW; LAMBERT-BEER LAW; BEER-BERNARD-LAMBERT-BOUGUER LAW

An exponential law of absorption of the electromagnetic spectrum in some absorbing medium is expressed for a particular wavelength. The flux density is a function of the characteristics of the absorbing medium for the particular wavelength, and distance as negative exponent of e:

$$I = I_0 \, e^{-kx}$$

where $I$ = flux density
$I_0$ = flux density before absorption
$k$ = constant characteristic of absorbing medium
$x$ = distance

**Keywords:** absorbing, electromagnetic, exponential, spectrum, wavelength
BOUGUER, Pierre, 1698-1758, French physicist and mathematician
**Sources:** Denny, R. C. 1973. 1982; Honig, J. M. 1953; Lapedes, D. N. 1974; Menzel, D. H. 1960; Thewlis, J. 1961-1964.
*See also* BEER; LAMBERT

### BOUGUER NUMBER, Bou OR $N_{Bou}$

A dimensionless group that relates the radiant heat transfer to dust laden gas through which heat is transferred:

$$Bou = 3C_D \, \lambda_r / 4\rho_0 \, R$$

where $C_D$ = mass of dust per bed volume
$\lambda_r$ = mean path length for radiation transfer
$\rho_0$ = mass density of the dust
$R$ = mean particle radius

**Keywords:** dust, heat transfer, radiation
BOUGUER, Pierre, 1698-1758, French physicist and mathematician
**Sources:** Bolz, R. E. and Tuve, G. L. 1970; Land, N. S. 1972; Parker, S. P. 1992; Potter, J. H. 1967; Thewlis, J. 1961-1964.

### BOUSSINESQ LAW

For atmospheric calculations, approximations can be made if one neglects the change of density except when coupled with gravity to produce buoyancy forces.

**Keywords:** atmosphere, buoyancy, density, gravity
BOUSSINESQ, Joseph Valentin, 1842-1929, French mathematician and physicist (mechanics)
**Source:** Houghton, J. 1986.

## BOUSSINESQ NUMBER, Bous OR $N_{Bous}$

A dimensionless group that is a modified Froude number, that represents waves in an open channel, and is the ratio of the inertia force and gravity force:

$$Bous = V/(2\ g\ R_H)^{1/2}$$

where V = velocity
g = gravitational acceleration
$R_H$ = hydraulic radius (cross-sectional area divided by the wetted perimeter)

**Keywords:** channel, gravity, inertial, waves
BOUSSINESQ, Joseph Valentin, 1842-1929, French mathematician and physicist (mechanics)
**Sources:** Bolz, R. E. and Tuve, G. L. 1970; Land, N. S. 1972; Parker, S. P. 1992; Potter, J. H. 1967.
*See also* FROUDE

## BOWDITCH LAW

The contraction of the heart muscle will not be increased by a stimulus greater than the minimum required to produce a contraction.

**Keywords:** contraction, heart, nerves, physiology, response, stimulus
BOWDITCH, Henry Pickering, 1840-1911, American physiologist
**Sources:** Friel, J. P. 1974; Landau, S. I. 1986; Stedman, T. L. 1976.
*See also* ALL-OR-NONE

## BOYLE-CHARLES LAW

A law pertaining to gases that combines both the Boyle (or Mariotte) law and the Charles law to represent the relationship between temperature, pressure, and volume of an ideal gas:

$$P_1V_1/T_1 = P_2V_2/T_2$$

where P = absolute pressure
T = absolute temperature
V = volume

Subscript 1 indicates initial condition; subscript 2 indicates final condition.

**Keywords:** gas, pressure, volume, temperature
BOYLE, Robert, 1627-1691, British physicist and chemist, born in Ireland
CHARLES, Jacques Alexandre, 1746-1823, French physicist
**Sources:** Considine, D. M. 1976; Lapedes, D. N. 1974; Hyne, N. J. 1991.
*See also* BOYLE; VAN DER WAALS

## BOYLE LAW (1660); BOYLE-MARIOTTE LAW; MARIOTTE LAW (1675)

The volume of a given quantity of confined ideal gas varies inversely as the absolute pressure, if the temperature remains unchanged. The generalization for perfect gases is that the product of the absolute pressure, P, and the volume, V, is constant if the temperature remains unchanged. The relationship was independently discovered by E. Mariotte.

**Keywords:** gas, pressure, quantity, temperature, volume
BOYLE, Robert, 1627-1691, British physicist and chemist, born in Ireland.
MARIOTTE, Edme, 1620-1684, French physicist

**Sources:** Considine, D. M. 1976; Isaacs, A. 1996; Krauskopf, K. B. 1959; Menzel, D. H. 1960; Parker, S. P. 1984; Thewlis, J. 1961-1964.
*See also* BOYLE-CHARLES; CHARLES; MARIOTTE; UNIVERSAL GAS

### BRAGG LAW (1912); BRAGG EQUATION; BRAVAIS LAW

The condition under which a crystal will reflect a beam of X-ray with maximum intensity is represented by:

$$n \lambda = 2 d \sin \theta$$

where $n$ = an integer
$\lambda$ = wavelength
$d$ = depth of penetration or distance of separation of layers of atoms in crystal expressed
$\theta$ = angle of rays with respect to crystal face

**Keywords:** crystal, reflection, wavelength, X-ray
BRAGG, Sir William Henry, 1862-1942, English physicist; Nobel prize, 1915, physics
BRAGG, Sir William Lawrence, 1890-1971, Australian English physicist; Nobel prize, 1915, physics (these two are father and son)
BRAVAIS, Auguste, 1811-1863, French physicist
**Sources:** Considine, D. M. 1976; Isaacs, A. 1996; Lapedes, D. N. 1974; Travers, B. 1996.
See also BRAVAIS

### BRAGG-GRAY PRINCIPLE OR LAW

The energy loss of ionizing radiation absorbed per unit mass of a given medium is related to the absorption is a small gas-filled cavity in the medium. The absorption is related to the product of the relative stopping power of the medium with respect to the gas, times the average energy required to produce an ion pair in the gas, times the number of ion pairs produced per unit mass of gas in the cavity.

**Keywords:** absorption, energy, ionization, physics, radiation
BRAGG, Sir William Lawrence, 1890-1971, Australian English physicist; Nobel prize, 1915, physics (shared with father)
GRAY, Alexander, 1884-1966, Austrian physician
**Source:** Hunt, V. D. 1979.

### BRAGG-PIERCE LAW

A relationship for determining an element mass absorption coefficient of the absorption of monochromatic X-rays as related to the wavelength of the beam and the atomic number of the material:

$$\mu = CZ^4 \lambda^{5/2}$$

where $\mu$ = mass absorption coefficient of monochromatic
$Z$ = atomic number of material
$C$ = constant representing material
$\lambda$ = wavelength of the beam

**Keywords:** absorption, atomic number, mass coefficient, X-rays
BRAGG, Sir William Lawrence, 1890-1971, Australian English physicist; Nobel prize, 1915, physics (shared with father)
PIERCE, John Robinson, twentieth century (b. 1910), American electrical engineer
**Sources:** Miller, D. 1996; Morris, C. G. 1992.

## BRAHE, TYCHO, OFTEN REFERRED TO SIMPLY AS TYCHO LAW

Although his work is referred to as a law, his data were used by J. Kepler to develop planetary laws.

BRAHE, Tycho, 1546-1601, Danish astronomer
**Sources:** Asimov, I. 1972; Bescancon, R. M. 1974; Encyclopaedia Britannica. 1961; Webster's Biographical Dictionary. 1959.
*See also* KEPLER (PLANETARY); TYCHO

## BRANDES LAW—SEE STORMS, LAW OF (1826)

## BRAVAIS LAW

In a metal structure, the crystals that form most frequently are those that have faces parallel to the planes of the smallest reticular area, like a net. The reticular area is inversely proportional to the interplanar spacing.

**Keywords:** crystals, faces, reticular area, interplanar spacing
BRAVAIS, Auguste, 1811-1863, French physicist
**Sources:** Ballentyne, D. W. G. and Lovett, D. R. 1972; Chemical Abstracts 2:1402, 1908; Gillispie, C. C.1981; Parker, S. P. 1984.
*See also* BRAGG

## BREAKDOWN VOLTAGE—SEE PASCHEN

## BRENNER LAW

The reaction of the auditory nerve to galvanic stimulation in which, with the cathode in the external auditory meatus (opening in passage), a sound is produced on closing the circuit, which is interrupted when the circuit is opened again. With the anode in the meatus, no reaction is experienced on closing the circuit, but a weak sensation of sound is experienced when the circuit is opened again.

**Keywords:** anode, auditory, cathode, circuit, reaction, sensation, sound
BRENNER, Rudolf, 1821-1884, German physician
**Source:** Landau, S. I. 1986.

## BRETON LAW

A parabolic psychological relationship that exists between the stimulus and just noticeable response is expressed by the formula:

$$S = (R/C)^{1/2}$$

where S = stimulus
R = response
C = constant

**Keywords:** medicine, parabolic, physiology, stimulus, response
BRETON, Andre, 1896-1966, French doctor and medical writer
**Sources:** Friel, J. P. 1974; Landau, S. I. 1986.

## BREWSTER LAW (1811)

The tangent of the angle of polarization of light for a substance is equal to the index of refraction. The polarizing angle is that angle of incidence for which the reflected polarized ray is at right angle to the refracted ray.

$$n = \tan \theta$$

where n = index of refraction
θ = polarizing angle

**Keywords:** angle, index, light, polarization
BREWSTER, Sir David, 1781-1868, Scottish physicist
**Sources:** Gray, D. E. 1963; Hodgman, C. D. 1952; Thewlis, J. 1961-1964.

### BRINKMAN NUMBER, Br OR $N_{Br}$ (BRINKMANN*)

A dimensionless group that relates the heat produced by viscous dissipation of friction heat generated and heat transported by molecular conduction, applicable to viscous flow:

$$Br = \mu V^2/kT$$

where μ = absolute viscosity
V = velocity
k = thermal conductivity
T = absolute temperature

**Keywords:** conduction, heat, molecular, viscous
BRINKMAN, H. C., wrote article in English in 1947.
**Sources:** Bolz, R. E. and Tuve, G. L. 1970: Cheremisinoff, N. P. 1986; Kakac, S. 1987; Land, N. S. 1972; Morris, C. G. 1992; *Parker, S. P. 1992; Potter, J. H. 1967.

### BROADBENT LAW

Lesions of the upper segment of the motor tract cause less marked paralysis of the muscles that habitually produce bilateral movements than of those that more commonly act independently of the opposite side.

**Keywords:** lesions, muscles, paralysis
BROADBENT, Sir William Henry, 1835-1907, English physician
**Sources:** Landau, S. I. 1986; Stedman, T. L. 1976.

### BRODY & PROCTER LAW—SEE BODY SIZE; RUBNER AND RICHET; SURFACE AREA

### BRÖNSTED CATALYSIS LAW OR RELATION (1923); BRÖNSTED-LOWRY LAW; LOWRY-BRÖNSTED LAW

For catalyzed reactions, the relationship between the catalyst effectiveness, as given by the rate constant, and the strength of the acid or base, ionization constant, is expressed by:

$$k = C K^a$$

where k = rate constant (or catalytic constant)
K = ionization constant
C and a = empirical constants governed by the type of reaction, temperature, and solvent

All base-acid reactions consist of the transfer of a proton from one base to another. An acid is the source of protons; a base is a species which can accept protons.

BRÖNSTED, Johannes Nicolaus, 1879-1947, Danish chemist
LOWRY, Thomas Martin, 1874-1936, English chemist
**Sources:** Chemical Heritage 16(1):24, 1998; Morris, C.G. 1992; Parker, S. P. 1987; Pelletier, P. A. 1994; Perry, R. and Green, D. W. 1984.

## BROWNIAN MOVEMENT, LAW OF (1827)

A given organic or inorganic colloidal or microscopic particle in a fluid, gas, or liquid will move through a very irregular and complicated random path, which was explained by Albert Einstein in 1905. R. Brown also identified the nucleus in living cells in 1831.

**Keywords:** colloidal, irregular, particle, path, random
BROWN, Robert B., 1773-1858, Scottish botanist
**Sources:** Brown, S. B. and L. B. 1972; World Book. 1989.

## BRUCKE LAW—SEE BEZOLD-BRUCKE

## BRUNAUER-EMMETT-TELLER ISOTHERM (1938)

The isotherm represents the relationship of the vapor pressure of the adsorbate in the liquid form. The adsorbed state builds up molecular layers of vapor:

$$V/V_M = KP/P_0/(1 - P/P_0)\ (1 - P/P_0 + KP/P_0)$$

where $P_0$ = vapor pressure of the adsorbate in liquid state
$V$ = volume of gas adsorbed per unit (STP) amount of solid
$V_M$ = volume adsorbed at saturation
$P$ = partial pressure of adsorbate
$K$ = constant based on material and conditions

**Keywords:** adsorbate, gas (vapor), isotherm, layers
BRUNAUER, Stephen, 1903-1986, Hungarian American chemist
EMMETT, Paul Hugh, 1900-1985, American chemist
TELLER, Edward, twentieth century (b. 1908), Hungarian American physicist
**Sources:** Hampel, C. A. and Hawley, G. G. 1973; Menzel, D. M. 1960.
*See also* ADSORPTION ISOTHERM; ISOTHERM

## BRUNS LAW—SEE BASTIAN-BRUNS

## BUBBLE NUMBER—SEE NUSSELT BUBBLE NUMBER

## BUCHELE LAWS OF AGRICULTURAL MACHINES (1969)

### First Law

Any operation performed by human hands can be performed by a machine or a series of machines.

### Second Law

Any mechanized operation performed by a machine or a series of machines can be done faster and/or at lower cost and/or with an improvement in quality of product by another machine or series of machines.

### Third Law

In an unlimited energy, material, and manpower economy, any uncontrolled, mechanized, profitably produced product or service will be in overproduction.

#### *First Corollary*

Any industry that does not control the supply or fails to manage its surplus of products or services will become bankrupt.

***Second Corollary***

The greater the quantity of product in permanent storage, the lower the buying price of the product.

**Keywords:** agriculture, cost, mechanized, product, quality, wages

BUCHELE, Wesley F., twentieth century (b. 1920), American agricultural engineer

**Source:** Buchele, W. F. 1969. No Starving Billions: The Role of Agricultural Engineering in Economic Development. Inaugural Lecture, University of Ghana, May 9. 35 p.

## BUCHELE LAWS OF PRODUCTION AGRICULTURE (1986)

**First Law**

One mechanized horsepower-hour of energy produces approximately one daily ration of food and fiber.

***First Corollary***

The cost of one daily ration of food and fiber is proportional to the cost of one mechanized horsepower-hour of energy.

***Second Corollary***

The daily wage earned by a third-world agricultural worker is proportional to the cost of one mechanized horsepower-hour of energy.

***Third Corollary***

The cost of one mechanized horsepower-hour of energy is equal to the cost of one tractor horsepower-hour of energy divided by the efficiency of the machine operated by the tractor.

$$m = c/f$$

where m = the cost of one mechanized horsepower-hour of energy

c = the cost of one tractor horsepower-hour of energy

f = the efficiency (decimal form) of the machine operated by a tractor

***Fourth Corollary***

The daily wage of a third world agricultural worker is proportional to the income from one daily ration of food and fiber.

**Second Law**

The daily income earned by a farmer is proportional to the income from one daily ration of food and fiber multiplied by the number of people supplied food and fiber.

**Third Law**

There is no relationship between the price of grain and the price of food.

**Fourth Law**

The agricultural policy that is best for a group of farmers is the policy that is worst for the individual farmer.

**Keywords:** agriculture, cost, food, mechanization, wages

BUCHELE, Wesley F., twentieth century (b. 1920), American agricultural engineer

**Source:** Buchele, W. F. 1986. Solving Industrial Problems of Agriculture. Paper presented at the Iowa Association of Electric Cooperatives, Des Moines, IA p. 16 and 17.

## BUCHELE LAWS OF SOIL-PLANT DYNAMICS

**First Law**
A good seed bed environment produces faster emergence, greater rates of emergence and greater total emergence of seedling than a poor environment.

**Second Law**
The finer the pulverization of the surface soil before a rain, the thicker and stronger the soil crust will be after a rain.

**Third Law**
The finer the pulverization of the surface soil the lower the water infiltration capacity of the soil.

**Fourth Law**
The finer the pulverization of soil, the greater the runoff of water.

**Fifth Law**
Tillage is the first step in the erosion process of cultivated soils, the soil is detached from soil mass by tillage.

**Sixth Law**
The greater the water runoff from cultivated soil, the greater the soil loss by water erosion.

**Seventh Law**
The greater the organic mulch on the surface of the soil, the greater the water infiltration capacity of the soil.

**Eighth Law**
The greater the organic mulch on the surface of the soil, the cooler the soil.

**Ninth Law**
The greater the organic mulch, the lower the available nitrogen in the surface soil.

**Tenth Law**
The greater the horizontal movement of the soil during tillage or planting, the greater the coverage of organic mulch.

**Eleventh Law**
Every plant creates its own bioskin to control wind and water erosion and soil temperature and minimize water run off. (The bioskin is composed of the canopy of the plant, the organic mulch of past growing seasons and root system.)

**Keywords:** infiltration, mulch, organic matter, plants, tillage, water
BUCHELE, Wesley, twentieth century (b. 1920), American agricultural engineer
**Source:** Buchele, W. F. 1979. ASAE Paper No. MC 79-303, St. Joseph, MI 15 p.

## BUCKINGHAM THEORY OR LAW (1914)

A method of dimensional analysis whereby the variables of some physical laws are grouped into one or more terms in which the dimensions cancel out, forming a dimensionless group or number. The approach can be used for modeling a system that can be scaled up or scaled down for making a prototype. The product of the dimensionless groups, sometimes called *pi*, with appropriate constants and exponents must equal zero. The variables can be represented by basic units such as mass, time, length, and temperature. If there are n variables and m basic units, the physical relationships are expressed in n minus m dimensionless groups. The ideas were first advanced by T. Simpson, a British mathematician, in 1757 and stated by E. Buckingham.

**Keywords:** dimensional analysis, dimensionless group, scaling, variables
BUCKINGHAM, Edgar, 1867-1940, American physicist
SIMPSON, Thomas, 1710-1761, English mathematician
**Sources:** Encyclopaedia Britannica. 1984; McKetta, J. J. 1976-1996; Parker, S. P. 1994.

## BUHL-DITTRICH LAW

The supposition that, in every case of acute military tuberculosis, there exists at least one old focus of causation in the body.

**Keywords:** causation, health, military tuberculosis
BUHL, Ludwig von, 1816-1880, German pathologist
DITTRICH, Franz, 1815-1859, German pathologist
**Sources:** Friel, J. P. 1974; Landau, S. I. 1986.

## BULK FLOW (FLUIDS), FIRST LAW OF

The first law equation for fluid flow (Fig. B.5) for a closed system, A – m, is:

$$E'' - (E' + e\,\delta\,m) = dQ + pv\,\delta\,m - dW_x$$

where $E'$ and $E''$ = initial and final energy inside $\sigma$, and e the energy per unit mass in $\delta$ m
p = pressure
v = specific volume of $\delta$ m
dWx = work delivered to the surrounding
dQ = heat entering the system

**Keywords:** energy, fluid, flow, heat, system, work
**Source:** Thewlis, J. 1962.

## BULK MODULUS NUMBER OR ELASTIC MODULUS

A numerical constant that describes the elastic properties of a solid or fluid when under pressure on all sides, expressed as a ratio of pressure on a body to its fractional decrease in volume:

$$\text{Bulk modulus number} = \text{pressure/strain} = p/[(V_o - V_n)/V_o]$$

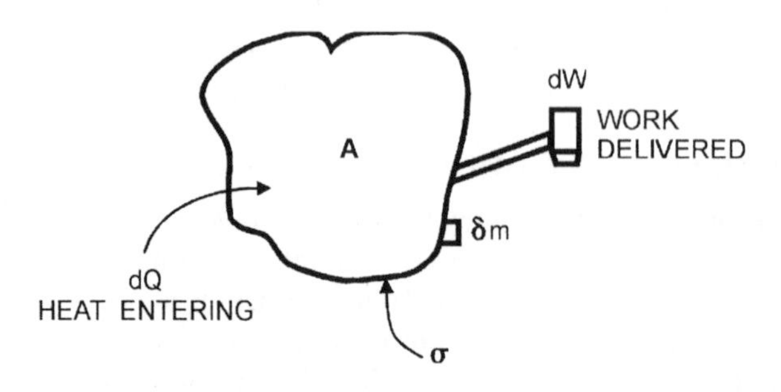


**Figure B.5** Bulk flow.

where $p$ = pressure applied to material
$V_o$ = original volume
$V_n$ = new volume after pressure applied

The compressibility is the reciprocal of the bulk modulus.

**Keywords:** bulk, compressibility, elastic, reciprocal
**Source:** Isaacs, A. 1996.

## BULYGIN NUMBER, Bu OR $N_{Bu}$ (BULYGEN*)

A dimensionless group that represents high intensity heat and mass transfer during evaporation. The number pertains to the amount of heat transfer expended to produce evaporation as related to the heat required to bring the liquid to boiling:

$$Bu = LH_v \, c_p p_o / c_q (t_c - t_o)$$

where $LH_v$ = latent heat of evaporation
$c_p$ = specific heat
$p_o$ = pressure in the material
$c_q$ = specific heat of the moist material
$t_c - t_o$ = temperature difference between the medium and material

**Keywords:** evaporation, heat, mass
**Sources:** Bolz, R. E. and Tuve, G. L. 1970; Land, N. S. 1972; Lykov, A. V. and Mikhaylev, Y. A. 1961; Morris, C. G. 1992; *Parker, S. P. 1992; Potter, J. H. 1967.
*See also* RAMZIN NUMBER

## BUNGE LAW

The secreting cells of the mammary gland take plasma mineral salts from the blood in proportion to the need for developing and strengthening of the offspring.

**Keywords:** blood, mammary, physiology, offspring
BUNGE, Gustav von, 1844-1920, Swiss physiologist
**Sources:** Chemical Abstracts 4:469; Friel, J. P. 1974; Landau, S. I. 1986.

## BUNSEN-ROSCOE LAW OR PHOTOGRAPHIC LAW; LAW OF RECIPROCITY (1862)

The time required for a reaction to a photochemical effect produced by a light source is inversely proportional to the intensity of illumination for a particular photochemical process. In other words, the duration times the intensity of illumination is constant for a short period of 50 milliseconds \or less.

**Keywords:** duration, illumination, intensity, photochemical, time
BUNSEN, Robert Wilhelm Eberhard von, 1811-1899, German chemist
ROSCOE, Sir Henry Enfield, 1833-1915, English chemist
**Sources:** Friel, J. P. 1974; Gennaro, A. R. 1979; Glasser, O. 1944; Honig, J. M.1953; Landau, S. I. 1986; Stedman; T. L. 1976; Thewlis, J. 1961-1964; Zusne, L. 1987.
*See also* BLOCH (K. E.); BLONDEL-REY; FERRY-PORTER; RECIPROCITY

## BUOYANCY, LAW OF—SEE ARCHIMEDES

## BUOYANCY NUMBER, Buoy OR $N_{Buoy}$ OR PARAMETER

A dimensionless group that relates the buoyant force and the viscous force in convection:

$$Buoy = L^2W\ \beta\ \Delta T/\mu\ V_o V$$

where L = characteristic length
W = weight
$\beta$ = coefficient of expansion
$\Delta T$ = difference in temperature
$\mu$ = absolute viscosity
$V_o$ = volume
V = velocity

The Buoyancy number equals the Grashof number divided by the Reynolds number squared.

**Keywords:** buoyant, convection, viscous
**Sources:** Bolz, R. E. and Tuve, G. L. 1970; Land, N. S. 1972; Parker, S. P. 1992.
*See also* ARCHIMEDES

### BURGERS MODEL—SEE VISCOELASTIC MODELS (MATERIALS)

### BUTLER-VOLMER LAW

For charge transfer control of an electrochemical reaction, the reaction rate depends mainly on the exchange current density and on the overpotential of the electrode, and is calculated by the following equation which gives the current given by half-cells in electrochemical reactions:

$$I = I_o[\exp(\alpha_a F\eta_s/RT) - \exp(\alpha_c F\ \eta_s/RT)]$$

where I = current density
$I_o$ = exchange current density
$\alpha_a$ = anodic charge transfer coefficient
$\alpha_c$ = cathodic charge-transfer coefficient
F = Faraday constant
$\eta_s$ = overpotential, $E - E_e$
R = gas constant, 8.314 J/mol-K
T = absolute temperature
$E_e$ = equilibrium electrode potential

The first term in the brackets represents the anodic (backward) reaction and the second term represents the cathodic (forward) reaction.

**Keywords:** charge transfer, chemical reaction, current density, overpotential
**Sources:** Chemical Engineering 102(1):103, January 1995; Grayson, M. and Eckroth, D. 1982; Kirk-Othmer. 1998; Lerner, R. G. and Trigg, G. L. 1991.
Also see NERNST EQUATION

### BUYS BALLOT LAW (1857); BARIC WIND LAW; BASIC WIND LAW

The relationship of the horizontal wind direction in the atmosphere to the pressure distribution is such that, if one stands with the back to the wind, the pressure to the left is lower than to the right in the northern hemisphere and vice versa in the southern hemisphere.

**Keywords:** atmosphere, pressure, wind
BUYS BALLOT, Christoph Hendrik Didericus, 1817-1890, Dutch meteorologist
**Sources:** Considine, D. M. 1967; Fairbridge, R. W. 1967; Huschke, R. E. 1959.
*See also* EGNELL; FAEGRI; FERREL; STORMS

## CAILLETET-MATHIAS LAW (1886); OR RECTILINEAR DIAMETER LAW

A linear function exists between the arithmetical average of the densities of a pure unassociated liquid and its saturated vapor and the temperature of the liquid:

$$(d_l + d_v)/2 = A + Bt$$

where $d_l$ and $d_v$ = densities of liquid and vapor, respectively
t = temperature, °C
A and B = constants of liquid, B is negative

This law was modified by S. Young in 1900. L. Cailletet was the first to liquefy oxygen, hydrogen, and nitrogen.

**Keywords:** density, liquid, temperature, vapor
CAILLETET, Louis Paul, 1832-1913, French physicist and technologist
MATHIAS, Emile, nineteenth century, French chemist
YOUNG, Sidney, 1857-1937, British chemist
**Sources:** Considine, D. M. 1957; Hix, C. F. and Alley, R. P. 1958; Honig, J. 1953; Morris, C. G. 1992; Perry, J. H. 1950; Thorpe, J. F., et al. 1947-1954.
See also RAMSAY AND YOUNG

## CAMERER LAW

Children of the same weight have the same food requirements independent of their ages.

**Keywords:** children, food, weight
CAMERER, John Friedrich Wilhelm, 1842-1910, German pediatrician
**Sources:** Friel, J. P. 1974; Landau, S. I. 1986.

## CAMPBELL LAW (1896)

The general law of the migration of a drainage divide in which the divide tends to migrate toward an axis of uplift or away from an axis of subsidence. When two streams that head opposite to each other are affected by an uneven lengthwise titling moment, the one whose declinivity is increased cuts down vigorously and growth in length headward at the expense of the other.

**Keywords:** drainage divide, migration, subsidence, uplift
CAMPBELL, Marius R., 1858-1940, American geologist
**Sources:** Bates, R. L. and Jackson, J. A. 1980. 1987; Morris, C. G. 1992.
*See also* BAER

## CAMP-MEIDELL CONDITION, INEQUALITY OR LAW (1921)

For determining the distribution of a set of numbers, the guideline stating that if the distribution has only one mode, if the mode is the same as the arithmetic mean, and if the frequencies decline continuously on both sides of the mode, then more than $1-(1/2.25\ t^2)$ of any distribution will fall within the closed range, $\overline{X} \pm \sigma$ where t is the number of items in a set, $\overline{X}$ is the average, and $\sigma$ is the standard deviation.

**Keywords:** arithmetic mean, distribution, frequencies, mode, numbers
CAMP, Burton Howard, twentieth century mathematician (b. 1880), American mathematician
**Sources:** Lapedes, D. N. 1974; NUC; Parker, S. 1994; Wilson, E. B. 1952 (called this a law).
*See also* PROBABILITY

## CANCELLATION LAW

If a + b and ab are real numbers, then a + c = b + c implies that a = b, ca = cb, and if c is not equal to zero, implies that a = b.

**Keywords:** algebra, mathematics, numbers, real
**Sources:** Good, C. V. 1973; Morris, C. G. 1992.

## CANNON LAW OF DENERVATION

The sensitivity of a structure to chemical stimulation increases as a result of denervation. When a structure is denervated, its irritability to certain chemical agents is increased.

**Keywords:** chemical, sensitivity, stimulation
CANNON, Walter Bradford, 1871-1945, American physiologist
**Sources**: Friel, J. P. 1974; Landau, S.I. 1986; Stedman, T. L. 1976.

## CAPACITANCE, LAW OF, IN SERIES—SEE SERIES RESISTANCE, LAW OF

## CAPILLARY ACTION, LAWS OF

A liquid in a capillary tube (1) rises in a tube that it wets and is depressed in a tube that it does not wet; (2) the amount of elevation or depression of a liquid in a tube is inversely proportional to the diameter of the tube; and (3) the amount of elevation or depression in a tube increases as the temperature decreases.

**Keywords:** depression, diameter, elevation, tube, wet
**Source:** Clifford, A. F. 1964.
*See also* POISUIELLE

## CAPILLARITY-BUOYANCY NUMBER, $K_F$

This number depends only on the value of g and physical properties and represents the effects of surface tension and acceleration in flowing media for two-phase flow:

$$K_F = g\,\mu^4/\rho\,\sigma^3 = (N_{We})^3/(N_{Fr})\,(N_{Re})^4$$

Where g = acceleration due to gravity
$\mu$ = dynamic viscosity
$\rho$ = density
$\sigma$ = surface tension

**Keywords:** acceleration, surface tension, two phase flow
**Sources:** Bolz, R. E. and Tuve, G. L. 1970; Parker, S. P. 1992; Potter, J. H. 1967.
*See also* POISEUILLE

## CAPILLARITY NUMBER, CAP OR $N_{Cap}$

A dimensionless group applied to two-phase flow which is the capillary force divided by filtration force:

$$Cap = \sigma\, k^{1/2}/\mu\, VL$$

where $\sigma$ = interfacial tension
$k$ = permeability
$\mu$ = absolute viscosity
$V$ = velocity
$L$ = characteristic length

This number is identified as Capillarity 1 by N. Land.*

**Keywords:** capillarity, filtration, force, two-phase flow
**Source:** *Land, N. S. 1972; Parker, S. P. 1992; Potter, J. H. 1998.
*See also* POISEUILLE

## CAPILLARY NUMBER, CA OR $N_{CA}$

A dimensionless group representing the viscous force divided by the surface tension force:

$$CA = \mu\, V/\sigma$$

where $\mu$ = absolute viscosity
$V$ = velocity of flow
$\sigma$ = surface tension

This number is identified as Capillarity 2 by N. Land.*

**Keyword:** surface tension, two-phase, viscous
**Sources:** Bolz, R. E. and Tuve, G. L. 1970; *Land, N. S. 1972; Parker, S. P. 1992; Potter, J. H. 1967.
*See also* POISEUILLE

## CARLSBAD LAW—SEE TWIN

## CARMICHAEL LAW

The anatomical behavior mechanisms may be elicited by experimental means at a time prior to that when the normal action of these patterns of behavior is essential in the adaptive life of the organism. He developed this law based on electrophysiological techniques.

**Keywords:** adaptive, anatomical, behavior, electrophysical, organism
CARMICHAEL, Leonard, 1898-1973, American psychologist
**Sources:** Greene, J. E. 1966-1968; Parker, S. P. 1980.

## CARNOT LAW (1824); CARNOT EFFICIENCY LAW; CARNOT NUMBER, Ca OR $N_{Ca}$

An engine is no more efficient than a reversible engine working between the same two temperatures, with the efficiency depending only on these two temperatures (Fig. C.1). The efficiency (eff.) of an engine working between $T_1$ and $T_2$ is:

$$\text{eff.} = (T_1 - T_2)/T_1 = \text{Carnot number} = Ca$$

where $T_1$ = absolute temperature of heat absorbed
$T_2$ = absolute temperature of heat given out

**Keywords:** efficiency, engine, heat, temperature, thermodynamics
CARNOT, Nicolas Leonard Sadi, 1796-1832, French physicist

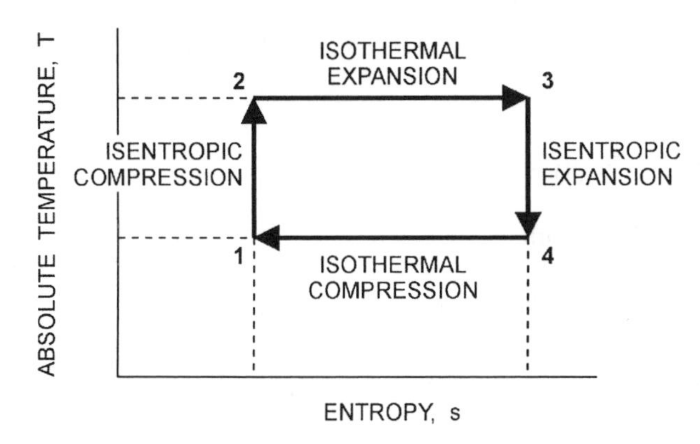


**Figure C.1** Carnot cycle.

**Sources:** Considine, D. M. 1974; Land, N. S. 1972; Layton, E. T. and Lienhard, J. H. 1988; Thewlis, J. 1961-1964.
*See also* THERMODYNAMICS

## CARNOT LAW OF IMPULSIVE MOTION

The deformation of impact of two bodies consist of two intervals: (1) the period of approach from time of contact to maximum deformation of a body; and (2) the restitution period which exists until separation (Fig. C.2). This law is represented by:

$$\Sigma \vec{P}_i \ V_i = \Sigma \, m_i V_i \ (\vec{V}_i - \vec{V}_{i0})$$

where i = number of rigid bodies
V = linear velocity
P = impulse
m = mass

**Keywords:** deformation, impact, restitution, separation
CARNOT, Nicholas Leonard Sadi, 1898-1832, French physicist
**Source:** Goldsmith, W. 1960.

## CARREL LAW; CIRCATRIZATION, MATHEMATICAL LAW OF

The rate of healing of a wound is dependent on the surface area of the wound and the age of the person, with the greater value of these requiring a longer time of healing.

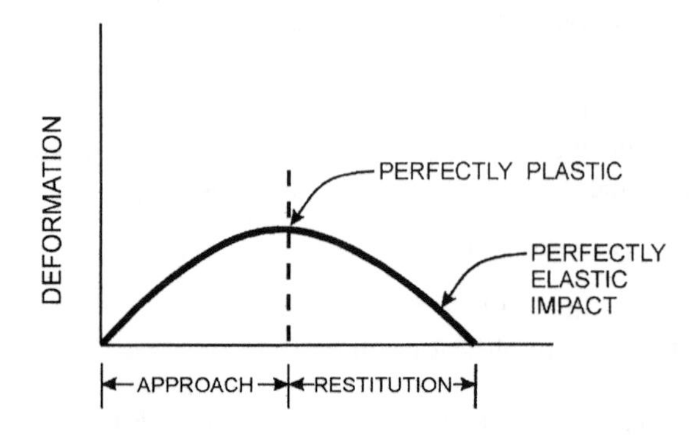


**Figure C.2** Impulsive motion.

**Keywords:** age, area, healing, time, wound
CARREL, Alexis, 1873-1944, French American surgeon; Nobel prize, 1912, physiology/medicine
**Source:** Schlesinger, B. S. and J. H. 1991.

## CASIMIR EFFECT (1948)

The motion of two parallel plates so close together that only small fluctuations count as a result of quantum fluctuations in a vacuum, and work to push the plates together. The effect was measured by S. K. Lamoreaux.

**Keywords:** fluctuation, parallel plates, quantum
CASIMIR, Hendrik B. C., twentieth century (b. 1909), Dutch physicist
LAMOREAUX, Steve K., twentieth century, American physicist (at Los Alamos National Lab.)
**Source:** Scientific American 277(6):84, December 1997.

## CASSINI LAWS (1721)

Three laws that describe the rotation of the Moon about its center of mass:

**Law 1**
The Moon rotates about an axis fixed within it, with constant angular velocity in a period of rotation equal to the mean sidereal period of revolution of the Moon about the Earth.

**Law 2**
The inclination of the mean plane of the Moon about the Earth to the plane of the ecliptic is constant.

**Law 3**
The poles of the lunar equator, the ecliptic and the Moon's orbital plane all lie on one great circle.

**Keywords:** Earth, inclination, Moon, rotation
CASSINI, Jacques, 1677-1756, French Italian astronomer
**Source:** Mitton, J. 1993.
*See also* KEPLER

## CASTIGLIANO LAW OR PRINCIPLE OF LEAST WORK

The partial derivative of the total elastic energy stored in a structure with respect to one of the loads, gives the displacement of the point of application of the load in the direction of the load.

**Keywords:** elasticity, energy, load
CASTIGLIANO, C. Alberto, 1847-1884, Italian engineer
**Sources:** Morris, C. G. 1992; Scott, J. S. 1993.
*See also* LAGRANGE PRINCIPLE; MAUPERTUIS

## CASTLE-HARDY-WEINBERG LAW—SEE HARDY-WEINBERG LAW

## CATALYSIS LAW—SEE BRÖNSTED

## CATEGORICAL JUDGMENT, LAW OF

A law of comparative judgment that should provide an equal-interval category scale that assumes that the psychological continuum of an individual can be divided into categories in which a category boundary is not a stable entity and the responses of the individual are predicated on the position of a category boundary so that the boundaries between adjacent categories behave like stimuli.

**Keywords:** boundary, comparative, judgment, psychological
**Source:** Wolman, B. B. 1989.
*See also* THURSTONE

### CAUCHY NUMBER, CA OR $N_{CA}$

A dimensionless group applicable in compressible flow that is the inertial force divided by the elastic force or compressibility:

$$CA = V^2/a^2$$

where V = velocity of flow
a = sonic velocity

**Keywords:** compressible flow, elastic force, inertial force
CAUCHY, Augustin Louis, Baron, 1789-1857, French mathematician
**Sources:** Bolz, R. E. and Tuve, G. L. 1970; Land, N. S. 1972; Parker, S. P. 1992; Perry, R. H. 1967; Potter, J. H. 1967.
*See also* MACH NUMBER; TENSOR TRANSFORMATION

### CAUCHY STRESS THEOREM

The force at a point on a surface of a body of fluid is given by sn where s is a symmetric second order Cartesian tensor and n is a unit vector normal to the surface.

**Keywords:** fluid, force, tensor, vector
CAUCHY, Augustin Louis, Baron, 1789-1857, French mathematician
**Source:** Bothamley, J. 1993.

### CAUSATION, LAW OF; CAUSALITY, LAW OF, OR LAW OF UNIVERSAL CAUSATION

Every fact or effect that has a beginning has a cause.

**Keywords:** cause, effect, fact
**Sources:** Brown, S. B. and Brown, L. B. 1972; Campbell, N. R. 1957.

### CAVINDISH LAW OF ELECTRICITY—SEE COULOMB

### CAVITATION NUMBER, $\sigma_c$

A dimensionless group applicable to cavitation excess of local static head over the vapor pressure head in pumps, nozzles, and hydraulic systems:

$$\sigma_c = (p - p_c)/q$$

where p = local static pressure
$p_c$ = fluid vapor pressure
q = dynamic pressure, $\rho V^2/2$
$\rho$ = mass density
V = velocity

**Keywords:** hydraulic, static head, vapor pressure
**Sources:** Bolz, R. E. and Tuve, G. L. 1970; Land, N. S. 1972; Parker, S. P. 1992; Potter, J. H. 1967.
*See also* LEROUX; THOMA

## CENTRAL LIMIT THEOREM (1812)—SEE ERRORS, LAW OF; PROBABILITY

## CENTRIFUGAL PUMP (FAN), AFFINITY LAWS OF

These laws apply to the design and testing of centrifugal pumps (fans) that relate the performance to speed:

1. The flow is directly proportional to speed
2. The head developed is proportional to speed squared
3. The horsepower is proportional to speed cubed
4. The efficiency remains approximately constant

**Keywords:** efficiency, flow, head, horsepower, speed
**Sources:** Considine, C. M. 1976; Hall, C. W. 1979.

## CENTRIFUGE NUMBER, Ce OR $N_{Ce}$

A dimensionless group representing the centrifugal force divided by the capillary force in slosh materials:

$$Ce = \rho R^2 h \omega^2/\sigma$$

where $\rho$ = mass density
R = tank radius
h = liquid depth
$\omega$ = angular velocity
$\sigma$ = surface tension

**Keywords:** capillary, centrifugal, slosh
**Source:** Land, N. S. 1972.
*See also* CORIOLIS

## CEPHEIDS, LAW OF—SEE LUMINOSITY PERIOD

## CHANCE, LAW OF—SEE PROBABILITY

## CHANNEL LENGTH, LAW OF—SEE STREAM LENGTH

## CHAPMAN LAW

The rate of production of ions in an ideal upper atmosphere (ionosphere) with only one constituent and of constant dimensions, for monochromatic incident radiation, is given a function of normalized height, and solar zenith angle, $\chi$, by:

$$q(z, \chi) = q(0,0) \exp\{1 - z - e^{-z} \sec \chi\}$$

where q = rate of production of ionization of upper atmosphere
q(0,0) = rate of production of ions at height, ho, when the sun is overhead, valid until $\chi$ is 80° angle
z = the normalized height, $(h - h_o)/H$, where
$h_o$ = a reference height
H = scale height over which the pressure changes by a factor of e
$\chi$ = solar zenith angle

**Keywords:** angle, atmosphere, ions, radiation, sun
CHAPMAN, Sydney, 1888-1970, American English geophysicist
**Sources:** NUC; Thewlis, J. 1967 (Supplement 2).

## CHARGE, LAW OF—SEE CONSERVATION OF CHARGE

## CHARGE CONJUGATION, LAW OF—SEE CONSERVATION OF SYMMETRY

## CHARGING LAW FOR ELECTRICAL CAPACITORS

The charging rate of electric capacitors that have no resistance or inductance is directly proportional to the rate of change of the potential and the electric capacitance, as follows:

$$I = C\, dv/dt \text{ and energy stored, } E = 1/2Cv^2$$

where I = current, amperes
v = potential, volts
C = capacitance, farads

**Keywords:** capacitance, charging, current, inductance, resistance, potential, volts
**Sources:** Fink, D. G. and Beaty, H. W. 1993; Michels, W. C. 1961; Parker, S. P. 1983 (physics).

## CHARLES LAW (1784*;1787); CHARLES-GAY-LUSSAC LAW; GAY-LUSSAC LAW (1802)

The volume, V, of a definite quantity of gas at a constant pressure is directly proportional to the absolute temperature, T. The relationship is

$$V = k\, T$$

where V = volume
k = constant
T = absolute temperature

The law was named after J. Charles even though J. Dalton first discovered it.

**Keywords:** gas, pressure, volume, temperature
CHARLES, Jacques Alexander Cesar, 1746-1823, French chemist/physicist
GAY-LUSSAC, Joseph Louis, 1778-1850, French chemist
**Sources:** Asimov, I. 1972; Considine, D. M. 1976; Hampel, C. A. and Hawley, G. G. 1973; Klauskopf, K. B. 1959; Layton, E. T. and Lienhard, J. H. 1988; Parker, S. P. 1989; Thewlis, J. 1961-1964; *Van Nostrand. 1947.
*See also* AMONTONS; BOYLE-CHARLES; DALTON; UNIVERSAL GAS

## CHARLES-GAY-LUSSAC LAW

Although expressed in a slightly different way from the above, the change in volume of a perfect gas is proportional to the change in temperature of a perfect gas:

$$t - t_o = 1/c\ \{(v - v_o)/v_o\}$$

where t = temperature
v = volume
c = coefficient of thermal expansion

So gases expand nearly equally at 1/273 of their volume at 0° C. J. Gay-Lussac obtained results from his work that established the authenticity of the work of J. Charles.

**Keywords:** gas, temperature, volume
CHARLES, Jacques Alexander, 1746-1823, French physicist
GAY-LUSSAC, Joseph Louis, 1778-1850, French chemist
**Sources:** Isaacs, A. 1996; Stedman, T. L. 1976.

*See also* CHARLES; UNIVERSAL GAS

## CHARPENTIER LAW

In the retinal fovea (small pit in the retina), the product of the area of a stimulus and its intensity is constant for stimuli at threshold intensities.

**Keywords:** retina, stimulus, threshold
CHARPENTIER, P. M. Augustin, 1852-1916, French biophysicist
**Source:** Bothamley, J. 1993; Zusne, L. 1987.

## CHATELIER—SEE LE CHATELIER

## CHAUVENET CRITERIA

The value of a data point is rejected if the probability of obtaining the particular deviation using all data, from the mean is less than 1/2 n. This criteria is more restrictive than the procedure of rejecting data if the probability for the observed deviation of a point is less than 1/n, another approach.

| Number of Readings, n | P(n) ≤ 1/2 n | Maximum Acceptable Deviation to Standard Deviation, $d_{max}/\sigma$ |
|---|---|---|
| 2 | .25 | 1.15 |
| 4 | .12 | 1.54 |
| 7 | .07 | 1.80 |
| 10 | .05 | 1.96 |
| 50 | .01 | 2.57 |
| 100 | .005 | 2.81 |

**Keywords:** data, deviation, probability, rejecting
CHAUVENET, William, 1820-1870, American mathematician
**Sources:** Hall, C. W. 1977; Webster's Biographical Dictionary. 1959.

## CHEMICAL AFFINITY, LAWS OF (1801)—SEE BERTHOLLET

## CHEMICAL COMBINATION, LAWS OF

A general classification of the laws of chemical combination include:

**Conservation of Mass (Matter), Law of (Lavoisier. 1774)**

**Definite (Constant) Composition, Law of (Proust. 1797)**

**Multiple Proportions, Law of (Dalton. 1804)**

**Combining Weights, Law of; or Equivalents, Law of; or Reciprocal Proportions, Law of**

**Sources:** Asimov, I. 1972; Brown, S. B. and Brown, L. B. 1972; Daintith, J. 1988; Isaacs, A. 1996; Uvarov, E. B. and Chapman, D. R. 1964.

## CHEMICAL EQUILIBRIUM, LAW OF; GULDBERG AND WAAGE LAW (1863)

A dynamic equilibrium is a dynamic condition in which the rate of forward reaction and the rate of reverse reaction are the same, as expressed by:

$$xA + yB + \ldots \Leftrightarrow mP + nQ + \ldots$$

The concentrations of the reactants A, B, etc., usually denoted by brackets [ ], satisfy the general equation, in which K is the equilibrium constant of a balanced reaction:

$$\frac{[P]^m \, [Q]^n \ldots}{[A]^x \, [B]^y \ldots} = K$$

**Keywords:** concentration, dynamic, equilibrium constant, reaction
GULDBERG, Cato Maximilian, 1836-1902, Norwegian chemist and mathematician
WAAGE, Peter, 1833-1900, Norwegian chemist
**Sources:** Honig, J. M. 1953; Isaacs, A 1996; Lowrie, R. S. 1965; Thewlis, J. 1961-1964.
*See also* GULDBERG AND WAAGE

## CHEMICAL REACTION, EXCHANGE, LAW OF

Two basic mechanisms are involved in a kinetic exchange reaction of an element or chemical grouping that takes place between two different species containing the element. The exchange is kinetically of the first order and dependent on the dissociation of AB,

1. $AB + B^* \Leftrightarrow [ABB^*] \Leftrightarrow AB^* + B$
2. $AB \Leftrightarrow A + B$, followed by $A + B^* \Leftrightarrow AB^*$

**Keywords:** dissociation, element, kinetic, order
**Source:** Thewlis, J. 1961-1964.

## CHEMICAL SPECIES—SEE CONSERVATION OF CHEMICAL SPECIES

## CHEZY FORMULA

The friction loss in water conduits is represented by:

$$V = C\,(R_H S)^{1/2}$$

where C = constant, depending on conditions, C = 1.486 R/r
r = roughness (Manning formula)
$R_H$ = hydraulic radius, cross-sectional area divided by wetted perimeter
S = energy loss per foot of length of conduit or drop in water surface

**Keywords:** conduits, energy, friction, hydraulic radius, loss
CHEZY, Antoine, 1718-1798, French civil engineer
**Sources:** Considine, D. M. 1976; Scott, J. S. 1993.
*See also* DARCY

## CHICK LAW

The number or organisms destroyed, or the bacterial die-away, by a disinfectant per unit of time is proportional to the number of organisms remaining.

**Keywords:** bacteria, death, disinfectant, organisms
CHICK, Harrietta, 1875-1977, English biochemist
**Source:** Holmes, F. L. 1990.

## CHILD LAW; CHILD-LANGMUIR LAW (1911); LANGMUIR-CHILD LAW

In a diode vacuum tube, the space charge in volts between two parallel plates, d centimeters apart, with one being an emitter with no initial electron velocity, and the other a collector, the space charge limited current is:

$$I = K\,V^{3/2}/d^2$$

where I = the current
V = space charge, volts
d = distance between two parallel plates
K = constant depending on the device and conditions

**Keywords:** charge, collector, current, electron, parallel plates, velocity
CHILD, Clement Dexter, 1868-1933, American physicist
LANGMUIR, Irving, 1881-1957, American chemist and engineer; Nobel prize, 1932, chemistry
**Sources:** Menzel, D. H. 1960; Thewlis, J. 1961-1964.
*See also* LANGMUIR-CHILD

### CHOSSAT LAW (1843)

An animal dies when it has lost 50 percent of its body weight, as represented by the relationship:

$$Y = 98\,e^{-.00835t}$$

where Y = weight in percent of weight at start of fast
t = length of fast, days

**Keywords:** animal, body, death, fast, loss of weight
CHOSSAT, Charles Jacques Etienne, 1796-1875, French physiologist
**Sources:** Kleiber, Max. 1961; Mendelsohn, E. 1964; National Union Catalogue. 1968.

### CHRISTIANSEN EFFECT OR LAW

Complete transparency of a liquid containing finely powdered materials that have the same index of refraction as the liquid can be obtained only with monochromatic light.

**Keywords:** illumination, light, liquid, refraction, transparency
CHRISTIANSEN, Christian, 1843-1917, Danish physicist
**Source:** Hodgman, C. D. 1952.
*See also* BROWNIAN

### CICATRIZATION, MATHEMATICAL LAW OF—SEE CARREL LAW

### CIRCUIT, LAW OF—SEE KIRCHHOFF

### CIRCUITAL LAW—SEE AMPERE

### CIRCULATION (BLOOD), GENERAL LAWS OF (1628)

#### Pressure, Law of

The pressure exerted by the blood on the blood vessel walls is determined by the cardiac rate (output per unit time) and by the resistance to circulation by the blood vessels.

#### Velocity, Law of

The velocity of the blood diminishes in the arteries as the distance from the heart is increased, is at a minimum in the capillaries, and increases on the venous side as the heart is approached.

**Volume Flow, Law of**

The amount or volume of blood passing through the cross-section of the circulation system during a given time is the same as that passing through a cross-section at any other point in the circulatory system.

Based on the early work of William Harvey.

**Keywords:** blood, flow, heart, medicine, pressure, velocity, volume
HARVEY, William, 1578-1657, English physician
**Source:** Thomson, W. 1984.

## CLAPEYRON EQUATION

The freezing point varies slightly with pressure, a small change ($\Delta p$) producing a change in temperature in the freezing point:

$$\Delta T = T/LH_c\ (v_2 - v_1)\Delta p$$

where $T$ = original freezing point
$LH_c$ = latent heat of freezing
$v_2$ = volume of liquid at T, K
$v_1$ = volume of solid at T, K

The boiling point of a liquid, where the liquid changes to a gas, is represented by:

$$dT_b/dp = T_b\ \Delta V_v/LH_v$$

where $T_b$ = boiling point
$\Delta V_v$ = volume change
$LH_v$ = latent heat of vaporization
$p$ = pressure on liquid

**Keywords:** freezing point, latent heat, pressure, volume
CLAPEYRON, Benoit Pierre-Emile, 1799-1864, French engineer
**Sources:** Isaacs, A. 1996; Meyers, R. A. 1992; Parker, S. P. 1998; Thewlis, J. 1961-1964.
*See also* CLAUSIUS

## CLAPEYRON LAW (1834)

A law of thermodynamics in which any action sets up a force that tends to counteract the action. The heat of condensation or the heat of vaporization may be obtained from the Clapeyron equation. Clapeyron put the ideas of Carnot into mathematical form. (See above equations.)

**Keywords:** physical chemistry, thermodynamics, vaporization
CLAPEYRON, Benoit Pierre-Emile, 1799-1864, French engineer
**Source:** Layton, E. T. and Lienhard, J. H. 1988.
*See also* CARNOT; CLAPEYRON EQUATION; CLAUSIUS; THERMODYNAMICS

## CLAUSIUS LAW (1850)

At constant volume, the specific heat of an ideal gas is independent of the temperature.

**Keywords:** constant volume, ideal gas, specific heat, temperature
CLAUSIUS, Rudolf Julius Emmanuel, 1822-1888, German physicist
**Sources:** Clifford, A. F.; 1964; Honig, J. M. 1953.
*See also* BERTHELOT; ENTROPY; THERMODYNAMICS

### CLAUSIUS-CLAPEYRON OR CLAPEYRON-CLAUSIUS EQUATION

This equation represents the change in vapor pressure of a liquid as the temperature changes, as follows:

$$\ln p_v = -\Delta LH_v/RT + B$$

where $p_v$ = vapor pressure of the liquid
$LH_v$ = latent heat of vaporization
R = universal gas constant
T = absolute temperature
B = constant related to material involved and conditions

**Keywords:** change, liquid, vapor pressure, temperature
CLAUSIUS, Rudolf Julius Emmanuel,1822-1888, German physicist
CLAPEYRON, Benoit Paul-Emile, 1799-1864, French engineer
**Sources:** Clifford, A. F. 1964; Honig, J. M. 1953; Isaacs, A. 1996; Lapedes, D. N. 1976.
*See also* GAS; THERMODYNAMICS

### CLAUSIUS-MOSSOTTI LAW OR EQUATION (1850; 1879)

A relationship between density and dielectric constant of a dielectric substance that is partially true for gases and approximately true for liquids and solids, that gives a constant, represented by:

$$(k - 1)/(k + 2)\ \rho = \text{a constant}$$

where k = dielectric constant
$\rho$ = density

that can also be stated as an equation:

$$\alpha = (3/4\ \pi\ N)/[(k - 1)/(k - 2)]$$

where $\alpha$ = polarizability of a molecule
N = number of molecules per unit volume
k = dielectric constant

**Keywords:** constant, density, dielectric constant
CLAUSIUS, Rudolf Julius Emmanuel, 1822-1888, German physicist
MOSSOTTI, Ottaviano Fabrizio, 1791-1863, Italian physicist
**Sources:** Clifford, A. F. 1964; Isaacs, A. 1996; Morris, C. G. 1992.
*See also* LORENTZ-LORENZ

### CLAUSIUS NUMBER, Cl OR $N_{Cl}$

A dimensionless group representing heat conduction in forced flow,

$$Cl = V^3 L\ \rho/k\ \Delta T$$

where V = velocity
L = characteristic length dimension
$\rho$ = mass density
k = thermal conductivity
$\Delta T$ = temperature difference

**Keywords:** conduction, forced flow, heat transfer

CLAUSIUS, Rudolf Julius Emmanuel, 1822-1888, German physicist
**Sources:** Bolz, R. E. and Tuve, G. L. 1970; Land, N. S. 1972; Parker, S. P. 1992; Potter, J. H. 1967.

### CLAUSIUS VIRIAL LAW

The equation of state for a solid can be obtained from the relationship that the mean kinetic energy of a system is equal to its virial. The virial (strength) of a system, in which an atom with coordinates x, y, z is acted upon by a force having components X, Y, Z parallel to these axes, is the average value with respect to the sum of the expression:

$$1/2(xX + yY + zZ)$$

**Keywords:** atom, kinetic energy, solid, virial
CLAUSIUS, Rudolf Julius Emmanuel, 1822-1888, German physicist
**Source:** Morris, C. G. 1992.

### CLOSURE, LAW OF (PSYCHOLOGICAL)

Brain activity follows the principle of equilibrium, and where there is a gap in electric current in brain activity, tensions on both sides of the gap and the electric current closes the gap as postulated by Koffka. The principle of closure is one of perception. When a figure is drawn, for example, with incomplete lines, the person perceiving the drawing completes or closes it in his or her mind and perceives it as complete.

**Keywords:** brain, drawing, equilibrium, perception
KOFFKA, Kurt, 1886-1941, German physiologist
**Sources:** Bothamley, J. 1993; Wolman, B.B. 1989.

### CLOSURE LAW OF ADDITION AND MULTIPLICATION

Any real numbers that are substituted for a and b, gives a + b as one and only one real number, and a x b as one and only one real number. The Closure law of addition for complex numbers is represented by:

$$(a + bi) + (c + di) = (a + c) + (b + d)i$$

**Keywords:** complex numbers, numbers, real numbers
**Source:** Gillispie, C. C. 1991.

### COEHN LAW

If two heterogeneous substances are in contact, both of which are dielectrics, the substance having the greater dielectric constant becomes charged to a potential higher than that of the other by an amount proportional to the difference in their dielectric constants.

**Keywords:** charge, dielectric, electricity, potential, proportional
COEHN, Alfred, 1863-1938, German chemist
**Sources:** Oesper, R. E. 1975; Pelletier, P. A. 1980.

### COERCION TO CULTURAL-GENETIC MEAN, LAW OF

There is a tendency of social pressure to force behavior to the existing central norm.

**Keywords:** behavior, central norm, social pressure
**Source:** Wolman, B. B. 1989.
*See also* BEHAVIOR

## COFFIN-MANSON LAW; OR COFFIN RELATIONSHIP

In tests in which a specimen is cycled under conditions of a constant plastic strain amplitude (the plastic strain is equal to the total strain less the elastic strain component) rather than under a constant stress amplitude the number, N, of cycles required to reach failure is given by this law or relationship:

$$N^{1/2} \varepsilon_p = C$$

where N = number of cycles
$\varepsilon_p$ = plastic strain amplitude
C = constant, approximately equal to $\varepsilon_s/2$

**Keywords:** amplitude, cycles, plastic, strain, stress
COFFIN, John Huntington, 1815-1890, American mathematician
MANSON, Samuel Stanford, twentieth century (b. 1919), American engineer
**Sources:** Lerner, Rita G. and Trigo, George L. 1981. 1991.

## COHEN LAW—SEE AUSTIN-COHEN

## COHESION, LAW OF (PSYCHOLOGY)

A principle that asserts that acts occurring close to each other tend to be integrated into more complex acts, as advanced by E. Tolman.

**Keywords:** acts, close, complex, integrated
TOLMAN, Edward Chase, 1886-1959, American psychologist
**Sources:** Bothamley, J. 1993; Goldenson, R. M. 1984.
*See also* COMBINATION

## COHN LAW

The bacteria with specific forms have a fixed and unchanging basis.

**Keywords:** bacteria, forms, unchanging
COHN, Ferdinand Julius, 1828-1898, German botanist and bacteriologist
**Source:** Millar, D. et al. 1996.

## COLBURN EQUATION

An equation that represents heat transfer in a fluid based on other dimensional groups:

$$St\ Pr^{2/3} = 0.023\ Re^{-0.2}$$

$$\text{or } Nu = 0.023\ Re^{-0.8}\ Pr^{1/3}$$

where St = Stanton number
Pr = Prandtl number
Re = Reynolds number
Nu = Nusselt number

**Keywords:** fluid, dimensionless groups, heat transfer
COLBURN, Allan Philip, 1904-1955, American chemical engineer
**Sources:** American Men of Science. 1955; Rohsenow, W. M. and Hartnett, J. P. 1973.
*See also* DITTUS-BOELTER; NUSSELT; PRANDTL; STANTON; REYNOLDS

## COLBURN NUMBER, Co OR $N_{Co}$

A dimensionless group relating momentum diffusivity to mass diffusivity, for diffusion in a flowing system

$$Co = \mu/\rho D$$

where $\mu$ = absolute viscosity
$\rho$ = mass density
D = mass diffusivity

A material property that is equal to Prandtl/Lewis number and equal to Schmidt number.

**Keywords:** diffusivity, mass, momentum
COLBURN, Allan Philip, 1904-1955, American chemical engineer
**Sources:** American Men of Science. 1955; Bolz, R. E. and Tuve, G. L. 1970; Land, N. S. 1972; Parker, S. P. 1992; Potter, J. H. 1967.
Also see LEWIS; PRANDTL; SCHMIDT

## COLES LAW OR EQUATION—SEE WAKE, LAW OF (HYDRAULICS)

## COLLES LAW (1837); COLLES-BAUMÈS LAW; BAUMÈS LAW

A child that is affected with congenital syphilitis, born from a mother showing no signs of the disease, will not infect its mother.

**Keywords:** birth, child, disease, infect, mother, syphilitis
COLLES, Abraham, 1773-1843, Irish medical doctor (surgeon)
BAUMÈS, Pierre Prosper Francois, 1791-1871, French physician
**Sources:** Friel, J. P. 1974; Landau, S. I. 1986; Stedman, T. L. 1976.

## COLLES-BAUMES LAW—SEE BAUMES LAW; COLLES LAW

## COLLINS LAW

After removal of a new or abnormal growth in infants and children, transfer of disease from one organ to another or recurrence does not occur for a period equivalent to the age of the patient plus nine months, and the possibility is slight.

**Keywords:** abnormal growth, children, disease, recurrence
COLLINS, Joseph, 1866-1950, American neurologist
**Sources:** Dorland, W. A. 1980; Friel, J. P. 1974; Landau, S. I. 1986.

## COLLIS AND WILLIAMS COOLING LAW (1959)—SEE COOLING LAWS

## COLOR MIXTURE, THREE LAWS OF

The results of mixing lights of the same brightness but of differing hues are represented by three laws (Fig. C.3):

### Law 1

All hues opposite each other on the color wheel combine to produce a gray color.

### Law 2

All hues not opposite each other on the color wheel fuse to produce different hues or blends.

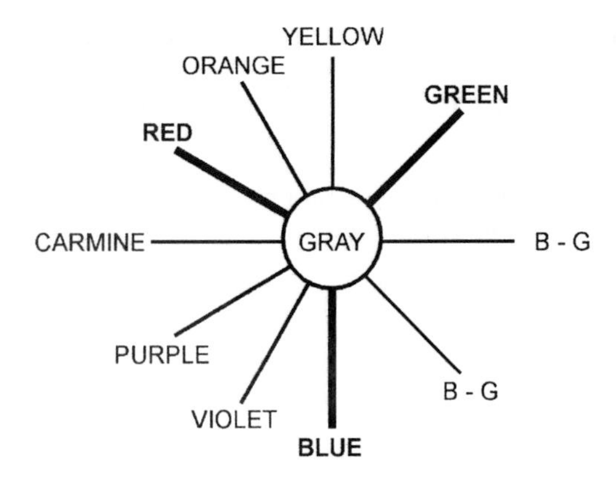


**Figure C.3** Color wheel.

**Law 3**

Secondary hues, that is those hues that are themselves produced by combining hues, mix or fuse according to Law 1 and Law 2.

**Keywords:** hues, light, mixing
**Sources:** Isaacs, A. 1996; Parker, S. P. 1983 (physics).
Also see YOUNG-HELMHOLTZ

## COLOR VISION, COEFFICIENT LAW OF

If the effect of a particular color adaptation or interpretation on any of three mathematical primaries of color is known, the effect may be determined for any combination of these, for any color.

**Keywords:** color, primary, vision
**Sources:** Michels, W. C. 1961; Thewlis, J. 1961-1964.
*See also* GRASSMANN

## COMBINATION, LAW OF (PSYCHOLOGY)

The principle that stimuli that occur simultaneously or in close proximity may produce a combined response or together when a stimulus causing either response is given.

**Keywords:** close, response, stimuli
**Source:** Goldenson, R. M. 1984.
*See also* COHESION

## COMBINATION LAW OF PROBABILITY

The probability of occurrence of either one or both of two events, is represented by:

$$P(\text{A or B or both}) = P(A) + P(B) - [P(A) \times P(B)]$$

where P = probability
A and B = events

**Keywords:** events, occurrence, probability, statistics
**Source:** Doty, Leonard. 1989.
*See also* ADDITIVE; COMPLEMENTARITY; CONDITIONAL; MULTIPLICATIVE

## COMBINATION PRINCIPLE—SEE RYDBERG AND SCHUSTER LAW

## COMBINED GAS LAW, OR GENERAL GAS LAW, OR UNIVERSAL GAS EQUATION

Boyle's Law and Charles' Law for the effect of pressure and temperature on a volume of gas can be incorporated into a combined gas law:

$$P V = K T, \text{ or } pV = MRT$$

where p = absolute pressure
V = volume
M = mass
R = constant of proportionality of ideal gas
T = absolute temperature

**Keywords:** gas, pressure, temperature, thermodynamics
**Sources:** Brown, S. B. and Brown, L. B. 1972; Isaacs, A. 1996.
*See also* BOYLE; CHARLES

## COMBINING VOLUMES, LAW OR PRINCIPLE OF; OR GAY-LUSSAC LAW OF COMBINING VOLUMES (1808)

In chemical reactions involving gases, the volumes of the reacting gases and those of the gaseous products are in the ratio of small whole numbers, provided that all measurements are made at the same temperature and pressure. The law is strictly true for ideal gases but is closely accurate at room temperature and atmospheric pressure. The work of F. W. H. von Humboldt (1805) helped establish this law.

**Keywords:** chemical reactions, gases, products, ratio
Gay Lussac, Joseph Louis, 1778-1850, French chemist and physicist
HUMBOLDT, Friedrick Wilhelm Heinrich von, 1769-1859, German naturalist
**Sources:** Brown, S. B. and Brown, L. B. 1972; Lapedes, D. N. 1974; Lagowski, J. J. 1997; Mandel, S. 1972; Parker, S. P. 1989. 1994; Webster's Biographical Dictionary. 1959. 1995.
*See also* AMAGAT-LEDUC; GAY-LUSSAC

## COMBINING WEIGHTS, LAW OF; RECIPROCAL PROPORTIONS, LAW OF; EQUIVALENTS, LAW OF

Elements combine chemically in ratio of their combining weights or in simple multiples of these weights.

**Keywords:** chemical, elements, weights
**Source:** Mandel, S. 1972.
*See also* EQUIVALENTS; RECIPROCAL PROPORTION

## COMBUSTION, LAWS OF

Combustion calculations are based on these laws:

1. Amagat Law
2. Avogadro Law
3. Combined Gas Law
4. Combining Weights, Law of
5. Conservation of Energy, Law of
6. Conservation of Mass (Matter), Law of (Lavoisier, Antoine)

7. Dalton Law
8. Ideal Gas Law

**Keywords:** combustion, heat, thermodynamics
**Sources:** Considine, C. M. 1976. 1995; Lorenzi, O. 1951.
*See also* SPEED OF COMBUSTION; THERMODYNAMICS

### COMMINUTION, LAWS OF—SEE BOND; KICK; RITTINGER

### COMMUNICATION, LAWS OF—SEE GRICE; HICKS; PROPAGATION OF RADIO WAVES

### COMMUTATIVE LAWS—SEE ALGEBRA OF PROPOSITION; BOOLEAN

### COMMUTATIVE LAW OF ADDITION; COMMUTATIVE LAW

The value of a mathematical expression is independent of the order of combination of numbers or terms of an equation in addition. If a + b and ab are real numbers, then a + b = b + a, and the order of addends does not affect the sum.

**Keywords:** addends, order, real numbers, sum
**Sources:** Considine, D. M. 1976; Good, C. V. 1973; James, R. C. and James, G. 1968; Lipschutz, S. and Schiller, J. 1995; Mandel, S. 1972; Newman, J. R. 1956.
*See also* ALGEBRA

### COMMUTATIVE LAW OF MULTIPLICATION; COMMUTATIVE LAW

The value of a mathematical expression is independent of the order of numbers or terms of an equation in multiplication. If ab are real numbers, then ab = ba, and the order of the factors in multiplication does not affect the product or xy = yx.

**Keywords:** multiplication, order, product, real numbers
**Sources:** Brown, S. B. and Brown, L. B. 1972; Good, C. V. 1973; James, R. C. and James, G. 1968; Lipschutz, S. and Schiller, J. 1995; Newman, J. R. 1956.
*See also* ASSOCIATIVE; DISTRIBUTIVE

### COMPARATIVE JUDGMENT, LAW OF—SEE FECHNER; THURSTONE

### COMPATIBILITY, LAW OF

In dimensional analysis, the condition in which the same number can be supplied for a letter in the model and the prototype without appreciably affecting the results.

**Keywords:** dimensional analysis, mathematics, model, prototype
**Source:** Gillispie, C. C. 1991.

### COMPETITIVE EXCLUSION PRINCIPLE—SEE GAUSE

### COMPLEMENT, LAW OF—SEE ALGEBRA OF PROPOSITIONS

### COMPLEMENTARITY, LAW OF—SEE BOOLEAN

### COMPLEMENTARITY LAW OF PROBABILITY

If $P_A$ is the probability than an event will occur then $1 - P_A$ is the probability that the event will not occur. The sum of the two equals 1.0.

**Keywords:** statistics, probability
**Source:** Doty, L. A. 1989.
*See also* ADDITIVE; COMBINATION; CONDITIONAL; MULTIPLICATIVE

### COMPONENT SUBSTANCES, LAW OF

Every material consists of one substance or a mixture of substances each of which has a specific set of properties independent of the other substances.

**Keywords:** material, mixture, properties, substances
**Source:** Morris, C. G. 1992.

### COMPOSITION, LAW OF—SEE DEFINITE COMPOSITION

### COMPOSITION OF FORCES, LAW OF

Equilibrium occurs if the state of a system is characterized by the fact that no spontaneous processes occur in it. Parallelogram of forces is a geometrical representation by which the sum or difference of two concurrent forces can be considered as the diagonals of a parallelogram. Polygon of forces is the figure obtained by considering force vectors as free vectors, then adding the vectors successively by placing the tail of each additional one at the head of the previous one. The contributions of Leonardo da Vinci were a forerunner in developing this law.

**Keywords:** equilibrium, parallelogram, polygon, vectors
LEONARDO da Vinci, 1452-1519, Italian artist and engineer
**Sources:** Freiberger, W. F. 1960; Honig, J. M. 1953; Scientific American 255(1):108-113, Sept. 1986.
*See also* EQUILIBRIUM; PARALLELOGRAM (OF FORCES)

### COMPOUND INTEREST LAW; EXPONENTIAL LAW

A condition where a function or magnitude, y, increases at a rate proportional to itself, where a and b are constants, and t is the time:

$$y = ae^{bt}$$

The equation representing the total sum after a number of years, at a particular interest rate, is:

$$A = P(1 + r)^n$$

where A = total sum
n = yearss
r = interest rate

**Keywords:** exponential, function, proportional, rate
**Source:** Gibson, C. 1981.

### COMPOUND PENDULUM, LAW OF

If y is the length of a simple pendulum which oscillates in the same time as a compound pendulum, the principle of the center of oscillation:

$$y = \Sigma\, mr^2/\Sigma\, mr$$

Its period T for small amplitudes:

$$T = 2\,\pi\,[\Sigma\, mr^2/g\, \Sigma\, mr]^{1\backslash 2}$$

where $r$ = the radius or perpendicular distance from the axis rotation to the center of gravity of the mass, m
$\Sigma mr^2$ = moment of inertia of the pendulum around the axis of oscillation

**Keywords:** amplitude, mechanics, oscillation, pendulum, period
**Source:** Northrup, E. T. 1917.

### COMPTON EFFECT OR LAW (1923)

Photons travel through space at the speed of light, which after X-ray are scattered from various materials, some of the scattered X-rays have an increased wavelength because X-ray photons may collide with electrons. The energy lost by the photon is,

$$h(\upsilon_1 - \upsilon_2)$$

where $h$ = Planck constant
$\upsilon_1$ and $\upsilon_2$ = frequencies before and after collision. $\upsilon_1$ is greater than $\upsilon_2$.

**Keywords:** electrons, photons, X-rays
COMPTON, Arthur Holly, 1892-1962, American physicist; Nobel prize, 1927, physics
DEBYE, Peter Joseph Wilhelm, 1884-1966, Dutch American engineer
**Sources:** Bynum, W. F. et al. 1981; Considine, D. M. 1976; Mandel, S. 1972; Parker, S. P. 1989.
*See also* RAMAN

### COMPTON-DEBYE, LAW OF—SEE COMPTON

### CONCRETE STRENGTH, LAW OF—SEE ABRAMS; FERET

### CONDENSATION NUMBER, Co OR $N_{Co}$

(1) A dimensionless group that relates viscous force to gravity force of condensing vapors:

$$Co_1 = h/k\ (\mu^2/\rho^2 g)^{1/3}$$

where $h$ = heat transfer coefficient
$k$ = thermal conductivity
$\mu$ = dynamic viscosity
$\rho$ = density
$g$ = gravitational acceleration

**Keywords:** condensation, gravity, vapors, viscous

(2) A dimensionless group applicable to condensation on vertical walls, sometimes called vapor condensation number:

$$Co_2 = L^3\ \rho^2\ g\ LH_v/k\ \mu\ \Delta t$$

where $L$ = length
$\rho$ = density
$g$ = gravitational acceleration
$LH_v$ = latent heat of vaporization
$k$ = thermal conductivity
$\mu$ = dynamic viscosity
$\Delta t$ = temperature difference across the liquid film

**Keywords:** condensation, gravity, vapors, vertical walls, viscous
**Sources:** Bolz, R. E. and Tuve, G. L. 1970; Land, N. S. 1972; McKetta, J. J. and Cunningham, W. 1971-1996; Parker, S. P. 1967; Potter, J. H. 1967.

## CONDITIONAL LAW OF PROBABILITY

Applies to dependent events, that is, in which two or more events are dependent such that the occurrence of one event affects the probability of the second, represented by,

$$P(A \text{ and } B) = P(A) \times P(B/A)$$

in which the last term is the probability of event B occurring if event A has already occurred.

**Keywords:** dependent events, mathematics, probability, statistics
**Source:** Doty, L. A. 1989.
*See also* ADDITIVE; COMBINATION; COMPLEMENTARITY; MULTIPLICATIVE

## CONDITIONING, LAW OF

When an animal learns to satisfy desires from something at the sounding of a noise, then when the noise is heard the animal will expect the object it enjoyed previously.

**Keywords:** animal, association, noise, psychology
**Source:** Scott, T. A. 1995.
Also see PAVLOV; REINFORCEMENT; SKINNER

## CONDUCTIVITY OF HEAT BY CRYSTALS, LAW OF—SEE EUKEN

## CONDUCTORS (ELECTRICAL), LAW OF—SEE ELECTROMOTIVE SERIES; VOLTA

## CONJUGACY, GENERAL LAW OF

Incident light coming from infinity with light rays parallel impinging on a curved surface, the term $n_1/f_1$ becomes zero and the relationship (Fig. C.4):

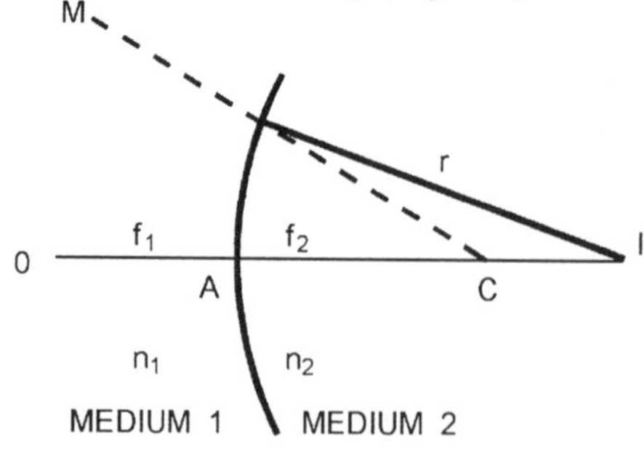


**Figure C.4** Conjugacy.

$$n_1/f_1 + n_2/f_2 = (n_2 - n_1)/r$$

with the general law becoming:

$$n_2/f_2 = (n_2 - n_1)/r$$

where $f_1$ = object distance
$f_2$ = image distance

n = index of refraction
r = radius of curvature

**Keywords:** incident, light, optics, physics
**Source:** Glasser, O. 1944.

## CONSERVATION, LAWS OF

The seven quantities concerned are:

1. Energy, including mass
2. Momentum
3. Angular momentum, including spin
4. Charge
5. Electron-family numbers
6. Muon-family numbers
7. Baryon-family numbers

**Keywords:** baryon, charge, electron, energy, momentum, muon
**Source:** Brown, S. B. and L. B. 1972.
*See also* PROHIBITION

## CONSERVATION LAW

The total magnitude of a physical property of a system (i.e., mass, energy, charge, etc.) remains unchanged, although there may be exchanges of that property between components of the system, considered as a closed system.

**Keywords:** change, physical, property, system
**Sources:** Clifford, A. F. 1964; Isaacs, A. 1996; Morris, C. G. 1992.
*See also* CONSERVATION (OF VARIOUS TOPICS)

## CONSERVATION LAWS (FLUID FLOW), DIFFERENTIAL EQUATIONS OF

The flow of a fluid is governed by laws which express the conservation of mass, momentum, and energy can be written as differential equations. A solution to the mass and momentum equation is:

$$u = U[1 - (y^2/L^2)]$$

where u = fluid velocity
y = any distance from center line of activity
U = maximum velocity
L = half distance across the flowing fluid

The solution of the energy relationship enables the computation of the fluid temperature at any selected position for given wall surface temperatures and fluid properties of viscosity and conductivity:

$$t - t_s = 1/3\ \mu\ U^2/k[1 - (y/L)^4]$$

where t = fluid temperature at a selected position
$t_s$ = temperature for given wall surface
y = selected position
$\mu$ = viscosity of fluid
k = conductivity of fluid

**Keywords:** energy, flow, fluid, mass momentum
**Sources:** Cheremisinoff, N. R. 1986; Flugge, W. 1962.

## CONSERVATION OF ANGULAR MOMENTUM, LAW OF

The total amount of angular momentum is constant in a closed system.

**Keywords:** closed system, mechanics, momentum
**Sources:** Isaacs, A. 1996; Besancon, R. M. 1974; Daintith, J. 1981; Gibson, C. 1981; Menzel, D. H. 1960; Parker, S. P. 1983 (physics); Parker, S. P. 1992.

## CONSERVATION OF AREAS, LAW OF—SEE MACH

## CONSERVATION OF BARYONS OR BARYON NUMBER, LAW OF

If one assigns an additive quantum number $B = +1$ to each baryon, $B = -1$ to each antibaryon, and $B = 0$ to all other particles, then all reactions conserve the total baryon number. A baryon is collective term for particles called nucleons and hyperons. Stated more simply, the law states that the net baryon number in any atomic process remains unchanged or is invariant.

**Keywords:** atomic, invariant, physics
**Sources:** Asimov, I. 1966 (III); Besancon, R. H. 1974; Brown, S. B. and L. B. 1972; Considine, D. M. 1976; Gray, D. E. 1963; Lerner, R. G. and Trigg, G. L. 1991.

## CONSERVATION OF CHEMICAL SPECIES, LAW OF

Molecular materials are conserved in the absence of chemical reactions and atomic materials are conserved in the absence of nuclear reactions.

**Keywords:** chemical, molecular, nuclear, physics
**Sources:** Clifford, A. 1964; Kakac, S. et al. 1987; Morris, C. G. 1952; Rohsenow, W. M. and Hartnett, J. P. 1973; Van Nostrand. 1947.

## CONSERVATION OF ELECTRIC CHARGE, LAW OF

The total amount of charge Q is conserved in all reactions. Reactions that would otherwise result in unbalanced charges will not occur, because there is a matching of negative charge and positive charge whenever possible. The charge is quantized in terms of +e or –e, where e is the magnitude of the charge on the electron. The total charge in any state is obtained by simply adding algebraically the charges on each of the particles in that state. This law was first clearly stated and demonstrated by B. Franklin.

**Keywords:** charge, electric, electron, negative, positive
FRANKLIN, Benjamin, 1706-1790, American statesman and scientist
**Sources:** Asimov, I. 1966 (III); Besancon, R. M. 1974; Brown, S. B. and Brown, L. B. 1972; Considine, D. M. 1976; Gray, D. E. 1972; Haus, H. and Melcher, J. R. 1989; Rothman, M. A. 1963.

## CONSERVATION OF ELECTRON FAMILY MEMBER

The electron family includes electrons, positrons, and neutrinos, as well antineutrinos, which is a part of maintaining the law of conservation.

**Keywords:** electrons, positrons, neutrinos
**Source:** Asimov, I. 1966 (v. III).

## CONSERVATION OF ELEMENTARY PARTICLES, LAW OF

The number of heavy particles (baryons) and the number of light particles (leptons) remain constant in any transformation process.

**Keywords:** baryons, leptons, light particles, transformation

**Source:** Besancon, R. M. 1974.

## CONSERVATION OF ENERGY, LAW OF (1695); OR FIRST LAW OF THERMODYNAMICS (1847)

The earliest precise statement was made by J. Bernoulli and named by G. Leibnitz in 1695 as kinetic energy. It was stated by Lagrange in 1788 in the relationship that kinetic energy plus potential energy is constant.

Energy cannot be created or destroyed. Helmholtz was among the first to state this law as a part of thermodynamics. J. Joule (1842) shared with Lord Kelvin and J. von Mayer the development of the conservation of energy.

**Keywords:** created, destroyed, energy, physics

BERNOULLI, Johannes, 1710-1790, Swiss mathematician

HELMHOLTZ, Hermann Ludwig Ferdinand von, 1821-1894, German physicist and physiologist

JOULE, James Prescott, 1818-1899, British physicist

KELVIN, Lord (THOMSON, William), 1824-1907, British physicist and electrical engineer

MAYER, Julius Robert von, 1814-1878, German physicist

LEIBNITZ, Gottfried Wilhelm, 1646-1710, German philosopher and mathematician

**Sources:** Brown, S. B. and Brown, L. B. 1972; Gibson, C. 1981; Krauskopf, K. B. 1959; Lagowski, J. J. 1997; Morris, C. G. 1992; Parker, S. P. 1992; Rothman, M. A. 1963; Thewlis, J. 1961-1964;

*See also* CONSERVATION; LAGRANGE PRINCIPLE; THERMODYNAMICS

## CONSERVATION OF FAMILIARITY, LAW OF—SEE PROGRAM EVOLUTION

## CONSERVATION OF HYPERCHARGE, LAW OF

In strong magnetic interactions of atomic elements, the hypercharge is conserved which is equal to twice the average charge of the members of a multiplet; hence, all 2I + 1 members of a given multiplet have the same hypercharge.

**Keywords:** atomic, hypercharge, multiplet, physics, magnetic, nuclear

**Sources:** Besancon, R. M. 1974; Parker, S. P. 1983; Thewlis, J. 1961-1964.

## CONSERVATION OF ISOSPIN, LAW OF

The isospin quantum number is conserved in strong interactions, which may or may not remain constant in other interactions.

**Keywords:** atomic, interactions, nuclear, physics, quantum

**Source:** Besancon, R. M. 1974.

## CONSERVATION OF KINETIC ENERGY, LAW OF—SEE CONSERVATION OF MOMENTUM

## CONSERVATION OF LEPTONS, LAW OF; LEPTON NUMBER, CONSERVATION OF LAW OF

If one assigns an additive quantum number = $\ell$ + 1 to each lepton $(\bar{e}, \bar{\mu}, \upsilon_e, \upsilon_n)$, and $\ell = -1$ to each antilepton $e^+$, $\mu^+$, $\upsilon_e$, $\upsilon_n$ and $\ell = 0$ to any other particle, in any physical process the total lepton number is conserved.

**Keywords:** atomic, nuclear, quantum, physics
**Sources:** Besancon, R. M. 1974; Bown, S. B. and Brown, L. B. 1972; Considine, D. M. 1976; Freiberger, W. F. 1960; Gray, D. E. 1972; Honig, J. M. 1953; Lerner, R. G. and Trigg, G. L. 1996.s

## CONSERVATION OF LINEAR MOTION, LAW OF—SEE NEWTON

## CONSERVATION OF MASS, LAW OF (1774)

In all ordinary changes in matter, the total mass is neither reduced nor enlarged, nor is it destroyed or created. The total mass of any system remains constant under all conversion. The principle of mass conservation has been modified into what is called the *principle of mass energy* based on the theory of relativity. Credit for early statement of this law is attributed to A. Lavoisier.

**Keywords:** changes, created, destroyed, enlarged, mass
LAVOISIER, Antoine Laurent, 1743-1794, French chemist
**Sources:** Brown, S. B. and Brown, L. B. 1972; Gibson, C. 1981; Lagowski, J. J. 1997; Leicester, H. M. and Klickstein, H. S. 1953; Mandel, S. 1972; Morris, C. G. 1992; Parker, S. P. 1992; World Book Encyclopedia. 1990.
*See also* CONSERVATION OF MASS-ENERGY

## CONSERVATION OF MASS-ENERGY, LAW OF; SOMETIMES CALLED RELATIVISTIC ENERGY, LAW OF

Energy and mass are mutually convertible from one physical state or form to another, and reversibly from one to the other, and the total matter and energy in the universe is constant, with the laws of conservation of mass and energy as special cases. The relationship of mass and energy is:

$$E = mc^2$$

where E = energy
m = mass
c = velocity of light

**Keywords:** conversion, energy, mass, relativistic
**Sources:** Gibson, C. 1981; Lerner, R. G. and Trigg, G. L. 1991; Mandel, S. 1972; Nourse, A. E. 1969; Parker, S. P. 1992.
*See also* CONSERVATION; EINSTEIN (RELATIVITY)

## CONSERVATION OF MASS FOR A CONTROL VOLUME, LAW OF

The rate of gain in mass inside a control volume (Fig. C.5) in terms of velocity and area vectors and specific volume where the fluid crosses the surface represented by:

$$\dot{m}_\sigma = -\Sigma[V_k \times n_k(a_k/\upsilon_k)]$$

where $m_\sigma$ = rate of increase of mass inside $\sigma$

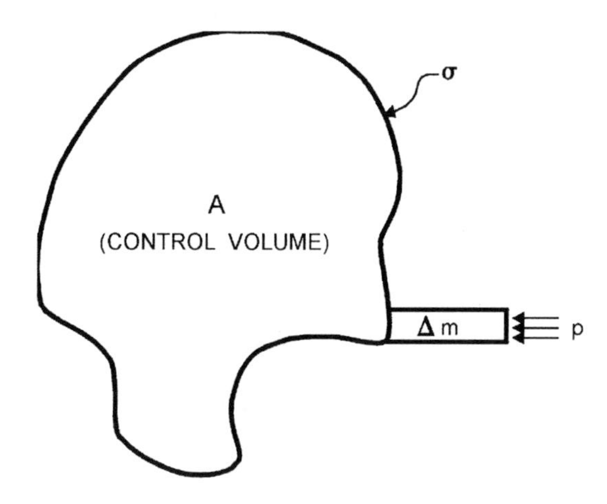


**Figure C.5** Control volume.

$n_k$ = an outward unit vector normal to the surface
$a_k$ = surface area
$V_k$ = velocity of change
$\upsilon_k$ = specific volume of the fluid at $a_k$

**Keywords:** fluid, gain, surface, velocity, volume
**Sources:** Honig, J. M. 1953; Kakac, S. et al. 1987.
*See also* NEWTON

## CONSERVATION OF MATTER, LAW OF

Matter cannot be created or destroyed. The total amount of matter in the universe is unaltered regardless of the changes which take place in the distribution of matter.

**Keywords:** changes, created, destroyed, distribution, universe
**Sources:** Brown, S. B. and Brown, L. B. 1972; Gibson, C. 1981; Krauskopf, K. B. 1959; Mandel, S. 1972; Nourse, A. E. 1969; Walker, P. M. B. 1988.
*See also* CONSERVATION OF MASS

## CONSERVATION OF MECHANICAL ENERGY, LAW OF (1693; 1788)

Related to what we now call kinetic energy, or the capacity to do work and was changed to other forms of energy. Thus, kinetic energy and potential energy equals a constant as determined by the initial conditions. This was stated by Lagrange in 1788.

**Keywords:** change, energy, kinetic, potential
LAGRANGE, Joseph Louis, 1736-1813, Italian French engineer
LEIBNIZ, Gottfried Wilhelm, 1646-1716, German philosopher and mathematician
**Sources:** Michels, W. C. 1956; Thewlis, J. 1961-1964.

## CONSERVATION OF MOMENTUM, LAW OF; NEWTON THIRD LAW

The vector sum of the momentum of each of the colliding bodies after collision is equal to the vector sum before collision. The total linear and total angular momentum of a system are each preserved in the system. Thus,

$$m_1u_1 + m_2u_2 = m_1v_1 + m_2v_2$$

where $m_1$ and $m_2$ = masses of two bodies

$v_1$ and $v_2$ = velocities of two bodies before impact
$u_1$ and $u_2$ = velocities of two bodies after impact

**Keywords:** collision, sum, vector
**Sources:** Daintith, J. 1981; Hodgman, C. D. 1952; Lerner, R. G. and Trigg, G. L. 1986. 1992; Mandel, S. 1972; Morris, C. G. 1992; Nourse, A. E. 1969; Parker, S. P. 1992; Rothman, M. A. 1964: Uvarov, E. S. and Chapman, D. R. 1964.
*See also* NAVIER-STOKES; NEWTON (THIRD)

**CONSERVATION OF MOTION, LAW OF—SEE NEWTON LAW (THIRD)**

**CONSERVATION OF MUON NUMBER, LAW OF**

Decay of the muon conserves the total lepton number and muon number, thus conserving the muon number. The usual mode of disintegration of a muon is

$$u^{+} \rightarrow e^{+} + \upsilon_e + \upsilon_\mu, \text{ or } u^{-} \rightarrow \bar{e} + \upsilon_e + \bar{\upsilon}_\mu$$

**Keywords:** atomic, decay, lepton, physics
**Sources:** Asimov, I. 1966 (III); Gray, D. E. 1972; Lerner, R. G. and Trigg, G. L. 1991.

**CONSERVATION OF NUCLEON CHARGE, LAW OF**

The total number of nucleon charges is constant in a system of particles.

**Keywords:** charge, nucleon, particles, system
**Source:** Gray, D. E. 1972.

**CONSERVATION OF ORGANIZATIONAL STABILITY, LAW OF—SEE PROGRAM EVALUATION (COMPUTER)**

**CONSERVATION OF PARITY, LAW OF, FOR WEAK INTERACTIONS (1956)**

In space, when a process is subjected to the operation of inversion or reflection such that all the signs of its coordinates are changed, the resulting process cannot be distinguished from the original one. If an event is possible, its reflection in a mirror could or might be an equally possible event. This law was proved by the findings of Tsung-Dao Lee and Chen Ning Yang, who received the Nobel prize in physics in 1957. Their work on parity laws led to many discoveries regarding elementary particles.

**Keywords:** inversion, reflection, space
LEE, Tsung-Dao, twentieth century (b. 1926), Chinese American physicist; Nobel prize, 1957, physics (shared)
YANG, Chen Ning, twentieth century (b. 1922), Chinese American physicist; Nobel prize, 1957, physics (shared)
**Sources:** Asimov, I. 1966 (III); Besancon, R. M. 1974; Brown, S. B. and Brown, L. B. 1972; Lapp, R. E. 1963; Schlessinger, B. S. and J. H. 1986; World Book Encyclopedia. 1990.

**CONSERVATION OF SPIN, LAW OF**

Now accepted as being a "strong" law of conservation in which, i.e., a pion that has no spin can produce two particles, are anti-muon, and a muon's neutrino in which the first spins one direction and second spins in the opposite direction, cancelling the effect of each other.

**Keywords:** anti-muon, pion, spin, strong law
**Source:** Nourse, A. E. 1969.

## CONSERVATION OF STRANGENESS, LAW OF

The strangeness number is a new quantum number which nuclear processes satisfy and which decay processes violate. "This law explains the inconsistency of long lifetimes and copious production in high energy collisions."

**Keywords:** atomic, decay, nuclear, quantum

**Sources:** Asimov, I. 1966 (III); Besancon, R. M. 1974; Considine, D. M. 1976; Parker, S. P. 1992.

*See also* MESON

## CONSERVATION OF SYMMETRIES, LAW OF

1. Charge Conjugation Symmetry
   When the sign of a charge of a particle is changed the particle should be converted into its corresponding antiparticle.
2. Charge Conjugation and Parity (Charge Conjugation Parity)
   Where both hold (CP) particularly for strong and electromagnetic interactions. CP means that laws of physics are unchanged by combinations of operational charge conjugation and space inversion.
3. Time Reversal
   Events on an atomic or smaller particle scale should be exactly reversible, a symmetry generally regarded as invariant.

**Keywords:** antiparticle, atomic, charge, electromagnetic, particle

**Sources:** Besancon, R. M. 1974; Parker, S. P. 1983 (physics); Rigden, J. S. 1996.

## CONSERVATION, UNIVERSAL LAW OF (1842)

Energy is neither created nor destroyed, but it can change states. Joule, Kelvin, and Mayer shared in the development of this law.

**Keywords:** creation, destroyed, energy, states

JOULE, James Prescott, 1818-1889, English physicist

KELVIN, William Thomson, 1824-1907, Scottish mathematician and physicist

MAYER, Julius Robert von, 1814-1878, German philosopher

**Sources:** Morris, C. 1992; Simpson, J. A. and Weiner, E. S. C. 1989.

*See also* CONSERVATION

## CONSOLIDATION PROCESS OF CLAY, LAW OF—SEE SOIL SETTLEMENT; TERZAGHI

## CONSTANCY, LAW OF (PSYCHOLOGICAL) OR PRINCIPLE OF CONSTANCY

All mental processes tend toward a state of equilibrium and the stability of the inorganic state, based on a suggestion by S. Freud.

**Keywords:** equilibrium, mental, stability

FREUD, Sigmund, 1856-1939, Austrian doctor (neurologist)

**Source:** Goldenson, R. M. 1984.

## CONSTANCY OF COMPOSITION, LAW OF—SEE CONSTANCY OF RELATIVE PROPORTIONS

## CONSTANCY OF HEAT, LAW OF—SEE HESS

## CONSTANCY OF INTERFACIAL ANGLES, LAW OF (1669)(1772); CONSTANT ANGLES, LAW OF (CRYSTALLOGRAPHY)

The angle between a pair of faces of a particular crystal is constant and is characteristic of the substance and not the size of the crystal. Was first clearly stated in 1772 by J. Rome de Isle, although first known in 1669 by N. Steno.

**Keywords:** angles, crystal, material, size
ROME DE ISLE, Jean B. L., 1736-1790, French mineralogist
STENO, Nicolaus, 1638-1686, Danish geologist
**Sources:** Bates, R. and Jackson, J. 1980; Gary, M. et al. 1972; Morris, C. G. 1992.
*See also* HAUY; RATIONAL INTERCEPTS

## CONSTANCY OF RELATIVE PROPORTIONS, LAW OF; OR LAW OF CONSTANCY OF COMPOSITION (1797) OR RULE OF CONSTANT PROPORTIONS

The ratios between the more abundant dissolved solids in sea water are virtually constant, regardless of the absolute concentration of the total dissolved solids. This law is attributed to L. Proust.

**Keywords:** concentration, dissolved, sea water, solids
PROUST, Louis Joseph, 1755-1826, French chemist
**Sources:** Bates, R. and Jackson, J. 1980; Gary, M. et al. 1972; Honig, J. M. 1953.

## CONSTANCY OF SYMMETRY, LAW OF

## CONSTANCY OF INTERFACIAL ANGLES, LAW OF RATIONAL INTERCEPTS, LAW OF

**Source:** Frye, K. 1981.

## CONSTANT ANGLES, LAW OF—SEE CONSTANCY OF INTERFACIAL ANGLES

## CONSTANT COMPOSITION, LAW OF (1831)

First stated explicitly by J. Berzelius based on the work of J. Dalton, in which each atom of oxygen and each hydrogen in each compound atom had precisely the same weight such that all samples of water contain the same elements in the same proportions by weight.

**Keywords:** atom, hydrogen, oxygen, water
BERZELIUS, Jons Jakob, 1779-1848, Swedish chemist
**Sources:** Brown, S. B. and L. B. 1972; Mandel, S. 1972.
*See also* DALTON; PROUST

## CONSTANT ENERGY CONSUMPTION, LAW OF—SEE RUBNER

## CONSTANT EXTINCTION, LAW OF—SEE VAN VALEN

## CONSTANT GROWTH QUOTIENT, LAW OF—SEE RUBNER

## CONSTANT HEAT FORMATION, LAW OF—SEE HESS

## CONSTANT HEAT SUMMATION, LAW OF—SEE HEATS OF REACTION; HESS

## CONSTANT, LAW OF—SEE THERMODYNAMICS (FIRST) (SCOTT, T. A. 1995)

## CONSTANT NUMBERS IN OVULATION, LAW OF—SEE LIPSCHUTZ

## CONSTANT PROPORTION(S), LAW OF; DEFINITE PROPORTIONS, LAW OF—SEE CONSTANT COMPOSITION; DEFINITE PROPORTION(S)

## CONSTANT RETURNS, LAW OF; CONSTANT RETURNS TO SCALE, LAW OF

An increase in the scale of production gives a proportionate increase in returns, as when there is a proportional increase in labor for the increased scale of production.

**Keywords:** increase, production, scale
**Sources:** Pass, C. et al. 1991; Pearce, D. W. 1992; Shim, J. K. and Siegel, J. G. 1995; Sloan, H. S. and Zurcher, A. 1970.
*See also* DIMINISHING RETURNS; INCREASING RETURNS

## CONTENT, LAW OF

The meaning of verbal material or mental processes is influenced by the conditions surrounding them and the degree of retention depends on the similarity between the learning situation and the retention condition.

**Keywords:** conditions, learning, meaning, mental, retention
**Source:** Wolman, B. B. 1989.

## CONTEXT, LAW OF

Words, phrases, or statements take on meaning in relation to the situation in which they are found or presented.

**Keywords:** meaning, phrases, statements, words
**Source:** Good, C. V. 1973.

## CONTIGUITY, LAWS OF; ARISTOTLE PRINCIPLE

One of Aristotle's laws of learning by association, that when two or more ideas are experienced at the same time the revival of one tends to revive the others. Learning depends on the proximity of stimulus and response in space and/or time.

**Keywords:** ideas, thoughts, memory
ARISTOTLE, 384-322 B.C., Greek philosopher
**Sources:** Angeles, P. A. 1981; Goldenson, R. M. 1984; Good, C. V. 1973; Stedman, T. L. 1974.
*See also* ASSOCIATION; CONTRAST

## CONTINUING CHANGE, LAW OF—SEE PROGRAM EVOLUTION (COMPUTERS)

## CONTINUITY, LAW OF

In steady uniform flow without accumulation or storage of material, the input at the upstream section must come out (emerge) at the downstream section, represented by:

$$w_1A_1V_1 = w_2A_2V_2$$

and for a liquid where $w_1 = w_2$:

$$A_1V_1 = A_2V_2 = Q \text{ (quantity)}$$

where w = weight
A = cross-sectional area
V = velocity

Leonardo's discoveries of nature became to be known as the law of continuity.

**Keywords:** flow, fluid, liquid
LEONARDO da Vinci, 1452-1519, Italian artist and engineer
**Source:** Guillen, M. 1995.
*See also* CONTINUITY

### CONTINUITY, LAW OF (NATURE) (GENERAL); LEIBNITZ PRINCIPLE

There is a continuous development or change in nature such that there is no break, and nothing passes from one state (stage) to another without passing through all the intermediate states (stages). In general, when the difference between two causes is diminished indefinitely, likewise is the difference between their effects.

**Keywords:** development, growth, state
LEIBNITZ, Gottfried Wilhelm, 1646-1716, German mathematician and philosopher
**Sources:** Angeles, P. A. 1981; Bothamley, J. 1993; Flew, A. 1984.
*See also* LEIBNITZ

### CONTRACTION, LAW OF—SEE PFLUGER

### CONTRADICTION, LAW OF

In logic, a proposition and its negation cannot both be true and false. This law was originally elucidated by Aristotle.

**Keywords:** false, logic, true
ARISTOTLE, 384-322 B.C., Greek philosopher
**Sources:** Good, C. V. 1973; Morris, C. G. 1992; Newman, J. R. 1956.
*See also* EXCLUDED MIDDLE; IDENTITY; THOUGHT

### CONTRARY INNERVATION, LAW OF—SEE MELTZER

### CONTRAST, LAW OF—SEE HERTZ

### CONTRAST, LAW OF; ARISTOTLE PRINCIPLE

The revival of one idea tends to effect the revival of its opposite in the thinking process. Original idea attributed to Aristotle and more recently expressed by T. Brown.

**Keywords:** idea, memory, thought
ARISTOTLE, 384-322 B.C., Greek philosopher
BROWN, Thomas, 1778-1820, Scottish physician
**Sources:** Angeles, P. A. 1981; Bothamley, J. 1993; Goldenson, R. M. 1984.
*See also* ASSOCIATION; CONTIGUITY

## COOLING, LAW OF; NEWTON LAW OF COOLING

For moderate temperature ranges the rate of cooling is proportional to the difference between the temperature of the cooling body and that of the surrounding medium, represented as:

$$t = t_m + (t_o - t_m)\, e^{-A\theta}$$

where $t_m$ = temperature of the surrounding medium
$t_o$ = temperature of the body at time, $\theta = 0$
$\theta$ = time

**Keywords:** cooling, heating, temperature
NEWTON, Sir Isaac, 1642-1727, English scientist and mathematician
**Source:** Dorf, R. C. 1995.

## COOLING LAWS

Relationships used in thermal anemometry to determine or estimate velocity, temperature, temperature difference, etc.:

**King Law (1914)**

$$I^2R/(T_s - T_f) = A_o + B_o\,(Re)^{0.5}$$

**Collis and Williams Law (1959)**

$$Nu\,(T_m/T_f)^{-0.17} = A_1 + B_1\,Re^n$$

**Kramer Law (1959)**

$$Nu = 0.42\,Pr^{0.20} + 0.57\,Pr^{0.33}Re^{0.50}$$

**Van der Hegge Zijnen**

$$Nu = 0.38\,Pr^{0.2} + (0.56\,Re^{0.5} + 0.001\,Re)\,Pr^{0.333}$$

where I = electric current
R = electric resistance
T = temperature (surface, film, medium)
Nu = Nusselt number
Pr = Prandtl number
Re = Reynolds number

Other letters are constants

**Keywords:** anemometry, velocity, temperature
KING, E. C., twentieth century, American engineer
HEGGE ZIJNEN, B. G. VAN DER, published thesis in Delft in 1924, Dutch engineer
**Sources:** Dorf, R. C. 1995; Jakob, M. 1957.
*See also* BIOT; NEWTON; NUSSELT; PRANDTL; REYNOLDS

## COPE LAW

Many types of organisms are originated by genera with little specialization; few biological variations are produced by highly specialized genera.

**Keywords:** biology, evolution, genera, growth, specialization
COPE, Edward Drinker, 1840-1897, American paleontologist
**Sources:** Friel, J. P. 1974; Landau, S. I. 1986.

**COPERNICUS THEORY OR LAW** (c. **1514;** c. **1535***)

The theory that the Earth and associated planets revolve around the sun. The theory replaces the 1500-year-old Ptolemaic system.

**Keywords:** Earth, planets, Ptolemaic, sun
COPERNICUS, Nicolaus, 1473-1543, Polish astronomer
**Sources:** Brown, S. B. and Brown, L. B. 1972; Encyclopaedia Britannica. 1961; *NYPL Desk Reference.
*See also* PTOLEMY

**COPE RULE (1887)**

Lineages of animals tend to increase in body size with evolution. Recent studies, in a large sample of North American mammals, imply that the body mass of new species is about 9 percent greater than for older species within a genera. This increase is not evident in small mammals.

**Keywords:** body, evolution, lineages, mass, size
COPE, Edward Drinker, 1840-1897, American paleontologist
**Source:** Science 280(5364):731-734. 1 May 1998.

**COPPET LAW (1871)**

Solutions that have the same freezing point are molecularly equivalent. The lowering of the freezing point of a solution below 0° C is proportional to the amount of solute dissolved. The same work was done earlier in 1771 by R. Watson.

**Keywords:** freezing, molecular, solutions
COPPET, Louis Casimir de, 1841-1911, French chemist and physicist
WATSON, Richard, 1737-1816, English chemist
**Source:** Stedman, T. L. 1976.
*See also* BLAGDEN; RAOULT

**CORIOLIS LAW OR EFFECT (1844)**

In an inertial system, the force exerted on a moving body to cause deflection of the body from a straight line to a curve. The coriolis force is:

$$F_c = 2\ m\ \omega\ \upsilon$$

where $F_c$ = force
$m$ = mass of the body
$\omega$ = constant angular velocity
$\upsilon$ = radial movement of body with constant speed

The above statement is based on the Newton second law.

**Keywords:** angular, force, mass, velocity
CORIOLIS, Gaspard Gustav de, 1792-1843, French physicist
**Sources:** Besancon, R. M. 1974; Considine, D. M. 1976; Rothman, M. A. 1963.

**CORIOLIS PARAMETER, $f_c$**

Similar to Coriolis effect, as applied to a planet:

$$f_c = 2\ \Omega \sin \phi$$

where $f_c$ = value of parameter
$\Omega$ = rotation rate
$\phi$ = latitude

**Keywords:** latitude, planet, rotation
CORIOLIS, Gaspard Gustave de, 1792-1843, French physicist
**Source:** Science 273, p. 335, 19 July 1996.

## CORRELATION OF FACIES (1893)

A principle in stratigraphy wherein, within a given sedimentary cycle, the same succession of facies that occurs laterally is also present in vertical succession.

First presented by J. Walther.

**Keywords:** facies, sedimentary, stratigraphy, succession
WALTHER, Johannes, 1860-1937, German geologist
**Sources:** Bates, R. L. and Jackson, J. A. 1980; Gary, M. et al. 1972.

## CORRESPONDENCE LAW OR THEORY (PSYCHOLOGY)

Whatever is true of molecular behavior is also true of molar behavior and that an unifying relationship can be found, due to the work of N. Bohr. The applicability to psychology is attributed to F. Logan.

**Keywords:** molar, molecular, unifying
BOHR, Niels, 1885-1962, Danish physicist
LOGAN, Frank Anderson, twentieth century (b. 1924), American psychologist
**Source:** Bothamley, J. 1993.

## CORRESPONDING STATES FOR THE CRYSTALLIZATION PROCESS, LAW OF; VON WEIMARN PRINCIPLE

Crystalline precipitates as a result of condensation or coalescence have the same in degree of dispersion and general physical appearance, irrespective of the chemical nature of these precipitates provided that the precipitation takes place under corresponding conditions,

$$V_i = k\ P_1/P_2$$

where $V_i$ = initial rate of condensation or coalescence
$k$ = constant depending on material and conditions
$P_1$ = condensation pressure
$P_2$ = condensation resistance

**Keywords:** condensation, crystallization, materials, precipitation
**Sources:** Clifford, A. F. 1964; Honig, J. M. et al. 1953; Michels, W. C. 1961; Morris, C. G. 1992.

## CORRESPONDING STATES, LAW OF (1881)

Real gases behave in a similar way in terms of their reduced variables, as for example, real gases in the same state of reduced volume and temperature have approximately the same reduced pressure. This relationship can be expressed as the van der Waals equation.

**Keywords:** gases, pressure, temperature, volume
VAN DER WAALS, Johannes Diderik, 1837-1923, Dutch physicist; Nobel prize, 1910, physics
**Source:** Bothamley, J. 1993; Lagowski, J. J. 1997; Morris, C. G. 1992.
*See also* VAN DER WAALS; VON WEIMARN

## CORRESPONDING STATES, LAW OF (GASES)

There is a single functional relationship for a gas relating the ratio of volume, $v_r$, at pressure, p, and temperature, T, to the volume at the critical state, $p_r$, which is the ratio of the pressure to the critical pressure, and $T_r$ is the ratio of the temperature t to the critical temperature. Thus, for two substances, if any two ratios of pressure, temperature, or volume to their respective critical properties are equal, the third ratio must equal the other two. These relationships have the form:

$$v_r = f\ (p_r, T_r)$$

where $v_r$ = ratio of volume
$p_r$ = ratio of critical pressure to pressure
$T_r$ = ratio of temperature to critical temperature

**Keywords:** critical, gas, pressure, temperature, volume
**Sources:** Parker, S. P. 1989; Rigden, J. S. 1996; Thewlis, J. 1961-1964.

## COSINE EMISSION LAW—SEE LAMBERT

## COSINE LAW (ANGLES); LAW OF COSINES

The square of a side of a triangle equals the sum of the squares of the other sides minus twice the product of these sides by the cosine of their included angle, thus (Fig. C.6):

$$a^2 = b^2 + c^2 - 2bc \cos A$$

for a plane triangle:

$$\cos A = (b^2 + c^2 - a^2)/2bc$$

where a, b, c, = the lengths of the sides
A = the angle opposite side a

for a spherical triangle:

$$\cos a = \cos b \cos c + \sin b \sin c \cos A$$

or

$$\cos A = \cos B \cos C + \sin B \sin C \cos a$$

**Keywords:** mathematics, plane, spherical, triangle, trigonometry
**Sources:** Baumeister, T. 1967; Isaacs, A. 1996; James, R. C. and James G. 1968; Karush, W. 1989; Morris, C. G. 1992; Parker, S. P. 1992; Tuma, J. J. and Walsh, R. A. 1998.
*See also* ANGLES; SINE; TANGENTS

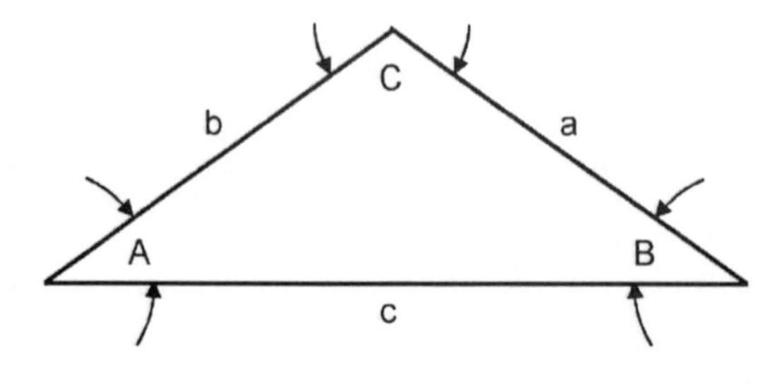


**Figure C.6** Cosine.

## COSINE LAW (GASEOUS MOLECULES)—SEE KNUDSEN

## COSINE LAW (LIGHT); LAMBERT COSINE LAW

The energy of light, per square centimeter, is proportional to a constant multiplied by the cosine of the angle made by a line connecting the source and the recipient, showing that light rays are most intense on parts at right angles to the line joining the light source and the recipient.

**Keywords:** angle, energy, light, radiation
**Source:** Glasser, O. 1944.
*See also* LAMBERT

## COSINE, LIGHT INCIDENCE, LAW OF—SEE PHOTOMETRIC

## COTTON-MOUTON LAW OR EFFECT

A dielectric aromatic organic liquid placed in a traverse magnetic field may become double refracting of light with retardation of the ordinary ray over the extraordinary ray. The retardation is related to the magnetic field strength and the characteristics of the liquid. The effect is small compared to the Faraday effect. This law is analogous to the electro-optical Kerr effect and follows the Faraday effect, discovered in 1845. The Faraday effect was the first indication of the interaction between light and magnetic field.

**Keywords:** dielectric, electricity, magnetic, refracting, retardation
COTTON, Aime August, 1869-1951, French physicist
MOUTON, Gabriel, 1618-1694, French mathematician and astronomer
**Sources:** Condon, E. U. and Odishaw, H. 1958; Gillispie, C. C. 1981; Morris, C. G. 1992; Parker, S. P. 1989; Thewlis, J. 1961-1964.
*See also* FARADAY; HAVELOCK; KERR

## COULOMB LAW (1785); LAW OF ELECTROSTATIC FORCE

The potential at any point in space resulting from a given configuration of discrete point electric charges is a summation of each charge divided by 4 $\pi$ r:

$$V = q/4\,\pi\,\varepsilon\,d^2$$

where V = the electrostatic force
q = each charge
$\varepsilon$ = the permittivity
d = the distance

**Keywords:** charge, electrostatic, permittivity, summation
COULOMB, Charles Augustin de, 1736-1806, French physicist
**Sources:** Clifford, A. F. 1964; Menzel, D. H. 1960; Peierls, R. E. 1956; Thewlis, J. 1961-1964.
*See also* AMONTONS; CHARGE

## COULOMB LAW OF FORCE OF TORSION OF A METAL WIRE OR ROD

The factors on which the force required to twist a wire or rod through a particular angle depend on length, diameter, angle in question, and modulus of rigidity.

**Keywords:** angle, diameter, force, rigidity, rod, twist, wire
COULOMB, Charles Augustine de, 1736-1806, French physicist
**Source:** Achinstein, P. 1971.

## COULOMB LAW OF FRICTION (1781)

The total friction that can be developed between sliding members is independent of the magnitude of the area of contact of the bodies, practically independent of the velocity, except for starting velocity, the total friction developed is proportional to the normal force:

$$F \propto N$$

and

$$F = \mu N$$

where F = force to overcome friction
μ = coefficient of friction
N = weight normal to surface

**Keywords:** area of contact, coefficient of friction, force, normal weight
COULOMB, Charles Augustine de, 1736-1806, French physicist
**Sources:** Considine, D. M. 1976; Thewlis, J. 1961-1964.
*See also* AMONTONS; FRICTION

## COULOMB LAW OF MAGNETIC FORCE BETWEEN TWO POLES

The magnetic force between two poles is proportional to the product of magnetic pole strengths and inversely proportional to the square of the distance between the poles:

$$F = m_1 m_2/\mu d^2$$

where F = magnetic force
$m_1$, $m_2$ = strengths of the magnetic poles
μ = permeability
d = distance between poles

The law is more generally stated as:

$$F = k_o q_1 q_2/d^2$$

where F = force (dynes)
q = electric charge
d = distance between charges
k = a constant

The law was found much earlier by H. Cavendish, 1731-1810, who did not publish his work, so the law was unknown until half a century later.

**Keywords:** electricity, force, magnetism, poles
CAVENDISH, Henry, 1731-1810, British chemist
COULOMB, Charles Augustine de, 1736-1806, French physicist
**Sources:** Honig, J. M. 1953; Landau, S. I. 1986; Lapedes, D. N. 1974; Morris, C. G. 1992; Parker, S. P. 1989; Peierls, R. E. 1956.

## COURVOISIER LAW

The gall bladder is usually not dilated and cannot be felt from jaundice caused by a stone blocking the common bile duct, whereas is dilated and can be felt from blocking by a tumor.

**Keywords:** dilation, gall bladder, medicine, physiology
COURVOISIER, Ludwig Georg, 1843-1918, Swiss surgeon

**Sources:** Friel, J. P. 1974; Landau, S. I. 1986; Rothenberg, R. E. 1982; Thomas, C. L. 1989. Stedman, T. L. 1976.

### COUTARD LAW

In radiotherapy, the point of origin of a mucous membrane tumor is the last site to heal following radiation treatment.

**Keywords:** mucous membrane, radiotherapy, tumor
COUTARD, Henri, 1876-1950, French radiologist in the U. S.
**Source:** Friel, J. P. 1974.

### COVERING LAW MODEL; OR DEDUCTIVE-NOMOLOGICAL MODEL OF EXPLANATION

An explanation that entails a phenomenon being explained in which the explanation may or may not be a necessary condition. The statement is credited to C. Hempel

**Keywords:** explanation, phenomenon
HEMPEL, Carl G., twentieth century (b. 1905), German philosopher
**Sources:** Achinstein, P. 1971; Bothamley, J. 1993; Flew, A. 1979.

### COWLING NUMBER, Cow OR $N_{Cow}$

A dimensionless group used in work on magnetohydrodynamics which is the ratio of the Alfven wave divided by the velocity of the fluid:

$$Cow = \beta/V(\rho\mu)^{1/2}$$

where $\beta$ = magnetic flux density
$\mu$ = magnetic permeability
$V$ = flux velocity
$\rho$ = mass density

The Cowling number is also equal to the reciprocal of the Mach Magnetic number, and also the reciprocal of the Alfven number.

**Keywords:** magnetohydrodynamics, velocity, wave
COWLING, T. G., twentieth century (b. 1906), English physicist
**Sources:** Bolz, R. E. and Tuve, G. L. 1970; Jerrard, H. D. and McNeill, D. B. 1992; Land, N. S. 1972; Parker, S. P. 1992; Potter, J. H. 1967.
*See also* ALFVEN NUMBER; MACH

### COWLING RULE (MEDICAL)

The dose of a drug for a child is obtained by multiplying the adult dose by the age of the child at its next birthday and dividing by 24.

**Keywords:** age, child, dose, drug
**Sources:** Friel, J. P. 1974; Stedman, T. L. 1976.
*See also* YOUNG

### CRAMER LAW (1942) (MEDICAL)

A hypothesis, now disproved, that in a given population, when the incidence of cancer of one particular organ is increased greatly, there is a compensating decrease in the incidence of cancer in a number of other organs. This hypothesis has been disproved by many specialists.

**Keywords:** cancer, incidence, organs
**Source:** Landau, S. I. 1986; Mettler, C. C. 1947.

### CR LAW; CAPACITOR RESISTANCE LAW

If a capacitor and resistance are in series, the ratio of the rise of the potential of the plates of the capacitor depends only on the product of the resistance and the capacitance.

**Keywords:** capacitor, potential, product, resistor
**Source:** Gray, H. J. and Isaacs, A. 1975.

### CRAYA-CURTET NUMBER, Ct OR $N_{Ct}$ (1952?)

A dimensionless group relating kinematic mean velocity and dynamic mean velocity in radiant heat transfer:

$$Ct = V_k/(V_d^2 - V_k^2)/2)^{1/2}$$

where $V_k$ = kinematic mean velocity
$V_d$ = dynamic mean velocity

**Keywords:** dynamic, kinematic, radiant heat transfer
CRAYA, A. E., published in 1952*; CURTET, published in Belgium
**Sources:** Bolz, R. E. and Tuve, G. L. 1970; *NUC; Parker, S. P. 1992; Potter, J. H. 1967.

### CRISPATION NUMBER OR GROUP, Cr OR $N_{Cr}$

A dimensionless group relating to cellular convection currents,

$$Cr = \mu D/\sigma d$$

where $\mu$ = viscosity
D = thermal diffusion
$\sigma$ = surface tension
d = fluid depth

Crispation refers to minute undulations on the surface of a liquid.

**Keywords:** cellular, convection, currents
**Sources:** Bolz, R. E. and Tuve, G. L. 1970; Land, N. S. 1972; Parker, S. P. 1992; Potter, J. H. 1967.

### CROCCO NUMBER, Cr OR $N_{Cr}$

A dimensionless group used for compressible flow relating the velocity of flow to the maximum velocity,

$$Cr = V/V_{max} = [1 + 2/(\gamma - 1)M^2]^{1/2}$$

where V = velocity
$V_{max}$ = maximum velocity of the gas when expanded to zero temperature
$\gamma$ = specific heat ratio
M = Mach number

**Keywords:** compressible, flow, velocity
CROCCO, Gaetano, A., 1877-1968, Italian aeronautics physicist/engineer
**Sources:** Bolz, R. E. and Tuve, G. L. 1970; Land, N. S. 1972; Morris, C. G. 1992; Parker, S. P. 1970; Pelletier, P. A. 1994; Potter, J. H. 1967; Wolko, H. S. 1981.
*See also* LAVAL NUMBER; MACH NUMBER

## CROSSCUTTING RELATIONSHIPS, LAW OF (GEOLOGY) (1669)

A stratigraphic principle whereby relative ages of rocks can be established in which a rock is younger than any other rock that it cuts.

**Keywords:** age, rocks, stratigraphic
**Sources:** Art, H. W. 1993; Bates, R. and Jackson, J. 1980; Gary, M., et al. 1972.

## CRYSTAL LAW—SEE CRYSTALS; MILLER

## CRYSTAL ZONES, LAW OF—SEE NEUMANN

## CRYSTALLOGRAPHIC LAWS

The following generalizations, called laws, made from the study of crystals:

**Constancy of Symmetry, Law of**

**Constancy of Interfacial Angles, Law of**

**Rational Intercepts, Law of**

*Source:* Frye, K. 1981.

## CRYSTALLOGRAPHY, LAWS OF—SEE CRYSTALS; NEUMANN

## CRYSTALS, THE LAW OF INTERFACIAL ANGLES (FUNDAMENTAL LAW OF CRYSTALLOGRAPHY OR FIRST LAW OF CRYSTALLOGRAPHY)

The inclination of the two definite crystal planes to each other, for the same substance, and measured at the same temperature, is constant and independent of the size and development of the planes. The law was formulated by N. Steno. The law is not quite true for all cases.

**Keywords:** crystals, materials, planes, size
STENO, Nicholaus, 1638-1686, Danish anatomist and geologist
**Sources:** Asimov, I. 1972; Frye, K. 1981.

## CUBE ROOT LAW

Used to predict the blast effect of an explosion, the approximate increase in magnitude and intensity is in ratio to the cube root of the energy released. Used in estimating the effect of a nuclear device, i.e., 1 kiloton weapon that produces a certain effect at 1 kilometer from the point of detonation, a megaton device would produce the same effect 10 kilometers from the point of detonation.

**Keywords:** distance, magnitude, nuclear
**Source:** Hogerton, J. F. 1963.
Also see CUBIC

## CUBIC LAW

Similar to the above law, stated for the explosive characteristics of grain dust explosions, for example, in which the ratio of the vent area required to limit the pressure rise to a specific value in two vessels is equal to the cube root of the volumes of the two vessels. It assumes that the same explosive mixture and ignition source are used in each vessel.

**Keywords:** grain dust, explosive, ignition, vent area, volume
**Source:** USDA Misc. Publ. No. 1375, 1979.

## COMMUTATIVE LAW OF ADDITION

Additions and subtractions can be performed in the order in which they occur or in any other order.

**Keywords:** addition, algebra, subtraction
**Sources:** Gillispie, C. C. 1991; Parker, S. P. 1992.
*See also* ADDITIVE; ALGEBRA

## CUNNERSDORF TWIN LAW (CRYSTALLOGRAPHY)

A rarely occurring relationship between normal twin crystals in feldspar in which the twin plane is (201).

**Keywords:** crystals, feldspar, twin
**Source:** Morris, C. G. 1992.

## CURIE LAW

Substances may be rendered radioactive by the influence of the emanations of radium, and these substances maintain their radioactivity longer when enclosed in an enclosure through which the emanations cannot pass. M. Curie used the term *radioactivity* for the first time to describe radiation produced by uranium and thorium.

**Keywords:** emanations, enclosures, radioactivity
CURIE, Marie Sklodowska, 1867-1934, Polish French chemist; Nobel prize, 1903, physics (shared with Pierre Curie and A. H. Becquerel); and Nobel prize, 1911, chemistry
**Sources:** Chemical Heritage 16(1):25, 1998; Daintish, J. 1981; Honig, J. M. 1953.

## CURIE LAW (1895); CURIE LAW FOR PARAMAGNETISM; CURIE LAW FOR MAGNETISM, INCLUDING CURIE CONSTANT

The paramagnetic susceptibility of a material is proportional to the Curie constant, $\psi = C/T$, and inversely proportional to the absolute temperature. Thus, in equation form,

$$\psi = N\,\mu^2/3k\,T = C/T$$

where C = Curie constant, $N\,\mu^2/3k$
C/T = susceptibility, $\psi$
$\mu$ = magnetic moment of each ion
k = Boltzmann constant
N = number of paramagnetic ions per unit volume

The Curie unit of radiation is named after Pierre Curie according to Isaacs* and named after Pierre and Marie Curie according to Mandel.**

**Keywords:** constant, electricity, magnetism
CURIE, Pierre, 1859-1906, French chemist; Nobel prize, 1903, physics (shared with Marie Curie and A. H. Becquerel) and 1911, chemistry
**Sources:** Besancon, R. 1974; Daintith, J. 198I (physics); *Isaacs, A. 1996; **Mandel, S. 1972; Rigden, J. S. 1996.
*See also* CURIE-WEISS

## CURIE-WEISS LAW (1907); CURIE-WEISS LAW FOR FERROMAGNETISM

A relation between electric and magnetic susceptibilities and the absolute temperature followed by magnetic material. Developed as the more general form of Curie Law, that the reciprocal of the magnetic susceptibility per gram molecule of a parametric substance, $\chi_m$, is proportional to the absolute temperature minus a constant or temperature difference:

$$\chi_m = C/(T - T_c)$$

where C = Curie constant
T = absolute temperature
$T_c$ = the Curie-Weiss temperature

**Keywords:** constant, electricity, magnetism
CURIE, Pierre, 1859-1906, French chemist; Nobel prize, 1903, physics (shared); 1911, chemistry
WEISS, Pierre Ernst, 1865-1940, French physicist
**Sources:** Besancon, R. 1974; Daintith, J. 1981. 1988; Kirk-Othmer. 1985; Parker, S. P. 1983.
*See also* BECQUEREL; CURIE; WEISS ZONE

## CUSHING LAW OR CUSHING PHENOMENON

An increase in intercranial tension causes an increase of blood pressure so as to cause an increase in pressure at a point slightly above the pressure exerted against the medulla.

**Keywords:** blood, intercranial, medical, pressure
CUSHING, Harvey, 1869-1939, American surgeon
**Sources:** Friel, J. P. 1974. 1985; Landau, S. I. 1986; Stedman, T. L. 1976.

## CYBERNETICS, LAW OF (1948)—SEE REQUISITE VARIETY

## DALE AND GLADSTONE—SEE GLADSTONE AND DALE

## D'ALEMBERT PRINCIPLE OR FORCE LAW (1743)

The resultant of external forces and kinetic reactions acting on a stationary body is zero. A force can be measured either by the momentum or by the energy given to a body. This is another way of expressing or writing Newton second law. Also applied to the condition for equilibrium of an electromagnetic system.

**Keywords:** energy, force, measure, momentum
D'ALEMBERT, Jean le Rond, 1717-1783, French mathematician and philosopher
**Sources:** Parker, S. P. 1989; Thewlis, J. 1961-1964.
*See also* HAMILTON; LAGRANGE; NEWTON

## DALE-FELDBERG LAW (1933)

An identical chemical transmitter is liberated at all the functional terminals of a single neuron.

**Keywords:** chemical, neuron, transmitter
DALE, Sir Henry Hallett, 1875-1968, British pharmacologist; Nobel prize, 1936, physiology/medicine
FELDBERG, Wilhelm, twentieth century (b. 1900), German English physiologist and pharmacologist
**Sources:** Bynum, W. F. et al. 1981; Landau, S. I. 1986; Stedman, T. L. 1976.

## DALTON LAW (1803) OR GIBBS-DALTON LAW OF PARTIAL PRESSURES

The pressure of a mixture of vapors or ideal gases, which do not react chemically, is equal to the sum of the partial pressures that each component would exert if only occupying the volume of the mixture at the temperature of the mixture. The equation representing this relationship is:

$$p = p_1 + p_2 + p_3 + \ldots = \Sigma p_i \text{ or } PV = V(p_1 + p_2 + p_3 + \ldots)$$

The maximum pressure exerted by a particular vapor of gas in a closed space depends only on the temperature of the gas and is independent of other vapors or gases that do not react chemically.

**Keywords:** gases, mixtures, pressure, vapors
DALTON, John, 1766-1844, British chemist
GIBBS, Josiah Willard, 1839-1903, American physicist
**Sources:** Clifford, A. F. 1964; Honig, J. M 1953; Layton, E. T. and Lienhard, J. H. 1988; Mendel, D. H. 1960; Parker, S. P. 1989. 1992; Thewlis, J. 1961-1964.
*See also* AMAGAT; AVOGADRO; GAY LUSSAC; MULTIPLE PROPORTIONS

## DALTON LAW OF ATOMIC THEORY OF MATTER (1803, 1808)

Each chemical element consists of a single type of atom. Although an amount of the element may contain many atoms, they are all identical in size, shape, and mass. The atoms of different elements always combine in the same ratios. Early work of Dalton of arranging elements in a table was a precursor of Mendeleev. Democritus proposed the first known atomic theory of matter in 460 B.C.

**Keywords:** atoms, chemical, combine, element, ratio
DALTON, John, 1766-1844, British chemist
DEMOCRITUS, fourth to fifth century B.C., Greek philosopher
**Sources:** David, Ian, et al. 1996; Isaacs, A. 1996; Layton, E. T. and Lienhard, J. H. 1988; Parker, S. P. 1992.
*See also* MENDELEEV

## DALTON LAW OF EVAPORATION

Although the pressure of a gas over a liquid decreases the rate of evaporation, the partial pressure of the vapor at equilibrium is independent of the presence of other gases and/or vapors.

**Keywords:** evaporation, gas, pressure, vapor
DALTON, John, 1766-1844, British chemist
**Sources:** Honig, J. M. 1953; Thewlis, J. 1961-1964.

## DALTON LAW OF SOLUBILITY OF GASES

The components of a mixture of gases are dissolved into solution independent of each other in the ratio to the partial pressure of each.

**Keywords:** components, gas, solution
DALTON, John, 1766-1844, British chemist
**Source:** Ballentyne, D. W. G. and Lovett, D. R. 1970.

## DALTON-HENRY LAW

When absorbing a mixture of gases, a fluid will absorb as much of each gas as would be absorbed of either gas separately.

**Keywords:** absorption, fluid, gas, mixture
DALTON, John, 1766-1844, British chemist
HENRY, Joseph, 1797-1878, American physicist
**Sources:** Friel, J. P. 1974; Dorland, W. A. 1988; Stedman, T. L. 1976.

## DAMKÖHLER FIRST NUMBER, $Da_I$ OR $N_{DaI}$ (1936); ALSO CALLED DAMKÖHLER RATIO

A dimensionless group applicable to chemical reactions, momentum and heat transfer, introduced for combustion, that represents the ratio between the reaction rate and flow rate of fluids:

$$Da_I = UL/Vc$$

where U = reaction rate
L = characteristic length dimension
V = velocity
c = concentration

**Keywords:** chemical, flow, fluids, heat transfer, momentum
DAMKÖHLER, Gerhard von, twentieth century (b. 1884), German engineer
**Sources:** Bolz, R. E. and Tuve, G. L. 1970; Land, N. S. 1972; Layton, E. T. and Lienhard, J. H.1988; Meyers, R. A. 1992; Morris, C. G. 1992; NUC; Parker, S. P. 1992; Perry, R. H. 1967; Potter, J. H. 1967.
*See also* DEBORAH

## DAMKÖHLER SECOND NUMBER, $Da_{II}$ OR $N_{DaII}$

A dimensionless group relating the reaction rate and diffusion rate in chemical reactions with mass transfer:

$$Da_{II} = UL^2/Dc$$

where U = reaction rate
L = characteristic length
D = mass diffusivity
c = concentration

**Keywords:** chemical, diffusion, mass, reaction
DAMKÖHLER, Gerhard von, twentieth century (b. 1884), German engineer
**Sources:** Bolz, R. E. and Tuve, G. L. 1970; Land, N. S. 1972; NUC; Parker, S. P. 1992; Perry, R. H. 1967; Potter, J. H. 1967.

## DAMKÖHLER THIRD NUMBER, $Da_{III}$ OR $N_{DaIII}$

A dimensionless group relating heat liberated and heat transported in a chemical reaction and heat transfer:

$$Da_{III} = QUL/c_p \rho VT$$

where Q = heat liberated per mass
U = reaction rate
L = characteristic length
$c_p$ = specific heat at constant pressure
V = velocity
$\rho$ = mass density
T = temperature

**Keywords:** chemical, heat, transfer
DAMKÖHLER, Gerhard von, twentieth century (b. 1884), German engineer
**Sources:** Bolz, R. E. and Tuve, G. L. 1970; Land, N. S. 1972; NUC; Parker, S. P. 1992; Perry, R. H. 1967; Potter, J. H. 1967.

## DAMKÖHLER FOURTH NUMBER, $Da_{IV}$ OR $N_{DaIV}$

A dimensionless group relating heat liberated and heat conducted in chemical reaction and heat transfer:

$$Da_{IV} = QUL^2/kT$$

where Q = heat liberated per unit mass
U = reaction rate
L = characteristic length
k = thermal conductivity
T = temperature

**Keywords:** chemical reaction, conduction, heat, liberated, transfer
DAMKÖHLER, Gerhard von, twentieth century (b. 1884), German engineer
**Sources:** Bolz, R. E. and Tuve, G. L. 1970; Land, N. S. 1972; NUC; Parker, S. P. 1992; Perry, R. H. 1967; Potter, J. H. 1967.

## DAMKÖHLER FIFTH NUMBER, $Da_V$ OR $N_{DaV}$

A dimensionless group relating the inertia force and the viscous force of the flow of fluids,

$$Da_V = \rho\ VL/\mu$$

where $\rho$ = mass density
$V$ = velocity
$L$ = characteristic length
$\mu$ = absolute (dynamic) viscosity

The Damköhler Fifth number is the same as Reynolds number.

**Keywords:** flow, fluids, inertia, viscous
DAMKÖHLER, Gerhard von, twentieth century (b. 1884), German engineer
**Sources:** Bolz, R. E. and Tuve, G. L. 1970; Land, N. S. 1972; NUC; Parker, S. P. 1992; Potter, J. H. 1967.
*See also* REYNOLDS NUMBER

## DARBOUX THEOREM—SEE MONODROMY

## DARCY (D'ARCY) LAW (1856)

The velocity of flow in a porous medium is directly proportional to the pressure gradient, which is particularly true for very low Reynolds numbers. The relationship is expressed as:

$$V = -K\ (\Delta H/L)$$

where $V$ = velocity of flow
$K$ = hydraulic conductivity
$\Delta H$ = loss of head
$L$ = distance

**Keywords:** flow, porous, pressure, Reynolds number
D'ARCY, Henri P., 1805-1858, French engineer
REYNOLDS, Osborne, 1842-1912, English engineer
**Source:** Thewlis, J. 1961-1964.
*See also* CHEZY; REYNOLDS

## D'ARCY LAW FOR PACKED BEDS

A specialized application of the basic Darcy law for fluids, relating the flow of a liquid, in which the velocity of flow in the x-direction is proportional to the permeability, of the cross-section divided by the viscosity, times the pressure gradient:

$$V = (\xi/\mu)dp/dx$$

where $V$ = velocity of flow in x-direction
$x$ = direction of flow
$\xi$ = permeability
$\mu$ = viscosity
$dp/dx$ = pressure gradient

**Keywords:** fluids, liquids, permeability, pressure gradient, viscosity
D'ARCY, Henri P., 1805-1858, French engineer
**Source:** Scott Blair, G. W. 1949.

## DARCY NUMBER, Da OR $N_{Da}$

A dimensionless group relating head loss, velocity head and diameter and length of pipe or a conduit through which a fluid flows:

$$Da = 2g\ H_l\ d/V^2L$$

where g = gravitational constant
$H_l$ = head loss in flow
d = diameter
V = velocity of flow
L = length of pipe or conduit

**Keywords:** diameter, head, length, loss, velocity
D'ARCY, Henri P., 1805-1858, French engineer
**Sources:** Bolz, R. E. and Tuve, G. L. 1970; Land, N. S. 1972; Parker, S. P. 1992; Potter, J. H. 1967.
*See also* FANNING NUMBER; REYNOLDS

## DARKENING, LAW OF (RADIATION)

In a semi-infinite atmosphere with a constant net flux, the greatest interest in radiative transfer is in angular distribution, known as the *law of darkening*.

**Keywords:** atmosphere, flux, radiation
**Source:** Morris, C. G. 1992 (partial)
*See also* RADIATION

## DARWIN—SEE EVOLUTION (1859); NATURAL SELECTION (1838) AND THEORY OF EVOLUTION (1859, ALTHOUGH PROPOSED IN 1842)

## DARWIN-EWALD-PRINS LAW

A complicated expression that can be used to describe the diffraction pattern formed by the reflection of X-rays from a crystal.

**Keywords:** crystal, diffraction, reflection, X-rays
DARWIN, Charles Galton, 1887-1962, English physicist
EWALD, Paul Peter, 1888-1985, German American physicist

## DASTRE-MORAT LAW

Constriction of the surface blood vessels of the body is usually accompanied by dilation of the splanchnic (relating to visceral) vessels, and vice-versa.

**Keywords:** blood, dilation, medicine, physiology, splanchnic, viscera
DASTRE, John Albert Francois, 1844-1917, French physician
MORAT, Pierre, 1846-1920, French physiologist
**Sources:** Landau, S. I. 1986; Stedman, T. L. 1976.

## DEAN NUMBER, De OR $N_{De}$

A dimensionless group representing the relationship between centrifugal force and inertial force for fluid flow in a curved channel or pipe:

$$De = (\rho VL/\mu)(l/2r)^{1/2}$$

or

$(L/2r)^{1/2}$ times Reynolds number

where $\rho$ = mass density
$V$ = velocity
$\mu$ = viscosity
$L$ = channel width or pipe diameter
$r$ = radius of bend curvature

**Keywords:** channel, flow, fluid, pipe
DEAN, Ernest W., 1888-1959, American chemical engineer (check in NC-50)
**Sources:** Bolz, R. E. and Tuve, G. L. 1970; Jerrard, H. G. and McNeill, D. B. 1992; Kakac, S. et al. 1987; Land, N. S. 1972; Parker, S. P. 1992; Potter, J. H. 1967.
*See also* DARCY NUMBER; REYNOLDS NUMBER

## DEBORAH NUMBER OR DEBORAH DIFFUSION NUMBER, Db OR $N_{Db}$

A dimensionless group applicable and fundamental in rheology that represents the ratio of the relaxation time and the characteristic diffusion time ($L^2/D$) for thixotropic flow and viscoelastic materials,

$$Db = \tau D/L^2$$

where $Db$ = time of relaxation divided by time of observation
$\tau$ = relaxation time
$D$ = diffusivity
$L$ = length dimension

**Keywords:** diffusion, relaxation, thixotropic, viscoelastic
DEBORAH. A name in the old testament, Judges 4 and 5, relating to flow of mountains used by Reiner to represent the number. Biblical reference may date back to 12th century B.C.
REINER, Markus, twentieth century (b. 1886), Israeli rheologist
**Sources:** American Institute of Chemical Engineering Journal 21(5):854, 1975; Bolz, R. E. and Tuve, G. L. 1970; Land, N. S. 1972; Physics Today 19(1):62, Jan. 1964; Parker, S. P. 1992; Potter, J. H. 1967; Reiner, M. 1960.
*See also* DAMKÖHLER FIRST NUMBER

## DEBYE-FALKENHAGEN EFFECT

An increase in inductance of an electrolyte occurs when measuring using high voltages, in the order of 100,000 volts.

**Keywords:** electrolyte, inductance, voltage
DEBYE, Peter Joseph Wilhelm, 1884-1966, Dutch American physical chemist; Nobel prize, 1936, chemistry
**Sources:** Honig, J. M. 1953; Parker, S. P. 1984.

## DEBYE-HÜCKEL LIMITING LAW (1923)

The activity coefficient of a dilute solution of a strong electrolyte is proportional to the valence factor, and to the reciprocal of the square root of the ionic strength of the solution:

$$\log \alpha(+ \text{ or } -) = -A \mid z_+ z_- \mid I^{-1/2}$$

where $\alpha$ = activity coefficient
$I$ = ionic strength of solution
$\mid z_+ z_- \mid$ = valence factor

**Keywords:** activity, chemistry, ionic, solution, valence

DEBYE, Peter Joseph Wilhelm, 1884-1966, Dutch American physical chemist; Nobel prize, 1936, chemistry

HÜCKEL, Erich, 1895-1980, German chemist

**Sources:** Bothamley, J. 1993; Bynum, W. F. et al. 1981; Considine, D. M. 1976; Hampel, C. A. and Hawley, G. G. 1973; NUC; Parker, S. P. 1989; Thewlis, J. 1961-1964.

*See also* OSTWALD DILUTION

## DEBYE NUMBER, Dy OR $N_{Dy}$

A dimensionless group applied to electrostatic probing in ionized gases that relates the Debye length and probe radius:

$$Dy = \lambda_D/a = (k_\varepsilon T/e^2 n)^{1/2}/a$$

where $a$ = probe radius
$\lambda_D$ = Debye length
$k$ = Boltzmann constant
$\varepsilon$ = permittivity of free space
$e$ = electron charge
$n$ = number of electrons per volume

**Keywords:** electrostatic, ionized gases, probe

DEBYE, Peter Joseph Wilhelm, 1884-1966, Dutch American physical chemist; Nobel prize, 1936, chemistry

**Source:** Land, N. S. 1972.

## DEBYE $T^3$ LAW (1912)

The heat capacity of a substance attributable to molecular (lattice) vibration is proportional to the third power of the absolute temperature, which in equation form:

$$c_v = (12\,\pi^4\,R/5\,\theta_D^3)\,T^3 = 12\,\pi^4\,R/5\,(T/\theta_D)^3$$

where $\theta_D$ = the Debye temperature, and $(T/\theta_D)$ varies from 0 to 1.0

**Keywords:** heat, thermal, temperature, vibration

DEBYE, Peter Joseph Wilhelm, 1884-1966, Dutch American physical chemist; Nobel prize, 1936, chemistry

**Sources:** Parker, S. P. 1983; Thewlis, J. 1961-1964.

## DECAY—SEE FORSTER; RADIOACTIVE EXPONENTIAL DECAY

## DECAY LAW FOR RADIOACTIVITY (1902)

If the number of active atoms of a given radioactive nuclide at time zero is a given quantity, the number of active atoms at a later time is given by:

$$N(t) = N(0)\exp(-\lambda\,t)$$

where $N(t)$ = activity at time, t
$N(0)$ = activity at time, 0
$\lambda$ = decay constant

which was formulated by E. Rutherford and F. Soddy.

**Keywords:** atoms, decay, nuclide, radioactive, time

RUTHERFORD, Ernest, 1871-1937, British physicist
SODDY, Frederick, 1877-1956, British chemist
**Source:** Thewlis, J. 1961-1964.
*See also* FORSTER; RADIOACTIVE

### DECLINING PRODUCTIVITY, LAW OF

The increase in any economic variable (land, labor, capital, etc.), while other variables remain fixed, will reach a limit beyond which a further increase diminishes productivity.

**Keywords:** diminishes, economic, increase, variable
**Source:** Sale, K. 1980.

### DECLIVITY, LAW OF (1877)

Formulated by G. Gilbert, the law states that, where homogeneous rocks are maturely dissected by consequent streams, all hillside slopes of the valleys cut by the streams tend to develop at the same slope angles, providing symmetric profiles of ridges, spurs, and valleys.

**Keywords:** hillside, homogeneous rocks, slopes, streams, profile
GILBERT, Grove Karl, 1843-1918, American geologist
**Source:** Gary, M., et al. 1972.
*See also* EQUAL DECLIVITIES

### DE COPPET—SEE COPPET

### DECREASING BASICITY, LAW OF (1896?)

A general statement relating the order of crystallization of minerals from igneous magnus, in which the first minerals to crystallize are ore minerals followed by iron, calcium, and alkali silicates, and other quartz.

**Keywords:** alkali, crystallization, igneous magnus, minerals, quartz
**Source:** Tomkeieff, S. I. 1983.
*See also* HILT; POLARITY; SUPERPOSITION

### DECREMENT, LAWS OF

Named by R. Hauy. After studying a variety of crystals, he reported that such crystals could be reduced to six diverse shapes. The building blocks of crystals were the tetrahedron, triangular prism, and the parallelepipedon. With the discovery of isomorphism and polymorphism by E. Mitscherlich in 1819 and 1822, Hauy's approach was abandoned.

**Keywords:** building blocks, crystals, shapes
HAUY, Rene Just, 1743-1822, French mineralogist
MITSCHERLICH, Eilhard, 1794-1863, German chemist
**Source:** Frye, K. 1981.
*See also* HAUY

### DEFINITE COMPOSITION LAW; OR CONSTANT COMPOSITION, LAW OF; DEFINITE ACTION, LAW OF; ALSO KNOWN AS DEFINITE PROPORTIONS LAW

A given pure chemical compound always contains the same elements in the same fixed proportion by weight.

**Keywords:** compound, elements, proportion, weight

**Sources:** Brown, S. B. and L. B. 1972; Honig, J. M. 1958; Krauskopf, K. B. 1959; Lapedes, D. N. 1974.
*See also* PROUST

## DEFINITE PROPORTIONS LAW—SEE CONSTANCY OF RELATIVE PROPORTIONS; DEFINITE COMPOSITION LAW; PROUST

## DEGRADATION OF ENERGY, LAW OF

Another name for the second law of thermodynamics, derived from the statement that the entropy of an isolated system is increased by irreversible processes, and the sum of available energy decreases.

**Keywords:** energy, entropy, second law, thermodynamics
**Source:** Michels, W. C. 1956.
*See also* THERMODYNAMICS

## DE HAAS-VAN ALPHEN EFFECT OR LAW (1930)

Refers to the oscillatory behavior of the magnetic moment of a pure metal crystal with changes in the applied magnetic field. A quasi-periodic variation of susceptibility, which may oscillate between diamagnetism and paramagnetism when plotted against 1/H, where H is the magnetic intensity, at low temperatures. The effect is derived from the quantization of the orbits of conduction electrons in a magnetic field.

**Keywords:** crystal, magnetic, oscillatory [complicated equation omitted]
DE HAAS, Wander J. de, 1878-1960, Dutch physicist
VAN ALPHEN, P. M., twentieth century, Dutch physicist
**Sources:** Besancon, R. M. 1974; Parker, S. 1983; 1994; Thewlis, J. 1961-1964.

## DEITERS LAW (1865)

The doctrine that all nerve fibers are long processes of nerve cells.

**Keywords:** cells, fibers, processes
DEITERS, Otto Friedrich Karl, 1834-1864, German anatomist
**Source:** Landau, S. I. 1986.

## DELAROCHE—SEE BIOT LAW; DELAROCHE AND BERARD

## DELAROCHE AND BERARD, LAW OF

For all elementary diatomic gases approximately in the perfect state, and for gas compounds formed without condensation and approximately in the perfect state, the product of the molecular weight and the specific heat at constant pressure has the same value.

**Keywords:** gases, molecular weight, physical chemistry, specific heat
DELAROCHE, Francois, 1775-1813, French scientist
BERARD, Jacques E., 1789-1869, French chemist
**Sources:** Nernst, W. 1923; Northrup, E. F. 1917; Truesdell, C. A. 1980.

## DE LA RUE AND MILLER LAW

For a field between two parallel plates, the sparking potential of a gas is the product of the gas pressure and the sparking distance only:

$$d = \ln f(E/pd) - \ln\phi\,(E/pd)$$

where d = distance between plates
E = sparking potential of gas
p = pressure of gas
f = for plate 1; ϕ is for plate 2

**Keywords:** distance, plates, sparking, voltage
DE LA RUE, Warren, 1815-1889, British astronomer and inventor
MILLER, William Allen, 1817-1870, English astronomer
**Sources:** Ballentyne, D. W. G. and Lovett, D. R. 1970; Ballentyne, D. W. G. and Walker, L. E. Q. 1961; Carter, E. F. 1976; Morris, C.G. 1992.

## DELTA LAW (ΔL) (1929)

All geometrically similar crystals of the same material suspended in the same solution grow at the same rate, based on an increase in length of crystals. For ΔL, the increase in length of one crystal will be the same for other crystals of the same geometrically corresponding distance. The law applies to existing crystals, does not apply to nucleation, and is based on the work of W. McCabe.

**Keywords:** crystals, growth, length, solution
McCABE, Warren Lee, 1899-1982, American chemical engineer
**Source:** Perry, J. H. 1950.

## DEMAND, LAW OF (ECONOMICS)

Other things being equal, more of a commodity will be bought at a lower price, and less will be bought at a higher price. This states a general expectation and is not a "law" in the true since, since it is not always clear what "other things" being equal means.

**Keywords:** amount, good(s), price
**Source:** Pearce, D. W. 1981. 1986.
*See also* DIMINISHING RETURNS

## DEMOCRITUS—SEE DALTON LAW OF ATOMIC THEORY

## DE MORGAN LAWS

May be stated as $(a + b)' = a'b'$ and $(ab)' = a' + b'$, and it may be used to transform conjunctions to disjunctions or vice versa. In logic, the law may be stated as: "If and only if all inputs are true, then the output is true" is logically equivalent to "if at least one input is false, then the output is false."

**Keywords:** false, logic, mathematics, true
DE MORGAN, Augustus, 1806-1871, British Indian mathematician and logician
**Sources:** Gray, P. 1967; Ito, K. 1987; James, R. C. and James, G. 1968; Lipschutz, S. and Schiller, J. 1995.
*See also* ALGEBRA OF PROPOSITION; BOOLEAN

## DENERVATION, LAW OF—SEE CANNON LAW

## DERIVATION, LAW OF

The operation of deducing one function from another according to some fixed relationship, such as finding the derivative of a function $f(x)$ as the coefficient of $f(x + h)$ in its expansion as a power series of h.

**Keywords:** coefficient, derivative, function
**Source:** Thewlis, J. 1961-1964 (partial).

## DERYAGIN NUMBER, De OR $N_{De}$

A dimensionless group applicable to coating that relates the film thickness and capillary length of path:

$$De = L\ (\rho g/2\sigma)1/2$$

where $L$ = film thickness
$\rho$ = mass density
$g$ = gravitational acceleration
$\sigma$ = surface tension

**Keywords:** coating, capillary, film
DERYAGIN, Boris Vladimirovich, twentieth century (b. 1902), Soviet chemist
**Sources:** Bolz, R. E. and Tuve, G. L. 1970; Land, N. S. 1972; Morris, C. G. 1992; Parker, S. P. 1992; Pelletier, P. A. 1994; Potter, J. H. 1967; Turkevich, J. 1963.
*See also* BOND; EÖTVÖS; GOUCHER.

## DESCARTES LAWS OF REFRACTION (1637)

The sine of the angle of incidence of light has a constant relation to the sine of the angle of refraction for two given media.

1. Incident and refracted rays are in the same plane with the normal to the surface.
2. Incident and refracted rays lie on the opposite sides of the normal to the surface.
3. The sines of the inclinations to the normal of incident and refracted rays are at a constant ratio to one another, the ratio depending only on the two media involved, not on the angles with the normal.

**Keywords:** angle, incidence, light, refraction
DESCARTES, Rene, 1596-1650, French philosopher
**Sources:** Considine, D. M. 1976; Stedman, T. L. 1976.
*See also* REFRACTION; SNELL

## DESIGN—SEE BUCKINGHAM THEORY; DIMENSIONLESS GROUPS; ENGINEERING DESIGN (AXIOMS)

## DESLANDRES LAWS

### First Law

In band spectra of the sun, the oscillation frequencies of the line starting from one head form arithmetical series. More than one such series can proceed from the same head.

### Second Law

The differences in frequencies of the heads of the bands in each group form an arithmetical series, but the arrangement of the heads is reversed from that of the lines forming each band.

**Keywords:** arithmetical series, band spectra, heads of bands, sun
DESLANDRES, Henri Alexandre, 1853-1948, French astrophysicist
**Sources:** Clifford, A. F. et al. 1964; Honig, J. M. 1953.

## DESMARRES LAW

In vision, when the visual axes are crossed, the images are uncrossed; when the axes are uncrossed or diverging, the images are crossed.

**Keywords:** axes, images, vision
DESMARRES, Louis A., 1810-1882, French ophthalmologist
**Sources:** Dorland, W. A. 1980; Friel, J. P. 1974; Landau, S. I. 1986; Stedman, T. L. 1976.

## DESPRETZ LAW

The temperature of water at maximum density is lowered below temperatures at maximum density below 4° C by the addition of a solute in proportion to the concentration of the solution.

The ratio of the latent heat of vaporization divided by the difference in volume a saturated vapor and liquid is approximately the same for different substances at their boiling points under equal pressures and is represented by:

$$LH_v/(v_v - v_L) = \text{constant}$$

where $LH_v$ = latent heat of vaporization
$v_v$ = volume of saturated vapor
$v_L$ = volume of liquid

**Keywords:** boiling, concentration, freezing, latent heat, liquid, solutions, vaporization
DESPRETZ, Cesar Mansuete, c. 1791-1863, French physicist
**Sources:** Bennett, H. 1939; Lewis, R. J. 1993.

## DESTRIAU, LAW OF OR EFFECT (1936)

Certain phosphorescent inorganic materials, when suspended in a dielectric field, may be excited to luminescence if subjected to the action of an alternating electric field. Also known as *electroluminescence*.

**Keywords:** dielectric, inorganic, luminescence, phosphorescent
DESTRIAU, Georges G., twentieth century, French scientist
**Sources:** Morris, C. G. 1992; Parker, S.P. 1987. 1992; Travers, G. 1996.

## DETACHMENT, LAW OF (HORNEY MODEL)

The Horney model typifies the law, in which the neurotic striving for safety is accomplished by exaggerating one of the three characteristics of anxiety: helplessness, aggressiveness, and detachment. The neurotic solution of detachment is typified by avoiding intimate or even casual contacts with others by moving away from them. These people strive to be completely and perhaps unrealistically self-sufficient.

**Keywords:** anxiety, argument, logic, neurotic
HORNEY, Karen, 1885-1952, German American psychologist and psychiatrist*
**Sources:** Corsini, R. J. 1994; Goldenson, R. M. 1984; *Reber, A. S. 1985; Sheehy, N. 1997.

## DETERMINISM, LAW OF; OR DETERMINISTIC LAW

The principle that nature follows exact laws so that what will happen in the future is based on the conditions and state of nature at any given moment in the past. It is generally accepted that all events of the world are not predetermined, although some behavior can be predicted the limits are not well defined. Based on the work by Laplace.

**Keywords:** condition, future, nature, past

LAPLACE, Pierre-Simon Marquis de, 1749-1827, French mathematician
**Sources:** Flew, A. 1984; Parker, S. P. 1992.

## DEUTSCH EQUATION (1922); DEUTSCH-ANDERSON EQUATION

A basic equation used in the design of electrostatic precipitators (ESP), relating the efficiency of separation:

$$\eta = 1 - e^{-w(A/Q)}$$

where $\eta$ = efficiency of separation
A = collector plate surface area
Q = exhaust gas volumetric flow rate
w = partial drift velocity
e = base of the natural logarithm

**Keywords:** design, efficiency, precipitators, separation
DEUTSCH, Walther, twentieth century, German physical chemist
**Sources:** Chemical Engineering 105(10):141, 1998; Dallavalle, J. M. 1948; Stevenson, L. H. and Wyman, B. C. 1991.
*See also* ELECTROVISCOUS

## DE WAELE-OSTWALD LAW; WAELE-OSTWALD LAW (1923, 1925)

A power law in rheology:

$$\nu = K\ pm$$

where $\nu$ = the kinematic viscosity
K and m = constants, with m greater than 1

de Waele himself later refused to recognize the equation as appropriate.

**Keywords:** power, rheology, viscosity
DE WAELE, A., twentieth century, German rheologist, published in German and English, 1920s to 1940s
OSTWALD, Wo., twentieth century, German rheologist, published in German, 1920s to 1940s
**Sources:** Reiner, M. 1960; Scott Blair, G. W. 1949.

## DE WAELE POWER LAW (1923)

A modification of the Newton equation relating shear and strain:

$$s = n\ \overline{v}^{\phi}$$

where s = strain
n = constant, not viscosity
$\overline{v}$ = rate of shear
$\phi$ = a constant exponent

**Keywords:** rate, shear, strain
DE WAELE, A., twentieth century, German rheologist, published in German and English, 1920s to 1940s
**Source:** Scott Blair, G. W. 1949.
*See also* NEWTON (STRESS PROPORTIONALITY); POWER LAWS

## DIDAY LAW

A woman whose first pregnancy results in stillbirth may later have a child with active syphilis or a child with latent syphilis or one entirely unaffected.

**Keywords:** birth, pregnancy, woman, syphilis
DIDAY, Paul Edouard, 1812-1894, French doctor
**Sources:** Garrison, F. H. 1929; Mettler, C. C. 1947.

## DIETERICI RULE (1899)

An equation representing the equation of state of real gases,

$$RT = p(V - b)\, e^{a/RTV}$$

where R = gas constant
T = temperature
p = pressure
V = volume
a and b = constants

This rule is related to the van der Waals equation, and the rule considers the influence of molecules with the liquid the near the boundary of the gas.

**Keywords:** gas, pressure, temperature, volume
DIETERICI, Conrad Heinrich, nineteenth century, (b. 1858), German chemist
**Sources:** Bothamley, J. 1993; Glasstone, S. 1958.
*See also* ARRHENIUS TEMPERATURE; VAN DER WAALS

## DIFFRACTION, LAW OF—SEE BABINET; DARWIN-EWALD-PRINS; FRAUNHOFER; FRESNEL

## DIFFUSE REFLECTION IN THE CONSERVATIVE CASE, LAW OF

The intensity of diffuse reflection, $I(0, \mu)$, is a function of the viscosity of the fields through which light passes, represented mathematically, for z-directed electromagnetic waves:

$$I(0, \mu) = 1/4E\ \mu_0/(\mu + \mu_0)\ H(\mu)\ H(\mu_0)$$

where $\mu$ = viscosity
E = electric intensity in x-direction
H = magnetic intensity in y-direction

**Keywords:** electromagnetic, intensity, viscosity, wave
**Source:** Parker, S. P. 1983 (physics).
*See also* LOMMEL-SEELINGER; REFLECTION

## DIFFUSION, FICK LAWS OF

### 1. Fick First Law

The flow, $s_{ix}$, during diffusion is proportional to the gradient of concentration:

$$s_{ix} = -D_i\ \partial c_i/\partial x$$

where D = proportionality of diffusion such that
$D_i = kT/Q_i\ \mu_i$
c = concentration

**2. Fick Second Law**

Within a region of length dx between x to x + dx, the increase in concentration per unit time (rate) is the excess of material diffusing into the region over that diffusing out, divided by the volume, dx:

$$\partial c/\partial t = D\ (\partial^2 c/\partial x^2)$$

**Keywords:** concentration, diffusion, input, output
FICK, Adolph Eugen, 1829-1901, German physiologist
**Sources:** Honig, J. M. 1953; Perry, R. and Green D. W. 1984.
*See also* GRAHAM; WAGNER

## DIFFUSION, LAW OF (MEDICAL)

A process that is established in the nerve center affects the entire organism by a process of diffused activity.

**Keywords:** activity, nerve, organism, physiology
**Sources:** Friel, J. R. 1974; Landau, S. I. 1986.

## DIFFUSION OF GASES—SEE GRAHAM

## DIFFUSIVITY, LAW OF—SEE EXNER

## DILUTION, LAW OF—SEE OSTWALD

## DILATANCY, RULE OF OR LAW OF OR RELATIONSHIP OF (1885)

Earlier, in 1611, Kepler wrote on the stacking of spheres. As applied to spheres in a pile in which the spheres are piled on top of each other, as predicted by O. Reynolds, they occupy a volume of $(2)^{1/2}$ that of a close-packed pile of the same spheres. The word *dilatancy* is also used in rheology writings.

**Keywords:** packed, spheres, volume
REYNOLDS, Osborne, 1842-1912, English engineer
**Sources:** Gillispie, C. C. 1981; Morris, C. G. 1992; Science News 154(7)103, August 15, 1998. See also KEPLER

## DIMENSIONLESS GROUPS AND NUMBERS

Acceleration; Alfvén; Archimedes; Arrhenius; Bagnold; Bansen; Bingham; Biot; Blake; Bodenstein; Boltzmann; Bond; Bouguer; Boussinesq; Brinkman(n); Bulygin; Buoyancy; Capillarity; Capillary; Carnot; Cauchy; Cavitation; Centrifuge; Clausius; Colburn; Condensation; Cowling; Craya-Curtet; Crispation; Crocco; Damköhler; Darcy; Dean; Deborah; Debye; Deryagin; Drew; Dulong; Eckert; Einstein; Ekman; Elasticity; Electroviscous; Ellis; Elsasser; Enthalpy recovery; Eötvös; Euler; Evaporation; Evaporation-elasticity; Explosion; Fanning friction; Fedorov; Fliegner; Flow; Fluidization; Fourier; Froude; Galileo; Gay Lussac; Goucher; Graetz; Grashof; Gravity; Gukhman; Hall; Hartmann; Heat transfer; Hedstrom; Hersey; Hodgson; Homochronous; Hooke; Jakob; J-factor; Jeffrey; Joule; Kármán; Kirpichev; Kirpitcheff; Knudsen; Kossovich; Kronig; Kutateladze; Lagrange; Laval; Leroux; Leverett; Lewis; Lorentz; Luikov; Lukomskii; Lundquist; Lyashchenko; Lykoudis; McAdams; Mach; Magnetic force; Magnetic Mach; Magnetic pressure; Magnetic Reynolds; Marangoni; Margoulis; Margulis; Mass ratio; Merkel; Miniovich; Momentum; Mondt; Morton; Naze; Newton; Number of velocity heads; Number for similarity; Nusselt; Ocvirk; Ohnesorge; Particle; Péclet; Phase-change; Pipeline; Plasticity; Poiseuille; Poisson; Pomerantsev; Posnov; Power;

Prandtl; Predvoditelev; Pressure; Psychrometric; Radiation; Rayleigh; Reech; Regier; Reynolds; Richardson; Romankov; Rossby; Russell; Sach; Sarrau; Schiller; Schmidt; Semenov; Sherwood; Slosh time; Smoluckowski; Sommerfeld; Spalding; Specific heat ratio; Specific speed; Stanton; Stefan; Stewart; Stokes; Strouhal; Suratman; Surface viscosity; Taylor; Thiele; Thoma; Thomson; Thring radiation; Toms; Truncation; Valensi, Velocity; Viscoelastic; Weber; Weissenberg; Zhukovsky.

## DIMINISHING RETURNS, LAW OF (ECONOMICS)

An increase of input into a system (such as labor, capital, energy, fertilizer, etc.) applied beyond a certain point causes a less than proportionate output (increase in the production) from the unit to which the additional input is applied.

**Keywords:** input, output, returns
**Sources:** Moffat, D. W. 1976; Pearce, D. W. 1986; Sloan, H. S. 1970.
*See also* DEMAND; DIMINISHING UTILITY; INCREASING RETURNS; SUBSTITUTION

## DIMINISHING UTILITY, LAW OF (ECONOMICS)

A consumer obtains considerable satisfaction from the first of an item received, but regardless of the desire for the items, the consumer is not longer interested after receiving a certain quantity.

**Keywords:** consumer, desire, quantity, satisfaction
**Source:** Moffat, D. W. 1976.
*See also* DIMINISHING RETURNS; SUBSTITUTION

## DIRAC MODEL (1930)

In symmetry with the electron, there is an opposite and equal charge known as the positron. This prediction was made by P. Dirac in 1930 and was confirmed by C. D. Anderson in 1932.

**Keywords:** charge, electron, positron
DIRAC, Paul Adrien Maurice, 1902-1984, English physicist; Nobel prize, 1933, physics (shared)
ANDERSON, Carl David, 1905-1991, American physicist; Nobel prize, 1936, physics (shared)
**Sources:** Millar, D. et al. 1996; Rothman, Milton A. 1963.

## DIRECT CURRENT (ELECTRICAL) PHYSICAL LAWS OF CIRCUIT ANALYSIS—SEE

1. Ohm Law
2. Kirchhoff Laws

## DIRECTION OF INDUCED CURRENT, LAW OF—SEE LENZ LAW

## DISHARMONY LAW

The logarithm of a dimension of any part of an animal is proportional to the logarithm of the dimension of the whole animal.

**Keywords:** animal, dimension, part, proportional
**Source:** Gray, P. 1967.

## DISPERSION OF LIGHT, LAW OF—SEE BIOT

## DISPLACEMENT, LAW FOR COMPLEX SPECTRA—SEE KOSSEL AND SOMMERFELD

## DISPLACEMENT, LAW OF—SEE WIEN

## DISPLACEMENT (NUCLEAR DISINTEGRATION), LAW OF (1913)

On the emission of an alpha particle, the atomic number of an atom decreases by two units, and the mass number decreases by four units. On the emission of an electron, the atomic number of the atom increases by one unit, and the mass number remains unchanged.

**Keywords:** alpha, atom, electron, mass
**Source:** Thewlis, J. 1961-1964.
*See also* DECAY; RADIATION; SODDY AND FAJANS

## DISPLACEMENT LAWS—SEE FAJANS; RUSSELL, FAJANS, SODDY; SODDY-FAJANS; SOMMERFELD AND KOSSEL OR KOSSEL AND SOMMERFELD

## DISPLACEMENT OF EQUILIBRIUM, LAW OF—SEE LE CHATELIER

## DISSIPATION OF ENERGY, LAW OF

In every energy conversion, some of the original energy is always changed to heat energy, which is not available for further conversion of energy.

**Keywords:** availability, conversion, energy, heat
**Source:** Krauskopf, K. B. 1959.
*See also* THERMODYNAMICS, SECOND LAW OF

## DISSOCIATION LAWS FOR NEW STATISTICS

The ratio of the mass of the system to that of the electron can be expressed as a constant, $\beta_m$.

For each component, m, of an assembly, there is an appropriate value of $\lambda_m$ according to the temperature and density, which are identified by rather complicated relationships for classical nonrelativistic and relativistic conditions:

$$\beta_m = m_m/m_e$$

is the ratio of the mass of the system ($m_m$) to that of the electron ($m_e$).

**Keywords:** density, electron, mass, temperature
**Source:** Menzel, D. H. 1960.

## DISSOLUTION, LAW OF—SEE JACKSON

## DISTRIBUTION LAW (CHEMICAL) (1872)

When a substance is placed between two immiscible solvents, the substance is distributed so that the concentration ratios in the two solvents are nearly constant, which is also approximately equal to the ratio of the solubilities of the substance in each of the two solvents, due to M. Berthelot.

**Keywords:** concentration, immiscible, solubilities, solvent
BERTHELOT, Marcellin Pierre Eugene, 1827-1907, French chemist

**Source:** Hodgman, C. D. 1952.
*See also* PARTITION

### DISTRIBUTION LAW (ELECTROMAGNETIC)—SEE ELECTROMAGNETIC RADIATION

### DISTRIBUTION LAW (SEVERAL MOLECULAR SPECIES)

When several molecular species are in both vapor form and in solution, the single and double molecules evaporate at constant temperature from a common solvent into a fixed vapor-space; the ratio of the concentration in the vapor-space of any one molecular species to its concentration in the solvent is constant. This distribution is independent of the presence of other molecular species, even when these are chemically reactive with the former.

**Keywords:** chemistry, concentration, evaporation, physical chemistry, vapor
**Sources:** Perry, J. H. 1950; Honig, J. M. 1953; Mandel, S. 1972.
*See also* HENRY; NERNST; PARTITION

### DISTRIBUTION LAW OF EXTRACTION

The behavior of solutes distributed between two immiscible solvents is described in terms of distribution ratio, D, the ratio of two phases:

$$D = |A_2|/|A_1|$$

**Keywords:** phases, ratio, solutes, solvents
**Source:** Parker, S. P. 1992.

### DISTRIBUTION LAW, POISSON—SEE SMALL NUMBERS

POISSON, Simeon Denis, 1781-1840, French mathematician

### DISTRIBUTION LAWS—SEE

1. BOLTZMANN DISTRIBUTION LAWS—relate to probability and entropy
2. MAXWELL DISTRIBUTION LAW—covers the kinetic energies of molecules
3. PLANCK DISTRIBUTION LAW—describes the energy of harmonic oscillation
4. BOSE-EINSTEIN DISTRIBUTION LAW
5. POISSON DISTRIBUTION LAW—see SMALL NUMBERS

**Keywords:** energy, entropy, harmonic, kinetic, oscillation, statistics
**Source:** Mandel, D. H. 1960.

### DISTRIBUTIVE—SEE BOOLEAN; ALGEBRA OF PROPOSITIONS

### DISTRIBUTIVE LAW OF MULTIPLICATION OVER ADDITION

If a number multiplies a sum, the total is the same as the sum of the separate products of the multiplier and each of the addenda. Thus, if a, b, c are real numbers, $a(b + c) = ab + ac$. For vectors $\alpha(a + b) = \alpha a + \alpha b$ and $(\alpha + \beta)a = \alpha a + \beta a$.

**Keywords:** real numbers, products, vectors
**Sources:** Considine, D. M. 1976; Good, C. V. 1973; James, R. C and James, G. 1968; Lipschutz, S. and Shiller, J. 1995; Mandel, S. 1972.
*See also* ASSOCIATIVE; COMMUTATIVE

## DISUSE, LAW OF

A law of learning, proposed by E. Thorndike, that the less frequently a connection is made between a situation and a reuse is exercised, the more difficult it is to recall the connection, with other things being equal.

**Keywords:** learning, recall, reuse
THORNDIKE, Edward Lee, 1874-1949, American psychologist
**Source:** Good, C. V. 1973.
*See also* EXERCISE; FREQUENCY; THORNDIKE

## DITTRICH—SEE BUHL-DITTRICH

## DITTUS-BOELTER EQUATION (1930)

As introduced in 1942 by W. H. McAdams, the relationship that represents turbulent flow heat transfer in a smooth tube, based on dimensionless numbers. After going through several forms the following equation is generally accepted:

$$Nu = 0.023\ Re^{0.8}\ Pr^{0.4}$$

where Nu = Nusselt number
Re = Reynolds number
Pr = Prandtl number

**Keywords:** dimensionless, flow, heat transfer, tube
DITTUS, Frederick William, 1897-1987, American mechanical engineer
BOELTER, Lewellyn Michael Kraus, 1889-1966, American engineer
McADAMS, William Henry, 1892-1975, American chemical engineer
**Sources:** International Journal of Heat and Mass Transfer 41(4,5)1998; Rohsenow, W. M. and Hartnett, J. P. 1973; Thewlis, J. 1961-1964.
*See also* COLBURN; PRANDTL; REYNOLDS

## DNA MODEL—SEE WATSON AND CRICK

## DOEBEREINER LAW OF TRIADS (1817)

Chemical elements that resembled each other often occurred in groups of three; that there was often a close relationship with the arithmetic mean of the atomic numbers of the lightest and heaviest elements. Thus, lithium, sodium, potassium, calcium, strontium, and barium are examples.

**Keywords:** atomic numbers, elements, heaviest, lightest, mean
DOEBEREINER, Johann Wolfgang, 1780-1849, German chemist
**Source:** Bothamley, J. 1993.

## DOLLO LAW OR RULE, OR DOLLO LAW OF IRREVERSIBILITY

Evolution is reversible in that features, structures, or functions gained can be lost, but irreversible in that features, structures, or functions lost cannot be regained.

**Keywords:** biology, development, evolution, organisms, reversibility
DOLLO, Louis Antoine Marie Joseph, 1857-1931, Belgian paleontologist
**Sources:** Friel, J. P. 1974; Gray, P. 1967; Henderson, L. E. 1963. 1989; Knight, R. L. 1948; Landau, S. I. 1986.
*See also* ARBER LAW.

## DONDERS LAW OR METHOD (1846)

Every oblique position of the eyeball is associated with a constant amount of torsional movement. The rotation of the eyeball, and hence the torsion, is determined by the distance of the object from the median plan and the line of the horizon. The position of the eyes in looking at an object is independent of the eyes to that position.

**Keywords:** distance, eyeball, horizon, torsional movement
DONDERS, Franciscus Cornelis, 1818-1889, Dutch ophthalmologist
**Sources:** Friel, J. P. 1974; Landau, S. I. 1986; Reber, A. S. 1985; Stedman, T. L. 1976; Wolman, B. B. 1989.
*See also* EMMERT; FERRY-PORTER; YOUNG-HELMHOLTZ

## DOPPLER EFFECT OR LAW OR PRINCIPLE (1842)

As the source of light (or sound or other waves) moves toward or away from the observer, or vice versa, there is an apparent change in wavelength and frequency of the light. The Doppler red shift provides a means of determining the dimensions of an expanding universe. The radial velocities of other galaxies away from our galaxy are proportional to their distances. The same principle applies when a sounding body approaches the ear such that the note perceived is higher than the true one, but if the source recedes the note is perceived as lower. This relationship was verified by C. Buys Ballot in 1845.

**Keywords:** distance, ear, frequency, light, wavelength
DOPPLER, Christian Johann (Johann Christian), 1803-1853, Austrian physicist
BUYS BALLOT, Christoph Hendrik Didericus, 1817-1890, Dutch meteorologist
**Sources:** Mandel, S. 1972; Parker, S. P. 1989; Thewlis, J. 1961-1964; Whitaker, J. C. 1996.
*See also* RED SHIFT

## DOSE-RESPONSE—SEE SHELFORD

## DOUBLE HELIX—SEE WATSON-CRICK MODEL

## DOVE LAW

In the world weather pattern, wind generally shifts in direction with the sun.

**Keywords:** meteorology, weather, wind
DOVE, Heinrich Wilhelm, 1803-1879, German meteorologist
**Source:** Webster's Biographical Dictionary. 1959.
*See also* GYRATION; STORMS

## DRAG FOR HIGH REYNOLDS NUMBERS, LAW OF

Although not as precise as at low Reynolds numbers, an approximate relationship exists at high Reynolds numbers such that the drag of a fluid is proportional to the density, to the square of the size, and to the square of the velocity of a body in a fluid. High Reynolds numbers are considered to be above 4000, although they can be higher, depending on the fluid and the characteristics of the carrying mechanism, following a transition from laminar to turbulent flow.

**Keywords:** density, fluid, size, velocity
REYNOLDS, Osborne, 1842-1912, English engineer
**Source:** Shapiro, A. H. 1961.
*See also* REYNOLDS

## DRAG FOR LOW REYNOLDS NUMBERS, LAW OF; ALSO STOKES LAW

For low Reynolds numbers, the drag is proportional to the speed, to the viscosity, and to the size of body in a fluid, and flow is dominated by the viscosity of the fluid. Low Reynolds numbers below 2000 represent laminar or streamline flow.

**Keywords:** fluid, size, speed, velocity, viscosity
REYNOLDS, Osborne, 1842-1912, English engineer
**Source:** Shapiro, A. H. 1961.
*See also* REYNOLDS; STOKES

## DRAPER EFFECT

An increase in volume when under constant pressure, hydrogen and chlorine react to create hydrogen chloride as a result of increased temperature resulting from an exothermic reaction.

**Keywords:** chloride, exothermic, hydrogen, pressure, temperature
DRAPER, John William, 1811-1882, English American chemist
**Sources:** Morris, C. G. 1992; NAS, Biographical Memoirs. 1886.

## DRAPER LAW (1841)

The only radiation energy that is effective in causing a chemical change is that which is absorbed by the photochemical substance. Thus, in the photochemical reaction of hydrogen and chlorine, the reaction rate is proportional to the intensity of absorbed light and to the pressure of the hydrogen, in the absence of other photochemical reactions. Renamed Grotthuss-Draper, based on similar findings by Grotthuss in 1817. Draper discovered infrared rays in 1847.

**Keywords:** absorption, chemical, energy, light, photochemical
DRAPER, John William, 1811-1882, English American chemist
**Sources:** Clifford, A. F. 1964; Friel, J. P. 1974; NAS, Biographical Memoirs. 1886; Stedman, T. L. 1976; Thewlis, J. 1961-1964,
*See also* GROTTHUSS-DRAPER LAW

## DRAPER LAW OF VISIBILITY (1847)

All bodies begin to become visible by self-emission of light at about the same temperature of 525° C. As the temperature of an incandescent body rises, it emits light of increasing refrangibility.* This statement was later shown not to be true.

**Keywords:** emission, light, temperature, visibility
DRAPER, John William, 1811-1882, English American chemist
**Sources:** *Layton, E. T. and Lienhard, J. H. 1988; NAS, Biographical Memoirs. 1886; Northrup, E. T. 1917.
*See also* STEFAN

## DREW NUMBER, D OR $N_D$

A dimensionless group applied to boundary layer mass transfer rates and drag coefficients for binary system:

$$D = [Z_A(M_A - M_B) + (M_B/Z_A - Y_{AW})(M_B - M_A)] \ln (M_v/M_w)$$

where $M_A$ = molecular weight of component A
$M_B$ = molecular weight of component B
$M_v$ = molecular weight of the moisture in vapor

$M_w$ = molecular weight at the wall
$Y_{AW}$ = mole fraction of A at the wall
$Z_A$ = mole fraction of A in diffusing stream

**Keywords:** boundary, diffusion, drag, mass transfer, molecular weight, vapor
DREW, Thomas Bradford, twentieth century (b. 1902), American chemical engineer
**Sources:** Bolz, R. E. and Tuve, G. L. 1970; Parker, S. P. 1992; Potter, J. H. 1967.

## DRUDE LAW

The spectral emissivity of metals is related to the square root of their resistivity as:

$$e_\lambda = 0.365\ (\rho_\lambda)^{1/2}$$

where $e_\lambda$ = spectral emissivity
$\rho$ = resistivity

Another statement of the Drude law is that the specific rotation for polarized light of an active material is related to the wavelength of the incident light:

$$\alpha = k/(\lambda^2 - \lambda_0^2)$$

where $\alpha$ = specific rotation
$k$ = rotational constant of material
$\lambda$ = wavelength of incident light
$\lambda_0^2$ = dispersion constant of material

**Keywords:** emissivity, resistivity, spectral
DRUDE, Paul Karl Ludwig, 1863-1906, German physicist
**Sources:** Ballentyne, D. W. G. and Lovett, L. E. Q. 1961; Thewlis, J. 1961-1964.

## DRUDE THEORY OF CONDUCTION

The specific resistance of a substance is:

$$1/\rho = Ne^2\tau/m$$

where $\rho$ = the specific resistance
$N$ = number of free electrons per unit volume
$e$ = charge of electrons
$m$ = mass of electrons
$\tau$ = time for each free electron to move between successive collisions

This theory is based on classical mechanical considerations, and has been replaced by an approach of quantum mechanics.

**Keywords:** collisions, electrons, resistance
DRUDE, Paul Karl Ludwig, 1863-1906, German physicist
**Source:** Ballentyne, D. W. G. and Lovett, L. Q. 1961.

## DUALISTIC THEORY (CHEMISTRY) (1811)—SEE BERZELIUS

## DUANE-HUNT LAW (C.1915); DUANE AND HUNT LAW; ALSO PLANCK-EINSTEIN EQUATION; QUANTUM-ENERGY LAW

The Planck constant multiplied by the maximum frequency in a beam of X-rays emitting from a tube is equal to the energy acquired by one of the cathode electrons in transversing the tube. There is a definite limiting frequency for X-rays produced by electrons of a given energy, the product of energy and maximum frequency being nearly Planck constant. This law is sometimes called the *inverse photoelectric equation.*

**Keywords:** electrons, energy, frequency, X-rays
DUANE, William, 1872-1935, American physicist
HUNT, Franklin Livingston, 1883-1973, American physicist
**Sources:** Considine, D. M. 1976; Holmes, F. L. 1990.

### DU BOIS-REYMOND LAW

The variation of current density, and not the absolute value of the current density at any given time, acts as a stimulus to a muscle or motor nerve.

**Keywords:** current, muscle, nerve, stimulus
DU BOIS-REYMOND, Emil Heinrich, 1818-1896, German physiologist
**Sources:** Friel, J. P. 1974; Landau, S. I. 1986; Stedman, T. L. 1976.
*See also* EXCITATION

### DU BOIS SURFACE AREA (PHYSIOLOGY), LAW OF; OR DU BOIS FORMULA

A method of calculating the surface of the body of an individual based on the height and weight of the person. The equation representing these relationships is:

$$A = 71.84W^{0.425}H^{0.725}$$

where $A$ = area, $cm^2$
$W$ = weight, kg
$H$ = height, cm

**Keywords:** body, height, weight
DU BOIS, Eugene F., 1882-1959, American physiologist
**Source:** Landau, S. I. 1986.
*See also* KLEIBER

### DUHRING RULE (1878)

The principle that the temperature at which one liquid exerts a given vapor pressure is a linear function of the temperature at which a second liquid exerts the same vapor pressure. The rule relates to the vapor pressures of similar substances at different temperatures. The rule is similar to, but less accurate than, the Ramsay and Young relationship but is more convenient to use.

**Keywords:** linear function; liquid; vapor pressure
DUHRING, W., nineteenth century, German chemist
**Sources:** Glasstone, S. 1948; Morris, C. G. 1992; Perry, J. H. 1950.
*See also* CLAUSIUS-CLAPEYRON; RAMSAY AND YOUNG

### DULONG AND PETIT COOLING LAW; 5/4TH POWER LAW

When a body is cooling by natural air convection, the loss of heat is proportional to:

$$(T - T_s)^{5/4}$$

where $T$ = temperature of the body
$T_s$ = temperature of the surroundings

**Keywords:** body, convection, cooling, natural, temperature
DULONG, Pierre Louis, 1785-1838, French chemist
PETIT, Alexis Therese, 1791-1820, French physicist
**Sources:** Bothamley, J. 1993; Thewlis, J. 1961-1964.
*See also* COOLING; NEWTON (COOLING)

## DULONG AND PETIT LAW (1817; 1821*); DULONG-PETIT LAW; LAW OF DULONG- PETIT

The atomic weight (A.W.) of a solid element multiplied by the specific heat (S.H.) gives an approximate constant of atomic heat (A.H.) of about 6.2-6.4, such that:

$$A.W. = 6.4/S.H.$$

The atoms of all elements have the same capacity for heat; the specific heats of several elements are inversely proportional to their atomic weight, with the atomic heats of solid elements as constant and approximately equal to 6.2 to 6.4, applicable to elements with atomic numbers below 35. Thus:

$$A.W. = A.H./S.H.$$

The quantity of heat contained in a solid substance is:

$$Q = 3NkT$$

where Q = the quantity of heat
N = number of atoms
k = Boltzmann constant
T = absolute temperature
3Nk = the specific heat (S.H.)

**Keywords:** atomic heat, atomic weight, element, specific heat
DULONG, Pierre Louis, 1785-1838, French chemist
PETIT, Alexis Therese, 1791-1820, French physicist
**Sources:** Daintith, J. 1981; Friel, J. P. 1974; Landau, S. I. 1986; Layton, E. T. and Lienhard, J.H.1988; Mascetta, J. A. 1996; Parker, S. P. 1983. 1992; Peierls, R. E. 1956; Stedman, T. L. 1961-1964; *Van Nostrand. 1947; Walker, P. M. B. 1988.
*See also* KOPP

## DULONG NUMBER, Du OR $N_{Du}$

A dimensionless group used to represent compressible flow that relates kinetic energy and thermal energy of the fluid,

$$Du = V^2/c_p(T_2 - T_1)$$

where V = velocity
$c_p$ = specific heat at constant pressure
$T_2 - T_1$ = temperature difference

This number is also identified as the Eckert number.

**Keywords:** compressible, energy, flow, fluid, kinetic, thermal
DULONG, Pierre Louis, 1785-1838, French
**Sources:** Land, N. S. 1972; Parker, S. P. 1992; Potter, J. H. 1967.
*See also* ECKERT

## DURAND RULE—SEE AREA

## DYAR LAW

The increase in head width between successive stages in the life of butterflies and moths (Lepidoptera–large order of insects) show a regular geometrical progression.

**Keywords:** butterflies, geometrical, Lepidoptera, head, width
DYAR, Harrison G., 1866-1929, American biologist
**Source:** Gray, P. 1976.

## DYNAMIC EFFECT OR LAW (PSYCHOLOGY)

Specific human behaviors become more habit-formed in relation to facilitating a particular goal. This law was proposed by R. B. Cattell.

**Keywords:** behavior, goals, habit, human
CATTELL, Raymond Bernard, twentieth century (b. 1905), English American psychologist
**Source:** Bothamley, J. 1993.

## DYNAMIC EQUILIBRIUM—SEE PREVOST

## DYNAMIC POLARIZATION, LAW OF

The conduction of nerve impulses occurs in one direction only, a postulate formulated as a law based on studies of S. Raman y Cajal.

**Keywords:** direction, impulses, neural
RAMAN Y CAJAL, Santiago, 1852-1934, Spanish neurohistologist
**Source:** McMurray, E. J. 1995.

## DYNAMIC SIMILARITY PARAMETERS—SEE FROUDE NUMBER (GRAVITY); PRANDTL NUMBER (HEAT CONDUCTION); REYNOLDS NUMBER (VISCOSITY)

## DYNAMIC SIMILITUDE, LAW OF; DYNAMIC(AL) SIMILARITY PRINCIPLE

For a physical situation where the Reynolds numbers of two or more different models or prototypes are equal, for example, the drag coefficients will be equal. Other dimensionless groups provide similar representations. Dynamical similarity does not necessarily mean geometric similarity.

**Keywords:** drag, coefficients, Reynolds
**Sources:** Bothamley, J. 1993; Parker, S. P. 1989; Thewlis, J. 1961-1964.
*See also* DIMENSIONLESS GROUPS AND NUMBERS

# E

## EARNSHAW THEOREM ON STABILITY (1842)

If a charged body is placed in an electric field and is altogether free to move, it is always in unstable equilibrium with respect to translational motion.

**Keywords:** charge, electricity, equilibrium, field, physics
EARNSHAW, Samuel, 1805-1888, Dutch physicist
**Sources:** Bothamley, J. 1993; Thewlis, J., 1961-1964.

## EBBINGHAUS LAW OR PRIMACY LAW

With an increase in the amount of material to be learned, there is a disproportionate increase in time to learn it.

**Keywords:** learn, material, time
EBBINGHAUS, Hermann von, 1850-1909, German psychologist
**Sources:** Bothamley, J. 1993; Good, C. V. 1973; Reber, A. S. 1985.

## ECKERT NUMBER, Ec OR $N_{Ec}$

A dimensionless group used in fluid flow that relates the temperature rise to temperature difference between the wall and fluid at the boundary layer in adiabatic flow or as otherwise stated the kinetic energy and thermal energy:

$$Ec = V^2/c_p \, \Delta T$$

where V = velocity at a great distance from wall
$c_p$ = specific heat at constant pressure
$\Delta T$ = temperature difference between wall and fluid at edge of boundary layer

The Eckert number equals the Dulong number.

**Keywords:** compressible, flow, kinetic energy, thermal energy
ECKERT, Ernst Rudolf George, twentieth century (b. 1904), German American mechanical engineer
**Sources:** Bolz, R. E. and Tuve, G. L. 1970; Kakac, S. et al. 1987; Land, N. S. 1972; Parker, S. P. 1992; Potter, J. H. 1967; Rohsenaw, W. M. and Hartnett, J. P. 1973.
*See also* DULONG NUMBER

## ECCENTRIC SITUATION OF THE LONG TRACTS, LAW OF—SEE FLATAU

## ECOGEOGRAPHIC RULES AND LAWS—SEE ALLEN LAW; BERGMANN RULE; GLOGER RULE; HEART-WEIGHT RULE; HOPKINS BIOCLIMATIC RULE; RENSCH LAWS (FOUR)

**Sources:** Art, H. W. 1993; Lincoln, R. J., et al. 1982.

## ECOLOGICAL VALENCY, LAW OF RELATIVITY OF; OR LAW OF RELATIVITY OF ECOLOGICAL VARIETY

The range of tolerance of a species is not constant throughout its entire range but varies according to the variations in environmental factors from locality to locality.

**Keywords:** constant, environmental, locality, species, tolerance
**Source:** Lincoln, R. J., et al. 1982.

## ECONOMY, LAW OF; OR ECONOMY PRINCIPLE—SEE OCKHAM; PARSIMONY

## EDDINGTON THEORY OR LAW (1923)

When describing a physical system, one must consider the observer, the nature of the observations, and the measurement processes. As related to relativity and quantum theory, A. Eddington is credited with the phrase, "the arrow of time."

**Keywords:** measurements, observations, observer, physical system
EDDINGTON, Sir Arthur Stanley, 1882-1944, English astronomer
**Sources:** Bothamley, J. 1993; Gillispie, C. C. 1981.

## EDDY CURRENTS—SEE SKIN EFFECTS

## EDINGER LAW

A gradual increase in the use and function of the neuron causes increased growth at first, but if irregular and excessive, atrophy and degeneration results.

**Keywords:** atrophy, degeneration, growth, neuron, physiology
EDINGER, Ludwig, 1855-1918, German anatomist and neurologist
**Sources:** Friel, J. P. 1985; Landau, S. I. 1986.

## EDISON EFFECT OR LAW (1883)

The phenomenon of electrical conduction between an incandescent filament and a separate independent cold electrode in the same envelope when the electrode is positive as related to the filament. The effect was at first ignored then revived decades later as the basis of the vacuum tube. This effect was explained by J. A. Fleming.

**Keywords:** conduction, electricity, filament, incandescent
FLEMING, Sir John Ambrose, 1849-1945, English electrical engineer
EDISON, Thomas Alva, 1847-1931, American inventor
**Sources:** Hodgman, C. D. 1952; Mandel, S. 1972; Speck, G. E. 1965.

## EDUCATION—SEE ASSOCIATIVE SHIFTING; BELONGINGNESS; EFFECT; EXERCISE; READINESS

## EFFECT, LAW OF (1899); OR THORNDIKE LAW OF EFFECT (1911); EMPIRICAL LAW OF EFFECT

The general observation that behaviors beginning as random activities that are rewarded tend to be repeated at the expense of behaviors that are not rewarded. The connections are strengthened between stimuli and responses when those responses are rewarded with a satisfying activity and are weakened with unsatisfying activity.

**Keywords:** behavior, stimulus, response
THORNDIKE, Edward Lee (1874-1949), American educator and psychologist
**Sources:** American Scientist 82(1):30, 1994; Deighton, L. C. 1971; Harre, R. and Lane, R. 1983; Morris, C. G. 1992; Walker, P. M. B. 1988; Wolman, B. B. 1989.
*See also* LEARNING; PAVLOV; SKINNER; THORNDIKE

## EFFLUX—SEE TORRICELLI

## EFFUSION OF GASES—SEE GRAHAM

## EGNELL LAW

The velocity of straight or nearly straight winds in the upper half of the troposphere, above a fixed location, increases with the height and is roughly proportional to the decrease in density of the air.

**Keywords:** atmospheric, density, troposphere, velocity, wind
EGNELL, A., twentieth century, meteorologist
**Sources:** Huschke, R. E. 1959; Morris, C. G. 1992; NUC
*See also* DOVE; STORMS

## EHRENFEST ADIABATIC LAW OR THEOREM (1907)

A model that can be derived from Schrödinger equation that shows that the laws of probability could produce an average trend toward equilibrium, even though the behavior of the model is reversible in time, and energy of one of its states would eventually recur. Thus, for slow alteration of the coupling conditions, the quantum numbers of the system do not change, and the number of terms do not change. This theorem contributed to adiabatic understanding and is related to black body radiation.

**Keywords:** adiabatic, coupling, probability, quantum
EHRENFEST, Paul, 1880-1933, Dutch Austrian physicist
**Sources:** Ballentyne, D. W. G. and Lovett, D. R. 1980; Holmes, F. L. 1980; Thewlis, J. 1961-1964.

## EINSTEIN-DE HAAS—SEE RICHARDSON

## EINSTEIN LAW OF GRAVITATION

Gravitation is assumed to be a physical effect produced by the curvature of four-dimensional space time. Twenty parameters contain the intrinsic characteristics of the gravitational field, those which would not depend on the motion of the observer. The generalization of Newton's gravitational potential is the metric tensor in terms of four-dimensional distance. The gravitational field equations are ten linear combinations of the curvature.

**Keywords:** gravitation, mass, motion, stress
EINSTEIN, Albert, 1879-1955, German Swiss American physicist; Nobel prize, 1921, physics
**Source:** Parker, S. P. 1992.
*See also* RELATIVITY (GENERAL)

## EINSTEIN LAW OF MOBILITY OF AN ION; EINSTEIN RELATION OF MOBILITY OF AN ION

The mobility (u) of an ion is related to the diffusion coefficient by:

$$\mu kT = eD$$

where $\mu$ = mobility of ion
$k$ = Boltzmann constant
$T$ = absolute temperature
$e$ = electron charge
$D$ = diffusion coefficient

**Keywords:** diffusion, ion, mobility
EINSTEIN, Albert, 1879-1955, German Swiss American physicist; Nobel prize, 1921, physics
**Sources:** Ballentyne, D. W. G. and Lovett, D. R.; Michels, W. C. 1956.

## EINSTEIN LAW OF PHOTOCHEMICAL EQUIVALENCE—SEE STARK-EINSTEIN LAW

## EINSTEIN LAW OF PHOTOELECTRIC EFFECT (QUANTUM) (1905)

A quantum of light striking a metal surface is absorbed by an electron; the total energy of the quantum can be calculated as the energy of the emitted electron plus the energy required to get the electron out of the metal surface.

**Keywords:** electron, energy, light, quantum
EINSTEIN, Albert, 1879-1955, German Swiss American physicist; Nobel prize, 1921, physics
**Sources:** Michels, W. C. 1956; Thewlis, J. 1961-1964.
*See also* STARK-EINSTEIN

## EINSTEIN LAW OF RELATIVITY (SPECIAL,1905; GENERAL, 1915); MASS-ENERGY EQUIVALENCE LAW(1905)

Every mass has a certain amount of energy connected with it, and all energy has a certain amount of mass associated with it, and these are related:

$$E = mc^2$$

where E = energy
m = mass
c = velocity of light

The special law of relativity expresses the notion that the laws of physics remain unchanged in systems moving at constant velocity in relation to each other. The special law does not include gravitation. The general law of relativity is based on the bending of light in a gravitational field, and it extended the special law to systems having relative motion to each other at any velocity. The general law includes the effects of acceleration in addition to uniform velocity and is more complicated than the special theory.

**Keywords:** energy, light, mass
EINSTEIN, Albert, 1879-1955, German Swiss American physicist; Nobel prize, 1921, physics
**Sources:** Asimov, I. 1994; Nourse, A. E. 1969; Parker, S. P. 1989.

## EINSTEIN LAW OF SPECIFIC HEAT (1907); EINSTEIN EQUATION OF SPECIFIC HEAT; EINSTEIN THEORY OF HEAT CAPACITY

The specific heat of a gas at constant volume ($c_v$) is simply but not precisely (because of temperature relations) related to the gas constant R:

$$c_v = 3R$$

Einstein proposed:

$$c_v = 3R(h\gamma/kT)^2 \exp h\gamma/kT/(\exp h\gamma/kT - 1)^2$$

where $\gamma$ = Einstein frequency
h = Planck constant
R = gas constant
k = Boltzmann constant

Also included in the relationship is:

$$E_i = nhf$$

where $E_i$ = energy of each oscillation in quantum mechanics
n = integer values
h = Planck constant
f = frequency

**Keywords:** constant volume, gas constant, Planck's constant, specific heat
EINSTEIN, Albert, 1879-1955, German Swiss American physicist; Nobel prize, 1921, physics
**Sources:** Ballentyne, D. W. G. and Lovett, D. R. 1970; Bothamley, J. 1993; Thewlis, J. 1961-1964.

## EINSTEIN NUMBER, Ei OR $N_{Ei}$

A dimensionless group used in magnetohydrodynamics that represents the ratio between fluid velocity and the velocity of light:

$$Ei = V/c$$

where V = fluid velocity
c = velocity of light

The Einstein number is the same as the Lorentz number.

**Keywords:** fluid, light, magnetodydrodynamics, velocity
EINSTEIN, Albert, 1879-1955, German Swiss American physicist; Nobel prize, 1921, physics
**Sources:** Land, N. S. 1972; Parker, S. P. 1992; Potter, J. H. 1967.
*See also* LORENTZ NUMBER

## EINSTEIN-STARK LAW—SEE STARK-EINSTEIN

## EINTHOVEN LAW OR EINTHOVEN EQUATION (1903)

When taking an electrocardiogram (EKG), or telegrams from the heart, the potential difference in lead II is equal to the sum of the potential difference of leads I and III. The electrocardiogram potential of any wave or complex in lead II is equal to the sum of the potential of leads I and II. The first recording of a human electrocardiogram was by A. D. Waller in 1887 (Fig. E.1).

**Keywords:** electrocardiogram, leads, potential

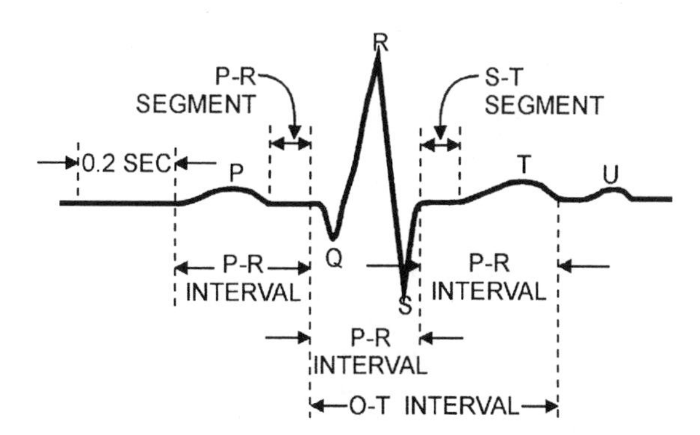


Figure E.1 Einthoven (EKG, ECG).

EINTHOVEN, Willem, 1860-1927, Dutch physiologist; Nobel prize, 1924, physiology/medicine
WALLER, Augustus Desire, 1856-1922, English physiologist
**Sources:** Arey, L. B. 1988; Barr, E. S. 1973; Friel, J. P. 1974; Garrison, F. H. 1929; Glasser, O. 1944; Landau, S. I. 1986; Stedman, T. L. 1976; Thomas, C. L. 1989.

## EKMAN NUMBER, E OR $N_E$

A dimensionless group used to relate variables of a fluid in a spinning tank or in magneto-fluid dynamics, it is the ratio of viscous force to centrifugal force (Coriolis force):

$$E = \upsilon/2\omega\, L^2 \text{ or } (\mu/2\, \rho\omega\, L^2)^{1/2}$$

where $\upsilon$ = kinematic viscosity
$\omega$ = angular velocity
L = characteristic length
$\rho$ = mass density

The Ekman number is equal to the Rossby number divided by the square root of Reynolds number.

**Keywords:** centrifugal, fluid, magneto-fluid dynamics, viscous
EKMAN, Vagn Walfrid, 1874-1954, Swedish scientist
**Sources:** Bolz, R. E. and Tuve, G. L. 1970; Land, N. S. 1972; Parker, S. P. 1992; Potter, J. H. 1967; Science 273(5277):942, 16 August 1996.
*See also* REYNOLDS NUMBER; ROSSBY NUMBER

## ELASTIC DISTORTION, LAW OF; HOOKE LAW OF ELASTIC DISTORTION

The shear of a square inscribed in a rectangle is produced by shearing stress resulting from the action of forces acting on all sides of the square. These forces which act in the longitudinal direction have tangential components. Equations are available for representing Hookean solids and for incompressible materials. It is found as follows:

$$\tau_{n(o)} = 2\mu\, e_{n(o)}$$

where $\tau$ = stress
n = for any direction to the normal by any section, regardless of orientation
$\mu$ = modulus of rigidity or shear modulus
e = strain

Subscript zero $_{(o)}$ refers to distortions.

**Keywords:** forces, longitudinal, shear, strain, stresses
HOOKE, Robert, 1635-1703, English physicist
**Source:** Reiner, M. 1960.
*See also* HOOKE; YOUNG

## ELASTICITY, GENERALIZED LAW OF

The relationship is given between the tensor of stress and the tensor of deformation, in which the mechanical condition of an element of a body is characterized, while the mechanical properties of the material are determined by relationships between these tensors. The state of stress at a given point of the body is determined by the state of deformation at this point, with small deformations and displacements of particles of the material (Fig. E.2).

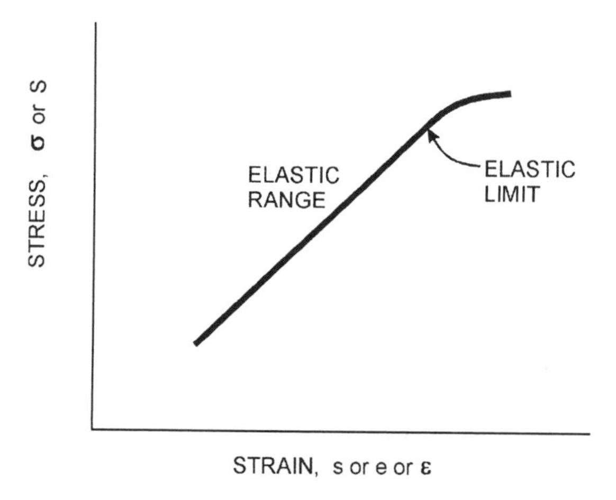


**Figure E.2** Elasticity.

**Keywords:** deformation, mechanics, properties, stress

**Sources:** Reiner, M. 1960; Scott Blair, G. W. 1949.

## ELASTICITY, LAW OF

In classical theory, the stress conditions of a body are assumed to be uniquely determined by the deformation of the body. Stress-strain relationships (curves or plots) are used to represent materials. If the elongation by tension on a body is proportional (straight line plot), the body is described as fitting the Hooke law. The ratio of the stress to strain in the elastic range is called the modulus of elasticity, represented by E, known as the Young modulus. As examples, the modulus of elasticity, E, is approximately: for steel, 200 $GN/m^2$; aluminum, 70; Douglas fir wood, 11; oak wood, 10; and lightweight concrete, 3.

**Keywords:** deformation, elastic, strain, stress

HOOKE, Robert, 1635-1703, English physicist

YOUNG, Thomas, 1773-1829, English physicist

**Sources:** Eirich, F. R. 1969; Reiner, M. 1960; Scott Blair, G. W. 1949.

*See also* HOOKE

## ELASTICITY 1 NUMBER, $El_1$ OR $N_{El(1)}$

A dimensionless group for viscoelastic flow based on the ratio of elastic force and inertia force:

$$El_1 = t_r \mu / \rho r^2$$

where $t_r$ = relaxation time

$\mu$ = absolute viscosity

$\rho$ = mass density

$r$ = pipe radius

**Keywords:** elastic, force, inertia, viscoelastic

**Sources:** Bolz, R. E. and Tuve, G. L. 1970; Land, N. S. 1972; Parker, S. P. 1992; Perry, R. H. 1967; Potter, J. H. 1967.

## ELASTICITY 2 NUMBER, $El_2$ OR $N_{El(2)}$

A dimensionless group representing the effect of elasticity in flow processes depending on physical properties:

$$El_2 = c_p/\beta\ a^2$$

where $c_p$ = specific heat at constant pressure
$\beta$ = coefficient of volume expansion
a = sonic speed

Also defined as the product of the Gay Lussac number and the Hooke number divided by the Dulong number.

**Keywords:** expansion, flow, volume, pressure
**Sources:** Bolz, R. E. and Tuve, G. L. 1970; Land, N. S. 1972; Parker, S. P. 1992; Potter, J. H. 1967.
*See also* DULONG; ELASTICITY 3; GAY LUSSAC; HOOKE

## ELASTICITY 3 NUMBER, $El_3$ OR $N_{El(3)}$

A dimensionless group representing a fluid property and the effect of elasticity on flow:

$$El_3 = \rho\ c_p/\beta\ E$$

where $\rho$ = mass density
$c_p$ = specific heat at constant pressure
$\beta$ = coefficient of bulk expansion
E = modulus of elasticity

**Keywords:** fluid, elasticity, expansion, pressure
**Sources:** Land, N. S. 1972; Potter, J. H. 1967.
*See also* ELASTICITY 2

## E-LAYER—SEE HEAVISIDE

## ELECTRICAL CONTACT TIME OF IMPACTING STEEL SPHERES, LAW OF

When two equal steel spheres come together, the time in millionths of a second during which they are in contact equals 74.7 times the diameter divided by the fifth root of their velocity of approach. Thus:

$$T_u = 74.7\ D/V^{1/5}$$

where $T_u$ = time (microseconds)
D = sphere diameter (centimeters)
V = velocity of contact (cm/sec)

**Keywords:** contact, impact, spheres, time, velocity
**Source:** Northrup, E. P. 1917.

## ELECTRICAL NUSSELT NUMBER, $NU_e$

The convection current as a result of an electric field:

$$Nu_e = V\ L/D^*$$

where V = voltage field
L = length dimension

$D^* = 1/2(D^+ + D^-)$

$D^+$ and $D^-$ = diffusion coefficients of ions

**Keywords:** convection, current, electric, field

NUSSELT, Ernst Kraff, 1882-1959, German physicist

**Source:** Parker, S. P. 1992.

*See also* NUSSELT; PÈCLET

## ELECTRIC CHARGES, LAW OF (1733); ALSO CALLED COULOMB LAW

Like electrical charges repel, and unlike electrical charges attract. This law was originally formulated by C. Cisternay du Fay.

**Keywords:** attract, charges, like, unlike, repel

COULOMB, Charles Augustin de, 1736-1806, French physicist

CISTERNAY DU FAY, Charles Francis de, 1698-1793, French physicist

**Sources:** Morris, C. G. 1992; Vecchio, A. del. 1964.

*See also* CONSERVATION OF ELECTRIC CHARGES

## ELECTRIC DISPLACEMENT—SEE ELECTRIC INTENSITY

## ELECTRIC FIELDS—SEE MAGNETIC AND ELECTRIC FIELDS

## ELECTRIC INTENSITY AND ELECTRIC DISPLACEMENT IN A MATERIAL, LAW OF

The pressure, temperature, and electric field strength fixes the stable equilibrium state of a material, similar to the manner in which for a single system, pressure and temperature, independent of each other, will fix the stable equilibrium state.

**Keywords:** electric, equilibrium, system

**Source:** Holmes, F. L. 1980. 1990.

## ELECTRICITY, LAW OF—SEE COULOMB

## ELECTRIC NETWORKS, LAWS OF—SEE KIRCHHOFF

## ELECTRIC POWER, LAW OF

The power in an electric system is equal to the product of the current and the potential difference:

$$W = EI$$

where W = power (watts)

E = potential difference, volts (or V)

I = current (amperage)

On the basis of electrical resistance of the system:

$$W = I^2 R$$

where R = electrical resistance (ohms)

**Keywords:** amperes, current, potential difference, resistance

**Source:** Bunch, B. 1992.

## ELECTRIC TRANSPORT PROPERTIES OF SOLIDS REPRESENTED BY

1. Electric conductivity (ohmic condition)
2. Mobility (drift mobility of charge carriers)
3. Hall effect
4. Magnetoresistance
5. Seebeck effect (thermoelectric power)
6. Thomson effect
7. Peltier effect
8. Kelvin relations
9. Nernst effect
10. Ettinghausen effect
11. Righi-Leduc effect
12. Thermal conductivity
13. Thermionic emission

**Source:** Gray, D. E. 1972.

## ELECTRICITY, LAW OF—SEE COULOMB

## ELECTROCARDIOGRAM—SEE EINTHOVEN

## ELECTROLYSIS, FARADAY LAWS OF (1833)

Each time a chemical bond is broken by electrolysis in an electrolyte, a certain quantity of electricity traverses the electrolyte, which is always the same.

**First Law**
"The amount of chemical change, as for example, chemical composition, dissolution, deposition, oxidation, reduction, produced by an electric current is directly proportional to the quantity of electricity passed through the solution."

**Second Law**
"The amounts of different substances decomposed, dissolved, deposited, oxidized, or reduced are proportional to their chemical equivalent weights. A chemical equivalent weight of an element or a radical is given by the atomic or molecular weight of the element or radical divided by it valence; the valence depends on the electrochemical reaction involved."

**Keywords:** chemistry, current, electricity, electrolyte
FARADAY, Michael, 1791-1867, English physicist

**Sources:** Besancon, R. 1990; Considine, D. M. 1976; Isaacs, I. 1996; Mandel, S. 1972.
*See also* FARADAY

## ELECTROLUMINESCENSE—SEE DESTRIAU

## ELECTROMAGNETIC ATTRACTION—SEE FARADAY

## ELECTROMAGNETIC ENERGY, LAW OF

The energy of an electromagnetic field is equal to the product of Planck constant and the frequency. Thus, $E = h f$. When applied to a photon, the energy of the photon follows the same law, which when combined with the speed of light in a vacuum, c, gives the relationship:

$$E = h c/\lambda$$

where E = energy of an electromagnetic field, ergs/s
h = Planck constant ($6.6 \times 10^{-27}$)
c = speed of light
$\lambda$ = wavelength of light

**Keywords:** energy, frequency, light, Planck, photon
**Source:** Bunch, B. 1992.

## ELECTROMAGNETIC FIELD—SEE FARADAY

## ELECTROMAGNETIC INDUCTION, LAW OF—SEE FARADAY

## ELECTROMAGNETIC LAW—SEE COULOMB

## ELECTROMAGNETIC RADIATION, DISTRIBUTION OF, LAW OF

### First Law

At any point at a distance from the center of disturbance, there is an electric and magnetic disturbance at right angles to a line drawn from the center of the disturbance. The electric force is on a tangent to a great circle of a sphere with the oscillation at its center and which has its poles at the intercepts of the axis of the oscillation produced; and the magnetic disturbances are on tangents to circles parallel to a plane perpendicular to the oscillation.

### Second Law

The electric and magnetic vibrations are transverse, as in light, and perpendicular to the direction of propagation of the wave front. The amplitude of these vibrations varies inversely as the distance. Their intensity varies inversely as the square of the distance from the oscillation. The vibrations maintain a constant direction as do those of polarized light.

**Keywords:** electromagnetic, magnetic, physics, radiation, wave
**Source:** Parker, S. P. 1983.

## ELECTROMAGNETIC REACTION, LAW OF—SEE MAXWELL-FARADAY

## ELECTROMAGNETIC SYSTEMS, LAW OF

Any two circuits carrying electric currents tend to distribute themselves in such a way that the flux of magnetic induction linking the two will be a maximum, and every electromagnetic system tends to change its configuration so that the flux of induced magnetism will be maximum.

**Keywords:** electricity, electromagnetic, flux, induction, magnetism
**Source:** Parker, S. P. 1983.
*See also* AMPERE

## ELECTROMAGNETIC THEORY, FUNDAMENTAL LAWS OF

These include (1) Faraday law; (2) Ampere law; (3) Conservation of charge. The equations for Faraday, Ampere, and Gauss laws are called Maxwell equations.

**Keywords:** ampere, charge, conservation, equations
**Source:** Menzel, D. H. 1960.
*See also* AMPERE; CONSERVATION (CHARGE); FARADAY; GAUSS; MAGNETIC; MAXWELL

## ELECTROMAGNETIC THEORY OF LIGHT

Light consists of electric oscillations and magnetic field oscillations propagating at right angles to each other and passing through space.

**Keywords:** electric, field, magnetic, oscillations
**Source:** Besancon, R. 1985.
*See also* EINSTEIN (PHOTOELECTRIC EFFECT)

## ELECTROMOTIVE SERIES, LAW OF (1800)

Working with various elements, zinc and silicon, with a salt solution between them, gave the largest voltage difference of the various elements.

**Keywords:** elements, salt solution, silicon, zinc
VOLTA, Count Alessandre, 1745-1827, Italian physicist
**Sources:** Holmes, F. L. 1980. 1990; Porter, R. 1994.

## ELECTROSTATIC ATTRACTION, LAW OF—SEE COULOMB

## ELECTROSTATIC REPULSION, LAW OF—SEE COULOMB

## ELECTROVISCOUS NUMBER, Elv OR $N_{Elv}$

A dimensionless group representing charged particles in flowing gas, as in electrostatic precipitators:

$$Elv = (\rho_p/2\pi\,\varepsilon)^{1/2}\,L^2/\upsilon\;q/m_p$$

where $\rho_p$ = mass density of particle cloud
$\varepsilon$ = permittivity of free space
$L$ = characteristic length
$\upsilon$ = kinematic viscosity of gas
$q$ = charge
$m_p$ = particle mass

**Keywords:** charged, flowing gas, particles, precipitator
**Source:** Land, N. S. 1972.
*See also* DEUTSCH

## ELEMENTARY PARTICLES—SEE CONSERVATION OF

## ELLIOTT LAW

The activity of epinephrine (a colorless weak basic compound produced by the adrenal gland, or synthetically, that regulates blood pressure and related activities) is caused by the stimulation of the endings of the sympathetic nerve.

**Keywords:** medicine, nerve, physiology, stimulus
ELLIOTT, Thomas R., 1877-1861, English medical doctor
**Sources:** Friel, J. P. 1974; Landau, S. I. 1986; Stedman, T. L. 1976.

## ELLIS NUMBER, El OR $N_{El}$

A dimensionless group that represents non-Newtonian flow of liquids:

$$El = \mu_o\,V/2\tau_{1/2}\,R$$

where $\mu_o$ = zero shear viscosity
$V$ = velocity
$\tau_{1/2}$ = shear stress when $\mu = \mu_o/2$
$R$ = tube radius

**Keywords:** flow, liquids, non-Newtonian
ELLIS, Samuel Benjamin, twentieth century, (b. 1904), American physical chemist
**Sources:** Bolz, R. E. and Tuve, G. L. 1970; Land, N. S. 1972; Parker, S. P. 1992; Potter, J. H. 1967.

## ELSASSER, El OR $N_{El}$

A dimensionless group used in magneto fluid dynamics:

$$El = \rho/\mu \sigma \mu_m$$

where $\rho$ = mass density
$\mu$ = absolute viscosity
$\sigma$ = electrical conductivity
$\mu_m$ = magnetic permeability

**Keywords:** electrical, magnetic, permeability, viscosity
ELSASSER, Walter M., twentieth century (b. 1904), German American physicist
**Sources:** Bolz, R. E. and Tuve, G. L. 1970; Land, N. S. 1972; Parker, S. P. 1992; Potter, J. H. 1967.

## EMBRYOGENESIS, LAW OF

That phase of prenatal development involved in the establishment of the characteristic configuration of the embryonic body; in humans, embryogenesis is usually regarded as extending from the end of the second week, when the embryonic disk is formed, to the end of the eighth week, after which the conceptus is usually spoken of as a *fetus*. von Baer states that the embryos of various species are alike but that the adults differ.

**Keywords:** characteristic, embryo, time
BAER, Karl Ernst von, 1792-1876, Estonian naturalist and embryologist
**Sources:** Dulbecco, R. 1991; Encyclopaedia Britannica, 1975; Stedman, T. L. 1990.
*See also* MÜLLER AND HAECKER

## EMISSION, LAW OF—SEE LIGHT EMISSION; RICHARDSON

## EMISSIVE POWER—SEE KIRCHHOFF-CLAUSIUS

## EMMERT LAW

The magnitude of the after-image visual sensation is directly proportional to the distance that it is projected. The after-image tends to increase in size in proportion to the projection distance.

**Keywords:** after-image, physiology, seeing, sight, visual
EMMERT, E., 1844-1911, Swiss scientist
**Sources:** Bothamley, J. 1993; Good, C. V. 1973; Landau, S. I. 1986; Wolman, B. B. 1989.
*See also* BLOCH

## ENCODING LAW

The law that defines the relative values of the quantum steps used in quantizing and encoding signals (to encode is to convert data by use of a code), frequently consisting of binary signals, such that reconversion to the original form is possible.

**Keywords:** code, data, quantizing, reconversion
**Source:** Weik, M. H. 1996.
Also see INVERSE SQUARE; SHANNON

## END-SQUARED LAW—SEE LANCHESTER

## ENERGY GENERATION LAW (STAR)

The energy output from a star or other celestial body is proportional to the emissivity, density, and absolute temperature exponents (or components), as follows:

$$E \propto \varepsilon(\rho, T, A)$$

where E = energy
ε = emissivity
ρ = density
T = temperature (absolute)
A = surface temperature

**Keywords:** celestial, energy, star, space
**Source:** Brown, S. B. and Brown, L. B. 1972.
*See also* STELLAR STARS

## ENERGY, LAWS OF

**I. Laws of motion**
- a) Universal gravitation, law of
- b) Inertia, law of

**II. Law of conservation of matter and energy**
- a) Conservation of matter-energy, law of
- b) Conservation of momentum, law of
- c) Conservation, other, such as electric charge, nucleon charge, symmetry, laws of

**III. Laws of thermodynamics**

**IV. Theories (laws) of relativity**
- a) Special
- b) General

**V. Theories (laws) of quantum mechanics**

**VI. Laws of forces and fields**
- a) Gravitational
- b) Electromotive force
- c) Weak nuclear binding forces
- d) Strong nuclear binding forces

**Source:** Nourse, A. E. 1969.

## ENGEL LAW (ECONOMICS) (1857)

**Law 1 (1857)**
As the income increases the proportion of income spent on food decreases.

**Law 2**
The law asserting the passage of quantity into quality after a "city reaches a threshold of population, and it passes to a qualitatively different type of urban organization."

**Keywords:** city, population, quality, urban
ENGEL, Ernst, 1821-1896, German statistician (probably wrong person, should be economist, etc.)
**Sources:** Bothamley, J. 1993; Ellul, J. 1964.

## ENGINEERING DESIGN, AXIOMS OF

The axioms provide initial assumptions that can be used as a basis for developing a rigorous procedure for decision making leading to design. The axioms can be the basis for selecting existing or development of procedures for decision making.

Axiom 1. Deterministic decision making
Axiom 2. Ordering of objectives
Axiom 3. Reduction of random choices
Axiom 4. Continuity
Axiom 5. Substitutability
Axiom 6. Transitivity
Axiom 7. Monotonicity
Axiom 8. Reality of engineering design

**Keywords:** continuity, decision-making, deterministic, objectives
**Sources:** Barrow, J. D. 1991; Hazelrigg, G. A. 1999. An axiomatic framework for engineering design. J. of Mechanical Design (ASME); Suh, N. P. 1990.

## ENTHALPY RECOVERY FACTOR, $r_h$

A dimensionless number expressed as follows:

$$r_h = (h_{ave} - h_1)/(h_{01} - h_1)$$

where $h_{ave,}$ $h_1$, and $h_{01}$ are flow enthalpies corresponding to temperatures $T_{ave,}$ $T_1$, and $T_{01}$.

**Keywords:** enthalpy, temperature
**Source:** Begell, W. 1983.

## ENTROPY, LAW OF (1865); ALSO KNOWN AS CLAUSIUS LAW (1850, 1854)

There is a tendency for organization to turn into disorganization or for the quantity of information available pertaining to a system to become smaller as time proceeds. The dissipated energy in a process is not destroyed but is merely unrecoverable or perhaps wasted but not destroyed. The entropy of a thermally related, closed system cannot decrease.

R. Clausius proposed the term *entropy* instead of the then used term *equivalence value*. Also, R. Clausius reconciled the work of J. Joule and N. Carnot to state the two independent laws of thermodynamics.

**Keywords:** disorganization, information, organization
CLAUSIUS, Rudolf Julius Emmanuel, 1822-1888, German physicist
**Sources:** Brown, S. B. and Brown, L. B. 1972; Harmon, P. M. 1952; Thewlis, J. 1961-1964.
*See also* BOLTZMANN; CARNOT; CLAUSIUS; JOULE; THERMODYNAMICS

## ENZYME—SEE FISCHER; MICHAELIS-MENTEN; NORTHROP; SCHÜTZ; WILHELMY

## EÖTVÖS LAW OR RULE (1886)

The molar free surface energy of a liquid is proportional to the difference between its temperature, surface area, and the critical temperature and to a universal constant. Expressed in equation form, the molecular surface energy:

$$S(M/\rho)^{2/3} = k(T_c - T - 6)$$

where $k$ = constant, 2.12, for normal liquids
$S$ = surface tension
$T$ = temperature of the liquid
$T_c$ = critical temperature
$M$ = molar free surface energy
$\rho$ = mass density

**Keywords:** energy, liquid, molar, surface
EÖTVÖS, Roland von, 1848-1919, Hungarian physicist
**Sources:** Fisher, D. J. 1988. 1991; Thewlis, J. 1961-1964.

## EÖTVÖS NUMBER, Eo OR $N_{Eo}$

A dimensionless group used to represent capillary flow and sloshing that relates the gravity force and the surface tension force:

$$Eo = \rho L^2 g/\sigma$$

where $\rho$ = mass density
$L$ = characteristic length
$g$ = gravitational acceleration
$\sigma$ = surface tension

The Eötvös number is equal to the Bond number and also equals the Weber number divided by the Froude number.

**Keywords:** capillary, gravity, sloshing, surface tension
EÖTVÖS, Roland von, 1848-1919, Hungarian physicist
**Sources:** Bolz, R. E. and Tuve, G. L. 1970; Land, N. S. 1972; Parker, S. P. 1992; Potter, J. H. 1967.
*See also* BOND; FROUDE; WEBER

## EÖTVÖS-RAMSAY-SHIELDS LAW (1886)

The surface tension of a liquid is proportional to the absolute value of the critical temperature minus the absolute temperature of the liquid minus a constant temperature.

**Keywords:** critical temperature, liquid, surface tension
EÖTVÖS, Roland von, 1848-1919, Hungarian physicist
RAMSAY, Sir William, 1852-1916, British physical chemist
SHIELDS, John Bickford, twentieth century (b. 1907), American chemist
**Sources:** Gillispie, C. C. 1981; Honig, J. M. 1953; NUC; Parker, S. P. 1983.

## EQUAL AREAS, LAW OF—SEE NEWTON SECOND LAW

## EQUAL DECLIVITIES, LAW OF; GILBERT PRINCIPLE (1877)

In the presence of homogeneous rocks cut by consequent streams, all hillside slopes of the valleys cut by the streams tend to develop at the same slope angle, thus producing symmetric profiles of ridges, spurs, and valleys.

**Keywords:** geological, hillside, homogeneous, ridges, streams, valleys
GILBERT, Grove Karl, 1843-1918, American geologist
**Sources:** Bates, R. L. and Jackson, J. A. 1980. 1987; Gary, M. et al. 1957. 1972.
*See also* DECLIVITY

## EQUALITY, LAW OF (PSYCHOLOGY)

As several components of a perceptual system become more similar, they tend to be conceived as a unit; the relationship is attributed to M. Wertheimer.

**Keywords:** components, perception, system, unit
WERTHEIMER, Max, 1880-1943, German psychologist
**Sources:** Bothamley, J. 1993; Bynum, W. F. et al. 1981.
*See also* WERTHEIMER

## EQUAL VOLUMES, LAW OF—SEE LINDGREN

## EQUILIBRIUM FOR A RIGID BODY, LAWS OF

**First Law**
The vector sum of all the external forces acting on a rigid or fixed body must equal zero.

**Second Law**
The sum of the moments of all the external forces acting on a body must be zero around any axis perpendicular to the location at which the forces act.

**Keywords:** body, forces, moments
**Source:** Daintith, J. 1981.

## EQUILIBRIUM LAW, GENERAL (CHEMICAL)

This law is derived from law of mass action and the work of numerous scientists, including A. F. Horstmann, J. W. Gibbs, and J. H. van't Hoff. When the pure crystalline phase of a molecular species exists in a chemical equilibrium of a system at a constant temperature and constant external pressure, the vapor pressure (or fugacity) of that molecular species cannot be changed by a displacement of the chemical equilibrium. The equilibrium law applies to reactions in solution and can be expressed mathematically as:

$$K_c = ([C]^c\,[D]^d/[A]^a\,[B]^b)_{equil}$$

where $K_c$ = the equilibrium constant
[ ] = concentration of products
and exponents = coefficients for the products

**Keywords:** chemistry, crystalline, equilibrium, fugacity, vapor pressure
HORSTMANN, August Friedrich, 1842-1929, German chemist
GIBBS, Josiah Willard, 1839-1903, American physicist
VAN'T HOFF, 1852-1911, Dutch chemist
**Sources:** Bothamley, J. 1993; Isaacs, A. 1996.
*See also* GIBBS; GULDBERG AND WAAGE; MASS ACTION; VAN'T HOFF

## EQUILIBRIUM OF A PARTICLE, LAW OF

When a number of forces act on a particle and keep it in equilibrium, the resulting force is zero. When expressed in vector form with lines representing forces, the vectors placed end to end form a closed figure when the particle is in equilibrium.

**Keywords:** equilibrium, forces, vectors
**Sources:** Holmes, F. L. 1980. 1990; Parker, S. P. 1987. 1992.
*See also* PARALLELOGRAM; VECTORS

## EQUILIBRIUM OF FLOATING BODIES, LAW OF

A floating body must displace a volume of liquid whose weight equals that of the body; the center of gravity of the floating body must be in the same vertical line with that of the fluid displaced; and the equilibrium of a floating body is stable or unstable according to whether the metacenter is above or below the center of gravity.

**Keywords:** displacement, floating, gravity, mechanics, volume
**Sources:** Holmes, F. L. 1980. 1990; Parker, S. P. 1987.
*See also* ARCHIMEDES

## EQUIPARTITION, LAW OF OR THEOREM; LAW OF EQUIPARTITION; LAW OF PARTITION OF KINETIC ENERGY

All particles regardless of density or state, independent of chemical nature or form, always possess the same mass energy of translation at a given temperature.

Each particle contributes 3k/2 to the kinetic part of the specific heat. In a mixture of gases, all at the same temperature, the average kinetic energy of each molecule will be the same. This was originally proposed by J. J. Waterston.

**Keywords:** form, kinetic, nature, particles, specific heat, translation
WATERSTON, J. J., 1811-1883, British physicist
**Sources:** Physics Today 19(8), 1966; Porter, R. 1994; Thewlis, J. 1961-1964.
*See also* PARTITION

## EQUIPOTENTIALITY LAW (MEDICAL)

Any part of a cerebral cortical region acting as a nerve center can, with proper training, carry out the function of any other part of that center which may have been lost by damage to the tissue.

**Keywords:** cerebral, cortical, damage, nerve, training
**Source:** Landau, S. I. 1986.

## EQUIVALENCE THEOREMS—SEE SECANT

## EQUIVALENT PROPERTIES—SEE RECIPROCAL PROPERTIES

## EQUIVALENT PROPORTIONS, LAW OF; EQUIVALENT RATIOS, LAW OF

The law of equivalent proportions is that, when two or more elements combine to form a compound, the weights of these elements are proportional to their equivalent weights, so as such the two are interrelated. A standard weight of eight parts (as one portion) is chosen for oxygen.

**Keyword:** elements, oxygen, proportions, ratios, weights
**Sources:** Honig, J. M. 1953; Parker, S. P. 1992; Walker, P. M. B. 1988.
*See also* EQUIVALENT WEIGHTS; RECIPROCAL PROPERTIES (PROPORTIONS)

## EQUIVALENT WEIGHTS, LAW OF

The weights of two elements, A and B, which combine separately to form compounds with identical weights of another element C are either the weights in which A and B combine together or are related to them in the ratio of small whole numbers. A standard weight of eight parts is chosen for oxygen, and the atomic weight of all elements is equal to the equivalent weight times a whole number called the *valence* of the element. An element can have more than one valence, and thus more than one equivalent weight.

**Keywords:** chemistry, compound, element, weight
**Sources:** Considine, D. M. 1976; Honig, J. M. 1953; Isaacs, A. 1996.
*See also* EQUIVALENT PROPORTIONS

## EQUIVALENTS, LAW OF

Chemical elements or substances combine in the ratio of their combining weights.

**Keywords:** chemistry, combinations, elements, substances
**Source:** Honig, J. M. 1953 (partial).
*See also* COMBINING WEIGHTS

## ERRERA LAW (1886)

Enlarging cells, which can be considered as physical systems, tend to divide into equal parts by walls of minimum area and tend to intersect the old at right angles.

**Keywords:** area, cells, intersect, parts, walls
ERRERA, Leo Abram, 1858-1905, Belgian biologist
**Sources:** Brown, S. B. and L. B. 1972; Gray, P. 1967.

## ERROR FUNCTION

One of the complex functions used to solve problems in probability theory. The error function or Gauss error function is the power function of the test of a hypothesis, x:

$$\text{erf } x = 2/\pi^{1/2}\int_0^x e^{-t^2}\,dt$$

For small values of x, in series form:

$$\text{erf } x = 2/\pi^{1/2}[x - x^3/1!3 + x^5/2!5 - x^7/3!7 + \ldots]$$

where erf x = error function

Or, more generally:

$$E_n(t) = n!\int e^{-y}dy$$

Tables are available to determine the various values of the error function.

**Keywords:** hypothesis, power, series
**Sources:** Hodgman, C. D. 1952 ff; James, R. C. and James, G. 1968; Lide, D. R. 1996; Whitaker, J. C. 1996.
*See also* ERROR; NORMAL LAW

## ERROR, LAW OF; GAUSSIAN DISTRIBUTION; NORMAL DISTRIBUTION, LAW OF

The square of any random error varies as the logarithm of the frequency of the error. The Gaussian or normal distribution, also known as the law of error, involving errors of observation, may be written as:

$$f(x) = h/\pi^{1/2} \exp(-h^2 x^x)$$

where h = a measure of precision in making measurements assuming that the mean is zero.

**Keywords:** error, frequency, observation, random, statistics
GAUSS, Karl F., 1777-1855, German mathematician
**Sources:** Freiberger, W. F. 1960; Gillispie, C. C. 1996; Hall, C. W. 1977.
*See also* ERROR FUNCTION; LARGE NUMBERS; NORMAL DISTRIBUTION; STANDARD ERROR

## ERRORS, LAW OF; ALSO CENTRAL LIMIT THEOREM

The average of a large number of independent and identically distributed random variables is approximately normally distributed. A scatter of a set of readings obeys the theory of probability as incorporated by the standard deviation.

**Keywords:** average, distributed, independent, normally, random
**Sources:** Bothamley, J. 1993; Freiberger, W. F. 1960; Honig, J. M. 1953; Morris, C. G. 1992; Gillispie, C. C. 1990; Whitehead, A. N. 1949. 1954.
*See also* CENTRAL LIMIT THEOREM; NORMAL DISTRIBUTION; PROBABILITY; STANDARD ERROR

## ERRORS, LAWS OF

I. Development of concept by numerous people, depending on application
II. Gaussian distribution or normal distribution
III. Non-Gaussian distribution

**Source:** Kotz, S. and Johnson, N. 1982. 1988.

## ERRORS, PROPAGATION LAW OF

The variance formula obtained by the delta method is called the *law of propagation of errors.* For large sample sizes, the properties of an estimate $g(\hat{\theta})$ of g(θ) where $(\hat{\theta})$ is an estimator of (θ) often may be studied using the delta method.

The delta method involves functions g and estimators (0) and expansion of g(0) in a Taylor series (see reference).

**Keywords:** delta, estimate, propagation, properties
**Source:** Lerner, R. G. and Trigg, G. L. 1981. 1991.

## ESTERIFICATION, LAW OF—SEE MEYER (VICTOR)

## ETTINGSHAUSEN EFFECT OR LAW (1887) (OR VON ETTINGSHAUSEN)

An electric current flowing across the lines of flux of a magnetic field produces an electromotive force which is at right angles to both the primary current and the magnetic field, and a temperature gradient is produced opposite in direction to the Hall electromotive force.

**Keywords:** current, electricity, Hall, magnetic, temperature

ETTINGSHAUSEN, Albert von, 1850-1932, Austrian physicist
HALL, Edwin Herbert, 1855-1938, American physicist
**Sources:** Gray, D. E. 1972; Hodgman, C. D. 1957; Lerner, R. G. and Trigg, G. L. 1991.
*See also* NERNST; ETTINGSHAUSEN (VON)

## EUCKEN (OR EUKEN) LAW OF CONDUCTIVITY OF HEAT BY CRYSTALS

The thermal conductivity of a crystal is inversely proportional to the absolute temperature for temperatures larger than the Debye temperature, which is typical of dielectric crystals.

**Keywords:** conductivity, crystal, Debye, dielectric, thermal conductivity
EUCKEN, Arnold Thomas, 1884-1950, German physical chemist
**Source:** Thewlis, J. 1964. 1969.

## EULER LAWS OF MOTION

Postulated by L. Euler, these axioms or postulates state that the force acting on a small element of a fluid (1) is equal to the rate of change of its momentum and (2) its torque is equal to the rate of change of its angular momentum.

**Keywords:** fluid, force, momentum, rate of change, torque
EULER, Leonhard, 1707-1783, Swiss mathematician
**Source:** Bothamley, J. 1993.
*See also* BERNOULLI-EULER; NEWTON (MOTION)

## EULER NUMBER, Eu OR $N_{Eu}$

A dimensionless group in fluid dynamics that represents fluid friction in conduits and is the ratio of the pressure force and inertia force:

$$Eu = \Delta p/\rho\, V^2$$

where $\Delta p$ = local static pressure or pressure drop
$\rho$ = mass density
$V$ = velocity

The Euler number is equal to the Newton number

**Keywords:** fluid, friction, inertia, pressure
EULER, Leonhard, 1707-1783, Swiss mathematician and physicist
**Sources:** Bolz, R. E. and Tuve, G. L. 1970; Kakac, S. et al. 1987; Parker, S. P. 1992; Perry, R. H. 1967; Potter, J. H. 1967; Rohsenow, W. M. and Hartnett, J. P. 1973.
*See also* CHEZY; DARCY; FANNING; NEWTON; REYNOLDS

## EVAPORATION—SEE DALTON

## EVAPORATION 1 NUMBER, $K_{H(1)}$

A dimensionless group that represents evaporation processes:

$$K_{H(1)} = V^2/LH_v$$

where $V$ = velocity
$LH_v$ = heat of vaporization

**Keywords:** evaporation, heat, velocity
**Sources:** Bolz, R. E. and Tuve, G. L. 1970; Land, N. S. 1972; Parker, S. P. 1992; Potter, J. H. 1967.

## EVAPORATION 2 NUMBER, $K_{H(2)}$

A dimensionless group that represents evaporation processes:

$$K_{H(2)} = c_p/LH_v \beta$$

where $c_p$ = specific heat at constant pressure
$LH_v$ = heat of vaporization
$\beta$ = coefficient of volume expansion

Also expressed as the Gay Lussac number times the evaporation number divided by the Dulong number.

**Keywords:** evaporation, expansion, heat, volume
**Sources:** Bolz, R. E. and G. L. Tuve. 1970; Land, N. S. 1972; Parker, S. P. 1992; Potter, J. H. 1967.
*See also* DU LONG; GAY LUSSAC

## EVAPORATION-ELASTICITY NUMBER, Ev-El, $N_{Ev\text{-}El}$

A dimensionless group that represents evaporation processes and is the Evaporation number divided by the Hooke number:

$$\text{Ev-El} = a^2/LH_v$$

where Ev and El = evaporation and elasticity, respectively
a = sonic speed
$LH_v$ = heat of vaporization

**Keywords:** elasticity, evaporation, Hooke
**Sources:** Bolz, R. E. and Tuve, G. L. 1970; Land, N. S. 1972; Parker, S. P. 1992; Potter, J. H. 1967.
*See also* HOOKE

## EVOLUTION, LAW OF (1859)

In the presence of a surplus of available energy, with a marked advantage to be gained by any species that may develop abilities to utilize excess energy, such a species, with other things being equal, tends to grow in numbers, and the growth will further increase the flux of energy through the system.

Expressed by C. Darwin (1859) as the theory of evolution by which all species of plants and animals and man have come into existence by change from earlier forms through natural selection; the tendency for survival and reproduction of organisms is most likely to occur in those best adapted to the environment. A. Wallace proposed an identical theory at the same time.

**Keywords:** energy, growth, species
DARWIN, Charles Robert, 1809-1882, English naturalist
WALLACE, Alfred R., 1823-1913, English naturalist
**Source:** Darwin, C. 1859 ff.
*See also* DOLLOS; PROGRAM EVOLUTION

## EWALD LAWS

Involuntary motion of the eye, resulting from endolymph (fluid from a duct in the ear) currents in a semicircular canal, is in the direction parallel with the plane of that canal and opposite to the current; in horizontal canals, the amount of eye motor impulse derived from the canals whose hair cells are bent toward the utricle (sac in inner ear) is twice as great as from the short end, but in vertical canals the reverse is true.

**Keywords:** canal (medical), duct in ear, eye, medicine, physiology
EWALD, Karl Emil Anton, 1845-1915, German physician
**Sources:** Friel, J. P. 1976; Landau, S. I. 1986.
*See also* DARWIN-EWALD-PRINS

## EWING THEORY OF FERROMAGNETISM (1885)

Each molecule of a ferromagnetic substance acts as a small magnet. When unmagnetized, these small magnets are arranged in a closed unit or chain so that the net external effect of their poles is zero. This theory is an extension of the molecular theory of magnetism as proposed by W. E. Weber.

**Keywords:** ferromagnetic, magnet, molecule
EWING, James Alford, Sir, 1855-1935, Scottish engineer
WEBER, Wilhelm Eduard, 1804-1891, German scientist
*See also* WEBER

## EXCHANGE, THEORY OF—SEE PREVOST

## EXCITATION, LAW OF

A motor nerve responds with the contraction of its muscles because of alterations of the strength of an electric current and not according to the absolute strength of the current.

**Keywords:** current, electric, muscles, nerve, physiology
**Sources:** Friel, J. P. 1976; Landau, S. I. 1986; Stedman, T. L. 1976.
*See also* DUBOIS-REYMOND

## EXCLUDED MIDDLE, LAW OF OR PRINCIPLE OF

A law that operates at the level of object language or logic and states that a proposition is true, or is false, but cannot be both true and false at the same time. A thing *A* is either *A* or it is not *A*. This law was elucidated by Aristotle.

**Keywords:** false, logic, proposition, statements, true
ARISTOTLE, 384-322 B.C., Greek philosopher
**Sources:** Flew, A. 1984; James, R. C. and James, G. 1968; Morris, C. G. 1992.
*See also* CONTRADICTION; THOUGHT

## EXCLUSION—SEE PAULI EXCLUSION

## EXERCISE, LAW OF OR THORNDIKE LAW

In many cases, complete learning will not be completed in one trial, so many trials are needed for learning. The law of exercise supported common practices of drilling in many learning situations or repetition of an act promotes learning. Thus, the repeated use of a connection between a stimulus and a response (S-R) strengthens, and disuse weakens the S-R bond.

**Keywords:** learning, response, stimulus, use
THORNDIKE, Edward Lee, 1874-1949, American psychologist and educator
**Sources:** Corsini, R. J. et al. 1984. 1987; Deighton, L. C. 1971; Morris, C. G. 1992; Wolman, B. B. 1989.
*See also* DISUSE; EFFECT; LEARNING; THORNDIKE; USE

### EXNER LAW

Diffusivity is inversely proportional to the square root of the molecular weight.

**Keywords:** diffusivity, molecular weight, physical chemistry
EXNER, Sigmund, 1846-1926, Austrian physiologist
**Sources:** Eggenberger, D. I. 1973; Garrison, F. H. 1929; Gillispie, C. C. 1981.

### EXPANSION OF A PERFECT GAS, LAW OF

The adiabatic expression of a perfect gas of volume V is:

$$p = k/V^r$$

where p = absolute pressure
k = constant
r = 1.4025 for air and 1.33 for steam (100% quality)

**Keywords:** adiabatic, gas, thermodynamics
**Source:** Honig, J. M. 1953.
*See also* BOYLE; CHARLES

### EXPLOSION NUMBER, Ex OR $N_{Ex}$

A dimensionless group representing a blast wave growth from instantaneous energy release:

$$Ex = r/\{(E/\rho)^{1/5}\, t^{2/5}\}$$

where r = blast wave radius
E = explosive energy
ρ = mass density of medium
t = time

**Keywords:** blast, energy, instantaneous, release
**Sources:** Land, N. S. 1972; Potter, J. H. 1967.

### EXPONENTIAL—SEE ABSORPTION; COMPOUND INTEREST; EXPONENTS

### EXPONENTIAL DECAY—SEE RADIOACTIVITY EXPONENTIAL DECAY

### EXPONENTIAL FAILURE LAW

Exponential relationship for normal operating failure is (Fig. E.3) represented by:

$$R = e^{-\lambda t} = e^{-t/m}$$

where R = reliability or probability of successful operation for time, t
λ = failure rate
m = mean time to failure (MTTF), the inverse of failure rate, 1/λ

For a series system of components, $R(t) = P(x_1) + P(x_2) + \ldots P(x_n)$

For a parallel system of components, $R(t) = P(x_1 + x_2 + \ldots x_n)$

The failure of an operating system of components can also be expressed in terms of reliability.

**Keywords:** failure, probability, reliability
**Sources:** Hall, C. W. 1977; Rothbart, H. A. 1996; Ushakov, I. A. and Harrison, R. A. 1944.
*See also* PRODUCT LAW; RELIABILITY (SYSTEM)

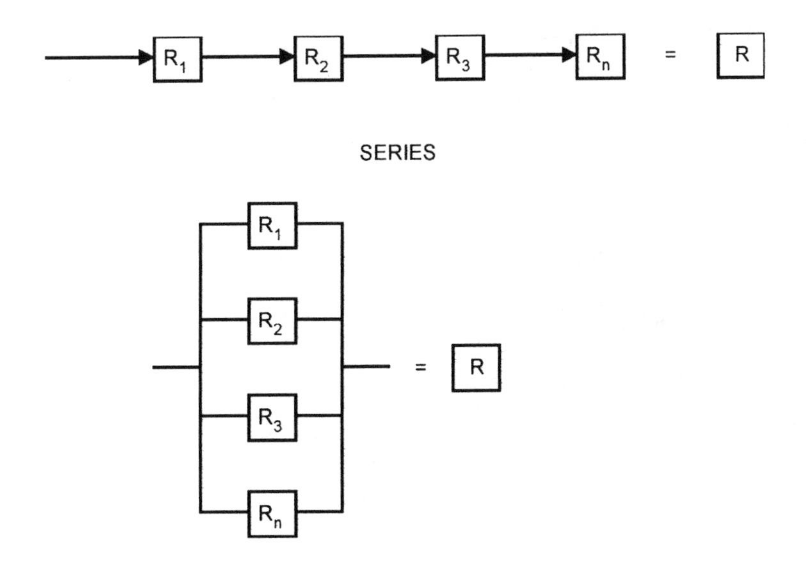


**Figure E.3** Exponential failure.

## EXPONENTS, LAWS OF

1. Multiplication
   In multiplying one term having an exponent by another term having an exponent, the product of the terms is obtained by adding exponents, such that:

$$n^a \text{ x } n^b = n^{a+b}$$

2. Division
   In dividing one term by another, the exponent of each letter in the quotient is the difference of exponents, such that:

$$n^a/n^b = n^{a-b}$$

3. Raising to a Power, or Involution
   A number having an exponent raised to a power has the value of that number with the product of the exponent and the power, such that:

$$(n^a)^b = n^{ab}$$

4. For Two Terms, m and n:

$$(m \text{ x } n)^a = m^a\ n^a$$

and

$$(m/n)^a = m^a/n^a$$

**Keywords:** algebra, division, involution, multiplication, numbers, power

**Sources:** James, R. C. 1992; Gibson, C. 1988; James, R. C. and James, G. 1968; Korn, G. A. and Korn, T. A. 1968; Morris, C. G. 1992; Tuma, J. J. and Walsh, R. A. 1998.

*See also* FRACTIONAL

## EXTENDED LAW OF MEAN—SEE MEAN

## EXTENDED TRANSONIC SIMILARITY LAW FOR GASES—SEE TRANSONIC SIMILARITY

## EXTINCTION—SEE PAVLOV; VAN VALEN

## EXTRACTION—SEE DISTRIBUTION

## EXTRALATERAL RIGHTS—SEE APEX

## EYRING THEORY OF LIQUID VISCOSITY

Molecules of a liquid flow over clusters of molecules, which in theory moves over an energy barrier, is a thermally activated process. The temperature-dependence of the viscosity is:

$$\ln \mu = A + B/T$$

and the pressure-dependence is:

$$\ln \mu = D + E\,p$$

where $\mu$ = viscosity
$T$ = absolute temperature
$p$ = pressure
$A, B, D, E$ = constants

**Keywords:** clusters, flow, molecules, pressure-dependence, temperature-dependence
EYRING, Henry, 1901-1981, American chemist
**Sources:** Bothamley, J. 1993; Thewlis, J. 1961-1964.

# F

## FABRE—SEE LOEB

## FACILITATION, LAW OF (MEDICAL)

Once an impulse has passed once through a certain set of neurons to the exclusion of others, the impulse will tend to take the same course on future occasions, with the ease of traversing the path increasing with each impulse.

**Keywords:** impulse, nerves, neurons, physiology
**Sources:** Critchley, M. 1978; Friel, J. P. 1974; Landau, S. I. 1986.

## FAEGRI LAWS OF CLIMATIC CHANGE (1950)

**First Law (1950)**
The shorter the duration of a climatic function, the smaller is the geographical area affected; likewise, the longer the cycle of climatic change, the greater the geographical area in which the effect is sensed.

**Second Law (1949, 1950)**
The longer the period of climatic change, the more complete is the reaction of all of the indicators of change.

Both of the above laws are based on the observations of H. C. Willett.

**Keywords:** area, change, duration, effect, indicators, period
FAEGRI, Knut, twentieth century, Scandinavian meteorologist
WILLETT, Hurd Curtis, twentieth century (b. 1903), American meteorologist
**Sources:** Fairbridge, R. W. 1967; NUC.
*See also* BUYS BALLOT; FERREL; STORMS

## FAGET LAW; FAGET SIGN

Human subjects with yellow fever have an increase in pulse initially but, as the temperature rises, the pulse shows a marked tendency to decrease.

**Keywords:** medicine, pulse, temperature, yellow fever
FAGET, Jean Charles, 1818-1884, French physician
**Sources:** Dorland, W.A. 1980; Friel, J. P. 1974; Landau, S. I. 1986.

## FAJANS LAW OR RULE (1923)

After emission of alpha rays from a radioactive substance, the product remaining has a valence decrease of two, and after the emission of beta rays has a valence increase of one.

**Keywords:** alpha, beta, radioactive, valence
FAJANS, Kasimir, 1887-1975, German Polish American chemist
**Sources:** Fisher, D. J. 1988. 1991; Friel, J. P. 1974.
*See also* RUSSELL, FAJANS, AND SODDY; SODDY-FAJANS

## FALL OF POTENTIAL, LAW OF (ELECTRICAL); POTENTIAL DROP, FALL OF

The decrease in electrical potential between two points on a conductor is proportional to the amount of electric energy transformed into other energy forms between the points, which for simple ohmic resistance is proportional to the resistance between the points.

**Keywords:** electricity, potential, resistance
**Sources:** Morris, C. G. 1992; Van Nostrand. 1947.

## FALLING BODY—SEE GALILEO

## FAN LAWS

The capacity of a fan varies directly as the change in speed ratio, the pressure varies as the square of the speed ratio, and the horsepower varies as the cube of the speed ratio.

The capacity of a fan varies as the cube of the size ratio, the static pressure as the square of the size ratio, and the power varies as the fifth power of the size ratio.

**Keywords:** capacity, horsepower, pressure, size, speed ratio
**Sources:** Baumeister, T. 1967; Hall, C. W. 1979; Perry, J. H. 1950.
*See also* FAN SOUND

## FANNING NUMBER, Fa OR $N_{Fa}$ (FANNING FRICTION FACTOR*)

A dimensionless group relating shear stress and dynamic pressure involved in fluid friction flow:

$$Fa = 2\,\tau/\rho V^2$$

where $\tau$ = shear stress
$\rho$ = mass density
$V$ = velocity of flow

**Keywords:** dynamic, friction, pressure, shear, stress
FANNING, John Thomas, 1837-1911, American hydraulic engineer
**Sources:** *Bolz, R. E. and Tuve, G. L. 1970; Land, N. S. 1972; Perry, R. H. 1967.
*See also* EULER; NEWTON

## FAN SOUND LAW

The noise generated by a fan is related to its size, static pressure, speed, and capacity is related by logarithm functions, with the change in sound power level varying as:

$$70 \log_{10} (\text{size}_2/\text{size}_1) = 50 \log_{10} (\text{speed}_2/\text{speed}_1)$$

$$20 \log_{10} (\text{size}_2/\text{size}_1) = 25 \log_{10} (\text{pressure}_2/(\text{pressure}_1)$$

$$10 \log_{10} (\text{capacity}_2/\text{capacity}_1) = 20 \log_{10} (\text{pressure}_2/\text{pressure}_1)$$

**Keywords:** capacity, noise, power, pressure, sound, speed
**Source:** Avallone, E. A. and Baumeister, T. 1996.

## FARADAY EFFECT (1846)

The rotation of the plane of polarization of a beam of light when it passes through matter in the direction of lines of focus of a magnetic field. This was the first indication of intersection between light and a magnetic field.

**Keywords:** beam, light, magnetic field

FARADAY, Michael, 1791-1867, English physicist
**Sources:** Daintith, J. 1981; Parker, S. P. 1989.
*See also* COTTON-MOUTON

## FARADAY LAW OF ELECTROMAGNETIC INDUCTION (1831)

The phenomenon whereby magnetism can be converted to electricity. A loop of wire turned in a magnetic field produced a flow of electric current. Faraday's work led to the concept of lines of force or flux in a magnetic field and cutting of these lines of force with a conductor to produce a current:

$$E \propto d\phi/dt$$

where $E$ = induced electromotive force
$d\phi/dt$ = rate of change of flux

This basic understanding led to the development of electric motors and dynamos. The principle of electromagnetism, the existence of a magnetic field surrounding a current-carrying conductor, was discovered by Oersted in 1820. J. Henry independently made the same discovery in 1831. The concept developed by Faraday was first expressed in mathematical terms by F. Neumann in 1845.

**Keywords:** electricity, lines of force, magnetic field
FARADAY, Michael, 1791-1867, English physicist
NEUMANN, Franz Ernst, 1798-1895, German physicist
OERSTED, Christian, 1777-1851, Danish scientist
**Sources:** Bynum, W. F. 1986; Daintith, J. 1981; Parker, S. P. 1989; Peierls, R. E. 1956; Thewlis, J. 1961-1964.
*See also* HENRY; MAXWELL; OERSTED

## FARADAY LAWS OF ELECTROLYSIS (1834)

### First Law

The quantity of an electrolyte decomposed by the passage of electric current is directly proportional to the quantity of electricity that passes through it. To deposit or to dissolve one gram equivalent of any material at an electrode requires the passage of 96,500 coulombs of electricity.

### Second Law

If the same quantity of electricity passes through different electrolytes, the weights of different ions deposited will be proportional to the chemical equivalents of the ions.

**Keywords:** chemical, electricity, electrolysis, electrolyte
FARADAY, Michael, 1791-1867, English physicist
**Sources:** Brown, S. B. and Brown, L. B. 1972; Daintith, J. 1981; Hodgman, C. D. 1952; Lederman 1993; Thewlis, J. 1961-1964.
*See also* ELECTROLYSIS

## FARADAY-NEUMANN—SEE NEUMANN

## FARR LAW OF EPIDEMICS

A property of all zymotic diseases (such as smallpox) in which the epidemic curve first ascends rapidly, then less rapidly to a maximum, and then descends more rapidly than the ascent, or a subsidence, until the disease attains a minimum density and remains stationary.

**Keywords:** disease, epidemic, subsidence
FARR, William, 1807-1883, English medical statistician
**Sources:** Friel, J. P. 1974; Landau, S. I. 1986; Stedman, T. L. 1976; Talbott, J. H. 1970.

### FARR LAW OF DEATH RATE

The death rate in a human population of a certain density is a logarithmic function as follows:

$$\log D = \log a + k \log d$$

where D = death rate
d = density of population
a, k = constants

**Keywords:** death, human, population
FARR, William, 1807-1883, English medical statistician
**Source:** Lotka, A. J. 1956.
*See also* GROWTH

### FATIGUE, LAW OF—SEE HOUGHTON

### FAUNAL ASSEMBLAGES, LAW OF

Similar assemblages of fossil organisms, faunas, and floras, in different parts of the world, indicate similar geologic ages for the rocks that contain them. Each geologic formation has a different aspect of life from that in the formations above it and below it.

**Keywords:** fauna, flora, fossil, geologic, rocks
**Source:** Bates, R. L. and Jackson, J. A. 1980.

### FAUNAL SUCCESSION, LAW OF

Fossil organisms succeed one another in a definite and recognizable order, each geologic formation having a different total aspect of life from that in the formations above and below it. The law was proposed by W. Smith.

**Keywords:** fossil, geologic, order, organisms
SMITH, William, 1769-1839, British geologist
**Sources:** Bates, R. L. and Jackson, J. A. 1980; Bothamley, J. 1993; Gary, M. et al. 1972; Schneer, C. J. 1960.

### FECHNER LAW (1860); ALSO CALLED RANGE OF SENSIBILITY

A physiological relationship in which the stimulus produces a sensation increasing in proportion to the logarithm of the intensity of stimulation, represented by:

$$S = k \log R$$

where S = sensation
k = constant depending on subject
R = stimulus

The division of stars into a scale of magnitude is based on this law. Fechner popularized Weber law which then became known as Weber-Fechner law.

**Keywords:** intensity, physiological, sensation, stars, stimulation
FECHNER, Gustav Theodor, 1801-1887, German psychologist and physicist

**Sources:** Abbott, D. 1984; Bothamley, J. 1993; Friel, J. P. 1974; Good, C. V. 1973; Landau, S. I. 1986; Morris, C. G. 1992; Thewlis, J. 1961-1964; Thomas, C. L. 1989;
*See also* WEBER-FECHNER

### FEDOROV NUMBER, Fe OR $N_{Fe}$

A dimensionless group that represents the relationship between the flow of particles and gas as in a fluidized bed:

$$Fe = d[(4g\,\rho^2/3\mu^2)\,(\lambda_M/\lambda_g - 1)]^{1/3}$$

where d = particle diameter
g = gravitational acceleration
$\rho$ = mass density of fluid
$\mu$ = absolute viscosity of fluid
$\lambda_M$ = specific gravity of particles
$\lambda_g$ = specific gravity of fluid

**Keywords:** fluidized bed, gas, particles
FEDOROV, Yevgenii Konstantinovich, twentieth century (b. 1910), Soviet geophysicist (Potter refers to B. I. Fedorov* for this number). Not sure which is correct.
**Sources:** Bolz, R. E. and Tuve, G. L. 1970; Land, N. S. 1972; Parker, S. P. 1992; *Potter, J. H. 1967; Turkevich, J. 1963.
*See also* ARCHIMEDES

### FENN EFFECT (PHYSIOLOGICAL) OR LAW

Any work accomplished by a muscle involves additional generation of metabolic energy, in addition to that work for shortening and activation. This phenomenon acts opposite the Abbott-Aubert effect.

**Keywords:** energy, muscle, work
FENN, Wallace Osgood, 1893-1971, American physiologist
**Source:** Parker, S. P. 1987.
*See also* ABBOTT-AUBERT EFFECT

### FERET LAW (1892*, 1896)

The strength of concrete is related to the ratio of cement to total solids content.

**Keywords:** cement, concrete, solids
FERET, nineteenth century, French engineer
**Sources:** Meyers, R. A. 1992; *Scott, J. S. 1993.
*See also* ABRAMS; NUTTING-SCOTT BLAIR

### FERMAT LAST THEOREM (1637)

If a, b, and c are integers greater than 0, and if n is an integer greater than 2, there are no solutions to the equation:

$$a^n + b^n = c^n$$

**Keywords:** equation, integers, solution
FERMAT, Pierre de, 1601-1655, French mathematician
**Sources:** American Scientist 82(2):144. 1994; Barrow, J. D. 1991.

## FERMAT LAW OR PRINCIPLE OF LEAST TIME (1637); OR PRINCIPLE OF STATIONARY TIME

Among a number of possible paths a ray of light might take between two points, the light ray will take the path which requires the least time than for any neighboring path (Fig. F.1). This law leads to the Snell law of refraction.

**Keywords:** least time, light, optics, physics

FERMAT, Pierre de, 1601-1665, French mathematician

**Sources:** Besancon, R. M. 1985; Honig, J. M. 1953; Mandel, S. 1972; Morris, C. 1992; Thewlis, J. 1961-1964.

*See also* SNELL

## FERMI-DIRAC DISTRIBUTION LAW (1926)

The distribution of electron energy, p, in a quantized solid is given by:

$$p = \frac{1}{e^{(W - W_F)/(kT)} + 1}$$

where $W_F$ = Fermi energy (critical energy)
T = absolute temperature
k = Boltzmann constant

The energy level when the function equals one-half is known as the Fermi level. The Fermi-Dirac distribution approaches the Boltzmann distribution law at high temperatures and low concentrations. It is particularly representative of electrons in metals.

**Keywords:** electron, energy, physics

FERMI, Enrico, 1901-1954, Italian American physicist; Nobel prize, 1938, physics

DIRAC, Paul Adrien Maurice, 1902-1984, English physicist; Nobel prize, 1933, physics (shared)

**Sources:** Ballentyne, D. W. G. and Lovett, D. R. 1980: Besancon, R. M. 1985; Bothamley, J. 1993.

*See also* BOISE-EINSTEIN; BOLTZMANN DISTRIBUTION

## FERMI INTERACTION LAWS

Regarding the absorption of the atomic electron by an unstable nucleus, with respect to positive beta decay, instead of emitting a positive electron and a neutrino, a nucleus may absorb from its own electron shell a negative electron and emit a neutrino.

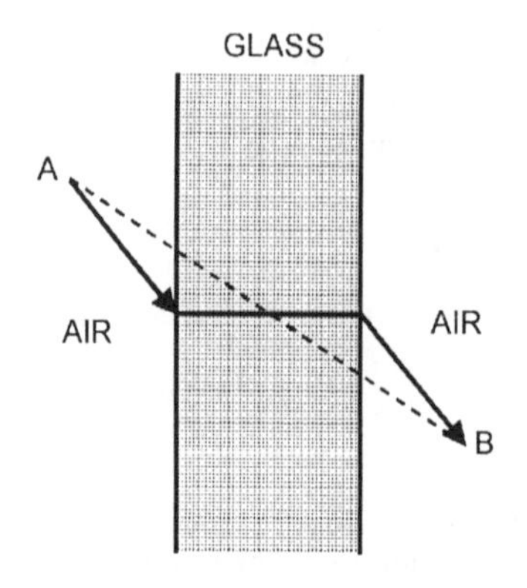


**Figure F.1** Fermat.

**Keywords:** absorption, beta, electron, decay, neutrino, nucleus
FERMI, Enrico, 1901-1954, Italian American physicist; Nobel prize, 1938, physics
**Sources:** Lerner, R. G. and Trigg, G. L. 1991; Parker, S. P. 1983 (physics).

## FERREL LAW (1856)

This law explains the phenomenon based on different surface velocities of the rotating Earth at different latitudes and the variations in centrifugal force at different latitudes. W. Ferrel was the first person to recognize the affect of the rotation of the Earth on the wind systems, followed independently by C. H. D. Buys Ballot a year later.

**Keywords:** centrifugal force, Earth, latitudes, velocities
FERREL, William, 1817-1891, American meteorologist
**Sources:** Fairbridge, R. W. 1967; Hunt, V. D. 1979; NUC; Speck, G. E. 1965.
*See also* BUYS BALLOT; FAEGRI; STORMS

## FERROMAGNETISM—SEE CURIE-WEISS

## FERRY FLOW LAW (1942)

An empirical relationship representing the flow behavior of polymeric systems at low rates of shear:

$$\dot{\gamma} = f(\tau) = (\tau/\mu)\,[1 + (\tau/G_i)]$$

where $\dot{\gamma}$ = rate of shear
$\tau$ = shear stress
$\mu$ = viscosity defined as the limiting value of $\tau/\dot{\gamma}$
$G_i$ = internal shear modulus

**Keywords:** flow, materials, polymer, stress, viscosity
FERRY, John Douglas, twentieth century (b. 1912), Canadian American chemist
**Sources:** Eirich, F. R. 1969; Scott Blair, G. W. 1949.

## FERRY-PORTER LAW (1890s)

The critical flicker frequency of retinal illumination is a logarithm function of the light intensity of the retinal illumination independent of the wavelength of light.

**Keywords:** flicker, illumination, light, retinal
FERRY, Edwin Sidney, 1868-1956, American physiologist
PORTER, Thomas Cunningham, 1860-1933, English scientist
**Sources:** Friel, J. P. 1974; Landau, S. I. 1986; Stedman, T. L. 1976; Thewlis, J. 1961-1964; Zusne, L. 1987.
*See also* BLOCH; BLONDEL-REY; BUNSEN-ROSCOE; DONDERS

## FICK LAW OF DIFFUSION (1856*, 1870)

For nonreacting systems of two compounds A and B, molecular diffusion for steady-state one dimensional transfer with constant c concentration is:

$$J_{Az} = -D_{AB}\, dc_A/dz \quad \text{or} \quad J_{Bz} = -D_{BA}\, dc_B/dz$$

where D = diffusion constant
A and B = two constants
c = concentration
z = direction perpendicular to x-y

There are two special cases of molecular diffusion: equimolal counterdiffusion, as in the case of distillation, and unimolal diffusion, where only one component A diffuses through B, such as in gas absorption.

**Keywords:** absorption, compounds, counterdiffusion, distillation, equimolal, unimolal
FICK, Adolf Eugen, 1829-1901, German physiologist
**Sources:** Besancon, R. M. 1974; Dallavalle, J. M. 1948; Kirk-Othmer. 1985; Landau, S. I. 1986; Thewlis, J. 1961-1964; *Webster Biographical Dictionary. 1959.
*See also* DIFFUSION; WAGNER

### FIELD-THEORY MODEL—SEE SCALING

### FIFTH POWER—SEE OWEN

### FILDES LAW

The presence of syphilitic reagin in the blood of the newborn is diagnostic of syphilis in the mother not of syphilis in the infant.

**Keywords:** birth, blood, medicine, syphilis
FILDES, Paul Gordon, active in twentieth century (b. 1882), English bacteriologist
**Sources:** Friel, J. P. 1974; Landau, S. I. 1986; NUC.

### FILIAL REGRESSION TO MEDIOCRITY, LAW OF (1869; 1889*)

Children whose parents deviate from the average of the population also deviate from the average in the same direction but regress about one-third of the parental deviation toward the mean. Galton first named this law and included in the consideration such traits as stature and intelligence, including general mental ability. This law has been discredited but continues to confound some psychologists.

**Keywords:** children, deviation, heredity, parents
GALTON, Sir Francis, 1822-1911, English geneticist
**Sources:** *Bothamley, J. 1993; Corsini, R. J. 1987; Landau, S. I. 1986; Schmidt, J. E. 1959.
*See also* GALTON

### FIRST DIGIT PROBLEM OR BENFORD LAW (1881)(1938)

Leading digits in tables or sets of data do not occur in numerical order with equal frequency but often conform to a distribution having a discrete probability density function equal to $\log_{10}\{(d + 1)/d\}$ for $d = 1, 2,\ldots, 9$. The law was formulated by S. Newcomb in 1881 but has since been named the Benford law.

**Keywords:** data, digits, frequency, order, probability
BENFORD, Frank, twentieth century, American mathematician
NEWCOMB, Simon, 1835-1909, Canadian American astronomer
**Source:** Kotz, S. and Johnson, N. 1988.
*See also* BENFORD

### FIRST LAW OF THERMODYNAMICS—SEE THERMODYNAMICS

### FIRST PRICE, LAW OF

The depression of prices in favor of the buyers, as a result of a need of raw materials by one country, of raw materials not needed by another country.

**Keywords:** depression, economics, international trade, need, prices, raw materials
**Source:** Toffler, A. 1980.

### FISCHER LIMITING LAW (1961)

For each protein molecule with a defined amino acid composition there is a corresponding number of nonpolar amino acid residues that the protein can accommodate, for which the defined *polarity ratio* is p in which:

$$p = V_e/V_t$$

where $V_e$ = volume of polar outer volume
$V_i$ = volume of polar inner volume

E. Fischer previously (1894) put forth the "lock and key" hypothesis of enzyme action, in which certain sugars are identical chemically but not observed to be the same under certain light transmission.

**Keywords:** amino acid, protein, polarity
FISCHER, Emil Hermann, 1852-1919, German chemist; Nobel prize, 1902, chemistry
**Sources:** Chemical Heritage 14(1):16. 1996-97; O'Daly, A. 1996.
*See also* METABOLISM; WILHELMY

### FITTS LAW (PSYCHOLOGY)

Motor movement time is related to the precision of time of movement and in turn to the distance to be covered:

$$mt = a + (b \log_2 2D/W)$$

where mt = motor movement time
a, b = constants
D = distance moved
W = width of target

The width of the target represents the precision of movement.

**Keywords:** distance, motor, movement, precision, time
FITTS, Paul Morris, 1912-1965, American psychologist
**Sources:** Bothamley, J. 1993; Reber, A. S. 1985; Sheehy, N. 1997; Zusne, L. 1987.

### FITZ LAW

When a previously healthy person is suddenly affected with violent pain in the upper middle section of the abdomen (epigastric), vomiting, and collapse, followed within 24 hours by epigastric swelling, distention of the abdomen due to gas or air in the intestine, or resistance, with slight elevation in temperature, the suspected cause is acute pancreatitis.

**Keywords:** medicine, pancreatitis, physiology
FITZ, Reginald Heber, 1843-1913, American physician
**Sources:** Friel, J. P. 1974; Landau, S. I. 1986.

### FIVE-FOURTHS LAW—SEE DULONG AND PETIT COOLING

### FLATAU LAW

The greater the length of the fibers of the spinal cord, the closer they are located to the periphery of the cord.

**Keywords:** fibers, length, spinal cord
FLATAU, Edward, 1869-1932, Polish neurologist
**Sources:** Dorland, W. A. 1980; Friel, J. P. 1974; Gennaro, A. R. 1979; Landau, S. I. 1986; Stedman, T. L. 1976.

## FLECHSIG MYELOGENETIC LAW

The taking on of the lipid substance that forms a sheath around the nerve fibers of the developing brain takes place in a definite sequence so that the fibers belonging to certain functional systems mature simultaneously.

**Keywords:** brain, nerves, medicine, physiology
FLECHSIG, Paul Emil, 1847-1929, German neurologist
**Sources:** Friel, J. P. 1974; Landau, S. I. 1986.

## FLEMING LAW OR RULE

The right-hand rule relates the flux, motion, and electromotive force of electricity for a generator, and the left-hand rule for a motor. The flux, electromotive force, and motion are represented by the forefinger, second finger, and thumb, at right angles to each other.

**Keywords:** electromotive, flux, generator, motion, motor, right hand
FLEMING, Sir John Ambrose, 1849-1945, English electrical engineer
**Sources:** Asimov, I. 1972; Hodgman, C. D. 1952.

## FLEXURE—SEE BERNOULLI-EULER

## FLICKER—SEE FERRY-PORTER

## FLIEGNER NUMBER, F OR $N_F$

A dimensionless group in compressible flow of fluids:

$$F = Q_m(c_pT)^{1/2}/A(p + \rho V^2)$$

where $Q_m$ = mass flow rate
$c_p$ = specific heat at constant pressure
$T$ = absolute temperature
$A$ = cross-section of flow area
$p$ = static pressure
$\rho$ = mass density
$V$ = velocity

**Keywords:** compressible, flow, fluid
FLIEGNER, Albert Friedrich, nineteenth century (b. 1842), German scientist
**Sources:** Bolz, R. E. and Tuve, G. L. 1970; Land, N. S. 1972; NUC; Parker, S. P. 1992; Potter, J. H. 1967.

## FLINT LAW (EMBRYOLOGY)

The development of the individual organism or organ is the evolutionary development of any species of its blood supply.

**Keywords:** blood, evolution, growth, organism
FLINT, Austin, 1836-1915, American physiologist
**Sources:** Friel, J. P. 1974; Landau, S. I. 1986.

## FLOATING BODIES—SEE ARCHIMEDES; EQUILIBRIUM

## FLOTATION, LAW OF

Following the Archimedes principle, an object floating in a liquid displaces a volume of fluid equal in weight to the object.

**Keywords:** displace, floating, liquid, volume, weight
ARCHIMEDES, c. 287-212 B.C., Greek mathematician and engineer
**Sources:** Daintith, J. 1981; Gibson, C. 1980; Morris, C. G. 1992.
*See also* ARCHIMEDES

## FLOURENS LAW (1824)

Stimulation of the semicircular canal causes spasmodic motion of the eyeball in the plane of the canal.

**Keywords:** canal, eye, spasmodic
FLOURENS, Marie Jean Pierre, 1794-1867, French physiologist
**Source:** Friel, J. P. 1974.

## FLOW LAW OF ICE—SEE GLEN LAW

## FLOW NUMBER, Fl OR $N_{Fl}$ (FLOW COEFFICIENT)

A dimensionless group used to represent flow in fans, turbines, rotary pumps, and similar devices:

$$Fl = Q_v/ND^3$$

where $Q_v$ = volume flow rate
N = rotational speed
D = impeller diameter

**Keywords:** fans, flow, fluid, turbines
**Sources:** Bolz, R. E. and Tuve, G. L. 1970; Land, N. S. 1972; Parker, S. P. 1992.
*See also* FAN; FROUDE

## FLUID DYNAMICS, GOVERNING LAWS OF

Disregarding quantum and relativistic effects, the universal governing laws applicable to all fluids are as follows:

1. Law of conservation of mass (matter)
2. Newton dynamical laws of motion (momentum and moment of momentum)
3. First law of thermodynamics
4. Second law of thermodynamics
5. Maxwell electrodynamic equation

**Keywords:** conservation, dynamics, electrodynamic, thermodynamics
**Source:** Streeter, V. L. 1961.
*See also* CONSERVATION; MAXWELL; NEWTON; THERMODYNAMICS

## FLUIDIZATION NUMBER, Fl OR $N_{Fl}$

A dimensionless group that relates the velocity of flow in a fluidized bed versus the initial velocity for fluidization:

$$Fl = V/V_{init}$$

where $V$ = velocity for fluidization
$V_{init}$ = velocity for initial fluidization

**Keyword:** fluidized bed, velocity of flow
**Sources:** Bolz, R. E. and Tuve, G. L. 1970; Parker, S. P. 1992; Potter, J. H. 1967.

## FLUORESCENCE, LAW OF

**First Law of Fluorescence is STOKES EMISSION LAW**

**Second Law of Fluorescence is LUMINESCENCE EQUIVALENCE LAW**

**Keywords:** emission, luminescence
STOKES, Sir George Gabriel, 1819-1903, British physicist and mathematician
*See also* EINSTEIN; LUMINESCENCE; STOKES

## FLUSH TOILET MODEL—SEE LORENZ HYDRAULIC

## FLUX—SEE CONSERVATION

## FLUX LAWS

1. Diffusive flux, simplified

$$G_{diff,j} = -\Gamma_j \text{ grad } m_j$$

2. Heat flux, Q

$$Q = -k \text{ grad } \Gamma$$

   related to Fourier law of heat conduction

3. Work flux, W
   W is dependent on shear stress and velocities, so at low Mach number, the influence is small, and is considered as zero

**Keywords:** diffusive, heat, work
**Source:** Potter, J. H. 1967 (vol. II).
*See also* DIFFUSION; FOURIER

## FOLICULAR CONSTANCY—SEE LIPSCHUTZ

## FORBUSH EFFECT

The intensity of cosmic rays is low during years of high solar activity and sunspot number, which follows an 11-year cycle.

**Keywords:** activity, cosmic, solar, sunspot
FORBUSH, Scott E., twentieth century (b. 1904), American physicist
**Sources:** Lerner, R. G. and Trigg, G. L. 1991; Morris, C. G. 1992; Parker, S. P. 1987.

## FORCE—SEE BOSCOVICH; D'ALEMBERT; CORIOLIS; PARALLELOGRAM LAW OF FORCES; YUKAWA INTERACTION

## FORCE BETWEEN TWO CHARGES—SEE COULOMB LAW

## FORCE OF TORSION OF A METAL WIRE—SEE COULOMB

## FÖRSTER DECAY LAW (1949)

Represented by the equation, as represented by photosynthesis process:

$$N(t) = N(0) \exp(-k_1 t - k_2 t^{1/2})$$

where $N(t)$ = number of excitations present at time, t
$k_1, k_2$ = constants

This is the equation for stationary or nondiffusing excitation in presence of a random distribution of exposures, in which case $k_1$ is an intrinsic decay rate, and $k_2$ is related to density of quenches and resonance transfer rate between excited molecule and the exposures.

**Keywords:** decay, density, molecules, photosynthesis, quenches
FÖRSTER, Thomas F., 1761-1825, German botanist
**Sources:** Barber, J. 1977; Smith, G. 1981.

## FORWARD DIRECTION, LAW OF (MEDICAL)

Stimulation of posterior roots of the spinal chord gives rise to nerve impulses in the anterior roots, but stimulation of the control end of cut anterior roots produces no activity in posterior roots.

**Keywords:** anterior, medical, nerve, posterior, spinal

## FOUNDER EFFECT—SEE MAYR

## FOURIER LAW OF HEAT CONDUCTION (1811); EQUATION (1822*)

The rate of heat flow through a wall is proportional to the temperature drop, proportional to the area, proportional to the thermal conductivity of the wall, k, and inversely proportional to the thickness of the wall:

$$dQ/dt = q = -k\,A\,\partial T/\partial x = -k \text{ grad } T$$

where $dQ/dt$ = area of conduction in x-direction
$\partial T/\partial x$ = temperature gradient normal to area
A = area perpendicular to direction of heat flow
x = thickness of wall
k = thermal conductivity of material

The work on flow of heat was first proposed by J. Biot but named after J. Fourier that inspired Ohm to do similar studies on the flow of electricity.

**Keywords:** conduction, heat, temperature, thermal, wall
BIOT, Jean Baptiste, 1774-1862, French physicist
FOURIER, Jean Baptiste Joseph, 1768-1830, French mathematician
**Sources:** *Bothamley, J. 1993; Layton, E. T. and Lienhart, J. H. 1988; Morris, C. G. 1987. 1991; Rigden, J. S. 1996; Thewlis, J. 1961-1964.
*See also* BIOT; HEAT CONDUCTION

## FOURIER NUMBER (HEAT TRANSFER), $Fo_h$ OR $N_{Fo(h)}$

A dimensionless group used to represent unsteady heat transfer:

$$Fo_h = kt/c_p\,\rho\,L^2$$

where $k$ = heat conductivity
$t$ = time
$c_p$ = specific heat at constant pressure
$\rho$ = mass density
$L$ = characteristic length (volume divided by surface area)

**Keywords:** heat transfer, unsteady
FOURIER, Jean Baptiste Joseph, 1768-1830, French mathematician
**Sources:** Bolz, R. E. and Tuve, G. L. 1970; Jerrard, H. G. and McNeill, D. B. 1992; Land, N. S. 1972; Parker, S. P. 19892; Perry, R. H.; Potter, J. H. 1967.

## FOURIER NUMBER (MASS TRANSFER), $Fo_m$ OR $N_{Fo(m)}$

A dimensionless group used to represent unsteady mass transfer:

$$Fo_m = Dt/L^2$$

where $D$ = the mass diffusivity
$t$ = time
$L$ = characteristic length

**Keywords:** mass transfer, unsteady
FOURIER, Jean Baptiste Joseph, 1768-1830, French mathematician
**Sources:** Bolz, R. E. and Tuve, G. L. 1970; Land, N. S. 1972; Parker, S. P. 1992; Potter, J. H. 1967.

## FOURTH POWER LAW—SEE STEFAN-BOLTZMANN

## FOWLER LAW (1786?)

"Tuberculosis frequently spreads from the apex of the dorsal lobe along the great fissure to the periphery and elsewhere."

**Keywords:** apex, dorsal, tuberculosis
FOWLER, Thomas, 1736-1801, British physician (probably)
**Source:** Critchley, M. 1978

## FRACTIONAL EXPONENTS, LAW OF

The same laws apply to fractional exponents as integral exponents except that one side of the equation may have to be multiplied by a suitable root to permit the appropriate manipulation.

If $a^{1/n}$ denotes the positive nth root of a, if there is one, and $a^{1/n} = (-1)^{1/n}(-a)^{1/n}$ for a negative, where $(-1)^{1/n}$ = a specified root of $-1$, then $a^{1/n}b^{1/n} \neq (ab)^{1/n}$ when a and b are both negative and n is even.

**Keywords:** exponents, mathematics, roots
**Sources:** Gibson, C. 1988; Gillispie, C. C. 1991; James, R. C. and James, G. 1968; Tuma, J. J. and Walsh, R. A. 1998.
*See also* EXPONENTS

## FRANCK-CONDON PRINCIPLE OR LAW (FRANCK, 1925; CONDON, 1928)

An electronic transition from one to another energy state in a molecule or crystal that takes place so quickly that the positions and the velocities of the nuclei are virtually unchanged in the process.

**Keywords:** crystal, electronic, molecule, nuclei, positions, velocities
CONDON, Edward Uhler, 1902-1974, American physicist; Nobel prize, 1925, physics (shared)
FRANCK, James, 1882-1964, German American physicist; Nobel prize, 1925, physics (shared)
**Sources:** Considine, D. M. 1976; Holmes, F. L. 1980. 1990; McMurray, E. D. 1995; Parker, S. 1989.

## FRANK LAW

The natural frequency of a registering manometer varies inversely to the square of the effective mass, and directly with the square root of the volume-elasticity coefficient of the system:

$$N \propto (E'/M')^{1/2}$$

$$N = 1/2\,\pi\,(E'/M')^{1/2}$$

where N = natural frequency
M′ = effective mass
E′ = elasticity coefficient

**Keywords:** frequency, manometer, natural frequency
FRANK, Otto, 1864-1944, German physiologist; Nobel prize, 1925, physics
**Source:** Wasson, T. 1987.

## FRANK-STARLING LAW OF THE HEART

The force developed by cardia contraction is proportional to the length of the myocardial (muscle tissue of the heart) fibers in diastole (period of dilation).

**Keywords:** cardiac, contraction, dilation, force, muscle tissue
FRANK, Otto, 1864-1944, German physiologist
STARLING, Ernest Henry, 1866-1927, English physiologist
**Source:** Friel, J. P. 1974.
*See also* HEART; STARLING

## FRASER-DARLING LAW OR EFFECT (1938)

The relative number of breeding individuals and young in a population of birds increases as the size of the population increased, with a corresponding shortening of the breeding season.

**Keywords:** birds, breeding, population, size
FRASER-DARLING, Sir Frank, 1903-1979, British zoologist
**Sources:** Allaby, M. 1992; Gray, P. 1967.

## FREEZING POINT AND SOLUBILITY LAW, GENERAL

Where the change of specific heat is so small that it can be neglected, the freezing point change is related to the solubility:

$$k_f = \Delta t/N$$

where N = moles of solute per 1000 g of solvent
$k_f$ = molal lowering for a normal dilute solution
$\Delta t$ = temperature decrease

The latent heat of fusion is:

$$LH_f = 1.99\, t_f^2/1000\, k_f$$

where $LH_f$ = latent heat of fusion, gcal/g
$t_f$ = freezing point of pure solvent, °C

**Keywords:** fusion, latent heat, molal, physical chemistry, solubility, thermal
**Source:** Gillispie, C. C. 1981.
*See also* COPPET; RAOULT

## FREQUENCY, LAW OF (ECOLOGY)

The generalization that when frequency indices of species in a stand are classified into five main classes, A, B, C, D, and E, a double peak occurs in homogeneous vegetation, i.e., attributed to C. Raunkaier.

$$A > B > C \gtrless D < E$$

**Keywords:** classes, ecology, homogenous, peak, vegetation
RAUNKAIER, Christen, 1860-1938, Danish botanist
**Sources:** Hanson, H. C. 1962; Merriam-Webster's Biographical Dictionary. 1995.

## FREQUENCY, LAW OF (LEARNING) (1886)

The more often an act or an association is repeated, the more rapid is the acquisition of learning of that act. Recall or recognition is concerned with learning attributed to the number of repetitions employed during the learning period, as exemplified by reliance on the drill method of instruction for some learning, formulated by T. Brown.

**Keywords:** drill, instruction, learning, recall, recognition, repeated
BROWN, Thomas, 1778-1820, Scottish physician
**Sources:** Corsini, R. J. 1987; Wolman, B.B. 1989.
*See also* EXERCISE; LEARNING

## FREQUENCY OF VIBRATING STRINGS—SEE VIBRATING STRINGS

## FRESNEL-ARAGO LAW (1811); FRESNEL LAW

Two rays of light polarized in the same plane interfere in the same manner as ordinary light; two rays at right angles do not interfere; two rays polarized at right angles from ordinary light and brought into the same plane of polarization do not interfere in the ordinary sense; two rays polarized at right angles interfere when brought into the same plane of polarization (Fig. F.2).

**Keywords:** light, interference, polarized, rays
ARAGO, Dominque Francois Jean, 1786-1853, French physicist

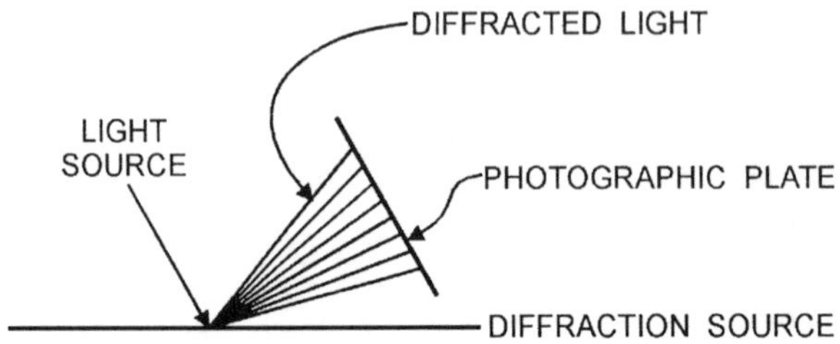

**Figure F.2** Diffraction.

FRESNEL, Augustin Jean, 1788-1827, French physicist
**Sources:** Asimov, I. 1966; Besancon, R. M. 1974; Considine, D. M. 1976; Thewlis, J. 1961-1964.
*See also* FRESNEL; HUYGENS-FRESNEL; REFLECTION

### FREUND LAW

Ovarian tumors change their position during growth such that, when pelvic tumors, they tend to grow downward behind the uterus, and when they have arisen out of the pelvis, they tend to pull forward toward the abdominal wall.

**Keywords:** medicine, pelvis, tumors, uterus
FREUND, Hermann Wolfgang, 1859-1925, German gynecologist
**Source:** Stedman, T. L. 1976.

### FREUNDLICH LAW OF ADSORPTION ISOTHERM; FREUNDLICH ISOTHERM MODEL (1909)

The relationship between a gas adsorbed and a hydroscopic material at constant temperature:

$$x/m = k\, p^{1/n}$$

where x = mass of gas adsorbed
m = mass of adsorbant
p = vapor pressure of gas
k, n = constant properties (material and gas)

**Keywords:** adsorbant, gas, vapor pressure
FREUNDLICH, Herbert M. F., 1880-1941, German chemist
**Sources:** Clifford, A. F. 1964; Considine, D. M. 1976; Menzel, D. H. 1960.
*See also* ADSORPTION ISOTHERM

### FRICTION, LAWS OF

Generally stated the frictional force between two bodies (Fig. F.3) is independent of the area of contact; the frictional force is proportional to the force holding the surfaces together; and in sliding friction the force is independent of the relative velocities of the surfaces. These were early established relationships. However, recent studies have revealed different relationships at the molecular or atomic level of contact.

**Keywords:** area, contact, force, sliding
**Sources:** Daintith, J. 1988; Gibson, C. 1981.
*See also* AMONTONS; COULOMB

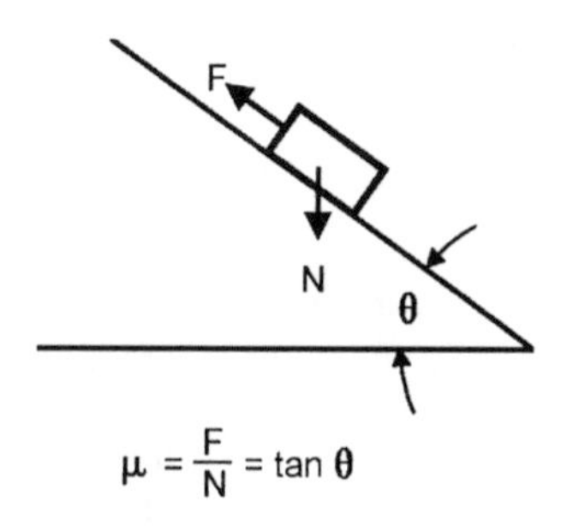


**Figure F.3** Friction forces.

## FRIEDEL LAW (1913)

The intensity of reflection in X-ray diffraction from opposite sides of the same set of crystal planes is the same, so X-rays cannot show the polarity of crystal structures.

**Keywords:** crystal, diffraction, X-rays
FRIEDEL, Georges, 1865-1933, French crystallographer
**Sources:** Besancon, R. M. 1974; Lerner, R. G. and Trigg, G. L. 1981; Thewlis, J. 1961-1964.

## FRORIEP LAW

The skull is developed by the annexation of true vertebrae, with the head growing at the expense of the neck.

**Keywords:** head, neck, skull, vertebra
FRORIEP, August von, 1849-1917, German anatomist
**Sources:** Friel, J. P. 1974; Landau, S. I. 1986.

## FROUDE LAW OF SKIN FRICTION (1852); LAW OF SIMILARITY

The resistance to flow of scale models of ships in water is represented by:

$$R = A\, v^n$$

where R = resistance
A = constant
n = usually about 1.85, not exceeding 2.0 for the roughest surface

or:

$$V/v = (L/\ell)^{1/2} = \lambda^{1/2}$$

where V = velocity of model
v = velocity of prototype
L = length of model
$\ell$ = length of prototype
$\lambda$ = linear relationship between prototype and model

The resistance to movement by similarly shaped ships are in the ratio of the cubes of their dimensions, and their speeds are in the ratio of the square roots of their dimensions.

**Keywords:** dimensions, flow, resistance, shape, ship
FROUDE, William, 1810-1879, English engineer
**Source:** Gillispie, C. C. 1981; Holmes, F. L. 1980. 1990.
*See also* SIMILITUDE

## FROUDE NUMBER, Fr OR $N_{Fr}$ (1869)

A dimensionless group, named after W. and R. Froude,* relating inertia force and gravity force applied to surface wave motion, surface ships, and gravity-affected motions:

$$Fr = V^2/g\,L \text{ or } Fr = V/(gL)^{1/2}$$

where V = velocity
g = gravitational acceleration or intensity of body force
L = characteristic length of body belonging to a family of geometrically similar bodies

In France called the Reech number (some references mistakenly use *Reece*). Also called the *Boussineq number* and the *Vedernikov number.* Some authors call this relationship the *Froude law.*

**Keywords:** gravity, inertia, motion, surface, wave
FROUDE, William, 1810-1879, English engineer, and his son Robert E.*
REECH, Ferdinand, nineteenth century, French engineer
**Sources:** Bolz, R. E. and Tuve, G. L. 1970; Land, N. S. 1972; Lerner, R. G. and Trigg, G. L. 1981; *Morris, C. G. 1992; Parker, S. P. 1992; Porter, R. 1994; Thewlis, J. 1961-1964.
*See also* BOUSSINESQ; DYNAMIC; REECH; RICHARDSON; VEDERNIKOV

### FRUEH NUMBER, Fru OR $N_{Fru}$ (1964)

A dimensionless group representing transonic wing flutter:

$$Fru = (b\omega_a/a)(\mu/C_{La})^{1/2}$$

where $b$ = wing half chord length
$\omega_a$ = first torsional natural frequency of wing
$a$ = sonic speed
$C_{La}$ = wing lift curve slope
$\mu$ = mass ratio, $m/\rho b^2$
$m$ = wing mass per length
$\rho$ = mass density of air

**Keywords:** flutter, transonic, wing
FRUEH, Frank J., twentieth century, American aeronauticist
**Source:** Land, N. S. 1972.
*See also* REGIER

### FUERBRINGER LAW (MEDICAL)

The developmental origin of a muscle can be determined from its nerve supply, based on the fundamental pattern of innervation of muscles established in the embryo.

**Keywords:** embryo, muscle, nerves, origin
FUERBRINGER, Paul Walther, 1849-1930, German physician
**Source:** Landau, S. I. 1986.

### FULLERTON-CATTELL LAW

The errors of observation and of just noticeable differences are proportional to the square root of the magnitude of the stimuli with accountable variations depending on the location of the stimuli.

**Keywords:** errors, magnitude, stimuli, square root
FULLERTON, G. S., 1859-1925, American psychologist
CATTELL, Raymond Bernard, twentieth century (b. 1905), American psychologist
**Sources:** Bothamley, J. 1993; Wolman, B. B. 1989.
*See also* WEBER

### FUNDAMENTAL LAWS OF PHYSICS

The following are considered among the most important fundamental laws of physics:

- Newton Law of Motion
- Newton Action-Reaction Law
- Newton Law of Gravitation
- Conservation of Energy
- The Second Law of Thermodynamics

- Huygen principle of Wave Propagation
- Coulomb Law of Electrostatic Force
- Ampere Law
- Faraday Law of Electromagnetic Induction
- Maxwell Postulate (Law)
- Principle of Relativity—Special and General Laws
- Quantum Principle
- Pauli Exclusion Principle
- Conservation of Elementary Particles

Numerous sources list what are called *fundamental laws,* often flavored by the author's background and experience. Fundamental laws of physics are listed by Considine, D. M. 1976, 1995; Menzel, D. H., 1960; Parker, S. P. 1987, 1994.

A categorization of the natural laws is given by Nourse (1969):

1. Laws of motion
   - Law of universal gravitation
   - Law of inertia
2. Laws of conservation
   - Law of conservation of matter-energy
   - Law of conservation of momentum
   - Other laws of conservation: electrical charge, nucleon charge, mirror symmetry, electrical charge
3. Laws of thermodynamics
   - Law of entropy
     Natural direction of heat from hot to cold
     Natural direction of heat from mechanical energy to heat
4. Laws of relativity
   - Special theory
   - General theory
5. Laws of quantum mechanics
6. Laws of forces and fields
   - Gravitational forces
   - Electromagnetic forces
   - Weak nuclear binding
   - Strong nuclear binding

**Sources:** Nourse, A. E. 1969; for a recent book of general interest explaining the current laws of forces and fields in physics, the reader is referred to LEDERMAN, Leon, 1993.

# G

## GAIA HYPOTHESIS (c. 1979)

The thesis that our planet acts like a giant living organism in which all living things interact to maintain a certain stability, through many feedback systems, that individuals and species unknowingly play a part in this interaction, and that living things are not passive but active agents that can change the environment. Gaia is the Greek goddess of the Earth. J. Lovelock and L. Margulis proposed the hypothesis, which has been widely discussed during the past 25 years.

**Keywords:** environment, feedback, interaction, organisms, species

LOVELOCK, James Ephraim, twentieth century (b. 1919), English medical doctor and scientist

MARGULIS, Lynn, twentieth century (b. 1938), British biologist

**Source:** Lederman, L. 1993.

## GALILEO LAW OF FALLING BODY (c. 1604); (1590*)

The velocity of descent of a falling body starting from rest is proportional to the time of falling, and the distance of descent is proportional to the square of the time of falling. The velocity, v, of a falling body is:

$$v = gt$$

where g = the acceleration produced by gravity

The average velocity is 1/2 v, and the distance is 1/2 vt or 1/2 $gt^2$.

$$d^2s/d^2t = g$$

$$s = 1/2\ gt^2 + A_1t + A_2$$

where s = the distance

$A_1$, $A_2$ = constants of integration and equal zero if the body starts from rest

**Keywords:** acceleration, distance, gravity, mechanics, velocity

GALILEO (Galilei), 1564-1642, Italian astronomer and physicist

**Sources:** Asimov, I. 1972; *NYPL Desk Reference; Parker, S. P. 1983 (physics); Webster's Biographical Dictionary. 1988.

## GALILEO NUMBER, Gal OR $N_{Gal}$

A dimensionless group relating gravity force and molecular friction force applicable to circulation of liquids in free flow, particularly for viscous fluids:

$$Gal = gL^3/\upsilon^2 \qquad \text{or} \qquad Gal = gL^3\ \rho^2/\mu^2$$

where g = gravitation acceleration

L = characteristic length

$\upsilon$ = kinematic viscosity

$\rho$ = density of fluid

$\mu$ = dynamic viscosity

The Galileo number is equal to the Grashof (Gr) number divided by the Gay Lussac (GaL) number.

**Keywords:** free flow, gravity, viscous

GALILEO (Galilei), 1564-1642, Italian scientist

**Sources:** Bolz, R. E. and Tuve, G. L. 1970; Land, N. S. 1972; Parker, S. P. 1992; Potter, J. H. 1967.

*See also* GAY LUSSAC; GRASHOF

## GALTON LAW; GALTON LAW OF ANCESTRAL HEREDITY (1897)

The average contribution to the makeup or composition of an individual is determined by the parents with of each parent contributing one-fourth (a total of one-half), of each grandparent is one-sixteenth (a total of one-fourth by the four grandparents), a sixty-fourth by each great grandparent, and so on.

R. A. Fisher wrote the classic paper pointing to the genetic theory of natural selection (1918) that decreased the antagonism between Mendelian and non-Mendelian factions.

**Keywords:** grandparents, contribution, heredity, individual, Mendelian, parents

GALTON, Sir Francis, 1822-1911, English scientist

FISHER, Ronald Aylmer, 1890-1962, English geneticist and statistician

**Sources:** Friel, J. P. 1974; James, R. C. and James, G. 1968; Landau, S. I. 1986; Stedman, T. L. 1976.

*See also* FILIAL REGRESSION; MENDEL

## GALTON LAW OF REGRESSION; GALTON LAW

Average parents tend to produce average children and the deviation from the average deviate in a negative or positive direction is by parents who produce like children in the same direction, but the offspring of extreme parents inherit in a less marked degree the extreme exhibited by the parents.

**Keywords:** children, deviation, heredity, inheritance

GALTON, Sir Francis, 1822-1911, English scientist

**Sources:** Friel, J. P. 1974; Gray, P. 1967; Landau, S. I. 1986; Stedman, T. L. 1976.

*See also* FILIAL REGRESSION; REGRESSION

## GALVANI LAW (1771)

The voltaic effect is a phenomenon of muscle contraction in legs of dead frogs when placed in a circuit containing dissimilar metals and an electrolyte.

**Keywords:** contraction, electrolyte, muscle

GALVANI, Luis, 1737-1798, Italian anatomist and physiologist

**Source:** Besancon, R. M. 1990.

*See also* ELECTROCHEMICAL; ELECTROMOTIVE; VOLTA

## GASKELL LAW OF PROGRESS

The nervous system has been the dominant factor in evolution.

**Keywords:** evolution, nervous, physiology, system

GASKELL, Walker H. 1847-1914, English physiologist

**Source:** Gennaro, A. R. 1979.

## GAS LAWS; COMBINED GAS LAW; IDEAL GAS LAW

The gas laws relate the pressure, temperature, and volume of a gas and are related and consistent with each other.

Boyle, R. (1622) related pressure and volume at a constant temperature, as $p_1V_1 = p_2V_2$ with T, temperature, a constant.
Charles, J. (1787) related volume and temperature with pressure as a constant, as $V_1/T_1 = V_2/T_2$ with p, pressure, a constant.
Dalton, J. (1804) related partial pressures of gases where the temperature is constant, as $p = p_1 + p_2$ with T, temperature.

The gas pressure law that can be deduced from Boyle law and Charles law; i.e., at constant volume the pressure of a gas, is in proportion to its absolute temperature:

$$p_1V_1/T_1 = p_2V_2/T_2$$

and can also be expressed as the Ideal gas law:

$$pV = nRT$$

where p = pressure
V = volume
n = number of moles
R = gas constant (universal constant consistent with units)
T = absolute temperature

**Keywords:** gas, temperature, partial pressure, pressure, volume
BOYLE, Robert, 1627-1691, British scientist
CHARLES, Jacques Alexandre Ceasur, 1746-1823, French physicist
DALTON, John, 1766-1844, British scientist
**Sources:** Brown, S. B. and Brown, L. B. 1972; Considine, D. M. 1976. 1995; Parker, S. P. 1989. 1992; 1994; Thewlis, J. 1961-1964.
*See also* BOYLE; CHARLES; CLAUSIUS-CLAPEYRON; DALTON; IDEAL GAS; VAN DER WAALS

## GASES, FUNDAMENTAL LAWS—SEE AVOGADRO; BERNOUILLI (KINETIC); BOYLE; CHARLES; DALTON; GRAHAM; JOULE; VAN DER WAALS

## GAUSE LAW OR EXCLUSION PRINCIPLE, HYPOTHESIS, RULE (GEOLOGY); OR GRINNELL AXIOM; OR COMPETITIVE EXCLUSION PRINCIPLE

Two species with the same ecological requirements cannot coexist in the same geographic place indefinitely; that is, they cannot form steady-state populations if they occupy the same ecological niche.

**Keywords:** ecological, geographic, niche, population, species
GAUSE, George Francis, twentieth century, German geneticist
**Sources:** Gray, P. 1967; King, R. C. and Stansfield, W. D. 1990; Landau, S. I. 1986; Lincoln, R. J. 1982.
*See also* COMPETITIVE EXCLUSION; GRINNELL

## GAUSS LAW

In electromagnetic theory, where there are many charges, one can consider a practically continuous distribution of charges so that the total lines of flux emanate from a given volume, such that the Gaussian representation is:

$$\int D\, dS = \Delta D = 4\, \pi\, q$$

$$\Delta\beta = 0 \text{ for magnetic induction}$$

where D = electric displacement
β = magnetic induction
q = electric charge
S = surface surrounding the volume containing charge

or the flux of an electric field through a surface is proportional to the charge inside the surface.

**Keywords:** charge, electric, electromagnetic, flux
GAUSS, Karl Friedrich, 1777-1855, German mathematician
**Sources:** Achinstein, P. 1971; Thewlis, J. 1961-1964.
*See also* COULOMB

## GAUSSIAN DISTRIBUTION LAW—SEE DISTRIBUTION; ERROR; NORMAL

## GAY LUSSAC LAW (1792, 1802); GAY-LUSSAC LAW

The coefficient of expansion of a gas under constant pressure is dependent on or related to the absolute temperature. The final volume of gas, $V_f$, as related to the original volume, $V_o$, and absolute temperature, $T_D$, under constant pressure is:

$$V_o/V_f = T_D/T_f$$

Also written as:

$$p_l = p_o(1 \pm \alpha\, t)$$

$$v_l = v_o(1 \pm \beta\, t)$$

where $p_l$ = gas pressure at temperature t
$p_o$ = gas pressure at 0° C
$v_l$ = gas volume at temperature t
$v_o$ = gas volume at 0° C
α = temperature coefficient of pressure increase
β = temperature coefficient of expansion
α = β = 1/T at low pressure

**Keywords:** expansion, gas, pressure, temperature
GAY LUSSAC, Joseph Louis, 1778-1850, French chemist
**Sources:** Bothamley, J. 1993; Hampel, C. A. and Hawley, G. G. 1973; Layton, E. T. and Lienhard, J. H. 1988; Thewlis, J. 1961-1964.
*See also* CHARLES; COMBINING VOLUMES; CHARLES-GAY LUSSAC; DALTON

## GAY LUSSAC LAW OF GASEOUS COMBINATION OR OF COMBINING VOLUMES (1809) OR LAW OF VOLUMES

When two ideal gases combine, a whole-number ratio exists between the volumes of the two gases, and between the volume of either one of the gases and that of the substance produced if it is a gas. This law is a modification of Charles law.

**Keywords:** chemistry, combination, gas, ratio
GAY LUSSAC, Joseph Louis, 1778-1850, French chemist
**Sources:** Hunt, V. D. 1979; Lagowski, J. J. 1997; Mandel, S. 1972.
*See also* AVOGADRO; CHARLES; GAS; OSTWALD

## GAY LUSSAC NUMBER, GaL OR $N_{GaL}$

A dimensionless group that represents the thermal expansion process due to temperature changes:

$$GaL = 1/\beta \, \Delta T$$

where $\beta$ = coefficient of bulk expansion
$\Delta T$ = temperature difference

**Keywords:** bulk, expansion, temperature, thermal
GAY LUSSAC, Joseph Louis, 1778-1850, French chemist
**Sources:** Bolz, R. E. and Tuve, G. L. 1970; Parker, S. P. 1992; Potter, J. H. 1967.
*See also* AVOGADRO

## GEARING—SEE TOOTHED GEARING, LAW OF

## GEIGER AND NUTTALL RULE (1911*; 1928)

There is a linear relationship between log of the wavelength and the energy of the alpha ray, with most alpha ray emitters in the domain of 2 to 10 MeV.

**Keywords:** alpha, emitters, energy, ray
GEIGER, Hans Wilhelm Johannes, 1882-1945, German physicist
NUTTALL, John Michael, 1890-1958, English physicist
**Sources:** *Bothamley, J. 1993; Parker, S. P. 1989; NYPL Desk Reference; Thewlis, J. 1961-1964.

## GEIGER LAW (1913)

The initial velocity, to the third power, $V_0^3$, of alpha particles is proportional to the range of alpha particles, R, represented by an equation in the form:

$$V_0^3 = a\,R$$

where $V_0$ = initial velocity of the alpha particle at the source, ft/sec
a = constant, $1.03 \times 10^{23}$
R = range of the alpha particle

The Geiger counter or Geiger-Müller (G-M) counter is based on the Geiger law and was invented in 1928.

**Keywords:** absorption, alpha particle, radiation, X-Ray
GEIGER, Hans Wilhelm Johannes, 1882-1945, German physicist
MÜLLER, Walther, late nineteenth-early twentieth century, German physicist
**Sources:** Besancon, R. M. 1974; Encyclopaedia Britannica. 1961; Hunt, V. D. 1979; Parker, S. P. 1992; Thewlis, J. 1961-1964; Travers, B. 1995; World Book Encyclopedia. 1990.
*See also* ABSORPTION; IONIZATION; 3/2 POWER

## GEIGER-NUTTALL LAW (1911; REV. 1921)

The relationship that the higher the energy of the emitted alpha particles, the shorter the life of the radioactive substance emitting them. Plotting on logarithmic scale the mean lifetimes of various elements against their alpha particle energies gave nearly a straight line from uranium with an alpha particle energy of 4.1 MeV and a mean half life of 4.5 billion years with an alpha particle energy of 7.7 Mev and a mean half life of 0.0002 sec.

**Keywords:** alpha, energy, lifetime, logarithmic
GEIGER, Hans Wilhelm Johannes, 1882-1945, German physicist
NUTTALL, John Michael, 1890-1958, English physicist
**Sources:** Gamow, G. 1970; Landau, S. I. 1986.

## GENERAL GAS—SEE COMBINED GAS

## GENETIC DRIFT—SEE SEWALL WRIGHT

## GEOCHEMICAL DISTRIBUTION OF ELEMENTS OF THE EARTH, LAW OF

1. That atomic structure was more important in explaining mineral similarities than their chemical composition
2. That binary crystal structures were explicable simply in terms of the ratio of two ions
3. That the most important parameters governing crystal chemistry was ionic charge and ionic polarization

**Keywords:** atomic, crystal, ionic charge, polarization
GOLDSCHMIDT, Victor Moritz, 1888-1947, Norwegian Swiss geophysicist
**Sources:** Asimov, I. 1972; Chem. Heritage 14(2):56. 1997; Dasch, E. J. 1996.

## GEODESICS MOTION, LAW OF

1. Newton law of gravitation-attractive force is proportional to masses and inversely to the square of the distance
2. Newton second law of motion

**Keywords:** attractive, distance, mass, motion
NEWTON, Sir Isaac, 1642-1727, English physicist
**Source:** Parker, S. P. 1992.

## GEOMETRIC PROGRESSION, LAW OF; GEOMETRIC SEQUENCE, LAW OF

A succession of terms, each of which, except for the first, is derived from the preceding term by multiplying the term by a constant, called the ratio. Thus, the sum of succession of terms is:

$$\Sigma S = a + ar + ar^2 + ar^3 + \ldots + ar^{n-1}$$

or sum of the terms is $a(1 - r^n)/(1 - r)$

where a = first term
r = ratio, which can be less or greater than 1; for example, 1, 3, 9, 27, 81,..., for r = 3

**Keywords:** algebra, mathematics, progression, ratio
**Sources:** Isaacs, A. 1996; James, R. C. and James, G. 1968; Taylor, J. G. 1975.
*See also* ARITHMETICAL PROGRESSION

## GEOMETRICAL OPTICS, FUNDAMENTAL LAWS OF

The fundamental laws of geometrical optics are:

- Rectilinear Propagation of Light, Law of
- Mutual Independence of the Component Parts of a Light Beam, Law of
- Rectangular Reflection, Law of
- Regular Refraction, Law of

**Sources:** Considine, D. M. 1976; Menzel, D. H. 1960.

## GEOMETRODYNAMIC LAW

A combined law of geometry and dynamics which states that forces on all space have this foam-like character, that is, permeated everywhere with wormholes.

**Keywords:** foam-like, forces, dynamics, geometry, space
**Source:** Taylor, J. G. 1975.

## GERBER LAW

An empirical law (Fig. G.1) regarding fatigue of materials whereby the alternating stress for a given endurance plotted on the ordinate and the mean static tensile strength plotted on the abscissa, gives the shape of a parabola:

$$S_A = S[1 - (S_M/u_t)^2]$$

where $S_A$ = alternating stress
$S_M$ = tensile strength
$u_t$ = ultimate tensile strength

**Keywords:** alternating stress, fatigue, strength, tensile
GERBER, John Gottfried Heinrich, 1832-1912, German engineer
**Source:** Thewlis, J. 1961-1964.
*See also* GOODMAN

## GERHARDT LAW; GERHARDT-SEMON LAW; SEMON-ROSENBACH LAW

In a progressive destructive lesion of the motor nerve supplying the largyneal muscles, the abductor mechanism is affected before that of adduction. Various peripheral and central lesions affecting the recurrent largyneal nerve cause the vocal cord to move to an intermediate position between where the muscle draws away from the main axis (abduction) and where the muscle draws toward the main axis (adduction), and the paralysis of the parts is incomplete.

**Keywords:** laryngeal, lesions, nerve, throat, vocal
GERHARDT, Adolf Christian Jacob, 1833-1902, German physician
SEMON, Sir Felix, 1849-1921, English physician
ROSENBACH, Ottomar, 1851-1907, German physician
**Sources:** Critchley, M. 1978; Friel, J. P. 1974; Landau, S. I. 1986; Stedman, T. L. 1976.

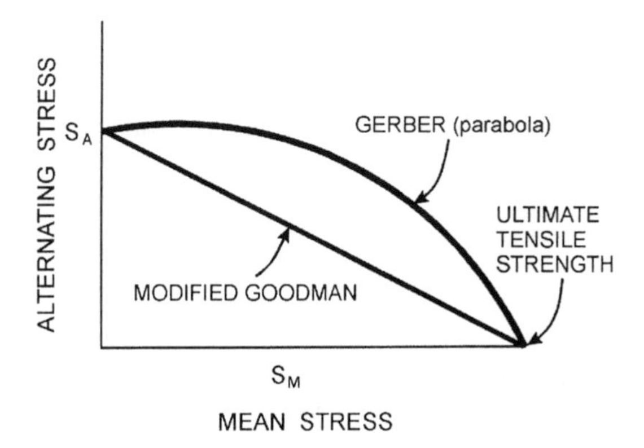


**Figure G.1** Gerber.

## GESTALT LAWS OF PERCEPTUAL ORGANIZATION

An emphasis on psychology, founded in Germany, that represents the procedure of looking at a broader or more holistic view than normally considered. The concept is based on the maxim that the whole is greater than the parts. The people who laid the background for this approach were M. Wertheimer, W. Kohler, and K. Koffka.

**Keywords:** holistic, psychology
WERTHEIMER, Max, 1880-1943, German American psychologist
KOHLER, Wolfgang, 1887-1967, German psychologist
KOFFKA, Kurt, 1886-1941, German psychologist
**Sources:** Bothamley, J. 1993; Encyclopaedia Britannica. 1961.
*See also* GOOD CONTINUATION; GOOD SHAPE; WERTHEIMER

## GIBBS ADSORPTION LAW (1878)

Any substance (solute) that lowers the surface or interface and lowers the surface tension of the solution. This relationship was found separately by Thomson.

**Keywords:** adsorption, interfacial, tension
GIBBS, Josiah Willard, 1839-1903, American physicist and engineer
THOMSON, Joseph John, 1856-1940, British physicist
**Sources:** Bothamley, J. 1993; Glasser, O. 1944; Glasstone, S. 1948.

## GIBBS FREE ENERGY OR GIBBS FUNCTION, G (1902)

The equation which is most useful in analyzing systems at constant pressure and temperature:

$$\Delta G = E + pv - TS$$

where $\Delta G$ = a measure of maximum attainable work
$E$ = internal energy
$p$ = pressure
$v$ = volume
$T$ = absolute temperature
$S$ = entropy

**Keywords:** attainable work, energy, entropy, pressure, temperature
GIBBS, Josiah Willard, 1839-1903, American physicist and engineer
**Sources:** Layton, E. T. and Lienhard, J. H. 1988; Parker, S. P. 1989.
*See also* HELMHOLTZ

## GIBBS-DALTON LAW—SEE DALTON

## GIBBS-DONNAN LAW

The distribution of diffusible ions between plasma and interstitial fluid should conform to:

$$r_{pf} = [H+]_f/[H+]_p$$

where $r$ = ratio
$f$ = fluid
$p$ = plasma

**Keywords:** diffusible, fluid, interstitial, ions, plasma
GIBBS, Josiah Willard, 1839-1903, American physicist and engineer
DONNAN, Frederick George, 1870-1956, English physical chemist
**Source:** Glasser, O. 1944.

## GIBBS-HELMHOLTZ EQUATION (GIBBS, 1875; HELMHOLTZ, 1882)

A thermodynamic relationship useful in calculating changes in the energy or enthalpy (heat content) of a system, using other applicable data. The two useful general forms of the equation are:

$$\Delta A - \Delta E = T\left(\frac{\partial \Delta A}{\partial T}\right)_v$$

$$\Delta F - \Delta H = T\left(\frac{\partial \Delta F}{\partial T}\right)_p$$

where A = the work function
E = energy of the system
T = absolute temperature
p = pressure
v = volume
F = free energy or thermodynamic potential
H = heat content of the system

**Keywords:** energy, enthalpy, heat, system, thermodynamic
GIBBS, Josiah Willard, 1839-1903, American physicist and engineer
HELMHOLTZ, Hermann Ludwig Ferdinand von, 1821-1894, German physicist/physiologist
**Source:** Honig, J. M. 1953.

## GIBBS PHASE RULE OR LAW (1875)

The number of degrees of freedom, F, of a system is the number of variable factors of the components, each of which must be arbitrarily fixed to properly define the system:

$$F = C - P - M + 2$$

where F = number of degrees of freedom
C = number of chemical species at equilibrium
P = number of phases
M = number of independent chemical reactions

**Keywords:** components, degrees of freedom, variable
GIBBS, Josiah Willard, 1839-1903, American physicist and engineer
**Sources:** Hodgman, C. D. 1952; Parker, S. P. 1989; Van Nostrand. 1947.

## GIBBS-THOMSON EQUATION—SEE OSTWALD-FREUNDLICH

## GIBRAT LAW OF PROPORTIONATE GROWTH (ECONOMICS) (1931)

A formulation that describes how a process of random growth can produce a log normal distribution of firm sizes. The significance of such a distribution is that it conforms with that most typically found in real-world industries, i.e., positively skewed with a few large firms and many small firms, justifying use of a lognormal distribution as indices of market conditions.

**Keywords:** firm, growth, sizes
GIBRAT, Robert Pierre Louis, 1904-1980, French economist
**Source:** Pearce, D. W. 1981. 1986.

## GILBERT PRINCIPLE—SEE EQUAL DECLIVITIES

## GIRAUD-TEULON LAW

The intersection of the primary and secondary axes of projection mark the location of the binocular retinal images.

**Keywords:** binocular, eye, images, retina, sight
GIRAUD-TEULON, M. A. L. Felix, 1816-1887, French ophthalmologist*
**Sources:** Friel, J. P. 1974; *Garrison, F. H. 1967; Landau, S. I. 1986; NUC.

## GLADSTONE AND DALE LAW (1858)

When a substance is compressed or its temperature is altered, the relationship that relates the variation in density and the refractive index is constant:

$$k = (n+1)/\rho \text{ is constant}$$

where $n$ = refractive index
$\rho$ = density

**Keywords:** compressed air, density, temperature, refractive index
GLADSTONE, John Hall, 1827-1902, English chemist
DALE, Thomas Pelham, 1821-1892, English scientist
**Sources:** Gillispie, C. C. 1981; Honig, J. M. 1953; NUC; Thewlis, J. 1961-1964.
*See also* LORENTZ-LORENZ FORMULA

## GLEN LAW OR FLOW LAW OF ICE (1955)

An experimental law for the plastic creep of ice (Fig. G.2) or a normal plastic substance when the cube size is doubled the collapse time is halved; but for ice, when the cube size is doubled the collapse time is one eighth. Ice weakens as the shear increases. For secondary creep over the range usually found in glaciers (50 to 200 kPa), the relation between shear stress and strain rate has the form:

$$\dot{e}_{xy} = A\tau^{n}{}_{xy}$$

where $\tau_{xy}$ = shear stress
$\dot{e}_{xy}$ = deformation rate or strain rate
$n$ = a constant, usually a value of 3 for ice
$A$ = a constant that depends on the temperature, impurities, and other factors

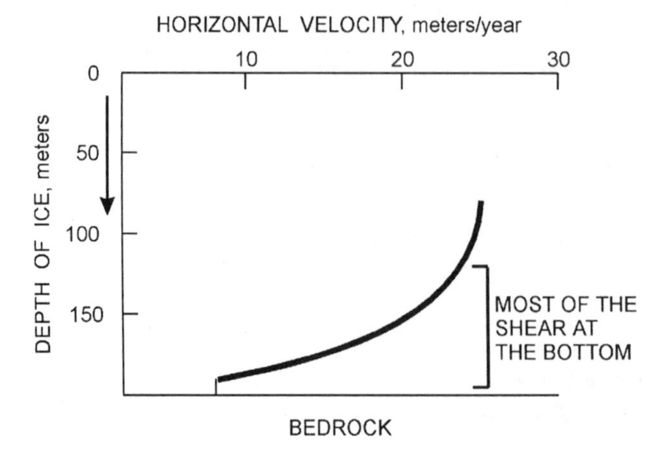


**Figure G.2** Movement of glacier ice.

**Keywords:** collapse, cube, glaciology, ice, plastic
GLEN, J. W., twentieth century, British geologist
**Sources:** Gary, M. 1973; Hoyle, F. 1981.

## GLOGER LAW

Skin pigmentation is related to climate. Cold reduces melanins, and warmth and humidity increase melanins. In southern races, warm-blooded animals tend to be dark colored, and in northern races, warm blooded animals tend to be light colored.

**Keywords:** animals, color, melanins, northern, southern
GLOGER, Constantine Wilhelm Lambert, 1803-1863, German zoologist
**Sources:** Gray, P. 1967; Landau, S. I. 1986.
*See also* ADAPTATION; ALLEN; BERGMANN

## GÖDEL THEOREM (1931) OR GÖDEL INCOMPLETENESS THEOREM

No single system of laws (axioms and rules) can be strong enough to prove all true statements of arithmetic without at the same time being so strong that it also proves the false ones. Equivalently, there is no single algorithm for distinguishing true arithmetical statements from false ones.

**Keywords:** arithmetic, false, statements, true
GÖDEL, Kurt, 1906-1978, Austrian American mathematician and logician (born in Czechoslovakia*)
**Sources:** Barrow, J. D. 1991; *Reber, A. S. 1985; Scientific American 273(10):116, April 1995.

## GODÉLIER LAW

Tuberculosis of membrane lining the abdominal wall is invariably associated with tuberculosis of gland membrane investing the lungs and lining of the thoracic cavity.

**Keywords:** abdominal, gland, tuberculosis
GODÉLIER, Charles Pierre, 1813-1877, French physician
**Sources:** Friel, J. P. 1974; Landau, S. I. 1986; Stedman, T. L. 1976.

## GOLDBERG LAW—SEE GULDBERG-WAAGE

## GOLDSCHMIDT LAW OR RULE (1929)

A fundamental law of crystal chemistry in which the structure of a crystal is described by the ratio of the numbers representing the ratios of sizes and properties of polarization of its structural units, based on the rock's temperature and pressure when formed.

**Keywords:** crystal, polarization, properties, structural
GOLDSCHMIDT, Victor Moritz, 1888-1947, Swiss Norwegian geochemist
**Sources:** Ballentyne, D. W. G. and Lovett, D. R. 1980; Clifford, A. F. 1964.

## GOLGI LAW

The severity of a malarial attack depends on the number of parasites in the blood.

**Keywords:** blood, malaria, parasites
GOLGI, Camillo, 1843-1926, Italian histologist, pathologist, and physiologist; Nobel prize, 1906, physiology/medicine (shared)
**Sources:** Friel, J. P. 1974; Landau, S. I. 1986; Reber, A. S. 1985.

## GOMPERTZ LAW; GOMPERTZ MODEL; OR LAW OF MORTALITY

The risk of dying increases geometrically and is equal to a constant multiple of a power of a constant, the exponent being the age for which the mortality is being determined, for a particular disease (Fig. G.3). One of the growth curves, in which the useful form is:

$$y = A \exp[-j \exp(-kx)]$$

where y = growth, vertical axis
x = the time, horizontal axis
A = amplitude of the sigmoidal oscillation
j, k = constants

**Keywords:** age, death, disease, mortality
GOMPERTZ, Benjamin, 1799-1865, English mathematician
**Sources:** Chemical Engineering 104(2);115-119. 1997; Friel, J. P. 1974; James, R. C. and James, G. 1968; Landau, S. I. 1986; Reber, A. S. 1985.
*See also* MAKEHAM

## GOOD CONTINUATION, LAW OF

A gestalt law of perception in which lines that are seen as broken or overlapped belong together as long as they result in a straight line or gentle curve, attributed to M. Wertheimer.

**Keywords:** gestalt, lines, perception
WERTHEIMER, Max, 1880-1943, German American psychologist
**Sources:** Bothamley, J. 1993; Reber, A. S. 1985; Wolman, B. B. 1983.
*See also* GESTALT; GOOD SHAPE; WERTHEIMER

## GOODELL LAW (SIGN)

The physical condition of the cervix indicates whether pregnancy exists such that, if the cervix is soft, a pregnancy is probable and, if hard, there is no pregnancy.

**Keywords:** birth, cervix, pregnancy
GOODELL, William, 1829-1894, American gynecologist
**Sources:** Friel, J. P. 1974; Landau, S. I. 1986.

## GOODHART LAW (ECONOMICS)

Whichever money aggregate is chosen as a target variable becomes disturbed by the very act of targeting it.

**Keywords:** aggregate, money, target

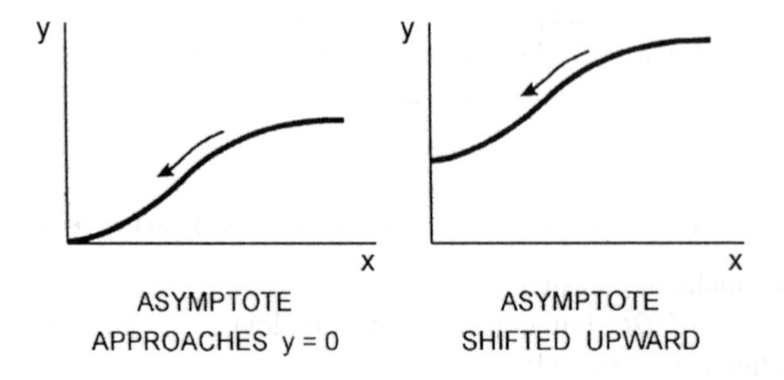


**Figure G.3** Gompertz.

GOODHART, Charles A. E., twentieth century (b.1936), English economist
**Source:** Blaugh, M. 1986.

### GOODMAN LAW (MODIFIED); GOODMAN MEAN STRESS LAW

An empirical law regarding fatigue of materials whereby the alternating stress for a given endurance plotted on the ordinate and the mean static tensile strength plotted on the abscissa gives a straight line:

$$S_A = S[1 - S_M/u_t]$$

where $S_A$ = safe alternating stress
$S$ = alternating fatigue stress
$S_M$ = tensile strength
$u_t$ = is ultimate strength

**Keywords:** alternating, endurance, static, tensile
GOODMAN, Laurence Eugene, twentieth century (b. 1920), American physicist/engineer
**Sources:** Eirich, F. R. 1956; Thewlis, J. 1961-1964.
*See also* GERBER

### GOOD SHAPE, LAW OF

A gestalt law, expressed by M. Wertheimer, that we generally perceive stimulus units and patterns in a methodical and organized way.

**Keywords:** patterns, perceptions, stimulus
WERTHEIMER, Max, 1880-1943, German American psychologist
**Source:** Bothamley, J. 1993.
*See also* GESTALT; GOOD CONTINUATION; WERTHEIMER

### GORTER-CASIMIR LAW (1911)

The relationships in a two-fluid model of superconductivity that regards the conduction electrons as two interpenetrating, noninteracting electronic fluids. One fluid does not interact and offers no resistance to flow, which is a condensed state at zero entropy. The other fluid responds and acts as normal conduction electrons, responsible for a-c resistance and heat conduction.

**Keywords:** electrons, entropy, fluid, resistance, superconductivity
GORTER, Cornelis J., twentieth century (b. 1907), Dutch physician
CASIMIR, Hendrik P. G., twentieth century (b. 1909), Dutch physicist
**Source:** Parker, S. P. 1983 (physics)
*See also* $T^3$ LAW

### GOUCHER NUMBER, Go OR $N_{Go}$

A dimensionless group applicable to coatings on wire or fibers that relates the gravity force and the surface tension force:

$$Go = r\,(\rho\, g/2\sigma)^{1/2}$$

where $r$ = radius of the wire
$\rho$ = mass density
$g$ = gravitational acceleration
$\sigma$ = surface tension

**Keywords:** coating, gravity, surface, tension
**Sources:** Bolz, R. E. and Tuve, G L. 1970; Land, N. S. 1972; Parker, S. P. 1992; Potter, J. H. 1967.
Also see BOND; DERYAGIN; EÖTVÖS

### GOVERNMENT SIZE, LAW OF

"Economic and social misery increase in direct proportion to the size and power of the central government of a nation or state."

**Keywords:** central government, economic, social, size
**Source:** Sale, K. 1980.

### GRAETZ NUMBER, Gz OR $N_{Gz}$

A dimensionless group used in representing conductive heat transfer in streamline flow that relates the thermal capacity of the fluid and the conductive heat transfer:

$$Gz = Q_m c_p/kL = [(L/d)/Re\ Pr]$$

$$Gz = \pi\ d\ Re\ Pr/4\ L \text{ in a pipe}$$

where $Q_m$ = mass flow rate
$c_p$ = specific heat at constant pressure
k = thermal conductivity
d = diameter of pipe
L = characteristic length

**Keywords:** conduction, heat transfer, thermal, streamline
GRAETZ, Leo P., 1856-1941, German physicist
**Sources:** Bolz, R. E. and Tuve, G. L. 1970; Jerrard, H. G. and McNeill, D. B. 1992; Land, N. S. 1972; Parker, S. P. 1992; Perry, R. H. 1967; Potter, J. H. 1967; Rohsenow, W. M. and Hartnett, J. P. 1973.
*See also* PRANDTL; REYNOLDS

### GRAHAM LAW OF DIFFUSION (1831)

The rates of diffusion of various gases vary inversely as the square roots of their densities or their molecular weights, at constant temperature and pressure:

$$r_1/r_2 = d_2^{1/2}/d_1^{1/2} = M_2^{1/2}/M_1^{1/2}$$

where r = rate of diffusion
d = density
M = molecular weight

**Keywords:** densities, diffusion, gases, molecular weight
GRAHAM, Thomas, 1805-1869, Scottish physical chemist
**Sources:** Honig, J. M. 1953; Landau, S. I. 1986; Mandel, S. 1969; Stedman, T. L. 1976; Thewlis, J. 1961-1964.

### GRAHAM LAW OF EFFUSION (1829)

The relative rates of effusion of different gases under the identical temperature and pressure are inversely proportional to the square roots of their densities.

**Keywords:** densities, effusion, gases, pressure, temperature

GRAHAM, Thomas, 1805-1869, Scottish physical chemist
**Source:** Bothamley, J. 1993.

## GRAIN PRESSURES, LAW OF LATERAL—SEE JANSSEN; KETCHUM; RANKINE

## GRASHOF NUMBER, Gr OR $N_{Gr}$ (1875)

A dimensionless group used in representing free convection heat and fluid transfer that relate inertia force and buoyant force to the viscous force:

$$Gr = \rho^{2} g L^{3} \beta \, \Delta T/\mu^{2}$$

where $\rho$ = mass density
$g$ = gravitational acceleration
$L$ = characteristic length
$\beta$ = temperature coefficient of volume expansion
$\Delta T$ = temperature difference
$\mu$ = absolute viscosity

Originally expressed by F. Grashof in 1875; named after him in 1934. The Grashof number = (the Reynolds number)$^2$ divided by the Gay Lussac number times the Froude number.

**Keywords:** buoyancy, convection, fluid, free, inertia, viscous
GRASHOF, Franz, 1826-1893, German mechanical engineer
**Sources:** Bolz, R. E. and Tuve, G. L. 1970; Land, N. S. 1972; Layton, E. T. and Lienhard, J. H. 1988; Parker, S. P. 1992; Perry, R. H. 1967; Potter, J. H. 1967; Rohsenow, W. M. and Harnett, J. P. 1973.

## GRASSET OR LANDOUZY-GRASSET LAW

Where there is lesion of one cerebral hemisphere, the head is turned to the side of the brain lesion if there is paralysis and to the side of the affected muscles if there is spasticity.

**Keywords:** brain, cerebral, head, medicine, muscles
GRASSET, Joseph,1849-1918, French physician
LANDOUZY, Louis, Theophil J., 1845-1917, French neurologist and physician
**Sources:** Friel, J. P. 1974; Landau, S. I. 1986; Webster's Biographical Dictionary. 1959.
Also see LANDOUZY-GRASSET

## GRASSMANN (GRASSMAN*) LAW OF COLOR VISION (1853); LAW OF COLOR MIXTURE

If a light is composed of known amounts of three components, called primaries, and is equivalent in color to an unknown light, the known amounts may be used as a color specification for this light. The amounts are called tristimulus specifications. The law states that "(1) When equivalent lights are added to or subtracted from equivalent lights, the sums or differences are equivalent are equivalent; (2) lights that are equivalent to the same light are equivalent to each other; and (3) for every color there is a complementary color that, when mixed with it in the correct proportion, gives white or gray, although the mixture of noncomplementary colors gives an intermediate color.

"Further, any beam of light, whether it originates from a self-luminous body, or comes, by transmission or reflection, from a non-self-luminous object, may be analyzed as the sum of a large number of portions of the spectrum."

**Keywords:** complementary, color, light, mixture

GRASSMANN, Hermann Gunther, 1809-1877, German Polish mathematician
**Sources:** Abbott, D. 1985; *Glasser, O. 1944; Physics Today 19(3):35. 1966; Thewlis, J. 1961-1964; Zusne, L. 1987.
*See also* ABNEY

## GRASSMANN LAW (1845) (MATHEMATICS)

Grassmann replaced Ampere's fundamental law for the reciprocal effect of two infinitely small current elements with a law requiring less arbitrary assumptions. In a broader sense, he developed the idea of an algebra in which symbols representing points, lines, or planes are manipulated according to certain rules that are easier to use than those at the time, and in which coordinates were used in three dimensions for space.

**Keywords:** ampere, current, reciprocal, space
GRASSMANN, Hermann Gunther, 1809-1877, German Polish mathematician who laid the foundation for modern vector analysis
**Sources:** Abbott, David. 1985; Morris, C. G. 1992.

## GRAVITATION—SEE EINSTEIN LAW OF GRAVITATION; GALILEO; KEPLER; NEWTON UNIVERSAL LAW OF GRAVITATION

## GRAVITATIONAL INVERSE-SQUARE LAW—SEE NEWTON

## GRAVITY NUMBER—Grav OR $N_{Grav}$

A dimensionless group representing two-phase fluid flow in porous media that relates the gravity force and filtration force:

$$Grav = k\ g\ \Delta\rho/\mu V$$

where k = permeability
g = gravitational acceleration
Δp = mass density difference between fluids
V = reference velocity
μ = absolute viscosity

**Keywords:** filtration, flow, fluid, gravity, porous, two-phase
**Sources:** Land, N.S. 1972; Potter, J. H. 1967.

## GRAY—SEE BRAGG-GRAY

## GREAT NUMBERS, LAW OF

All of our phenomena are but an envelope and the result of an immense number of molecular and electronic phenomena that cannot be observed but are represented by great numbers that enter into life. In mathematical statistics, relationships of probability, p, are expressed, one of which is Law of Great Numbers, which can be expressed by saying that, if n is taken large enough, almost all n-field selections have α-ratios nearly equal to p, where n is a number increasing without limit.

**Keywords:** life, phenomenon, probability, statistics
**Source:** Braithwaite, R. B. 1953. 1960. 1953.
*See also* LARGE NUMBERS

### GRESHAM LAW

In planning, activity that is programmed tends to drive out unprogrammed activity. In the flow of coins, there is a tendency, when two or more coins are equal in debt-paying power but unequal in intrinsic value, for the one having the least intrinsic value to remain in circulation and for the other to be kept or hoarded.

**Keywords:** coins, flow, planning, programmed
GRESHAM, Sir Thomas, 1519-1579, English financier
**Sources:** Pearce, D. W. 1986; Sloan, H. S., et al. 1970.

### GRICE MAXIMS

A set of maxims relating to the understanding of the intended meaning in communication between human beings. These are used as guidelines in the formulation of computer-generated messages for reducing misunderstanding by the user.

1. The maxim of *quantity* states that a message should be informative but say only what is necessary for understanding and no more.
2. The maxim of *quality* states that messages should include only information known to be true.
3. The maxim of *relevance* states that a message should make its context clear to the receiver.
4. The maxim of *manner* states that messages should be brief and unambiguous, using simple grammatical structures.

**Keywords:** communications, manner, quality, quantity, relevance
GRICE, H. Paul, twentieth century (b. 1913), American philosopher
**Source:** Watters, C. 1992.

### GRIMM LAW (1822)

A statement of the regular changes undergone by the mute consonants of the early Indo-European languages in their development into sister languages, such as g into k.

**Keywords:** consonants, European, language, speech
GRIMM, Jacob, 1785-1863, German philologist
**Sources:** Bothamley, J. 1993; Webster's Biographical Dictionary. 1959.
*See also* VERNER

### GRIMM-SOMMERFELD RULE

The absolute valence of an atom is numerically equal to the number of electrons "engaged" in attracting the remaining electrons.

**Keywords:** atom, electrons, physical chemistry, valence
GRIMM, Hans August George, twentieth century (b. 1887), German physicist
SOMMERFELD, Arnold Johannes Wilhelm, 1868-1951, German physicist
**Sources:** Gillispie, C. C. 1981; NUC; Webster's Biographical Dictionary. 1959.

### GRINNELL AXIOM—SEE GAUSE LAW

### GROSCH LAW (1949)

The economics of scale in computer operation is that the computing power increases as the square of the cost:

$$p = kc^2$$

where p = the computing power
k = constant
c = system cost

There is a question whether this relationship still holds, but it can be used as guide.

**Keywords:** computers, economics, power, set, scale
GROSCH, Herbert R. J., twentieth century, American computer businessman
**Source:** Ralston, A. and Reilly, E. D. 1993.
*See also* MOORE; ZIPF

## GROTTHUSS-DRAPER LAW (1817, 1841) (GROTTHUS*)

There is a pigment material in a photoreceptor which is affected by light, described as *photosensitive*, and to react with visible light it must absorb these wavelengths. Simply stated, only light that is absorbed is effective in producing chemical change. This is a basic law of photochemistry.

**Keywords:** light, photoreceptive, photosensitive, pigment
GROTTHUSS, Christian Johann Dietrich von (also called Theodor), 1785-1822, German Russian (Lithuanian) chemist
DRAPER, John William, 1811-1882, English American chemist
**Sources:** *Ballentyne, D. W. G. and Lovett, D. R. 1980; Considine, D. M. 1976; Friel, J. P. 1974; Gillispie, C. C. 1981; Glaser, O. 1944; Honig, J. M. 1953; Thewlis, J. 1961-1964.
*See also* DRAPER; STARK-EINSTEIN

## GROUP DISPLACEMENT LAW

The emission of an alpha particle in a radioactive change produces a product that occupies a position two places to the left of its parent in the periodic table, and the emission of a beta particle results in a product that occupies a position one place to the right of its parent in the periodic table.

**Keywords:** alpha, atomic, beta, nuclear, periodic
**Sources:** Hazewinkel, M. 1987; Honig, J. M. 1953.

## GROWTH—SEE GOMPERTZ; LINEAR GROWTH CURVE

## GROWTH OF POPULATION, LAW OF

The fundamental system of equations, $dx/dt = F(x)$, represents the change in population, based on the human population, x, over time, t.

**Keywords:** human, population
**Source:** Lotka, A. J. 1956.
*See also* VERHULST-PEARL LAW

## GRÜNEISEN LAW (1908) OR GRÜNEISEN SECOND RULE

The ratio of the coefficient of thermal expansion of an isotropic solid or metal to its specific heat at constant pressure is approximately independent of temperature, or is a constant at all temperatures. Thus:

$$\beta = \Gamma Z_0 C_v/V$$

where $\beta$ = volume coefficient of expansion
$\Gamma$ = dimensionless quantity
$Z_0$ = isothermal compressibility at 0° K
$C_v$ = molecular specific heat at constant volume
V = molecular volume of the solid

The first Grüneisen rule was used as a basis of this law. The first rule of Grüneisen, the equation of state for a solid is:

$$pV + G(V) = \gamma E$$

where p = pressure
V = volume
G(V) = interatomic potential energy
E = energy of atomic vibrations
$\gamma$ = dimensionless quantity

**Keywords:** expansion, solid, specific heat, thermal
GRÜNEISEN, Eduard, 1877-1949, German physicist
**Sources:** Thewlis, J. 19661-1964; Uvarov, E. and Chapman, D. R. 1964 (first rule).
*See also* DEBYE

### GUDDEN LAW

In the division of a nerve, degeneration in the proximal portion is toward the nerve cell. (The degeneration of the proximal end of a divided nerve is cellulipetal.)

**Keywords:** degeneration, nerve
GUDDEN, Berhard Aloys von, 1824-1886, German neurologist
**Sources:** Friel, J. P. 1974; Landau, S. I. 1986; Thomas, C. L. 1989.

### GUDDEN-POHL EFFECT

When an electric field is applied to a phosphorescent material, the electrons are pulled out more rapidly and used to produce luminescence, and decay is accelerated.

**Keywords:** electric field, electron, luminescence
GUDDEN, Bernhard Aloys von, 1824-1886, German neurologist
**Sources:** Holmes, F. L. 1980. 1990; Morris, C. G. 1992; Parker, S. 1987.

### GUEST LAW OF MAXIMUM SHEARING STRESS (1900); ALSO CALLED COULOMB THEORY

The theory predicts the failure of a material under a combination of loads when the maximum shearing stress is equal to one-half the normal stress.

**Keywords:** failure, material, maximum, normal, stress
GUEST, James J., early twentieth century, English scientist
**Sources:** Higdon, et al. 1985; NUC; Philosophical Magazine. 1900; Thewlis, J. 1961-1964.
*See also* COULOMB; RANKINE; ST. VENANT

### GUKHMAN NUMBER, GU OR $N_{Gu}$ (1951?)

A dimensionless group used as a thermodynamic criterion of evaporation under adiabatic conditions and convective heat transfer from a moist surface:

$$Gu = (T_g - T_s)/T_g$$

where $T_g$ = temperature of ambient gas flowing over moist surface
$T_s$ = wet bulb temperature at moist surface

**Keywords:** adiabatic, convective, evaporation, heat transfer, moist surface
GUKHMAN, Aleksandr Adolfovich, twentieth century, Soviet engineer
**Sources:** Bolz, R. E. and Tuve, G. L. 1970; Land, N. S. 1972; Morris, C. G. 1992; NUC; Parker, S. P. 1992; Potter, J. 1967.

### GULDBERG AND WAAGE LAW (1864*; 1867); WAAGE-GULDBERG LAW; LAW OF MASS ACTION; OR LAW OF CHEMICAL KINETICS; OR LAW OF CONCENTRATION EFFECT

The velocities of the forward and reverse homogeneous chemical reactions are proportional to the active concentrations of the reacting substances, each raised to a power equal to the molecules appearing in the balanced equation. This law was independently formulated by J. H. van't Hoff. G. N. Lewis modified and improved the accuracy of this law.

**Keywords:** active masses, chemical, concentration, reaction
GULDBERG, Cato Maxmilian, 1836-1902, Norwegian chemist and mathematician
LEWIS, Gilbert Newton, 1875-1946, American chemist
WAAGE, Peter, 1833-1900, Norwegian chemist
VAN'T HOFF, Jacobus Henricus, 1852-1911, Dutch physical chemist
**Sources:** Hampel, C. A. and Hawley, G. G. 1973; Hellemans, A. 1988; Mandel, S. 1972; Thewlis, J. 1961-1964; Thorne, J. O. 1984; *Van Nostrand. 1947; Walker, P. M. B. 1988.
*See also* MASS ACTION; VAN'T HOFF

### GULDBERG AND WAAGE LAW (1863) OF CHEMICAL EQUILIBRIUM

For an infinitesimal displacement from equilibrium, with a constant temperature or pressure, of a reaction in a closed system:

$$aA + bB + \ldots \Leftrightarrow xY + yY + \ldots$$

for which, at equilibrium, the change of free energy must vanish. Chemical equilibrium originated with the approximately idealized law of mass action and can be derived from F. Trouton.

**Keywords:** equilibrium, free energy, mass action, system
GULDBERG, Cato Maxmilian, 1836-1902, Norwegian chemist and mathematician
WAAGE, Peter, 1833-1900, Norwegian chemist
**Sources:** Friel, J. P. 1974; Hampel, C. A. and Hawley, G. G. 1973; Isaacs, A. 1996; Millar, D. et al. 1996; Parker, S. P. 1989.
*See also* CHEMICAL EQUILIBRIUM; MASS ACTION; TROUTON

### GULDBERG LAW OR RULE (1890)

The law describes the relationship between the boiling point and critical temperature, the point above which the gas cannot be liquefied by additional pressure, on the absolute scale, represented by:

$$T_b/T_c = 2/3$$

where $T_b$ = boiling point temperature
$T_c$ = critical temperature

The law was discovered independently by P.-A. Guye.

**Keywords:** boiling point, critical temperature, pressure
GULDBERG, Cato Maxmilian, 1836-1902, Norwegian chemist and mathematician
GUYE, Phillippe-Auguste, 1862-1922, Swiss chemist and physicist
**Sources:** Daintith, J., et al. 1981; Fisher, D. J. 1988. 1991; Gillispie, C. C. 1981; Honig, J. M. 1953; Ralston, A. and Reilly, E. D. 1993.
*See also* LORENZ; TROUTON

### GULLSTRAND LAW

When a patient turns his head while fixing his sight on a distant object, the corneal reflex from either eye moves in the direction in which the head is turning; it moves toward the weaker muscle.

**Keywords:** corneal, eye, medical, sight
GULLSTRAND, Allvar, 1862-1930, Swedish ophthalmologist; Nobel prize, 1911, physiology/medicine
**Sources:** Friel, J. P. 1974; Landau, S. I. 1986.

### GULL-TOYNBEE LAW; TOYNBEE LAW

In ear inflammation (otitis) media, the lateral sinus and cerebellum are liable to involvement in mastoid disease, and the cerebrum may be attached when the roof of the middle ear (tympanum) becomes decayed (carious).

**Keywords:** disease, mastoid, medical, sinus
GULL, Sir William Whitney, 1816-1890, English physician
TOYNBEE, Joseph, 1815-1866, English otologist
**Sources:** Friel, J. P. 1974; Landau, S. I. 1986.

### GUNN LAW

When treating a limb dislocation, the limb must be placed in the same position as at the time of injury and force exerted on the displaced bone in the reverse direction to that which caused the dislocation.

**Keywords:** bone, dislocation, injury, limb
GUNN, Moses, 1822-1887, American surgeon
**Sources:** Friel, J. P. 1974; Koenigsberg, R. 1989; Landau, S. I. 1986.

### GYRATION, LAW OF—SEE DOVE

# H

## HACK LAW (1957)

An approach for optimal channel networks for water flow based on energy minimization. The law relates the length of the longest stream in the drainage region (measured from any site to the edge of the subbasin) to the drainage area of the basin (the number of upstream sites):

$$d_L = (1 + H)(2 - \tau)$$

where L = length of the longest stream in drainage region
H = number less than 1 that relates length vs width of basin
$\tau$ = the exponent representing the characteristic scale below which the distribution is algebraic

**Keywords:** basin, channel, drainage, energy minimization, network
HACK, John Tilton, twentieth century (b. 1915), American geologist
**Source:** Science 272(5264): 984-986, 17 May 1996.

## HAECKEL LAW; ALSO BIOGENETIC LAW; RECAPITULATION LAW

Developing from the ovum, an organism goes through the same changes as the species in developing from the lower to higher forms of animal life. Haeckel is credited to being the first to use the term *environment.*

**Keywords:** evolution, organism, ovum, species
HAECKEL, Ernst Heinrich, 1834-1919, German naturalist
**Sources:** Dulbecco, R. 1991; Friel, J. P. 1974; Gray, P. 1967; Lapedes, D. N. 1974; Stedman, T. L. 1976; Thomas, C. L. 1989.
*See also* BIOGENETIC LAW; MÜLLER-HAECKEL; RECAPITULATION; VON BAER (BAER)

## HAGEN-POISEUILLE LAW (1839; 1840); POISEUILLE LAW

An expression for the laminar flow of fluid through a capillary (circular) whereby the flow is proportional to the pressure gradient and the fourth power of the radius, and inversely to the length of the capillary:

$$Q = \pi\, p\, R^4/8\, \mu L$$

where Q = volume of flow per unit time
$\mu$ = viscosity
L = length
R = radius
p = pressure

Each person discovered the law independently, and it is generally named after both.

**Keywords:** capillary, flow, fluid, viscosity
HAGEN, Gotthilf Heinrich Ludwig, 1797-1884, German hydraulic engineer
POISEUILLE, Jean Leonard Marie, 1799-1869, French physician and physicist

**Source:** Layton, E. T. and Lienhard, J. H. 1988.
*See also* POISEUILLE; (de) WAELE-OSTWALD

## HAHNEMANN RULE—SEE SIMILARS, LAW OF

## HALDANE LAW OR RULE (1932)

In first-generation (F1) hybrids between two species, in which one sex does not occur, or is rare or sterile, that sex is the heterogametic. J. Haldane devised the "polygraph."

**Keywords:** generation, hybrids, sex, species
HALDANE, John Burdon Sanderson, 1892-1964, Scottish geneticist
**Sources:** Gray, P. 1967; Knight, R. L. 1948; Lincoln, R. J. et al. 1982.

## HALE LAW (1908)

A characteristic magnetic polarity related to the magnetic cycle of the sun. The sun's magnetic pattern is continuously changing in a random nature.

**Keywords:** cycle, magnetic, polarity, sun
HALE, George Ellery, 1868-1938, American astronomer
**Source:** Lapedes, D. N. 1971.
*See also* SPÖRER LAW

## HALL EFFECT (1879)

When a steady current is flowing in a steady magnetic field, electromotive forces (voltages) are developed at right angles both to the magnetic force and to the current, and these are proportional to the product of the intensity of the current, the magnetic force, and the sine of the angle between the directions of these quantities (Fig. H1).

**Keywords:** current, electromotive forces, magnetic force, sine, voltages
HALL, Edwin Herbert, 1855-1938, American physicist
**Sources:** Encyclopaedia Britannica. 1984; Hodgman, C. D. 1952; Mandel, S. 1972.
*See also* NERNST

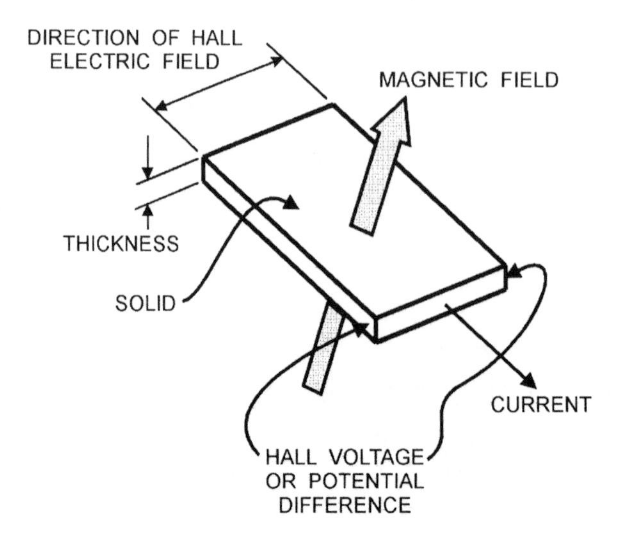


**Figure H.1** Hall effect.

## HALL NUMBER, H OR $N_H$ (HALL COEFFICIENT)

A dimensionless group relating to magnetohydrodynamics in cyclotron operation:

$$H = \omega\tau$$

where $\omega$ = cyclotron frequency, 1/t
$\tau$ = average free path divided by the average velocity

**Keywords:** cyclotron, magnetohydrodynamics
HALL, Edwin Herbert, 1855-1938, American physicist
**Sources:** Bolz, R. E. and Tuve, G. L. 1970; Land, N. S. 1972; Parker, S. P. 1992; Potter, J. H. 1967.

## HALLION LAW

Extracts from an organ injected into the body exert a stimulating influence on the same organ, a law no longer accepted.

**Keywords:** extracts, injection, organ, physiology
HALLION, Louis, 1862-1940, French physiologist
**Sources:** Friel, J. P. 1974; Landau, S. I. 1986.

## HALSTED LAW

Transplanted tissue will grow only if there is a lack of tissue in the host.

**Keywords:** growth, tissue, transplanted
HALSTED, William Stewart, 1852-1922, American surgeon
**Sources:** Friel, J. P. 1974; Stedman, T. L. 1976.

## HAMBURGER LAW

When the blood is alkaline, albumens and phosphates pass to the red corpuscles from the serum, and chlorides pass to the serum from the cells, with the reverse occurring when the blood is acid.

**Keywords:** acid, alkaline, blood, physiology, red corpuscles, serum
HAMBURGER, Hartog Jacob, 1859-1924, Dutch physiologist
**Sources:** Lamdau, S. I. 1986; Stedman, T. L. 1976.

## HAMILTON PRINCIPLE OR LAW (1835)

A body under the influence of a field or force must follow a path that at all times keeps the least value of kinetic energy less the potential energy.

**Keywords:** electricity, energy, kinetic energy, potential energy
HAMILTON, Sir William Rowan, 1805-1865, British mathematician
**Source:** Asimov, I. 1972.
*See also* D'ALEMBERT

## HANAU LAW OF ARTICULATION

A set of physical laws that define the contacting surfaces of the teeth when chewing, and the arrangement of artificial teeth to serve the functions of the natural teeth for chewing and speaking to provide balanced action.

**Keywords:** arrangement, contacting, teeth
HANAU, Rudolph L., published book on dental engineering in 1921

**Sources:** Friel, J. P. 1974; Dorland, W. A. 1980; Landau, S. I. 1986; NUC.
*See also* ORBITALCANINE; ORBITALGNATHION

### HARDY (1900)-SCHULZE LAW (1882); HARDY-SCHULTZE* LAW; SCHULZE-HARDY RULE OR LAW

The most effective precipitants are ions that have a charge of opposite sign to that carried by the suspended particle, with the higher valence materials having greater precipitating action.

**Keywords:** charge, ions, precipitants, valence
HARDY, W. B.; and SCHULZE, H.,** German chemist
SCHULZE, Max Johann Sigismund, 1825-1894, German biochemist
**Sources:** Bennett, H. 1939; *Bothamley, J. 1993; **Fischer, D. J. 1991; Gillispie, C. C. 1981; *Morris, C. G. 1987.

### HARDY-WEINBERG LAW OR PRINCIPLE; CASTLE-HARDY-WEINBERG LAW (1908)

The genetic constitution of a population tends to remain stable, even in the absence of selection, since the frequency of the pairs of allelic genes is an expansion of a binomial equation. The law provides a simple standard for comparing a natural population. In a large population in which the males and females mate at random with respect to the genes in question and produce equal numbers of equally fertile offspring, one would get roughly the same ratios of homozygotes and heterozygotes, expressed simply by the binomial equation:

$$p^2 + 2pq + q^2 = 1$$

where $p^2$ = frequency of one homozygote
$q^2$ = frequency of the other homozygote
$2pq$ = frequency of the heterozygote

**Keywords:** genetic, heterozygote; homozygote, population
CASTLE, William Ernest, 1867-1962, American biologist
HARDY, Godfrey Harold, 1877-1947, English mathematician
WEINBERG, Wilhelm, 1862-1937, German physician
**Sources:** Bothamley, J. 1993; Friel, J. P. 1974; Gray, P. 1967; Landau, S. I. 1986; Lapedes, D. N. 1974; McMurray, E. J. 1995.
*See also* MENDEL

### HARKINS-JURA ISOTHERM MODEL (1946)

A method of representing the amount of moisture adsorbed on hygroscopic materials, derived on the basis of adsorbed material is a condensed monolayer or moisture:

$$1/V^2 = A - B \ln P$$

where $V$ = volume of gas adsorbed per unit amount of solid
$P$ = partial pressure of adsorbate
$A, B$ = appropriate constants

**Keywords:** adsorption, condensed monolayer, hygroscopic, moisture
HARKINS, William D., 1873-1951, American physical chemist
JURA, George, 1886-1951, American chemist
**Sources:** Menzel, D. 1960; NYT Obit., June 23, p. 15, 1951.
*See also* ISOTHERM

## HARMONIC LAW OR PERIODIC LAW

The movement of a small body or particle along a straight path and through a central point so that its acceleration toward the point is proportional to its direction from the point. This is a fundamental harmonic or parabolic relationship of natural science, which is fundamental to all periodically recurring phenomenon, represented by

$$y = a \sin nx$$

where y = distance on the scale at right angles to x
x = distance on the horizontal scale
a = amplitude on the y-scale
n = amplitude on the x-scale

**Keywords:** nature, parabolic, periodic
**Sources:** Mandel, S. 1972; Parker, S. P. 1983.

## HARMONIC PROGRESSION OR SEQUENCE, LAW OF

The terms a, b, c, etc. form a harmonic series if their reciprocals, 1/a, 1/b, 1/c, etc., form an arithmetical series.

**Keywords:** arithmetical, reciprocals, series
**Sources:** Freiberger, W. F. 1960; Gibson, G. 1981; Gillispie, C. C. 1991; Parker, S. P. 1983.

## HARTLEY LAW

Separations of components in any one series of doublet or triplet spectral lines, given frequencies or wave numbers are equal.

**Keywords:** components, frequencies, spectral, wave numbers
HARTLEY, Sir Walter Noel, 1846-1913, English chemist (spectroscopy)
**Source:** Bennett, H. 1939.

## HARTMANN NUMBER, Ha OR $N_{Ha}$ (1957)

A dimensionless group used in magnetohydrodynamics that relates magnetic force and viscous force on flow of conducting fluids:

$$Ha = \beta\, \sigma^{1/2}\, L/\mu^{1/2}$$

where $\beta$ = magnetic induction
$\sigma$ = electrical conductivity
L = characteristic length
$\mu$ = absolute viscosity

**Keywords:** electric, magnetic, viscosity
HARTMANN, R. A.,* twentieth century, German scientist
**Sources:** Bolz, R. E. and Tuve, G. L. 1970; Jerrard, H. G. and McNeill, D. B. 1992; Kakac, S. et al. 1987; *Land, N. S. 1972; Morris, C. 1992; Parker, S. P. 1992; Potter, J. H. 1967; Rohsenow, W. M. and Hartnett, J. P. 1973.

## HAÜY LAW (1784)

The relative development of the different faces of two crystals of the same material may be widely different, but the interfacial angles of corresponding faces of the two crystals of the same material are all equal and are characteristic of that substance.

**Keywords:** angles, crystal, faces, lattice
HAÜY, Abbe Rene Just, 1743-1822, French mineralogist
**Sources:** Clifford, A. F. 1964; Perry, J. H. 1950; Webster's Biographical Dictionary. 1959.
*See also* DECREMENT; RATIONAL INDICES

### HAVELOCK LAW

The relationship between the refractive index and wavelength of a material is represented by:

$$k = B\,\lambda\, n/(n-1)^2$$

where $k$ = material constant
$B$ = Kerr constant of material, approximately proportional to the absolute temperature
$\lambda$ = wavelength
$n$ = refractive index

The dispersion of the Cotton-Mouton effect is given by the Havelock law.

**Keywords:** Kerr, refractive index, wavelength
HAVELOCK, Sir Thomas Henry, 1877-1968, British mathematician
KERR, John, 1824-1907, Scottish physicist
**Sources:** Dictionary of National Biography; Parker, S. P. 1983; Pelletier, P. A. 1994.
*See also* COTTON-MOUTON; KERR

### HAZEN LAW

An estimate of permeability for relatively uniform sand and clay is represented by:

$$k = (D_{10})^2/100$$

where $k$ = permeability, such as m/s
$D_{10}$ = the 10% size, or effective size, such as mm, with typical values of k for coarse sand of $1 \times 10^{-2}$ and clay of $1 \times 10^{-10}$ m/s

**Keywords:** clay, permeability, sand
HAZEN, Allen, 1869-1930, American civil engineer
**Sources:** Blake, L. S. 1994; Who Was Who. 1976.

### HEART, LAW OF THE; OR STARLING LAW

For each contraction of the heart, there is a simple function relating the energy set free and the length of the fibers composing the muscular walls.

The distension of a complete ventricle by increasing the pressure of the liquid in it produces a greater amplitude of contraction and therefore a greater output of liquid.

**Keywords:** energy, physiology, ventricle
STARLING, Ernest Henry, 1866-1927, English physiologist
**Sources:** Friel, J. P. 1974; Landau, S. I. 1986; Morris, C. G. 1992; Thomas, C. L. 1989.
*See also* FRANK-STARLING; STARLING

### HEART-WEIGHT RULE OR HESSE RULE

The ratio of body weight to heart weight increases in races of animals found in cold regions compared with races from warm regions, attributed to the necessity of maintaining a greater temperature differential between body and environment.

**Keywords:** body, cold, heart, warm, weight
**Sources:** Lincoln, et al. 1982; Pelletier, P. A. 1994.

## HEAT CONDUCTION, LAW OF

The quantity of heat transferred by conduction is proportional to the cross-sectional area, temperature difference, and the thermal conductivity of the material, and inversely proportional to the thickness of the material:

$$q = k\,A\,\Delta t/\Delta x$$

where $q$ = heat transferred, Btu/hr
$A$ = cross-sectional area, $ft^2$
$\Delta t$ = temperature difference, °F
$\Delta x$ = length of path of heat transfer
$k$ = thermal conductivity, Btu-$ft^2$ ft hr °F

This relationship was introduced by J. Biot but is often credited to J. Fourier.

**Keywords:** area, conductivity, heat, length, temperature, thermal
**Sources:** Parker, S. P. 1983 (physics); Parker, S. P. 1989.
*See also* BIOT; FOURIER

## HEAT EFFECT OF AN ELECTRIC CURRENT, LAW OF—SEE JOULE

## HEAT EXCHANGE, LAW OF

The quantity of heat lost by a substance is equal to the quantity of heat gained by a cooler substance when two substances are in contact:

$$Q_{lost} = Q_{gained}$$

where $Q$ = quantity of heat

**Keywords:** gain, heat, loss, thermal
**Source:** Mandel, J. 1972.

## HEAT FLOW, LAW OF FLOW FOR STEADY STATE

Heat flow by conduction is proportional to the temperature difference for the particular material:

$$Q = k\,\Delta t\,A\,\theta$$

where $Q$ = total heat
$k$ = thermal conductivity of the material
$\Delta t$ = temperature difference
$A$ = area perpendicular to heat flow
$\theta$ = time

**Keywords:** area, conduction, heat transfer, temperature difference, time
**Sources:** Perry, J. H. 1950. 1973.
See HEAT CONDUCTION; HEAT EXCHANGE

## HEAT LIBERATION LAW; OR SINUSOIDAL HEAT LIBERATION LAW

When the heat is applied in a sinusoidal fashion by a heating element, the heat liberated from the surface of the heating section:

$$q = q_{max} \sin \frac{\pi(\chi + x/L)}{\chi + 1}$$

where $q$ = heat given off, kcal/m$^2$ hr
$\chi$ = constant
$x$ = distance from heated surface
$L$ = length of heated section

**Keywords:** heating, liberated, surface
**Source:** Kutateladze, S. S. and Borishanskii, V. M. 1966.

## HEAT OF REACTION, LAW OF (CHEMICAL SYSTEM)

In a chemical system, the sum of the heat produced in a reaction and external work performed is called the *heat of reaction*. This heat of reaction (or *heat of formation*), may be positive or negative, which represents the total energy of the chemical system. The total heat generated in a chemical reaction is entirely independent of the steps followed in passing from initial to final state of the system, and this principle, *the law of constant heat summation,* makes it possible to calculate the heat of formation for steps which are chemically impractical.

**Keywords:** chemistry, heat of formation, heat summation
**Sources:** James, A. M. 1976; Thewlis, J. 1961-1964.
*See also* HEAT SUMMATION

## HEAT RADIATION—SEE KIRCHHOFF; PLANCK; RAYLEIGH-JEANS; STEFAN-BOLTZMANN; WIEN

## HEAT SUMMATION, LAW OF

The difference in energy between the identical conditions of a system must be the same, independent of the path through which the system is transferred from one condition to the other.

**Keywords:** energy, heat, path, thermal
**Source:** James, A. M. 1976.
*See also* HESS

## HEAT TRANSFER, GENERAL LAWS OF

1. Conservation of mass, law of
2. First law of thermodynamics
3. Second law of thermodynamics
4. Newton second law of motion

**Source:** Kakac, S. et al. 1987.
*See also* CONSERVATION; NEWTON; THERMODYNAMICS

## HEAT TRANSFER, SPECIFIC MODES, LAWS OF

***Conduction*. The basic law is credited to J. Fourier based on J. Biot:**

$$dQ/dt = -k\,A\,d\theta/dx$$

where $dQ/dt$ = rate of heat flow
$A$ = area perpendicular to the direction of heat flow
$d\theta/dx$ = temperature flow in the direction of heat flow
$k$ = thermal conductivity, a property of the material

*Convection.* **Based on the use of heat transfer coefficients defined by I. Newton:**

$$Q = h\ A\ t\ \Delta\theta$$

where $Q$ = quantity of heat flow over an area, A, in time, t
$h$ = convection heat transfer coefficient
$A$ = area from which heat is transferred
$t$ = time involved
$\Delta\theta$ = temperature difference between the hot body and the fluid stream (certain conditions define use of log mean temperature difference)

For natural convection, the Grashof number, $N_{Gr}$, defines the relationship:

$$N_{Gr} = \beta\ \Delta\ t(N_{Ga})$$

where Ga = the Galileo number

For forced convection, the Nusselt number, $N_{Nu}$, defines the relationship:

$$N_{Nu} = h\ d/k$$

***Radiation.*** **Based on the Stefan-Boltzmann law:**

$$q = \sigma\ T^4$$

where $q$ = rate of heat transfer by radiation per unit area per unit time
$\sigma$ = Stefan-Boltzmann constant
$T$ = absolute temperature of the radiating surface

**Keywords:** conduction, convection, radiation
**Source:** Kakac, S. et al. 1987.
*See also* BIOT; FOURIER; GALILEO; GRASHOF; NUSSELT; STEFAN-BOLTZMANN

## HEAT TRANSFER NUMBER, $K_H$ OR $N_{KH}$

A dimensionless group used to represent heat transfer in a stream:

$$K_H = Q/\rho\ V^3L^2$$

where $Q$ = rate of heat flow
$\rho$ = mass density
$V$ = velocity
$L$ = characteristic length

**Keywords:** heat transfer, stream
**Sources:** Bolz, R. E. and Tuve, G. L. 1970; Land, N. S. 1972; Parker, S. P. 1992; Potter, J. H. 1967.

## HEAVISIDE LAYER, LAW OF; ALSO CALLED KENNELLY-HEAVISIDE LAYER

There is an ionized layer in the air, concentric with the Earth's surface, that might serve as a reflecting surface that would confine radiation between it and the earth.

**Keywords:** air, atmosphere, Earth, ionized, radiation
HEAVISIDE, Oliver, 1850-1925, English electrical engineer and physicist
KENNELLY, Arthur E., 1861-1939, American electrical engineer
**Sources:** Gillispie, C. C. 1981; Mandel, S. 1972; Parker, S. P. 1989; Pelletier, P. A. 1994.

## HEBB LAW (1949)

Learning is postulated to be dependent on cell assemblies in the brain that form the neural basis for concept formation. The cell assemblies build coordinated groups of cells in the cortex of the brain. Networks learn by experience.

**Keywords:** brain, cells, experience, neural
HEBB, Donald Olding, twentieth century (b. 1904), Canadian physiologist
**Source:** Stewart, I. 1998.
*See also* HOPFIELD

## HECKER LAW

The weight of the infant at each successive childbirth usually exceeds that of the preceding infant by about 150 to 200 grams.

**Keywords:** birth, child, medical, weight
HECKER, Karl von, 1827-1882, German obstetrician
**Sources:** Landau, S. I. 1986; Stedman, T. L. 1976.

## HEDSTROM NUMBER, He OR $N_{He}$ (1952)

A dimensionless group that represents the flow of Bingham type plastics:

$$He = \sigma_o \, \rho \, L^2/\mu_p^2$$

where $\sigma_o$ = stress at elastic yield
$\rho$ = mass density
L = characteristic length
$\mu_p$ = absolute viscosity in plastic state

The Hedstrom number is the product of the Reynolds number and the Bingham number. This number is sometimes called the Hedstrom 1 number, and the Bingham number is sometimes called the Hedstrom 2 number.

**Keywords:** elastic, flow, plastic, yield
**Sources:** Bolz, R. E. and Tuve, G. L. 1970; Land, N. S. 1972; Parker, S. P. 1992; Potter, J. H. 1967.
*See also* BINGHAM; REYNOLDS

## HEIDENHAIN LAW

The secretion by a gland always results in a change in the structure of the gland.

**Keywords:** gland, medical, physiological, secretion
HEIDENHAIN, Rudolf Peter, 1834-1897, German physiologist
**Sources:** Friel, J. P. 1974; Landau, S. I. 1984; Stedman, T. L. 1976.

## HEISENBERG UNCERTAINTY PRINCIPLE OR LAW (1927) OR INDETERMINANCY PRINCIPLE

One cannot accurately determine the position and velocity of a particle simultaneously, because in making measurements you disturb the particle. In locating the position of a moving particle, such as an electron, the more precisely one attempts to locate the position, the less knowledge one has of its velocity.

$$\Delta x \, \Delta p \approx h/4\pi$$

where x = uncertainty in position of the particle
p = uncertainty in momentum
h = Planck constant, or constant of proportionality

This law describes the world as not completely deterministic.

**Keywords:** momentum, particle, position, uncertainty, velocity
HEISENBERG, Werner Karl, 1901-1976, German physicist; Nobel prize, 1932, physics
**Sources:** Barrow, J. D. 1991; Daintith, J. M. et al. 1994; Lederman, L. 1993; Peierls, R. E. 1956; NYPL Desk Ref.
*See also* UNCERTAINTY PRINCIPLE

### HELLIN LAW (1895) OR HELLIN-ZELENY LAW

There is a statistical expectation with respect to natural multiple births that 1 in about 89 natural pregnancies results in the birth of twins, 1 in 892 pregnancies results in birth of triplets, and 1 in 893 pregnancies results in quadruplets, 1:N, 1:$N^2$, and 1:$N^3$.

**Keywords:** births, children, medical, multiple, pregnancies
HELLIN, Dyonizy, 1867-1935, Polish pathologist
ZELENY, Charles, 1878-1939, American zoologist
**Sources:** Friel, J. P. 1974; Gennaro, A. R. 1979; Good, C. V. 1973; National Cyclopaedia, v.42, 1958; Stedman, T. L. 1976; Thomas, C. L. 1989.

### HELMHOLTZ—SEE SMITH-HELMHOLTZ

### HELMHOLTZ ACCOMMODATION THEORY OF THE EYE (c. 1856)

The lens of the eye assumes a more spherical form by its own elasticity when the suspensory ligaments are tense. Previous work on the subject was done by T. Young (in 1793), and R. Descartes (in 1637).

**Keywords:** elasticity, eye, form, lens, ligaments
HELMHOLTZ, Hermann Ludwig Ferdinand von, 1821-1894, German physicist and physiologist
DESCARTES, Rene, 1596-1650, French scientist and philosopher
YOUNG, Thomas, 1773-1829, English physician
**Source:** Schmidt, J. 1959.

### HELMHOLTZ, CONSERVATION OF ENERGY—SEE CONSERVATION OF ENERGY; THERMODYNAMICS, FIRST LAW

### HELMHOLTZ FREE ENERGY FUNCTION, F

A thermodynamic function that is a measure of the ability of a system to do useful work in isothermal conditions:

$$F = U - TS$$

where U = internal energy
T = temperature
S = entropy

**Keywords:** energy, entropy, isothermal, work
HELMHOLTZ, Hermann Ludwig Ferdinand von, 1821-1894, German physicist and physiologist

**Sources:** Daintith, J. 1981; Parker, S. P. 1992.
*See also* GIBBS

### HELMHOLTZ-LAGRANGE, LAW OF

In electron optics, a mathematical expression that relates the size of apertures of the imaginary pencils, which are assumed to be small, and the magnification of the objects.

**Keywords:** electron, optics, physics
HELMHOLTZ, Hermann Ludwig Ferdinand von, 1821-1894, German physicist and physiologist
LAGRANGE, Joseph Louis, Comte de, 1736-1813, Italian French astronomer and mathematician
**Source:** Menzel, D. H. 1960.
*See also* LAGRANGE

### HELMHOLTZ-LAMB HYPOTHESIS OR LAW

Two dissimilar phases in contact become oppositely charged in a manner similar to the static charge gained by frictional electricity.

**Keywords:** charges, electricity, friction, static
HELMHOLTZ, Hermann Ludwig Ferdinand von, 1821-1894, German physicist and physiologist
LAMB, Horace Sir, 1849-1934, English applied mathematician and fluid dynamicist
**Sources:** NUC; Zeleny, P. O. 1990.

### HELMHOLTZ LAW (ELECTRIC CIRCUIT)

An equation that considers the inductance of an electric circuit as well as the resistance (as defined by Ohm law) as:

$$L\,dI/dt + R\,I = E$$

where I = current, amperes
R = resistance, ohms
L = inductance
E = electromotive force, or V for voltage

**Keywords:** circuit, electric, inductance, resistance, voltage
HELMHOLTZ, Hermann Ludwig Ferdinand von, 1821-1894, German physicist and physiologist
**Source:** Gillispie, C. C. 1981.
*See also* KIRCHHOFF; OHM LAW

### HELMHOLTZ, LAW OF; OR LAW OF THOMSON

In an electric cell, the heat of reaction is a direct measure of the electromotive force, in which chemical energy is converted to electrical energy.

**Keywords:** cell, chemical, electrical energy, electromotive, force
HELMHOLTZ, Hermann Ludwig Ferdinand von, 1821-1894, German physicist and physiologist
**Sources:** Considine, D. M. 1976; Honig, J. M. 1953.
*See also* THOMSON (THOMPSON)

### HELMHOLTZ NERVE IMPULSE LAW (1850)

The measurement of velocity of a nerve impulse was extended by means of a pendulum, extending thinking from non-living to living matter. In 1850, H. Helmholtz measured the speed of human nerve response.

**Keywords:** impulse, nerve, velocity

HELMHOLTZ, Hermann Ludwig Ferdinand von, 1821-1894, German physicist and physiologist

**Source:** Zeleny, R. O. 1990.

### HELMHOLTZ THEORY OF COLOR VISION—SEE YOUNG-HELMHOLTZ

### HELMHOLTZ-THOMSON RULE OR LAW

The chemical energy of a process is equal to the heat effects of the process when the process takes place without performing external work. In a voltaic cell's chemical reaction, the electrical energy produced is equal to the heat of reaction.

**Keywords:** chemical, electric, energy, heat

HELMHOLTZ, Hermann Ludwig Ferdinand von, 1821-1894, German physicist and physiologist

THOMSON, William (Lord Kelvin), 1824-1907, Scottish mathematician and physicist [Michel, W. C. 1956 has THOMPSON, Sir Benjamin (Count Rumford), 1753-1814, born in America, then moved to Great Britain, made a count in Austria*]

**Sources:** Gray, H. J. and Isaacs, A. 1975; *Michels, W. C. 1965.

### HELMHOLTZ VORTEX LAW

The vortex laws of motion state that a (vortex) filament is always composed of the same fluid particles, and its strength is constant, this being the flux of the vorticity through the cross-section. Utilizing the equations of motion, the vortex law relates the vorticity vector, assuming that the viscosity is constant, for incompressible and non-viscous flow:

$$D/Dt\ (\omega/\rho) = (\omega/\Delta p)\ V + \mu/\rho^2\ \nabla^2\ \omega$$

where $\omega$ = vorticity, the curl of vector function of space x and time t
$V$ = velocity
$p$ = pressure
$\rho$ = mass density
$\mu$ = coefficient of viscosity

For a barotropic fluid, that is where the $\rho$ is uniquely determined by p:

$$\omega/\rho \propto dL$$

**Keywords:** flow, incompressible, mechanics, non-viscous, vortex

HELMHOLTZ, Hermann Ludwig Ferdinand von, 1821-1894, German physicist and physiologist

**Sources:** Flugge, W. 1962; Lerner, R. G. and Trigg, G. L. 1981.

*See also* BIOT-SAVART; VORTEX MOTION

### HELM-OSTWALD INTENSITY LAW

In physico-chemical systems, the topographic parameter v (volume) is the capacity factor of energy, and associated with v is a conjugate parameter $p_i$ (pressure), which is the intensity factor of the energy in question, i.e., that factor which determines the direction of any changes in the capacity factor, independent of the nature of v and p:

$$p_i \Leftrightarrow p_e$$

where pi = conjugate pressure
$p_e$ = external

The variables dv and dt are conjugate parameters, in which dv can be greater than or equal to zero, and dt is less than or equal to zero.

**Keywords:** capacity, chemical, intensity, physics, volume
HELM, Jacob A., 1761-1831, German? medical doctor
OSTWALD, Friedrich Wilhelm, 1853-1932, Russian German chemist
**Sources:** Lotka, A. J. 1956; Pelletier, P. A. 1994; Talbott, J. H. 1970.
*See also* FECHNER; LE CHATELIER

## HEMIHEDRAL CORRELATION, LAW OF—SEE PASTEUR

## HENRY LAW (1830,* 1835) OR LAW OF ELECTRIC INDUCTION

Joseph Henry developed the electric relay that became the basis of the telegraph whereby messages could be sent through long distances of wire. The wire carried a current of sufficient strength to just lift a small iron key. The key, when lifted, closed a second circuit with a current from a battery, etc., and thereby a current could be sent over long distances as a result of the battery or self-induction. Henry discovered electric induction in 1831, the same year that M. Faraday is given credit.

**Keywords:** electricity, self-induction
HENRY, Joseph, 1797-1878, American physicist, Director, Smithsonian Inst., 1846-1878
**Sources:** *David, I. et al. 1996; Thewlis, J. 1961-1964.
*See also* FARADAY

## HENRY LAW OF SOLUBILITY (1801,1803)

At constant temperature, the solubility of a gas in a liquid with which it does not react chemically is proportional to the pressure. Another statement of the law is that the solubility of a gas in a liquid is directly proportional to the vapor pressure of the dilute component of the gas exerted above the liquid at equilibrium. The Henry law was later shown to be a special case of the more general H. Nernst law.

**Keywords:** gas, liquid, pressure, solubility, vapor
HENRY, William, 1774-1836, English chemist and physician
**Sources:** Gray, H. J. and Isaacs, A. 1975; Hodgman, C. D. 1952; Isaacs, A. 1996; Menzel, D. H. 1960; Perry, J. H. 1950; Stedman, T. L. 1976; Tver, D. F. 1981.
*See also* ABSORPTION OF GASES; BABO; NERNST; PARTITION; RAOULT

## HEREDITY—SEE MENDEL

## HERING LAW; LAW OF OCULAR MOVEMENTS; HERING LAW OF EQUAL INNERVATION (188x)

The two eyes are controlled by similar and equal nerve impulses such that one eye is never moved independently of the other, except for extreme lateral movements. According to this law, the two eyes move in a conjugate manner, because they receive identical signals from the brain or equal innervation. The clearness of any conception of the eye depends on the proportion existing between its intensity and the sum total of intensities of all the simultaneous conceptions.

**Keywords:** eye, medical, movement, physiological, seeing
HERING, Carl Ewald Konstantin, 1834-1918, German physiologist
**Sources:** Critchley, M. 1978; Friel, J. P. 1974; Science v. 248, p. 1118, 1 June 1990.

### HERSCHEL LAW (1800)

The infrared part (not visible to the eye) of the spectrum, so identified and named by W. Herschel, is the part of the spectrum responsible for the heat that interferes with the solar astronomers' vision. He discovered the planet Uranus in 1781.

**Keywords:** infrared, heat, spectrum, vision
HERSCHEL, Sir William, 1738-1822, English astronomer (born in Germany)
**Sources:** Carter, E. F. 1976; NUC; Today's Chemist at Work, p. 31, October, 1995; NYPL Desk Reference.

### HERSCHEL-BULKLEY LAW (1926)

The flow of a material under loading, a special case of the Bingham solid for n = 1, is represented by:

$$f(\tau) = 0 \qquad \text{for } \tau \leq \tau_f$$

$$f(\tau) = (1/k)(\tau - \tau_f)^n \qquad \text{for } \tau > \tau_f$$

which is a power law flow curve.

**Keywords:** Bingham, flow, loading, mechanics, rheology, solid
HERSCHEL, Winslow H., published in English and German, 1920s and 1930s
BULKLEY, R., published in English and German, 1920s and 1930s
**Sources:** Eirich, F. R. 1969 (v. 3); Scott Blair, G. W. 1949.
*See also* VISCOELASTIC

### HERSEY NUMBER, Her OR $N_{Her}$

A dimensionless group representing lubrication of bearings:

$$Her = F/\mu\, VL$$

where F = load on the bearing
μ = absolute viscosity
V = bearing surface speed
L = bearing length

Also known as the Truncation number.

**Keywords:** bearings, load, lubrication
HERSEY, Mayo D.,1886-1978, American? engineer
**Sources:** Bolz, R. E. and Tuve, G. L. 1970; Land, N. S. 1972; Nat. Ency. Am. Biog. Supp.; Parker1992; Potter, J. H. 1967.
*See also* OCVIRK NUMBER; SOMMERFELD NUMBER

### HERTZ CONTACT LAW; LAW OF CONTACT FOR IMPACT

A theory of local indentations in which two impacting bodies have contact, equivalent to the electrostatics problem (Fig. H.2). Stresses and deformations are described at the contact point

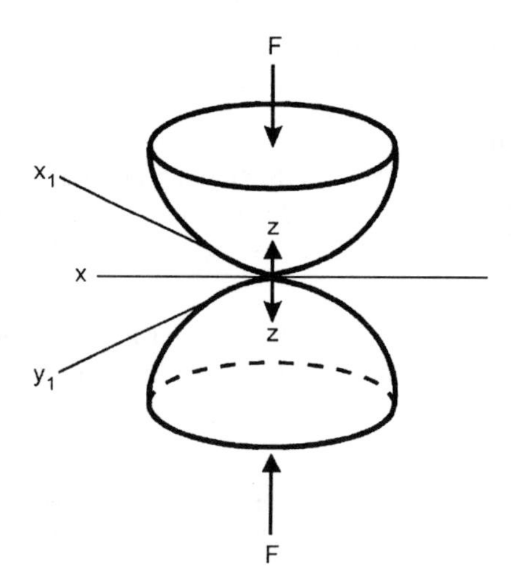


**Figure H.2** Hertz contact.

and surrounding area as a function of the geometrical and elastic properties of the bodies. The radius of contact between a sphere and a surface can be represented by an equation relating the ratio of the normal force times the radius of sphere divided by the Young modulus of the material, as follows:

$$r = 1.11\ (N\ a/E)^{1/3}$$

where r = radius of contact
a = radius of sphere
N = normal force exerted on the sphere
E = Young modulus of material of the sphere

The Hertz law is applicable in the elastic range.

**Keywords:** contact, radius, sphere, Young's modulus
HERTZ, Heinrich Rudolf, 1857-1894, German physicist and electrical engineer; Nobel prize, 1925, physics
**Sources:** Ballentyne, D. W. G. and Lovett, D. R. 1972; Goldsmith, W. 1960.
*See also* MEYER (MATERIALS)

## HERTZIAN WAVES (RADIO WAVES) LAW (1888)

Electromagnetic waves generated in space travel at the speed of light but have a greater wavelength than light. Hertz discovered electromagnetic waves and produced radio waves in the laboratory.

**Keywords:** electromagnetic waves, light, speed
HERTZ, Heinrich Rudolph, 1857-1984, German physicist and electrical engineer; Nobel prize, 1925, physics
**Source:** Daintith, J. 1981.

## HESS LAW OF HEAT SUMMATION (1840); LAW OF CONSTANCY OF HEAT

The heat evolved or absorbed in a given reaction must always be constant and independent of the particular manner in which the reaction takes place, and it is related to the first law of thermodynamics.

**Keywords:** heat, reaction, physical chemistry

HESS, Germain Henri, 1802-1850, Swiss Russian chemist

**Sources:** Asimov, I. 1972; Bothamley, J. 1993; Hodgman, C. D. 1952; Isaacs, A. 1996; James, A. M. 1976; Millar, D. et al. 1996; Morris, C. G. 1992; Perry, J. H. 1950; Thewlis, J. 1961-1964; Van Nostrand. 1947.

*See also* BLAGDEN; HEAT SUMMATION; THERMODYNAMICS; THERMONEUTRALITY

## HESSE RULE—SEE HEART-WEIGHT

## HEYMANS LAW

The threshold value of a visual stimulus to sight in relation to a simultaneously occurring inhibitory stimulus is:

$$T_a = T_o + K_a$$

where $T_a$ = threshold when it is raised by the occurrence of a second identified stimulus
$T_o$ = threshold of a given stimulus
$K_a$ = coefficient of inhibition, depending on conditions and subject

**Keywords:** sight, stimulus, strength, visual

HEYMANS, Corneille Jean Francois, 1892-1968, French Belgian physiologist; Nobel prize, 1938, physiology/medicine

**Sources:** Dorland, W. A. 1980; Friel, J. P. 1974; Landau, S. I. 1986; Muir, H. 1994; Dorland, B. B. 1989.

## HICK-HYMAN LAW (1950s)

The amount of information processed by a person can be represented by equation:

$$P_t = a + b\ H_t$$

where $P_t$ = processing time (or reaction time)
$H_t$ = amount of information processed
a and b = constants, depending on task

The reaction time increases as a function of information transmitted in making a response.

**Keywords:** data, information, processing, time

HICK, W. E., twentieth century, American psychologist

HYMAN, Ray, twentieth century (b.1928), American psychologist

**Sources:** Bothamley, J. 1993; Reber, A. S. 1985.

*See also* SHANNON

## HILDEBRANDT RULE (1915)

The entropy of vaporization, that is, the ratio of the heat of vaporization to the temperature at which it occurs, is a constant for many substances if it is determined at the same molal concentration of vapor for each substance.

**Keywords:** concentration, entropy, evaporation, vapor, vaporization

HILDEBRANDT, Georg Friedrich, 1764-1816, German chemist

**Sources:** Belzer, J. et al. 1975; Muir, H. 1994; Thewlis, J. 1961-1964.

## HILL POSTULATE OF A MODEL FOR RUNNING; HILL-KELLER THEORY (1920s)

Based on the Newton law of motion, a runner's acceleration is equal to the force exerted per unit mass as that time, minus the resistance per unit mass at a given speed, expressed as:

$$dv/dt = f(t) - R(v)$$

where dv/dt = acceleration
f(t) = force exerted per unit mass at that time
R(v) = resistance per unit mass at given speed

Hill postulated that the effective resistance experienced by a runner increases linearly with speed, so the resistance is a constant $R_0$ multiplied by the runner's speed. Wind resistance does not increase linearly with speed but constitutes only a small amount of the total resistance—about 3 percent. Combining the Hill postulate and Keller's four empirical constants, F, $R_0$, $E_0$, and $\sigma$-maximum propulsive force, resistance, the energy store, and the energy-replacement rate) gives the Hill-Keller theory of running.

**Keywords:** acceleration, force, motion, resistance, running, speed
HILL, Archibald, 1886-1977, British biologist; Nobel prize, 1922, physiology/medicine
KELLER, Joseph Bishop, twentieth century (b. 1923), American mathematician
**Source:** American Scientist, 82(6):546-550. Nov.-Dec. 1994.

## HILT LAW OR RULE (1873)

The volatile matter in coal decreases proportionately with the depth of the seam from the surface, which can be stated as *the coal at a deeper depth is of higher rank than that above.*

**Keywords:** coal, depth, mining, ore
HILT, C., nineteenth century, Belgian geologist
**Sources:** Francis, W. 1961; Raistrick, A. and Marshall, C. E. 1939. 1948.

## HILTON LAW

A nerve trunk that supplies the muscles of any given joint also supplies the muscles that move the joint and the skin over such muscles.

**Keywords:** joint, medical, muscles, nerve
HILTON, John, 1804-1878, English surgeon
**Sources:** Friel, J. P. 1974; Landau, S. I. 1986; Stedman, T. L. 1976; Thomas, C. L. 1989.

## HOBBES—SEE EFFECT

## HODGKIN LAW (1947); ALSO CALLED HODGKIN-HUXLEY LAW

The electrical potential of a nerve fiber during conduction of an impulse exceeds the potential of a fiber at rest. The activity of a nerve fiber depends on the fact that a large concentration of potassium ions is maintained inside the film, and a large concentration of sodium ions is in the surrounding medium.

**Keywords:** electrical potential, nerve fiber, impulse, potassium, sodium
HODGKIN, Sir Alan Lloyd, twentieth century (b. 1914), British physiologist; Nobel prize, 1963, physiology/medicine
**Source:** Encyclopaedia Britannica. 1984.

## HODGSON NUMBER, H OR $N_H$

A dimensionless group representing pulsating gas flow that relates the time constant of a system and the period of pulsation:

$$H = V\ f\ \Delta p/Q\ p$$

where V = volume of the system
f = pulsation frequency
$\Delta p$ = pressure drop
Q = volume flow rate
p = average static pressure

**Keywords:** flow, gas, pulsation
**Sources:** Bolz, R. E. and Tuve, G. L. 1970; Land, N. S. 1972; Parker, S. P. 1992; Potter, J. H. 1967; Talbott, J. H. 1970.

## HOFACKER-SADLER LAW OR HOFACKER AND SADLER LAW

When the mother is younger than the father, the ratio of male to female births is 113 to 100; for parents of equal age, the ratio is 93.5 males to 100 females; when the mother is older than the father, the ratio is 88.2 males to 100 females.

**Keywords:** birth, children, female, male, medicine
HOFACKER, Johann Daniel, 1788-1828, German obstetrician
SADLER, Michael Thomas, 1834-1923, English obstetrician
**Sources:** Friel, J. P. 1974; Gray, P. 1967; Landau, S. I. 1986; Stedman, T. L. 1976.
*See also* HELLIN

## HOFF LAW—SEE VAN'T HOFF

## HOMEOPATHY, LAW OF; OR FIRST PRINCIPLE OF HOMEOPATHY (1796)

Relates to the treatment of a disease with small amounts of drugs which, if used in high dosage, cause symptoms similar to those of the disease.

**Keywords:** disease, dosage, drugs, quantity, symptoms
HAHNEMANN, Christian Friedrich Samuel, 1755-1843, German physician
**Source:** Bynum, W. F., et al. 1981.
*See also* HELLIN; INFINITESIMALS; SIMILARS

## HOMOCHRONOUS NUMBER, Ho OR $N_{Ho}$

A dimensionless group that relates the duration of process to the time for a liquid to move through a distance, L, often used to select time scales for processing or mixing:

$$Ho = Vt/L$$

where V = fluid velocity
t = time for the liquid to move distance
L = characteristic distance

**Keywords:** duration, process, time scale
**Sources:** Bolz, R. E. and Tuve, G. L. 1979; Parker, S. P. 1992; Potter, J. H. 1967.

## HOMOGENEOUS CIRCUIT, LAW OF

An electric current cannot be sustained in a circuit of a single homogeneous metal by the application of heat only, even with varying cross-section.

**Keywords:** current, electricity, heat, metal
**Source:** Considine, D. M. 1972.
*See also* INTERMEDIATE CIRCUIT; SUCCESSIVE OR INTERMEDIATE TEMPERATURE

## HOMONYM LAW OR LAW OF HOMONYMY

A principle in taxonomy in which two different taxa cannot have the same name. A junior homonym of another name must be rejected and replaced.

**Keywords:** name, taxonomy
**Sources:** Bates, R. L. and Jackson, J. A. 1980; Gary, M. et al. 1972; Gray, P. 1967.
*See also* PRIORITY, LAW OF

## HOOKE LAW (1676); YOUNG MODULUS

Within the elastic limit of a material, the stress is proportional to the strain or change in length to which the material is subjected by the applied load (Fig. H.3). The ratio of the applied force to the deformation, elongation, or compression per unit length is a constant, e:

$$E = e = \text{stress/strain} = F/A/(\Delta L/L) = Y = \text{Young modulus}$$

where $F$ = applied force or load
$A$ = cross-sectional area perpendicular to the deformation

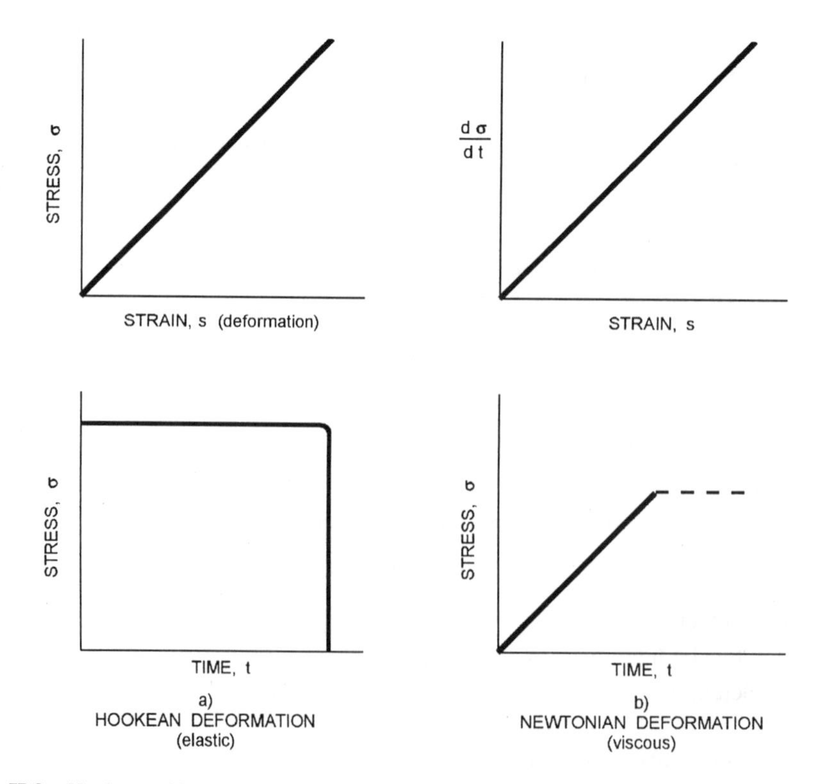


**Figure H.3** Hooke vs. Newton.

$\Delta L$ = the change of length or deformation
L = length

The Hooke law was observed independently by E. Mariotte in 1680.

**Keywords:** deformation, elastic, strain, stress, Young's modulus
HOOKE, Robert, 1635-1703, English physicist
MARIOTTE, Edme, 1620-1684, French physicist
YOUNG, Thomas, 1773-1829, English physicist and physician
**Sources:** Bynum, W. F., et al. 1981; Mandel, S. 1972; Perry, J. H. 1950; Rigden, J. S. 1996; Scott Blair, G. W. 1949; Thewlis, J. 1961-1964.
*See also* ELASTIC DISTORTION; ELASTICITY

### HOOKE NUMBER, Hk OR $N_{Hk}$

A dimensionless group used in representing the elasticity of a flowing medium, as in a compressible flow that relates the inertia force and the elastic force:

$$Hk = V^2/a^2$$

where V = velocity of flowing medium
a = speed of sound

The Hooke number equals the Mach number squared and also equals the Cauchy number.

**Keywords:** elasticity, fluid flow, inertia force
HOOKE, Robert, 1735-1703, English physicist
**Sources:** Bolz, R. E. and Tuve, G. L. 1970; Land, N. S. 1972; Parker, S. P. 1992; Potter, J. H. 1967.
*See also* CAUCHY NUMBER; MACH NUMBER

### HOORWEG LAW

In the provocation or stimulation of neuromuscular response, there is a duration of electric discharge above which time does not count, and a duration below which the time counts.

**Keywords:** electric discharge, neuromuscular, physiology, response, stimulus
HOORWEG, Jan Leendert, 1841-1919, Dutch biophysicist
**Sources:** Friel, J. P. 1974; Landau, S. I. 1986; NUC.

### HOPFIELD MODEL (1982)

A representation of an artificial neural system, such as by electrical components (resistors, capacitors, etc.), in which there is a feedback network of signals to content addressable memories. This concept has permitted building of models for associative recall to assist in the model learning from disturbances in the neural system. The Hebb law describes unsupervised learning where training occurs only if a processing element's output is active.

**Keywords:** artificial, feedback, learning, neural
HOPFIELD, John Joseph, twentieth century (b. 1933), American physicist
HEBBS, D. O., 1904-1985, American biophysicist
**Sources:** Am. Men of Science; Corsini, R. J. 1966; Meyers, R. A. 1992; Pelletier, P. A. 1994.
*See also* HEBB

### HOPKINS LAW; BIOCLIMATIC LAW OR RULE

A biotic event in temperature in North America will generally lag by 4 days for each degree of latitude, for 5 degrees of longitude, and with 500 feet altitude, northward, eastward, and upward, in spring and early summer. The reverse is true in late summer and fall and in the Southern Hemisphere.

**Keywords:** atmospheric, biotic, climate, ecology, meteorology, weather
HOPKINS, Andrew Delmas, 1857-1948, American entomologist
**Sources:** Gray, P. 1967; Landau, S. I. 1986; National Cyclopaedia, v. 41, 1956.
*See also* BIOTIC

### HOPKINSON LAW

A specific law of superposition that covers the dispersion effect of electrons dealing with a rather complex network of effective condensers and resistances in a nonhomogeneous dielectric.

**Keywords:** condensers, dielectric, electrons, resistances
HOPKINSON, John, 1849-1898, English electrical engineer (he patented 3-wire system of distribution in 1882)
**Source:** Thewlis, J. 1961-1964.

### HORNER LAW

Color blindness is transmitted from males to males through normal females.

**Keywords:** color-blind, eyes, physiology, sight
HORNER, Johann Friedrich, 1831-1886, Swiss ophthalmologist
**Sources:** Friel, J. P. 1974; Landau, S. I. 1986.

### HORNEY LAW—SEE DETACHMENT

### HOTELLING LAW (1929) (ECONOMICS)

Competitors in business differentiate their goods and services as little as possible to maximize demand from the public.

**Keywords:** business, competitors, differentiate, goods, services
HOTELLING, Harold, 1895-1973, American economist
**Source:** Bothamley, J. 1993.

### HOUGHTON LAW OF FATIGUE (PHYSIOLOGICAL)

When a muscle, or a group of muscles, is kept in constant activity until fatigued, the total work done multiplied by the rate of work is constant.

**Keywords:** fatigue, muscle, physiology, work
HOUGHTON, E. Mark, 1867-1937, American physician
**Sources:** Friel, J. P. 1974; Landau, S. I. 1986.

### HUBBLE LAW (1929)

Relates to the expansion of the universe in that galaxies are receding from each other at a rate nearly proportional to their distance from our galaxy. The motion of a galaxy is detected from its effect on the light it emits, with each wavelength shifting to the red end of the spectrum by an amount that increases with velocity, expressed in the galaxy's "red shift."

**Keywords:** distance, galaxy, light, receding, velocity, wavelength
HUBBLE, Edwin Powell, 1889-1953, American astronomer
**Sources:** Lederman, L. 1993; Lerner, R. G. and Trigg, G. 1991; Technology Review, p. 70, March, 1981.
*See also* DOPPLER

## HUBBLE-HUMANSON (RED SHIFT)—SEE HUBBLE

## HÜCKEL—SEE DEBYE-HÜCKEL LIMITING LAW

## HUES—SEE COLOR MIXTURE

## HUGONIOT—SEE RANKINE-HUGONIOT

## HUMAN MORTALITY—SEE MAKEHAM

## HUMBOLDT LAW (1817)

In nature, the upper tree line occurs at lower and lower elevations as one moves away from the equator. The tree line reaches sea level above the Arctic Circle.

**Keywords:** elevations, equator, sea level, tree line
HUMBOLDT, Friedrich William Heinrich Alexander von, 1769-1859, German natural scientist
**Sources:** Bothamley, J. 1993; Parker, S. P. 1989.

## HUME LAW

Hume asserted that facts alone cannot tell us what action should be taken. Any recommendation for what we or society ought to do embodies some ethical principles as well as factual judgments. As an example, to recommend policies if and only if their economic benefits exceed their costs would imply the ethical principle that increasing net economic benefits is the only worthy goal for society.

**Keywords:** action, ethical, facts, judgments
HUME, David, 1711-1776, Scottish philosopher
**Source:** Resources for the Future, n. 120, p. 5, Summer 1995.

## HUNT—SEE DUANE-HUNT

## HUYGENS LAW (1678)

In the impact of two bodies, the sum of the product of the mass with the square of the respective velocity of each body is the same before and after impact:

$$m_1v_1^2 + m_2v_2^2 = m_1v_3^2 + m_2v_4^2$$

where m = the mass of bodies 1 and 2
v = the velocity at times 1, 2, 3, and 4

**Keywords:** impact, mass, mechanics, velocity
HUYGENS, Christiaan, 1629-1695, Dutch physicist and astronomer
**Sources:** Asimov, I. 1972; Parker, S. P. 1989.
*See also* REFLECTION

## HUYGENS PRINCIPLE OF WAVE PROPAGATION (1678); HUYGENS-FRESNEL PRINCIPLE

Every point on a wavefront, such as a radiation field from an aperture, may be considered as a source of secondary disturbance that gives rise to spherical wavelets, with due regard to their wavelength differences, when they reach the point in question. There is a small amount of light beyond the point in question that is called *diffraction*.

**Keywords:** diffraction, light, physics, radiation, wavelets
HUYGENS, Christiaan, 1629-1695, Dutch physicist and astronomer
FRESNEL, Augustin Jean, 1788-1827, French physicist
**Sources:** Asimov, I. 1972; Gray, D. E. 1972; Parker, S. P. 1994; Thewlis, J. 1961-1964.

## HYDRATION NUMBER, Hy OR $N_h$

The average number of water molecules that move with an ion in solution is called the *hydration number* of the ion. Some values of the hydration number are estimated to be as shown below.

| ion | $Li^+$ | $Na^+$ | $K^+$ | $F^-$ | $Cl^-$ | $I^-$ |
|---|---|---|---|---|---|---|
| $N_h$ | 4.5 | 4 | 3 | 4 | 2 | 1 |

**Keywords:** ion, molecules, solution, water
**Source:** Levine, I. N. 1995.

## HYDRAULICS—SEE ARCHIMEDES

## HYDRODYNAMIC PARAMETER, Ho OR $N_{Ho}$

A dimensionless group that describes the rate of change of fluid flow with time:

$$Ho = Vt/L$$

where V = velocity
t = time
L = characteristic length

**Keywords:** fluid flow, rate of change
**Sources:** Gillispie, C. C. 1991; Parker, S. P. 1983.

## HYDRODYNAMIC PRESSURE—SEE BERNOULLI LAW OF

## HYDROSTATIC LAWS

In a mass of fluid at rest the pressure is equal at all points in any horizontal plane, and the change in pressure from one horizontal plane to another is equal to the weight of a vertical column of the fluid over the particular area.

**Keywords:** fluid, mass, mechanics, pressure
**Source:** Parker, S. P. 1983 (physics).
*See also* ARCHIMEDES; FLOTATION; HYDROSTATICS; PASCAL

## HYDROSTATICS, FIRST LAW OF; ARCHIMEDES LAW

When a body floats in a liquid, it appears to lose weight, and the apparent loss of weight is equal to the weight of liquid displaced.

**Keywords:** float, liquid, physics, weight
**Sources:** Isaacs, A. 1996; Mandel, S. 1972; Speck, G. E. 1965.
*See also* ARCHIMEDES; FLOTATION; HYDROSTATIC

## HYMAN LAW (1959)

Adults, particularly parents, are more likely to influence a decision by a child with respect to decisions on future aspirations—political, economic, educational—than the child.

**Keywords:** adults, aspirations, child, decisions, parents
HYMAN, Harold Melvin, twentieth century (b. 1924), American psychologist
**Source:** Bothamley, J. 1993; NUC.

## HYPERCHARGE—SEE I SPIN

# I

## I SPIN AND HYPERCHARGE—SEE ISOSPIN LAW

## ICOSACANTHIC LAW—SEE MÜLLERIAN LAW

## IDEAL DISTRIBUTION LAW—SEE DISTRIBUTION

## IDEAL GAS LAWS; IDEAL GAS, LAWS OF

The ideal gas law relates the conditions of pressure, volume, absolute temperature, and moles involved:

$$p\,V = n\,T$$

and for the gas constant:

$$p\,V = n\,R\,T$$

where p = pressure
V = volume
T = absolute temperature
n = the moles involved
R = universal gas constant

Another statement of the gas law is:

$$p\,v = (p_o v_o/273)\,T = R\,T$$

where p = atmospheric pressure
v = volume in liters of gas
T = temperature in degrees absolute, Celsius
R = gas constant, 0.08204

**Keywords:** gas, pressure, temperature, thermodynamics, volume
**Sources:** Considine, D. M. 1976. 1995; Parker, S. P. 1989. 1992. 1994.
*See also* GAS LAWS

## IDEAL MIXTURES, LAW OF

The property of a mixture of gases and some solids and some liquids is an additive function of the same property of the components, an assumption which in many cases is far from correct:

$$W\,A = W_1\,A_1 + W_2\,A_2 + W_3\,A_3 + \ldots$$

**Keywords:** gases, liquids, solids, materials, mixtures, properties
**Source:** Thewlis, J. 1961-1964.
*See also* ARRHENIUS; PERFECT GAS

## IDEMPOTENT—SEE ALGEBRA OF PROPOSITIONS; BOOLEAN

## IDENTICAL DIRECTION, LAW OF

When an object is fixed visually, the vision of the two eyes intersect at the fixation point, while the vision continues beyond the fixation point. The law was formulated by E. Hering.

**Keywords:** eyes, fixation point, vision
HERING, Ewald, 1834-1918, German physiologist
**Source:** Bothamley, J. 1993.

## IDENTITY, LAW OF—SEE ALGEBRA OF PROPOSITIONS; THOUGHT

## IDENTITY LAW FOR ADDITION—SEE ADDITIVE IDENTITY

## IDENTITY LAW FOR MULTIPLICATION—SEE MULTIPLICATIVE IDENTITY

## ILLUMINANCE, LAW OF

The incident luminous flux per unit area varies according to the inverse square law:

$$E = I/d^2$$

where E = illuminance
I = intensity of illuminous flux (illumination)
d = distance of point of illumination from plane surface

**Keywords:** flux, illumination, inverse square law
**Source:** Thewlis, J. 1961-1964.
*See also* ILLUMINATION; LAMBERT; PURKINJE

## ILLUMINATION—SEE LAMBERT (COSINE); PURKINJE

## ILLUMINATION (COSINE-CUBED), LAW OF

If a point source illuminates a plane, the illumination at any point P is proportional to the cube of the cosine of the angle between the ray from the source to P and the normal.

**Keyword:** angle, cosine, cubed, illumination, normal, source
**Source:** Thewlis, J. 1961-1964.
*See also* LAMBERT; LIGHT EMISSION

## IMPACT, LAW OF

When two bodies meet in impact, the impulse of the restitute equals the impulse of compression times a constant, the coefficient of elasticity, or coefficient of resilience:

$$I' = e\,I$$

where I′ = impulse of restitute
I = impact of compression
e = a constant, coefficient of elasticity or resilience

The laws of impact of elastic bodies are derivable from the Newton laws of motion and not vice versa.

**Keywords:** elasticity, impact, mechanics, restitution
**Source:** Jammer, M. 1957.
*See also* CARNOT; HERTZ CONTACT; MEYER; NEWTON LAW OF IMPACT

### IMPULSE-MOMENTUM LAW

Impulse is equal to the change of momentum. The impulse-momentum law is expressed as:

$$F \, \Delta t = (mv)_2 - (mv)_1$$

where F = force
Δt = time interval
m = mass
v = velocity

**Keywords:** force, impulse, mass, momentum, velocity
**Sources:** Besancon, R. M. 1975; Goldsmith, W. 1960.

### IMPULSIVE ACTION—SEE CARNOT

### INCIDENCE MASS (SOUND)—SEE MASS

### INCLINED PLANE, LAW OF

On an inclined plane or slope, where friction is neglected, the effort or force to move an object will be multiplied as many times as the length of the plane is times the height of the plane:

$$\frac{\text{weight (resistance)}}{\text{force (effort)}} = \frac{\text{length of plane}}{\text{height of plane}}$$

**Keywords:** force, friction, plane, slope
STEVINUS, Simon, 1548-1620, military and civil engineer
**Source:** Northrup, E. T. 1917.
*See also* MACHINES

### INCOMPLETENESS THEOREM—SEE GÖDEL

### INCREASING COMPLEXITY—SEE PROGRAM EVOLUTION (COMPUTER)

### INCREASING RETURNS TO SCALE, LAW OF

An increase in intensive cultivation or manufacturing gives an increase in product proportionately greater than the increase of cost. This law refers to all inputs being varied, in contrast to the law of diminishing returns (Fig. I.1).

**Keywords:** cost, manufacturing, production, return
**Source:** Link, A. N. 1933.
*See also* DIMINISHING RETURNS

### INCREASE OF THE RANDOM ELEMENT, LAW OF

The energy given off nonsimultaneously in multidirections and with increasing diffusion, which brings about an expanding physical universe.

**Keywords:** energy, diffusion, direction, nonsimultaneously, universe
**Source:** Fuller, B. 1975.

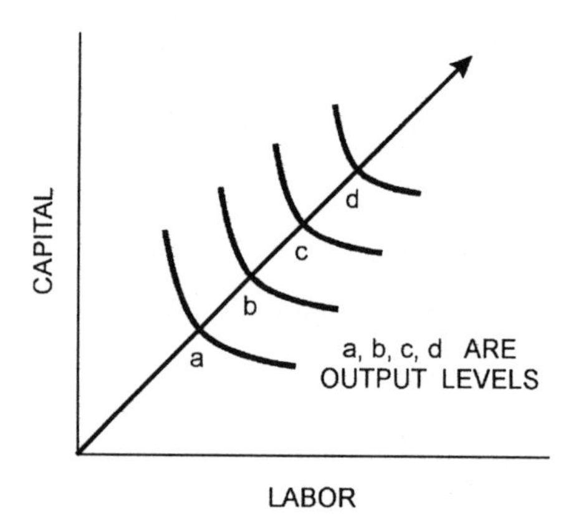


**Figure I.1** Increasing returns.

## INDENTATION, LAW OF—SEE IMPACT; MEYER

## INDEPENDENT ASSORTMENT, LAW OF; (GENETIC EXPRESSION OF MENDEL SECOND LAW)

In reproduction genetics, the members of gene pairs segregate independently during mitosis (differential segregation of replicated chromosomes in a cell nucleus preceding cell division).

**Keywords:** genetics, heredity, mitosis, reproduction
**Sources:** Friel, J. P. 1974; Parker, S. P. 1994; Stedman, T. L. 1976.
*See also* MENDEL; MENDELIAN

## INDEPENDENT MIGRATION OF IONS—SEE KOHLRAUSCH

## INDESTRUCTIBILITY OF MATTER; ALSO CALLED LAW OF CONSERVATION OF MATTER

Matter is neither created or destroyed in a mechanical or chemical reaction.

**Keywords:** chemical, created, destroyed, matter, mechanical
**Source:** Gibson, C. 1981.
*See also* CONSERVATION OF MATTER

## INDETERMINACY PRINCIPLE—SEE HEISENBERG; UNCERTAINTY PRINCIPLE

## INDEX LAWS (MATHEMATICS)—SEE ASSOCIATIVE; COMMUTATIVE; EXPONENTS

## INDUCED CURRENT—SEE LENZ

## INDUCTION—SEE MAGNETIC INTENSITY

## INERTIA, LAW OF (1638); ALSO KNOWN AS NEWTON FIRST LAW OF MOTION (1667)

Based on Galileo's statement, a body in motion will continue in motion if there is not force acting on the body or will remain at rest if at rest.

**Keywords:** body, force, mechanics, movement, rest
GALILEO (Galilei), 1564-1642, Italian astronomer and physicist
NEWTON, Sir Isaac, 1642-1727, English scientist and mathematician
**Source:** Brown, S. B. and Brown, L. B. 1972.
*See also* JACOBI-SYLVESTER; NEWTON FIRST LAW

## INFIDELITY, LAW OF—SEE LIKING, LAW OF; OR LAW OF ATTRACTION

## INFINITESIMALS (INFINITESIM), LAW OF; OR SECOND PRINCIPLE OF HOMEOPATHY (1810?)

A tenet of homeopathic medicine in which the more the active agent for a remedy is diluted, the stronger its effect. A 28× order of magnitude dilution of active agent represents a mixture that has 10 trillion quadrillion times as many inert as active ingredients.

**Keywords:** active agent, diluted, homeopathy
HAHNEMANN, Christian Friedrich Samuel, 1755-1843, German physician (founder of homeopathy)
**Source:** Bynum, W. F., et al. 1981.
*See also* HOMEOPATHY; SIMILARS

## INHERITANCE—SEE FILIAL REGRESSION; GALTON

## INITIAL VALUE (STIMULUS AND RESPONSE), LAW OF; WILDER LAW OF INITIAL WORK

With a standard stimulus and in a standard period of time, the extent and direction of response of a physiological function at rest depends to a large measure on its initial value, so that with the higher initial value, the smaller the response to function-increasing, and the larger the response to function-decreasing stimuli.

**Keywords:** physiological, response, stimulus
**Sources:** Landau, S. I. 1986; Stedman, T. L. 1976.
*See also* WILDER

## INOCULATION, PRINCIPLE OR LAW OF (1796)

Discovered by E. Jenner, it states that the use of material from sores could be used to treat a patient against the disease, first demonstrated with cowpox and then with smallpox. This principle was first rejected in 1796.*

**Keywords:** cowpox, disease, sores, treatment
JENNER, Edward, 1749-1823, English physician
**Sources:** Bendiner, J. and Bendiner, E. 1990; Garrison, F. H. 1929; *Hellemans, A. and Bunch, B. 1988; Morris, C. 1992; Morton, L. T. and Moore, R. J. 1989; Travers, B. 1995.

## INTEGRAL EXPONENTS, LAWS OF

The integral exponents may be appropriately manipulated when powers or exponents of the same base are multiplied or divided:

$$a^n a^m = a^{n+m}$$

$$a^m/a^n = a^{m-n}$$

$$(a^m)^n = a^{mn}$$

$$(ab)^n = a^n b^n$$

$$(a/b)^n = a^n/b^n, \; b \neq 0$$

**Keywords:** exponents, mathematics, powers
**Sources:** James, R.B. and James, G. 1968; Korn, G. A. and Korn, T. M. 1968; Morris, C. G. 1992.
*See also* EXPONENTS; FRACTIONAL EXPONENTS

## INTEGRATION—SEE OLIGOMERIZATION

## INTEGRATION, LAW OF (EVOLUTION)

A trend in evolution that is manifested by a reduction in the number of segments of a body.

**Keywords:** body, evolution, number, segments
**Source:** Lincoln, R. J., et al. 1982.
*See also* EVOLUTION

## INTENSITY, LAW OF; SEE FECHNER (PHYSIOLOGY); HELM-OSTWALD

## INTENSITY OF ILLUMINATION—SEE BIOT

## INTERACTION—SEE FERMI INTERACTION

## INTERFACIAL ANGLES, LAW OF (CRYSTALLOGRAPHY)—SEE CONSTANCY OF INTERFACIAL ANGLES

## INTERMEDIATE METALS, LAW OF

In any circuits made of solid conductors, the temperature is uniform through all the conducting matter, and the algebraic sum of the electromotive forces in the entire circuit is totally independent of intermediate matter and is the same as if two points separated by the conductor were placed in contact.

**Keywords:** circuit, conductors, electromotive, temperature
**Source:** Considine, D. M. 1976.
*See also* HOMOGENEOUS CIRCUIT; SUCCESSIVE OR INTERMEDIATE TEMPERATURES; THERMOCOUPLES

## INTERMEDIATE TEMPERATURES—SEE SUCCESSIVE TEMPERATURES

## INTERNAL PRESSURE PRODUCED BY AN ELECTRICAL CURRENT IN A CONDUCTOR, LAW OF

In every conductor that carries an electric current, a pressure is produced within the substance of the conductor that results from the mutual attraction of all the current-carrying elements of the conductor. The pressure per unit area, at any radial distance from the axis is:

$$g = I^2/\pi R^4 (R^2 - r^2)$$

where $g$ = pressure per unit area, dynes
$I$ = electric current, amperes
$R$ = radius of conductor, cm
$r$ = radial distance from the axis, cm

**Keywords:** conductor, current, electricity
**Source:** Thewlis, J. 1961-1964.

## INTESTINES, LAW OF

The presence of a bolus (round mass) in the intestine induces contraction above and inhibition below the point of stimulus, thereby producing a progression of intestinal contents.

**Keywords:** bolus, bowel, digestion, physiology
**Sources:** Friel, J. P. 1974; Landau, S. I. 1986; Stedman, T. L. 1976; Thomas, C. L. 1989.

## INVERSE LAW FOR ADDITION—SEE ADDITIVE INVERSE

## INVERSE LAW FOR MULTIPLICATION- SEE MULTIPLICATIVE INVERSE

## INVERSE PHOTOELECTRIC—SEE DUANE-HUNT

## INVERSE-SQUARE LAW; INVERSE-SQUARE, GENERALITY OF LAW

A relation between physical quantities of the form where x is proportional to $1/y^2$, where y usually represents a distance and x represents a flux or force.

An example is that the differential illumination varies as the square of the distance between source and receiving surface. Other examples that follow this relationship include sound, heat, and odor.

**Keywords:** flux, force, illumination, mathematical, physical, proportional
**Sources:** Bunch, B. 1992; Glasser, O. 1944; Peierls, R. E. 1956; Weik, M. H. 1996; Wolman, B. B. 1989.
*See also* NEWTON (GRAVITY); PHOTOMETRIC

## INVERSE-SQUARE LAW OF RADIATION

The intensity of radiation or illumination at any surface varies inversely as the square of the distance of that surface from the source of radiation or light.

**Keywords:** illumination, light, radiation
**Source:** Thewlis, J. 1961-1964.
*See also* KEPLER FIRST LAW

## INVERSE VARIATION, LAW OF (INTENSION/EXTENSION)

Words may be arranged in order of increased extension, i.e., animal, vertebrate, mammal, canine, dog, poodle, or vice-versa.

**Keywords:** arrangement, extension, intension, order
**Source:** Angeles, P. A. 1981.

## INVOLUTION—SEE BOOLEAN

## IONIZATION, THREE-HALVES POWER LAW; OR GEIGER LAW (1913)

The range that an alpha particle will travel in air is proportional to the cube of its velocity. The range of the alpha particle in air follows:

$$r = av^3 = b\,E^{3/2}$$

where $r$ = range
$a, b$ = constants
$v$ = velocity
$E$ = energy of the particle

**Keywords:** alpha, energy, particle, velocity
GEIGER, Hans, 1882-1945, German physicist
**Source:** Thewlis, J. 1961-1964.
*See also* GEIGER-NUTTALL

## ION MOBILITY—SEE EINSTEIN

## IRON LAW OF WAGES (1863)

Also known as the *subsistence theory of wages,* in which wages tend to keep down to subsistence level, it is believed that if the wages rise above that level it will lead to an increase in population, and as a result wages will fall again to subsistence level. The expression was first made by F. J. G. Lassalle but does not hold today.

**Keywords:** population, subsistence, wages
LASSALLE, Ferdinand J. G., 1825-1864, German economist
**Sources:** Bothamley, J. 1993; Hanson, J. L. 1986; Pearce, D. W. 1986; Sloan, H. S. and Zurcher, A. 1970.

## IRREVERSIBILITY—SEE DOLLO

## ISOBAR LAWS—SEE MATTAUCH

## ISOCHRONISM, LAW OF

A nerve and its innervated (stimulated) muscle have identical chronaxy values. A chronaxy value is the time interval required to stimulate a nerve electrically with twice the current needed to elicit a threshold response.

**Keywords:** muscle, nerve, physiology, response
**Sources:** Friel, J. P. 1974; Landau, S. I. 1986; Stedman, T. L. 1976.
*See also* LAPICQUE

## ISODYNAMIC LAW

In metabolism heat production in the body, different foods are interchangeable in accordance with their caloric values.

**Keywords:** calorie, food, heat, metabolism
**Sources:** Friel, J. P. 1974; Landau, S. I. 1986; Stedman, T. L. 1976.
*See also* BODY SIZE

## ISOLATED CONDUCTION, LAW OF

The nervous impulse that is passed through a neuron is never communicated to other neurons except at the ends.

**Keywords:** nerves, physiology, stimulus
**Sources:** Critchley, M. 1978; Friel, J. P. 1974; Landau, S. I. 1986.

## ISOMORPHISM, LAW OF (1817; 1819)

An equal number of atoms, if they are bound in the same way, produce similar crystal forms, and the crystal forms depend on the number and method of combination, not on the nature of the atoms. Isomorphs are elements that belong to the same group. The law was published in Sweden in 1821 based on a paper read in 1819.

**Keywords:** atoms, combination, crystals, elements, method, number
MITSCHERLICH, Eilhardt, 1794-1863, German chemist
**Sources:** Brown, S. B. and Brown, L. B. 1972; Lagowski, J. J. 1997; Walker, P. M. B. 1980.
*See also* MITSCHERLICH

### ISOSPIN LAW (OR ISOTOPIC SPIN) OR I SPIN

In the nuclear family are neutrons and protons that are in the atomic nucleus. Isospin is a concept, sometimes called a law, in which the nucleus is considered as a single object coming in two isospin states, neutron and proton. The pion comes in three isospin states, $\pi^+$, $\pi^-$, and $\pi^0$. Spin is a property, like charge and mass, and is a conserved quantity, which means that the total isotopic spin number remains constant.

**Keywords:** atomic nucleus, neutron, pion, property, proton, state
**Sources:** Besancon, R. M. 1974; Lederman, L. 1992.

### ISOTHERMAL LINES, LAW OF

Isothermal lines drawn on graphs and diagrams, such as P-V (pressure-volume) diagrams, join points of the same temperature. The temperature on one side of the line is higher than the temperature on the other side of the line.

**Keywords:** graphs, lines, temperature, thermal
**Source:** Thewlis, J. 1961-1964.

### ISOTHERM MODELS

Some of the representative models for pure-component isotherm relationships of adsorption of gas/liquid components are as follows:

| | |
|---|---|
| HENRY LAW | $n^* = K\ n_M\ C$ |
| LANGMUIR LAW (1916) | $n^* = K\ n_M\ C/1 + KC$ |
| FREUNDLICH (1909) | $n^* = AC^B$ |
| BRUNAUER-EMMETT-TELLER (BET) | $n^* = K\ n_M C_1/[1 + (K - 1)\ C_1](1 - C_1)$ |
| HARKINS-JURA (1946) | $1/V^2 = A - B \ln P$ |
| LANGMUIR-FREUNDLICH | $n^* = A\ n_M\ C^B/(1 + AC^B)$ |

where $n^*$ = loading
$n_M$ = monolayer loading
K = Henry law coefficient
V = volume of gas adsorbed per unit solid
P = partial pressure of adsorbent
C = concentration
$C_1$ = partial concentration, $C/C_{sat}$
A, B, D = constants, empirical parameters

Isotherms can also be represented graphically based on data obtained based on volumetric, gravimetric, or chromatographic means. These data may plot as Type I (concave downward), Type II (concave upward), or Types III, IV, and V (alternating convex and concave regimes) (Fig. I.2).

**Keywords:** adsorption, concentration, gas/liquid, isotherms
BRUNAUER, Stephen, twentieth century (b. 1903), Hungarian American chemist
EMMETT, Paul Hugh, twentieth century (b. 1900), American chemist
FREUNDLICH, Herbert M. F., 1880-1941, German chemist

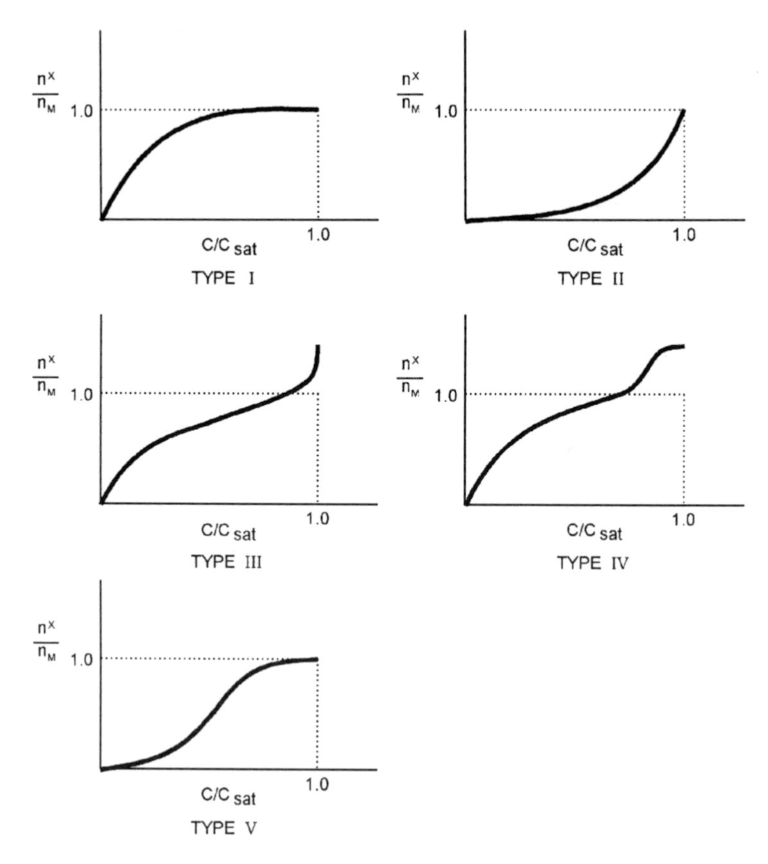


**Figure I.2** Isotherms.

HENRY, William, 1774-1836, English chemist
LANGMUIR, Irving, 1881-1957, American chemist
TELLER, Edward, twentieth century (b. 1908), Hungarian American physicist
**Sources:** Chemical Engineering 102(11):92-102, Nov. 1995; Menzel, D. H. 1960.
*See also* BRUNNER-EMMETT-TELLER; FREUNDLICH; HARKINS-JURA; HENRY; LANGMUIR

## ITERATED LOGARITHM, LAW OF THE

There are quite a few generalizations with regard to the expression and application of this law, many of which are quite complicated, and it is a modification of the strong law of large numbers. The law of integrated logarithms for the square of independent variables has served well as a starting point for the applicability to sequences of dependent random variables. The law deals with the sums of independent and ideally distributed random variables, xi, with i = 1, 2, 3, … with:

$$Ex_1 = \mu \text{ and } \operatorname{var} x_1 = E(x_1 - \mu)^2 = \sigma^2$$

where E = expected value
$\sigma$ = variance

**Keywords:** distributed, independent, random, variables
**Sources:** Hazewinkel, M. 1987; Kotz, S. and Johnson, N. L. 1983.
*See also* KOMOGOROV; LARGE NUMBER; STRONG LAWS

# J

## JACKSON LAW OR RULE; ALSO CALLED LAW OF DISSOLUTION

When a mentally disturbed person has lost memory functions, events immediately preceding the onset of the disturbances are most difficult to recall. Also, the nerve functions that are developed last are the first to be lost.

**Keywords:** medicine, memory, mentally, mind

JACKSON, John Hughlings, 1835-1911, British neurologist

**Sources:** Friel, J. P. 1974; Landau, S. I. 1986; Stedman, T. L. 1976.

*See also* JOST

## JACOBI-SYLVESTER LAW OF INERTIA—SEE SYLVESTER

## JAKOB NUMBER, JAK

A dimensionless group in heat transfer that relates the heat of evaporation and the sensible heat change:

$$\mathrm{JAK} = LH_v/c_p\ \Delta T$$

where $LH_v$ = heat of evaporation
$c_p$ = specific heat at constant pressure
$\Delta T$ = temperature difference

**Keywords:** evaporation, sensible heat, transfer

JAKOB, Max, 1879-1955, German American engineer and physicist

**Sources:** Grigull, U. et al. 1982; Land, N. S. 1972; Layton, E. T. and Lienhard, J. H. 1988; Parker, S. P. 1992; Potter, J. H. 1967.

## JAKOB NUMBER OR MODULUS, Ja

A dimensionless group is boiling heat transfer that relates the maximum bubble radius and the thickness of the superheated film:

$$Ja = (T_o - T_{sat})\ \rho_l\ c_p/LH_v\ \rho_v$$

where $T_o$ = bulk liquid temperature
$T_{sat}$ = saturation temperature
$c_p$ = specific heat at constant pressure
$\rho_l$ = mass density of liquid
$\rho_v$ = mass density of vapor
$LH_v$ = latent heat of vaporization

**Keywords:** boiling, bubble, heat transfer

JAKOB, Max, 1879-1955, German American engineer and physicist

**Sources:** Bolz, R. E. and Tuve, G. L. 1970; Grigull, U. et al. 1982; Land, N. S. 1972; Potter, J. H. 1967.

## JANSSEN LAW OF LATERAL PRESSURE (1895) OR LAW OF BIN LOADING

The lateral pressure of granular materials (Fig. J.1) is represented by:

$$L = w\ R/\mu'\ (1 - e^{-k\,\mu'\,h/R})$$

where L = lateral pressure
w = material density
R = hydraulic radius
$\mu'$ = coefficient of friction
h = distance from the top of the load
k = L/V or $\sigma_n/\sigma_y$

**Keywords:** granular, lateral, material, pressure

JANSSEN, H. A., late nineteenth century, German engineer

**Sources:** Cheremisinoff, N. P. 1986; Ketchum, M. S. 1919.

*See also* KETCHUM; RANKINE; SEMI-FLUIDS

## JEANS—SEE RAYLEIGH-JEANS

## JEFFREYS NUMBER, Je OR $N_{Je}$ (c. 1929)

A dimensionless group used to represent slow flow that relates gravity force and viscous force:

$$Je = \rho\ gL^2/\mu V$$

where $\rho$ = mass density
g = gravitational acceleration
L = characteristic length
$\mu$ = absolute viscosity
V = velocity

The Jeffreys number equals the Reynolds number divided by the Froude number. Also, the Jeffreys number is the reciprocal of the Stokes number.

**Keywords:** flow, gravity, slow, viscous

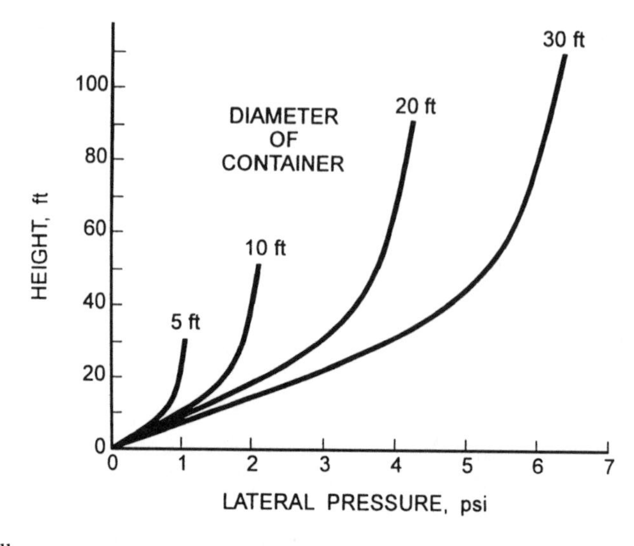


**Figure J.1** Wall pressure.

JEFFREYS, Sir Harold Jeffreys, 1891-1989, British geologist
**Sources:** Dallavalle, J. M. 1948; Land, N. S. 1972. Millar, D. 1966; Pelletier, P. A. 1994; Potter, J. H. 1967.
*See also* STOKES

## J-FACTOR (HEAT TRANSFER), $j_H$

A dimensionless group used in heat, mass, and momentum transfer:

$$j_H = h/c_pG\ (c_p\mu/k)^{2/3}$$

where $h$ = heat transfer coefficient
$c_p$ = specific heat at constant pressure
$G$ = mass transfer
$\mu$ = absolute viscosity
$k$ = heat conductivity

The J-factor is equal to the Nusselt number divided by the Reynolds number and Prandtl number to the one-third power.

**Keyword:** heat, mass, momentum, transfer
**Sources:** Bolz, R. E. and Tuve, G. L. 1970; Land, N. S. 1972; Parker, S. P. 1987. 1992; Potter, J. H.1967.
*See also* NUSSELT; PRANDTL; REYNOLDS

## J-FACTOR (MASS TRANSFER), $j_M$

A dimensionless group applicable to mass transfer:

$$j_M = k_c p/G\ (\mu/\rho\ D)^{2/3}$$

where $k_c$ = mass transfer coefficient
$\rho$ = mass density
$G$ = mass transfer
$\mu$ = absolute viscosity
$D$ = mass diffusivity

**Keywords:** density, diffusivity, mass transfer, viscosity
**Sources:** Bolz, R. L. and G. L. 1970; Land, N. S. 1972; Parker, S. P. 1992; Potter, J. H. 1967.
*See also* COLBURN; SCHMIDT

## JORDAN LAW OR RULE

### Law 1

Organisms in species that are closely related tend to occupy adjacent rather than identical or distant ranges.

### Law 2

Fish of a given species develop more vertebrae in a cold environment than those in warm waters.

**Keywords:** climate, cold, ecology, evolution, warm, species
JORDAN, David Starr, 1851-1931, American biologist (ichthyology)
**Sources:** Gray, P. 1967; Lapedes, D. N. 1974.

## JOST LAW

As applied to the results of continued practice on material initially equally well learned, the practice is more effective for recall or utilization of the older of two associations than for the more recently acquired; the older loses strength less rapidly with the passage of time.

**Keywords:** learning, psychology
JOST, Adolph, nineteenth twentieth century (b. 1874), German psychologist born in Austria
**Sources:** Gillispie, C. C. 1974; Goldenson, R. M. 1984; Landau, S. I. 1986; Reber, A. S. 1985; Wolman, B. B. 1989; Zusne, L. 1987.
*See also* EXERCISE (LEARNING); JACKSON

## JOUKOWSKY (ZHUKOVSKI)—SEE KUTTA-JOUKOWSKY

## JOULE FIRST LAW (1842); JOULE LAW OF ELECTRIC HEATING

The heat produced in an electrical conductor is proportional to the resistance of the conductor, to the time and to the square of the current and is represented by:

$$\text{Heat} = I^2Rt$$

where I = current
R = resistance
t = time
E = voltage or electromotive force

This relates to the Watt equation in which power = watts = $I^2R = EI$

**Keywords:** conductor, current, heat, resistance
JOULE, James Prescott, 1818-1889, English physicist
**Sources:** Daintith, J. 1981; Honig, J. M. 1953; Thewlis, J. 1961-1964.
*See also* OHM; WATT

## JOULE NUMBER, J OR $N_J$

A dimensionless group in magnetohydrodynamics that relates the heating energy to the magnetic field energy:

$$J = 2\ \rho c_p\ \Delta T/\mu H^2$$

where $\rho$ = mass density
$c_p$ = specific heat at constant pressure
$\Delta T$ = temperature difference
$\mu$ = magnetic permeability
H = magnetizing force

**Keywords:** energy, heating, magnetic
JOULE, James Prescott, 1818-1889, English physicist
**Sources:** Bolz, R. E. and Tuve, G. L. 1970; Land, N. S. 1972; Parker, S. P. 1992; Potter, J. H. 1967.

## JOULE SECOND LAW (1843); OR MAYER HYPOTHESIS (1842)

The internal energy of a given mass of gas is independent of its volume (not strictly true). The internal energy of a given mass of gas is a function of temperature alone and independent of the pressure and volume of the gas, and the molar heat capacity of a solid compound is equal to the sum of the atomic heat capacities of its component elements in the solid state, another aspect of the law of conservation of energy.

**Keywords:** energy, gas, volume
JOULE, James Prescott, 1818-1889, English physicist
MAYER, Julius Robert von, 1814-1878, German physicist
**Sources:** Daintith, J. 1981; Honig, J. M. 1953; Mandel, S. 1972; Thewlis, J. 1961-1964; Truesdell, C. A. 1980; Walker, P. M. B. 1988.
See also KOPP

## JOULE-THOMSON EFFECT (1852) OR LAW; JOULE-KELVIN EFFECT

When a highly compressed gas is allowed to expand in such a way that no external work is done, the cooling is inversely proportional to the square of the absolute temperature.

**Keywords:** compressed, cooling, gas, work
JOULE, James Prescott, 1818-1889, English physicist
THOMSON, William (Lord Kelvin), 1824-1907, Scottish mathematician and physicist
**Sources:** Mandel, S. 1972; Morris, C. G. 1992; Rigden, J. S. 1996.

## JUNGE POWER LAW

A simple function represents a natural aerosol from 0.1 to 20 μm when in equilibrium, which is:

$$dn/d \log r = C\, r^{-b}$$

where dn = number of particles in a logarithmic size interval
C = constant
r = radius
b = value of approximately 3

Experiments have shown that, at 75 to 95% relative humidity, from 0.5 to 0.9 of the submicron aerosol mass can be liquid water. The presence of air pollutants caused by combustion and other volatiles and particles greatly effect the constants and relationships.

**Keywords:** aerosol, equilibrium, logarithmic, particles
JUNGE, Gustav, twentieth century (b. 1879), German mathematician
**Sources:** Besancon, R. M. 1974; Lerner, R. G. and Trigg, G. L. 1981; NUC.

## JURIN LAW OF CAPILLARY ACTION (1750)

In the same liquid and at the same temperature, the mean height of the ascent of a column of liquid in a capillary glass tube is inversely related to the diameter of the tube. Thus, the diameter times height is a constant, in which:

$$h = 2\, \gamma \cos \alpha / \rho\, g\, a$$

where h = height of rise
a = internal radius of capillary tube
$\gamma$ = surface tension of the liquid
$\rho$ = density
g = acceleration due to gravity
$\alpha$ = angle of contact with the capillary

**Keywords:** capillary, diameter, flow, height, liquid

JURIN, James, 1684-1750, British physiologist and mathematician
**Sources:** Ballentyne, D. W. G. and Lovett, D. R. 1980; Clifford, A. F. 1964.
*See also* CAPILLARITY; CAPILLARY

# K

## KAHLER LAW

The ascending branches of the posterior (rear) roots of the spinal nerves pass within the cord in succession from the root zone toward the mesial (middle) plane.

**Keywords:** anatomy, medicine, medial, nerves, spinal
KAHLER, Otto, 1849-1893, Austrian physician
**Sources:** Friel, J. P. 1974; Landau, S. I. 1986.

## KARLSBAD—SEE CARLSBAD; TWIN

## KÁRMÁN NUMBER, Ka

A dimensionless group that is a measure of stream turbulence in fluid flow:

$$Ka = [(v_x^2 + v_y^2 + v_z^2)/3w_o^2]^{1/2}$$

where Ka = Kármán number
$v_x$ = stream velocity in x-direction
$v_y$ = stream velocity in y-direction
$v_z$ = stream velocity in z-direction
$w_o$ = flow velocity

**Keywords:** fluid, flow, stream, turbulence
KÁRMÁN, Theodor von, 1881-1963, Hungarian American engineer
**Source:** Hall, C. W. 1979.

## KÁRMÁN NUMBER 1, $Ka_1$ OR $N_{Ka(1)}$

A dimensionless group representing fluid flow in conduits with friction:

$$Ka_1 = \rho p d^3/\mu^2 L$$

where $\rho$ = mass density
p = pressure drop
d = pipe diameter
$\mu$ = absolute viscosity
L = characteristic length

**Keywords:** flow, fluid, friction
KÁRMÁN, Theodor von, 1881-1963, Hungarian American engineer
**Sources:** Bolz, R. E. and Tuve, G. L. 1970; Land, N. S. 1972; Parker, S. P. 1992; Perry, R. H. 1967; Potter, J. H. 1967.
*See also* REYNOLDS

## KÁRMÁN NUMBER 2, $Ka_2$ OR $N_{Ka(2)}$

A dimensionless group in magnetohydrodynamics relating flow speed and Alfvén wave speed:

$$Ka_2 = V(\rho\mu)^{1/2}/\beta$$

where V = flow velocity
$\rho$ = density
$\mu$ = magnetic permeability
$\beta$ = magnetic induction

The Kármán number 2 equals the Alfvén number and also the Magnetic Mach number.

**Keywords:** flow, fluid, magnetohydrodynamics, wave speed
KÁRMÁN, Theodor von, 1881-1963, Hungarian American engineer
**Sources:** Bolz, R. E. and Tuve, G. L. 1970; Land, N. S. 1972; Parker, S. P. 1992; Potter, J. H. 1967; Travers, B. E. 1996.
*See also* ALFVÉN; MAGNETIC MACH

## KASSOWITZ—SEE DIDAY

## KEITH LAW

Ligaments are never used for the continuous support of any body part such as a joint.

**Keywords:** anatomy, joints, ligaments, medical
KEITH, Sir Arthur, 1866-1955, Scottish anatomist
**Sources:** Morton, L. T. and Moore, R. J. 1989; Talbott, J. H. 1970.

## KELVIN EFFECT—SEE THOMSON EFFECT

## KELVIN LAW

If a system of rigid circuits does mechanical work while maintaining constant current conditions, the energy of the circuits increases at the same rate that work is done.

**Keywords:** current, electrical, mechanical, work
KELVIN, Lord (William Thomson or William Thompson), 1824-1907, Scottish mathematician and physicist
**Source:** Considine, D. M. 1976.

## KELVIN LAW OF CONDUCTORS

The most economical size of electric conductors to use in a transmission line is that for which the annual cost of the losses is equal to the annual interest and depreciation on that part of the capital cost of the conductor that is proportional to its cross-sectional area.

**Keywords:** area, conductor, economics, electricity, transmission
KELVIN, Lord (William Thomson or William Thompson), 1824-1907, Scottish mathematician and physicist
**Source:** Gillispie, C. C. 1974.

## KENNELLY—SEE HEAVISIDE

## KEPLER CONJECTURE ON STACKING SPHERES (1611)

The most compact method of packing spheres is the one exploited by nature to arrange atoms into crystals and by grocers to stack oranges into four-sided pyramids. The easiest way to create this pattern is to form a layer of spheres consisting of even vertical and horizontal rows in which the spheres in the next layer up nestle in the niches between each foursome of the layer below, as recently proved by computer modeling as well. The Kepler conjecture shows that the upper bound of the volume filled by spheres is approximately 74 percent, or pi divided by the square root of 18.

**Keywords:** compact, stacking, spheres
KEPLER, Johannes, 1571-1630, German astronomer
**Sources:** Science News 154(7):103, August 15, 1998; Scientific American 271(6): 32, December 1994.
*See also* DILATANCY

### KEPLER LAWS (PLANETARY) (1609-1619); THEORY OF PLANETARY MOTION (1609)

The three laws of planetary motion are:

**First Law (1609)**
The planets move about the sun in elliptical paths, with the sun at one focus of the ellipse.

**Second Law; Law of Areas (1609)**
The radius vector connecting each planet with the sun describes equal areas in equal times.

**Third Law; Harmonic Law (1619)**
The cubes of the mean distances of the planets from the sun are proportional to the squares of their times of revolution about the sun.

Kepler based much of his work on planetary action on the meticulous data obtained by Tycho Brahe.

**Keywords:** astronomy, ellipses, planets, sun
KEPLER, Johannes, 1571-1630, German astronomer
BRAHE, Tyco den, 1546-1601, Danish astronomer
**Sources:** Barrow, J. D. 1991; Besancon, R. M. 1974; Brown, S. B. and Brown, L. B. 1972; Daintith, J. 1981 (physics); Fairbridge, R. W. 1967; Hodgman, C. D. 1952; Mandel, S. 1972; NYPL Desk Ref. 1989; Thewlis, J. 1961-1964.
*See also* BRAHE; CASSINI

### KERR EFFECTS OR LAW (1875); KERR ELECTRO-OPTIC EFFECT

When polarized light is incident on the polished mirror surface of a pole of an electromagnet, the plane of polarization of the reflected light is not the same when the magnet energized as when it is not energized, with the direction of rotation opposite to that of the currents exciting the pole from which the light is reflected. Another Kerr effect was the creation of double refraction (birefringence) in a transparent substance by an electric field. A parallel activity is the Kerr magneto-optic effect.

**Keywords:** electromagnet, light, magnet, polarized
KERR, John, 1824-1907, Scottish physicist
**Sources:** Considine, D. M. 1976; Hodgman, C. D. 1952; Thewlis, J. 1961-1964.
*See also* COTTON-MOUTON; PRESTON; ZEEMAN

### KETCHUM LAW

The flow rate out of an orifice in a bin is independent of the head and varies as the cube of the orifice, independent of the pressure of grain, or aggregate, when stationary or in motion.

**Keywords:** bin, flow, rain, head, orifice, pressure
KETCHUM, Milo Smith, 1872-1934, American civil engineer

**Source:** Ketchum, M. S. 1919.
*See also* JANSSEN; RANKINE; SEMI-FLUIDS

### KEYNES LAW OF CONSUMPTION

At every level of income, a certain portion of that income is spend for consumptive goods, with the proportion decreasing as income increases.

**Keywords:** consumptive goods, income, portion
KEYNES, John Maynard, 1883-1946, English economist
**Source:** Sloan, H. S. and Zurcher, A. J. 1979.

### KHINSHIN—SEE WEAK LAW OF LARGE NUMBERS

### KICK LAW (1885)

The work done in crushing a particle is a function of the fractional reduction of size of the particle:

$$E = C \log L_1/L_2$$

where $E$ = work energy required or needed to provide the reduction
$L_1/L_2$ = size reduction ratio, where $L_1$ is the initial dimension
$C$ = proportionality constant based on material and reduction

**Keywords:** crushing, grinding, size, work
KICK, Friedrich, 1840-1915, German engineer
**Sources:** Brown, G. G. et al. 1950; Grayson, M. 1978; Honig, J. M. 1953; Kirk-Othmer. 1978; NUC; Perry, J. H. 1950.
*See also* BOND; RITTINGER

### KIND—SEE NATURE

### KINETICS—SEE NEWTON

### KING COOLING LAW (1914)—SEE COOLING LAWS

### KINNEY LAW

A person with normal speech and hearing who subsequently loses all hearing begins to show speech changes in a length of time that is directly proportional to the length of time that the person had been speaking.

**Keywords:** hearing, physiology, speech
KINNEY, Richard, twentieth century (b. 1924), American educator
**Source:** Zusne, L. 1987.

### KIRCHHOFF-CLAUSIUS ON THE RELATION BETWEEN EMISSIVE POWER AND THE MEDIUM, LAW OF; CALLED KIRCHHOFF LAW BY MANY

The emissive power of perfectly black bodies is proportional to the square of the index of refraction of the surrounding medium:

$$e = \mu^2 E$$

where $e$ = emissive power
$\mu$ = index of refraction
$E$ = emissive power of a perfectly black body

**Keywords:** black bodies, emissive, refraction

KIRCHHOFF, Gustav Robert, 1824-1887, German physicist

CLAUSIUS, Rudolf Julius Emmanuel, 1822-1888, German physicist

**Sources:** Gillispie, C. C. 1981; Morris, C. G. 1992.

*See also* HELMHOLTZ; STEWART AND KIRCHHOFF

## KIRCHHOFF LAW (FOR ELECTRICITY)—SEE KIRCHHOFF NETWORK LAWS

## KIRCHHOFF LAW (FOR RADIATION OF HEAT) (1859)

The values of emissivity and absorptivity of a body under equilibrium conditions are the same. The law is also stated as, "the ratio of the emissive power of a body to that of a black body at the same temperature is equal to the absorptivity."

**Keywords:** absorptivity, emissivity, heat

KIRCHHOFF, Gustav Robert, 1824-1887, German physicist

**Sources:** Daintith, J. 1988; Layton, E. T. and Lienhard, J. H. 1988; Parker, S. P. 1989; Thewlis, J. 1961-1964; Van Nostrand. 1947.

*See also* KIRCHHOFF(OPTICAL); STEFAN-BOLTZMANN; STEWART AND KIRCHHOFF

## KIRCHHOFF NETWORK LAWS (ELECTRICITY) (1845)

**First Law: Kirchhoff Current Law; Kirchhoff Junction Law**

At any point in an electric circuit where two or more conductors are joined, the sum of the currents directed toward the junction equals the sum of the currents directed away from the junction.

$$\Sigma I = 0$$

**Second Law: Kirchhoff Loop Law; Kirchhoff Voltage Law**

Around any closed path in an electric circuit, the algebraic sum of the potential difference equals zero.

$$\Sigma E - \Sigma I R = 0$$

**Keywords:** circuit, electricity, physics

KIRCHHOFF, Gustav Robert, 1824-1887, German physicist

**Sources:** Bronson, R. 1993; Daintith, J. 1988 (physics); Hodgman, C. D. 1952; Honig, J. M. 1953; Isaacs, A. et al. 1996; Mandel, S. 1972; Morris, C.G. 1992; Wells, D. E. and Slusher, H. S. 1983; Williams, T. 1994.

## KIRCHHOFF OPTICAL LAW; LAW OF KIRCHHOFF, OPTICAL

The relationship between the powers of emission and the powers of absorption for rays of the same wavelength is constant for all bodies at the same temperature. In 1859, G. Kirchhoff and R. Bunsen devised spectrum analysis.

**Keywords:** absorption, emission, rays, wavelength

KIRCHHOFF, Gustav Robert, 1824-1887, German physicist

**Source:** Honig, J. M. 1953.

*See also* BUNSEN-ROSCOE; NEWTON (OPTICS); STEWART AND KIRCHHOFF

## KIRPICHEV NUMBER (HEAT TRANSFER), $Ki_h$ OR $N_{Ki(h)}$

A dimensionless group for heat transfer that relates the intensity of external heat transfer to the intensity of internal heat transfer:

$$Ki_h = q^*L/k\ \Delta T$$

where $q^*$ = heat flux density
$L$ = length dimension on the surface
$k$ = thermal conductivity
$\Delta T$ = temperature difference

A different relationship is given by W. Begell* as follows:

$$Ki = \sigma_o\ T_o^3/\lambda k$$

where $\sigma$ = Stefan-Boltzmann constant
$T_o$ = characteristic absolute temperature of the medium
$k$ = thermal conductivity of medium
$\lambda$ = extinction factor of the medium, volumetric

**Keywords:** external, heat transfer, internal
KIRPICHEV, Mikhail Viktorovich, son of Victor Lvovich Kirpichev (1844-1913), Russian engineer
**Sources:** *Begell, W. 1983; Bolz, R. E. and Tuve, G. L. 1970; Land, N. S. 1972; Lykov, A. V. and Mikhaylovk Y. A. 1961; Parker, S. P. 1992; Potter, J. H. 1967.
*See also* NUSSELT

## KIRPICHEV NUMBER (MASS TRANSFER), $Ki_m$ OR $N_{Ki(m)}$

A dimensionless group for mass transfer, such as moisture evaporation from porous bodies, that relates the intensity of external mass transfer to the intensity of internal mass transfer:

$$Ki_m = GL/D\ \rho\ \Delta u$$

where $G$ = mass of moisture evaporated per area time
$L$ = length dimension on the body surface
$D$ = mass diffusivity of moisture in bed
$\rho$ = mass density
$\Delta u$ = change in moisture in body

**Keywords:** external, internal, mass transfer
KIRPICHEV, Mikhail Viktorovich, son of Victor Lvovich Kirpichev (1844-1913), Russian engineer
**Sources:** Begell, W. 1983; Bolz, R. E. and Tuve, G. L. 1970; Land, N. S. 1972; Lykov, A. V. and Mikhaylov, Y. A. 1961; Parker, S. P. 1992; Potter, J. H. 1967.

## KIRPITCHEFF NUMBER, Kir OR $N_{Kir}$

A dimensionless group used to represent flow around obstacles or over immersed bodies:

$$Kir = \rho\ F_R/\mu^2$$

where $\rho$ = mass density of fluid
$F_R$ = resistance force in flow over bodies
$\mu$ = absolute viscosity of fluid

**Keywords:** flow, immersed, obstacles

KIRPITCHEFF, Victor L., 1844-1913, Russian engineer (appears to be earlier translated name)

**Sources:** Begell, W. 1983; Bolz, R. E. and Tuve, G. L. 1970; Land, N. S. 1972; Parker, S. P. 1992; Potter, J. H. 1967.

### KIRSCHMANN LAW

The greatest contrast in color sensed by the eye is seen when the difference in luminosities is small.

**Keywords:** color, contrast, eye, luminosities, sensed

KIRSCHMANN. Biographical information not located.

**Source:** Critchley, M. 1978.

*See also* COLOR MIXTURE; COLOR VISION; YOUNG-HELMHOLTZ

### KJERSTAD-ROBINSON LAW (1919)

For human verbal learning, the amount of material learned during equal portions of learning time is the same for different lengths of material to be learned.

**Keywords:** learning, material, time, verbal

KJERSTAD, Conrad Lund, 1883-1967, American psychologist

ROBINSON, Edward Stevens, 1893-1937, American psychologist

**Source:** Zusne, L. 1987.

*See also* THORNDIKE

### KLEIBER LAW (1932)

In mammals, the 3/4th-power scaling of metabolic rate is the ratio of food consumption to basal metabolism, and maintenance is independent of body weight. Therefore, the excess food that may go for production processes is independent of body weight.

**Keywords:** body, food, metabolism, weight

KLEIBER, Max, 1893–c. 1976, Swiss American physiologist

**Sources:** BioScience 48(11):888, November 1998; Kleiber, M. 1961. 1975.

*See also* ALLOMETRIC SCALING; DU BOIS SURFACE AREA; SURFACE

### KNAPP LAW

There should be no difference in retinal image size in the correction of spherical axial anisometropia if the lenses are placed at the anterior focal point of the eye.

**Keywords:** anisometropia, anterior, eye, image, lens, retinal

KNAPP, Herman Jacob, 1832-1911, American ophthalmologist

**Sources:** Friel, J. P. 1974; Landau, S. I. 1986.

### KNIGHT-DARWIN LAW

The postulate that nature “abhors” perpetual self-fertilization.

**Keywords:** fertilization, nature

KNIGHT, Thomas Andrew, 1759-1838, English botanist

DARWIN, Charles Robert, 1809-1882, English naturalist

**Source:** Gray, P. 1967.

## KNUDSEN COSINE LAW

For a gas at rest at a uniform temperature, the number of molecules striking or leaving an area of the wall in a solid angle, at an angle with the normal, is expressed by the equation:

$$N = (dS/4\pi)\, n\, \bar{c}\, \theta\, dW$$

where N = number of molecules
dS = area of the wall
n = number of molecules per $cm^3$
$\bar{c}$ = mean velocity
dW = solid angle with the wall
$\theta$ = angle with the normal

**Keywords:** flow, fluid, gas, molecules
KNUDSEN, Martin Hans Christian, 1871-1949, Danish physicist
**Sources:** NUC; Parker, S. P. 1992; Thewlis, J. 1961-1964.

## KNUDSEN LAW

For molecular flow of a gas, where the molecules of a gas do not collide with one another, the rate of flow through a circular capillary is expressed by the equation:

$$N = 4/3(g_c/2\pi\, MRT)^{1/2}\, d/L(p'' - p')$$

where N = molecular flow, lb moles/$ft^2$ sec
$g_c$ = constant, 32.2
M = molecular weight
d = capillary diameter
L = capillary length
$p''$ = upstream pressure, psf
$p'$ = downstream pressure, psf

**Keywords:** capillary, flow, gas, molecular flow, rate
KNUDSEN, Martin Hans Christian, 1871-1949, Danish physicist
**Sources:** NUC; Thewlis, J. 1961-1964.

## KNUDSEN NUMBER, Kn OR $N_{Kn}$

A dimensionless group applicable to low-pressure gas flow or rarefied gas flow that relates the mean free path of gaseous diffusion, such as in drying, to the characteristic dimension:

$$Kn = \lambda/L$$

where $\lambda$ = length of mean free path
L = characteristic length dimension

There is also a Knudsen number that relates bulk diffusion and Knudsen diffusion as well as a separate representation of Knudsen number for diffusion.

**Keywords:** diffusion, gas, low pressure, mean free path
KNUDSEN, Martin Hans Christian, 1878-1943, Danish physicist
**Sources:** Bolz, R. E. and Tuve, G. L. 1970; Eckert, E. R. G. and Drake, R. M. 1972; Jerrard, H. G. and McNeill, D. B. 1972; NUC; Parker, S. P. 1992; Perry, R. H. 1967; Potter, J. H. 1967; Rohsenow, W. M. and Hartnett, J. P. 1973.

## KOCH LAW OR POSTULATES (1876) OR LAW OF SPECIFICITY OF BACTERIA

A test devised more than a century ago by Koch for proving that a disease is caused by a specific microbe. Koch maintained that, for causation to be established, it must be possible to isolate the microbe from an organism that has the disease. The microbe must then be given to a healthy host, where it causes the same disease. Then the microbe must be isolated again.

**Keywords:** cause, disease, microbe
KOCH, Robert, 1843-1910, German bacteriologist; Nobel prize, 1905, physiology/medicine
**Sources:** Friel, J. P. 1974; Landau, S. I. 1986; Parker, S. P. 1992; Science 266: 1647, 9 December 1994; Stedman, T. L. 1976; Thomas, C. L. 1989; Walter, P. M. B. 1983.

## KOENIG LAW—SEE KÖNIG

## KOHLRAUSCH LAW (1876)

When ionization of an electrolyte is complete, the conductivity is equal to the sum of the conductivities of the ions into which the substance dissociates.

**Keywords:** conductivity, electrolyte, ions
KOHLRAUSCH, Friedrich Wilhelm Georg, 1840-1910, German physicist
**Sources:** Bynum, W. F., et al. 1981; Hodgman, C. D. 1952; Thewlis, J. 1961-1964; Van Nostrand. 1947; Webster's Biographical Dictionary. 1959.
*See also* following KOHLRAUSCH

## KOHLRAUSCH LAW OF THE INDEPENDENT MIGRATION OF IONS (1875)

In an electrolyte, regardless of the kind of anion, there is a nearly constant difference between the ionic conductivities of potassium and sodium salts. The ionic conductivity of a solution is the sum of two independent terms, one representing the conductivity of the anion, and one representing the cation.

**Keywords:** anion, electrolyte, cation, chemistry
KOHLRAUSCH, Friedrich Wilhelm Georg, 1840-1910, German physicist
**Sources:** Considine, D. M.; 1976. 1995; Honig, J. M. 1953; Thewlis, J. 1961-1964. Webster's Biographical Dictionary. 1959.

## KOHLRAUSCH SQUARE-ROOT LAW

The equivalent conductance of a strong electrolyte in a very dilute solution gives a straight line when plotted against the square root of the concentration.

**Keywords:** chemistry, concentration, electrolyte, solution
KOHLRAUSCH, Friedrich Wilhelm Georg, 1840-1910, German physicist
**Sources:** Lide, D. R. 1995; Thewlis, J. 1961-1964; Webster's Biographical Dictionary. 1959.

## KOLMOGOROFF (KOLMOGOROV) (KOHMOGOROV)—SEE STRONG LAW OF LARGE NUMBERS

## KOLMOGOROFF–5/3 POWER LAW

Represents the dissipation of energy away from the atmospheric boundary:

$$E(\lambda) = \alpha_1 \, \varepsilon^{2/3} \, \lambda^{-5/3}$$

where $E(\lambda)$ = spectral energy
$\alpha_1$ = constant

$\varepsilon$ = dissipation of energy at rate E per unit mass
$\lambda$ = wavelength (slope of the line is –5/3)

**Keywords:** atmosphere, boundary, energy
KOLMOGOROFF, Andrei Nikolaevich, 1903-1987, Soviet mathematician
**Sources:** Abbott, D. 1986; Daintith, J. et al. 1994; Houghton, J. T. 1986.

## KOLMOGOROFF-OBUKHOV LAW (1941)

A relationship in turbulence, considered as a natural state of fluids in motion. The diffusive nature of turbulence is its most significant property. A pioneer in the field of turbulence was G. I. Taylor. His papers were written in 1921 and later. He pointed out the tendency of turbulence toward isotropy and that the mean dissipation rate per unit mass depends on the characteristic velocity and length scale, but is independent of the molecular viscosity, which developed into the equation:

$$\varepsilon \approx U^3/L$$

where $\varepsilon$ = mean dissipation rate per unit mass
U = characteristic velocity
L = length scale

This became known as the Kolmogoroff-Obukhov law.

**Keywords:** dissipation, isotropy, length scale, velocity, viscosity
KOLMOGOROFF, Andrei Nikolaevich, 1903-1987, Soviet mathematician
OBUKHOV, Aleksandr, twentieth century (b. 1918), Soviet geophysicist
TAYLOR, Geoffrey Ingram, 1886-1975, English engineer and scientist (mathematician)
**Sources:** Daintith, J. et al. 1994; Lerner, R. G. and Trigg, G. L. 1981; Lewytzkj, B. 1984; Morris, C. G. 1992; Thewlis, J. 1961-1964.
*See also* TAYLOR

## KÖNIG LAW OF KINETIC ENERGY (1733)

The kinetic energy of a system of mass points is equal to the sum of the kinetic energy of the motion of the system relative to the center of gravity and the kinetic energy of the total mass of the system considered as a whole.

**Keywords:** energy, gravity, motion, system
KÖNIG, Johann Samuels, 1712-1757, German mathematician and physicist
**Source:** Gillispie, C. C. 1981.
*See also* MAUPERTUIS

## KONOVALOV LAWS

These laws represent the relationships that characterize the phase equilibrium in liquid-vapor systems.

### First Law

The mole fraction, $y_1$, in the vapor phase of a two component equilibrium system is greater than the mole fraction, $x_1$, in the liquid phase whenever an isothermal addition of this component into the liquid phase increases the total pressure.

### Second Law

A maximum or minimum on the curve of the isothermal phase diagram that shows the dependence of the total pressure of the saturated vapor on the composition of a two-component (two-phase) system corresponds to an azeotropic system.

**Keywords:** azeotropic, equilibrium, isothermal, liquid, phase, vapor
KONOVALOV, Evmenii Grigorevich, twentieth century (b. 1914), Soviet scientist
**Sources:** Ulicky, T. J. and Kemp, T. J. 1992; Lewytzkj, B. 1984.

## KOPP LAW OR RULE (1864); ALSO CALLED WOESTYN LAW OR RULE (1848)

The specific heat of a solid element at room temperature is the same whether it is free or part of a solid compound. Another statement of the law is that the specific heat of a solid at room temperature is approximately equal to the sum of the atomic specific heat capacities divided by the molecular weight. The specific heat of atomic carbon is 1.8; oxygen, 4.0; and silicon, 3.5

**Keywords:** element, physics, solid, specific heat
KOPP, Hermann Franz Moritz, 1817-1892, German physical chemist
WOESTYN, A. C., nineteenth century, German physicist
**Sources:** Ballentyne, D. W. G. and Walker, D. R. 1960; Baumeister, T. 1967; Honig, J. M. 1953.
*See also* DULONG AND PETIT; JOULE

## KOPP-NEUMANN LAW

The specific heat is inversely proportional to the molecular weight for compounds of similar composition, the constant of proportionality varying from one series of components to another.

**Keywords:** molecular weight, specific heat, thermal
KOPP, Hermann Franz Morita, 1817-1892, German physical chemist
NEUMANN, Franz Ernst, 1798-1895, German physicist and mineralogist [double check this]
**Sources:** Thewlis, J. 1961-1964; Ulicky, T. J. and Kemp, T. J. 1992.
*See also* BLADGEN; DE COPPET; HESS

## KOSCHMIEDER LAW; AIRLIGHT LAW OR FORMULA

The apparent luminance of a distant black object relating to visual range, such that the apparent luminance of the background sky above the horizon, and the extinction coefficient of the air layer near the ground, are represented by:

$$B_b = B_{sky}\,(1 - e^{-\sigma x})$$

where $B_b$ = apparent luminesce of black body
$B_{sky}$ = luminance of sky near horizon
$\sigma$ = extinction coefficient
x = distance from observer to object

**Keywords:** atmosphere, horizon, light, luminance, vision
KOSCHMIEDER, Hans Harald, twentieth century, German atmospheric physicist
**Sources:** Gillispie, C. C. 1981; Huschke, R. E. 1959; McIntosh, D. H. 1972.
*See also* KOPP; LUMINESCENCE; LUMINOSITY

## KOSCHMIEDER RELATIONSHIP

An equation that gives the visual range in the atmosphere using estimates for scattering and absorption of light by naturally occurring and pollutant gases and pollutants:

$$L_v = 3.92/b_{ext}$$

where $L_v$ = visual range or distance
$b_{ext}$ = extinction coefficient in terms of l/kilometers

**Keywords:** atmosphere, light, pollution, vision
KOSCHMIEDER, Hans Harald, twentieth century, German atmospheric physicist
**Sources:** Gillispie, C. C. 1981; Stevenson, L. H. and Wyman, B. C. 1991.
*See also* ABSORPTION; REFLECTION

## KOSSEL AND SOMMERFELD DISPLACEMENT LAW FOR COMPLEX SPECTRA

This law relates to the multiplet structure of the arc spectrum of any element, such that this arc spectrum is similar to the first spark spectrum (singly ionized atom spectrum) of the element one place higher in the periodic table or to the second spark spectrum (doubly ionized atom spectrum) of the element two places high, etc. Another statement is that an element with an odd atomic number that produces an arc spectrum shows an even multiplicity, and an element with an even atomic number that produces an arc spectrum shows odd multiplicity.

**Keywords:** arc, atom, element, ionized, spectrum
KOSSEL, Walter, 1888-1956, German physicist
SOMMERFELD, Arnold Johannes Wilhelm, 1868-1951, German physicist
**Sources:** Ballentyne, D. W. G. and Lovett, D. R. 1980; Thewlis, J. 1961-1964.

## KOSSOVICH NUMBER, Ko OR $N_{Ko}$

A dimensionless group for convective heat transfer during evaporation that relates the heat required for evaporating moisture in a material to heat required for heating the material from 0 to $T_c0$:

$$Ko = LH_v \, u/c \, \Delta T$$

where $LH_v$ = latent heat of vaporization
$u$ = moisture content, decimal
$c$ = specific heat of moist material
$\Delta T$ = temperature difference, $T_c - T_0$

The Kossovich number is the analog of the Biot heat transfer number but characterizes only the internal heat transfer without regard to the external process.

**Keywords:** convection, evaporation, heat transfer
**Sources:** Bolz, R. E. and Tuve, G. L. 1970; Lykov, A. V. and Mikhaylov, Y. A. 1961; Parker, S. P. 1992; Potter, J. H. 1967.
*See also* BIOT; RAMZIN

## KRAMER COOLING LAW (1959)—SEE COOLING LAWS

## KRANZBERG LAWS OF TECHNOLOGY (1997*)

**First Law**
Technology is neither good nor bad; nor is it neutral.

**Second Law**
Invention is the mother of necessity.

**Third Law**
Technology comes in packages, large and small.

**Fourth Law**
Although technology might be a prime element in many public issues, nontechnical factors take precedence in technology-policy decisions.

**Fifth Law**
All history is relevant, but the history of technology is most relevant. Technology's history can cast light on the most perplexing problems confronting us.

**Sixth Law**
Technology is a very human activity.

**Keywords:** history, invention, policy
KRANZBERG, Melvin, twentieth century (b. 1917), American educator and historian
**Source:** *STS Newsletter No. 114, Lehigh University, Winter 1997.

### KRIGAR-MENZEL LAW

When a string is bowed at any rational point, p/q, when p and q are primaries to each other, the part of the string immediately under the bow begins to move back and forth with two constant velocities whose ratio depends only on q and is:

$$1/(q - 1)$$

**Keywords:** bow, mechanics, string, velocities, vibration
KRIGAR-MENZEL, Otto, nineteenth century (b. 1861), German mathematician
**Sources:** NUC; Parker, S. P. 1997.
*See also* MERSENNE; PYTHAGOREAN

### KROGH LAW (1916)

The rate of a biological process is directly correlated to the temperature.

**Keywords:** biological, process, rate, temperature
KROGH, August, 1874-1949, Danish physiologist; Nobel prize, 1920, physiology/medicine
**Sources:** Gray, P., 1967; Lincoln, R. J. et al. 1982.
*See also* $Q_{10}$

### KRONIG NUMBER, Kr OR $N_{Kr}$

A dimensionless group in convective heat transfer that relates the electrostatic force and the viscous force:

$$Kr = 4L^2\beta\ \rho^2\ \Delta t\ E_S^2\ N[\alpha + 2/3(\rho_o^2/k\ \Gamma)]/\mu^2 M$$

where $E_s$ = electric field at surface
$N$ = Avogadro number
$\alpha$ = polarization coefficient
$\rho_o$ = molecular dipole moment
$k$ = Boltzmann constant
$M$ = molecular weight
$L$ = characteristic dimension
$\beta$ = coefficient of bulk expansion
$t$ = temperature
$\rho$ = density
$\Gamma$ = rate of change of temperature of medium (gamma)
$\mu$ = viscosity

**Keywords:** convective, electrostatic, heat transfer, viscous
KRONIG, Arthur Karl, 1822-1879, German scientist (physicist)
**Sources:** Bolz, R. E. and Tuve, G. L. 1970; Gillispie, C. C. 1981; Parker, S. P. 1992; Potter, J. H. 1967.

## KUNDT LAW

As an absorption band is approached from the red side of the spectrum, the refractive index is abnormally increased by the presence of the band whereas, with the approach from the blue side of the spectrum, the index is abnormally decreased.

**Keywords:** absorption, light, refractive, spectrum
KUNDT, August Adolph Eduard, 1839-1894, German physicist
**Sources:** Hodgman, C. D. 1952; Lide, D. R. 1980-1995.

## KUNDT LAW OF ABNORMAL DISPERSION

When the refractive index of a solution increases because of changes in composition and other causes, the optical absorption bands are displaced toward the red.

**Keywords:** absorption, light, physics, refraction, solution
KUNDT, August Adolph Eduard, 1839-1894, German physicist
**Source:** Thewlis, J. 1961-1964.

## KUSTNER LAW

A left-sided ovarian tumor causes torsion of its pedicle toward the right, and if right-sided, toward the left.

**Keywords:** medical, ovarian, tumor
KUSTNER, Otto Ernst, 1850-1931, German gynecologist
**Source:** Friel, J. P. 1974.

## KUTATELADZE NUMBER (1), Ku OR $N_{Ku}$

A dimensionless group used to represent electric arcs in gas streams:

$$Ku = I\,E\,L/\rho\,V\,\mu'$$

where $I$ = current density
$E$ = voltage
$L$ = characteristic length
$\rho$ = density of gas stream
$V$ = gas stream velocity
$\mu'$ = enthalpy

**Keywords:** arcs, electric, gas, streams
KUTATELADZE, Samson Semenovich, twentieth century (b. 1914), Soviet thermophysicist
**Sources:** Bolz, R. E. and Tuve, G. L. 1970; Lewytzkj, B. 1984; Parker, S. P. 1992; Potter, J. H. 1967.

## KUTATELADZE NUMBER (2), K OR $N_K$ (1951?)

A dimensionless group representing combined heat and mass transfer in evaporation:

$$K = LH_v/c_p\,(t_o - t_w)$$

where $LH_v$ = latent heat of vaporization
$c_p$ = specific heat
$t_o$ = stream temperature
$t_w$ = wall temperature

**Keywords:** evaporation, heat transfer, mass transfer
KUTATELADZE, Samson Semenovich, twentieth century (b. 1914), Soviet thermophysicist

**Sources:** Bolz, R. E. and Tuve, G. L. 1970; Lewytzkj, B. 1984; Parker, S. P. 1992; Potter, J. H. 1967.

## KUTTA-JOUKOWSKY (ZHUKOVSKI) LAW(EARLY 20TH CENTURY)

In aeronautics, the lift component perpendicular to the direction of motion exerted on a vehicle by the atmosphere, in an inviscid, incompressible fluid, is:

$$Li = \rho V_f G$$

where Li = lift component
$\rho$ = density of fluid
$V_f$ = velocity relative to fluid
G = circulation factor of air around airfoil

**Keywords:** aeronautics, air, air foil, fluid flow, lift

KUTTA, Wilhelm Martin (also listed as Martin), 1867-1944, German physicist

JOUKOWSKY (ZHUKOVSKI), Nikolair Jepowitch, 1847-1921, Soviet aeronautics scientist

**Sources:** Bothamley, J. 1993; Flugge, W. 1962; James, R. C. and James, G. 1968; Parker, S. P. 1989; Pelletier, P. A. 1994; Sneddon, I. N. 1976; Thewlis, J. 1961-1964; Wolko, H. S. 1981.

# L

## LADENBURG LAW

The speed of a photoelectron is proportional to the square root of the voltage of excitation.

**Keywords:** excitation, photoelectron, speed, voltage
LADENBURG, Rudolf, W. 1882-1952, German physicist
**Sources:** Clifford, A. F. et al. 1964; Gillispie, C. C. 1973; Honig, J. M. 1953.

## LAGRANGE LAW; ALSO KNOWN AS LAGRANGE-HELMHOLTZ; HELMHOLTZ-LAGRANGE; SMITH-HELMHOLTZ; HELMHOLTZ EQUATION

The relationship between linear and angular magnification at a spherical refracting interface is:

$$n_1\, y_1 \tan\theta_1 = n_2\, y_2 \tan\theta_2$$

where $y_1$ and $y_2$ = linear dimensions of object and image
$n_1$ and $n_2$ = refractive indexes of object and image
$\theta$ = angle between linear and angular magnification

**Keywords:** image, light, physics, refraction
LAGRANGE, Joseph Louis Comte, 1736-1813, Italian French mathematician
HELMHOLTZ, Herman Ludwig Ferdinand von, 1821-1894, German physicist and physiologist
SMITH, Robert, 1689-1768, English physicist
**Sources:** Becanson, R. M. 1974; Menzel, D. 1960; Parker, S. P. 1989.

## LAGRANGE NUMBER OR GROUP 1, $La_1$ OR $N_{La(1)}$

A dimensionless group used to represent agitation:

$$La_1 = \pi/\mu\, L^3N^2$$

where $\mu$ = dynamic viscosity
L = characteristic dimension of agitator
N = rate of rotation

**Keywords:** agitation, rotation, viscosity
LAGRANGE, Joseph Louis, 1736-1813, Italian French mathematician
**Sources:** Asimov, I. 1972; Bolz, R. E. and Tuve, G. L. 1970; Parker, S. P. 1992; Potter, J. H. 1967.

## LAGRANGE NUMBER 2, $La_2$ OR $N_{La(2)}$

A dimensionless group used in mass transfer in a turbulent system that relates combined molecular and eddy mass transfer rate to the molecular transfer rate:

$$La_2 = \frac{D + E_D}{D}$$

where D = molecular diffusivity
$E_D$ = eddy transfer rate

**Keywords:** eddy, mass, molecular, transfer
LAGRANGE, Joseph Louis, 1736-1813, Italian French mathematician
**Sources:** Bolz, R. E. and Tuve, G. L. 1970; Parker, S. P. 1992; Potter, J. H. 1967.

### LAGRANGE NUMBER 3, $La_3$ OR $N_{La(3)}$

A dimensionless group used in the representation of magneto fluid dynamics:

$$La_3 = \Delta P\ R/\mu V$$

where $\Delta P$ = pressure difference
R = radius
$\mu$ = dynamic viscosity
V = the velocity of flow

**Keywords:** dynamics, magneto-fluid, viscosity
LAGRANGE, Joseph Louis, 1736-1813, Italian French mathematician
**Sources:** Bolz, R. E. and Tuve, G. L. 1970; Parker, S. P. 1992; Potter, J. H. 1967.

### LAGRANGE PRINCIPLE OF LEAST ACTION

In the movement of bodies which interact so that the total energy remains constant, the sum of the products of the masses by the velocities and by the spaces described is a minimum. This was extended by Lagrange to systems of masses:

$$\delta\ \Sigma\ m \int vds = 0$$

where m = mass
$\delta$ = change in total energy
v = velocity
s = distance

**Keywords:** bodies, least action, masses, mechanics, moments, space
LAGRANGE, Joseph Louis Comte, 1736-1813, Italian French mathematician
**Sources:** (partial): Gillispie, C. C. 1981; Menzel, D. H. 1960; Parker, S. P. 1987.
*See also* CASTIGLANO; D'ALEMBERT; HAMILTON; MAUPERTUIS

### LAMB LAW

For thermal and hydraulic design of heat exchangers, and for very low Reynolds numbers, the drag coefficient of fluid flow is:

$$c = \frac{8\pi}{Re(2 - \ln Re)}$$

where c = the drag coefficient of flow, generally less than 1
Re = Reynolds number

**Keywords:** drag, flow, fluid, hydraulic, Reynolds, thermal
LAMB, Sir Horace, 1849-1934, English mathematician and fluid dynamicist
**Sources:** Kutateladze, S. S. and Borishanskii, V. M. 1966; NUC.

### LAMBERT-BEER LAW

Light absorption by colored solutions, based on concentration of particular solutions, intensify of incident light:

$$\log I_0/I = a\ b\ c$$

where $\log I_0/I$ = absorbance
$I_0$ = intensity of incident light
$I$ = intensity of transmitted light
$a$ = absorptivity
$b$ = length of path
$c$ = concentration of colored solution

**Keywords:** absorption, colored, concentration, incident, light
LAMBERT, Johann Heinrich, 1728-1777, German mathematician
BEER, Wilhelm, 1797-1850, German astronomer
**Source:** Considine, D. M. 1976.
*See also* ABSORBANCE; BEER

### LAMBERT-HOLZKNECHT LAW

A basic principle of radiotherapy stating that, to obtain a homogeneous dose on the surface of an area of the irradiated field, the source to skin distance must be at least equal to or greater than twice the greatest diameter of the field.

**Keywords:** distance, dose, irradiated, radiotherapy
LAMBERT, Johann Heinrich, 1728-1777, German mathematician
HOLZKNECHT, Guido, 1872-1931, Austrian radiologist
**Source:** Landau, S. I. 1986.

### LAMBERT LAW OF ABSORPTION (1760); LAMBERT-BEER-BOUGUER LAW; LBB LAW

When light is incident on a sample, part of the light is reflected, part is absorbed, and part is transmitted through the sample. The absorption of a material is expressed by the L-B-B law, and the intensity passing through a distance, x, can be written:

$$I = I_0\, e^{-a\,x} \qquad \text{or} \qquad I_0 \exp(-ax)$$

where $I$ = intensity of light at a distance
$x$ = distance into material
$I_0$ = intensity at the surface
$a$ = absorption coefficient

**Keywords:** absorbed, light, reflected, transmitted
LAMBERT, Johann Heinrich, 1728-1777, German mathematician
BEER, Wilhelm, 1797-1850, German astronomer
BOUGUER, Pierre, 1698-1758, French mathematician
**Sources:** Bothamley, J. 1993; Gray, D. E. 1972; Hodgman, C. D. 1952; Honig, J. M. 1953; Thewlis, J. 1969.
*See also* BEER-LAMBERT

### LAMBERT LAW OF EMISSION (LIGHT) (1760); LAMBERT COSINE EMISSION LAW

The law relates the emission of radiation of heat and light in different directions from a radiating surface. When the intensity of radiation from a surface that is luminous (whether by diffuse reflection or by temperature radiation) is measured, the apparent light intensity, as measured by a photometer, is greatest when the measurement is taken along a line perpendicular to the surface, and the intensity falls off as one moves from the perpendicular. The energy emitted in any direction is proportional to the cosine of the angle of that direction makes with respect to the normal.

**Keywords:** cosine, emission, light, normal
LAMBERT, Johann Heinrich, 1728-1777, German mathematician
**Sources:** Considine, D. M. 1976; Daintith, J. 1981; Layton, E. T. and Lienhard, J. H. 1988; Thewlis, J. 1961-1964.
*See also* ILLUMINATION (COSINE CUBED); KNUDSEN; LIGHT EMISSION; PHOTOMETRIC; STOKES

## LAMBERT LAW OF ILLUMINATION;LAMBERT COSINE ILLUMINATION LAW

The illumination of a surface upon which light falls is inversely proportional to the square of the distance of the source from the surface. If the source is inclined at an angle from the normal to the surface (direction of the rays), the illuminance is proportional to the cosine of the angle.

**Keywords:** angle, cosine, distance, inclined, light
LAMBERT, Johann Heinrich, 1728-1777, German mathematician
**Sources:** Friel, J. P. 1974; Glasser, O. 1944; Gray, D. E. 1973; Hodgman, C. D. 1952.
*See also* PHOTOMETRIC; STOKES

## LAMI THEOREM OR LAMY THEOREM

When a particle is acted upon by three forces, the necessary and sufficient condition for equilibrium is that the three forces shall be in one plane and each shall be proportional to the size of the angle between the other two.

**Keywords:** equilibrium, forces, mechanics, particles
LAMY, Bernard, 1640-1715, French mathematician
**Source:** Thewlis, J. 1961-1964.

## LAMONT LAW

The permeability of a material, such as steel, at a given flux density is proportional to a constant times the difference between the flux density and the saturation flux density:

$$\mu_o = k\,(\phi - \phi_s)$$

where $\mu$ = permeability
$\phi$ = flux density
k = constant for the material
$\phi_s$ = saturation flux density

**Keywords:** flux density, material, permeability, saturation
LAMONT, Johann von, 1805-1879, Scottish German astronomer
**Sources:** Gillispie, C. C. 1981; Pelletier, R. A. 1994.

## LANCEREAUX LAW OF THROMBOSIS

Clotting always takes place where the tendency to stagnation (stasis) is greatest, and especially at points where the influence of thoracic aspiration and cardiac propulsion is smallest.

**Keywords:** cardiac, clotting, physiology, thoracic
LANCEREAUX, Etienne, 1829-1910, French physician
**Source:** Landau, S. I. 1986.

## LANCHESTER LAW (MILITARY) OR END-SQUARED LAW (1914)

A law on which a military tactic is based, which states that if five aircraft meet seven opposing aircraft in battle, the aim should be to divide the seven into two groups of three and four, attacking first one group and then the other. Since $5^2 = 4^2 + 3^2$, the group of five has a better prospect of success than if the seven are attacked simultaneously.

**Keywords:** aircraft, battle, group, military
LANCHESTER, Frederick William, 1868-1946, British engineer
**Sources:** Dupuy, T.N. 1993; Johnson, J. E. 1980.

## LANCHESTER LAW (STATISTICS)

Assuming that $\Phi(z) = C_1 z$, the system effectiveness is a linear function of the number of successfully operating units. The expression for system effectiveness is:

$$E = C_1 N_n \pi r_i$$

when $1 \leq i \leq n$.

This formula leads to the fact that all variants of the system in Fig. L.1 are equivalent for a linear function $\Phi(z)$, where E is the system effectiveness, and $N_n$ is number of executing units.

**Keywords:** effectiveness, linear function, system, operating units
LANCHESTER, Frederick William, 1868-1946, British engineer
**Sources:** Parker, S. P. 1992; Ushakov, I. A. 1994.

## LANDOLT-OUDEMAN, LAW OF—SEE OUDEMAN

## LANDOUZY-GRASSET, LAW OF—SEE GRASSET

## LANE LAW

If a star contracts, the internal temperature of the star must increase. The general law is that gaseous bodies, from the generation of heat by contraction, may grow hotter independent of the source of energy, which assumes that the star behaves like a perfect gas.

**Keywords:** internal, solar, star, temperature

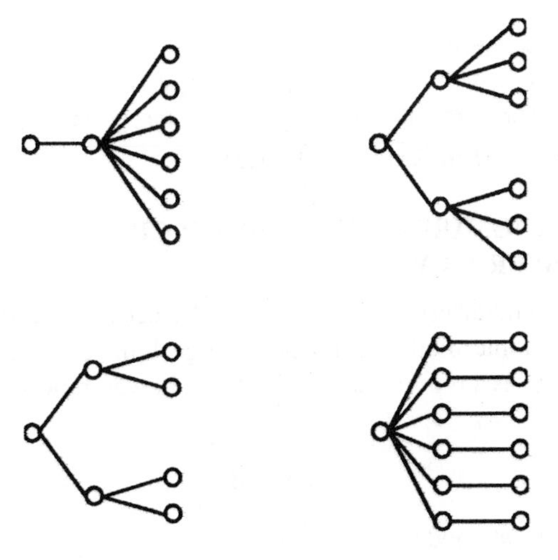

**Figure L.1** Lanchester.

LANE, Jonathan Homer, 1819-1880, American physicist
**Sources:** Ballentyne, D. W. G. and Lovett, D. R. 1980; Thewlis, J. 1961-1964.

### LANGEVIN-DEBYE LAW OF MAGNETIC SUSCEPTIBILITY (1905)

A law, or equation, from which the permanent dipole movements of polar molecules may be obtained from the temperature variation of measured values of polarization:

$$X = N\,p^2\,\mu_B^2/3kT + N\alpha$$

where X = static susceptibility of paramagnetic substance
N = number of magnetic dipoles per unit volume
p = effective magneton number
$\mu_B$ = Bohr magneton
k = Boltzmann constant
T = absolute temperature
$\alpha$ = temperature independent contribution of paramagnetism

**Keywords:** Bohr, dipole, magnetic, polarization, temperature
LANGEVIN, Paul, 1872-1946, French physicist
DEBYE, Peter Joseph Wilhelm, 1884-1966, Dutch American physical chemist; Nobel prize, 1936, chemistry
**Sources:** Freiberger, W. F. 1960; Gillispie, C. C. 1981; Thewlis, J. 1961-1964.

### LANGMUIR ISOTHERM MODEL (1916)

Based on the idea that adsorption of gas molecules are held to surface by bonds of the same character as covalent chemical bonds, termed chemisorption. The fraction of active site occupied by a monolayer of adsorbed molecule:

$$\frac{V}{V_M} = \frac{KP}{1 + KP}$$

where V = volume of gas adsorbed per unit amount of solid, STP
$V_M$ = volume adsorbed at saturation
P = partial pressure of adsorbate
K = constant depending on material and conditions

**Keywords:** adsorption, gas, molecules, surface
LANGMUIR, Irving, 1881-1957, American chemist; Nobel prize, 1932, chemistry
**Sources:** Clifford, A. F. 1964; Honig, J. M. 1953; Menzel, D. H. 1960; Parker, S. P. 1992.
*See also* ADSORPTION ISOTHERM; ISOTHERMS

### LANGMUIR LAW; LANGMUIR-CHILD LAW OR CHILD-LANGMUIR (1911); THREE-HALVES POWER LAW

The voltage and current characteristics of a gas-filled electron tube differ from those of a vacuum tube. In a thermionic diode with space-charge limited current, the anode current density in a vacuum tube is proportional to the three-halves power of the anode-cathode (plate) voltage:

$$j = K\,V^{3/2}/d^2$$

where j = current density at anode or electrode
K = constant for conditions

V = voltage (potential difference)
d = distance between two electrodes

It is also expressed as:

$$I_p = k_o V_p^{3/2}$$

where $I_p$ = plate current
$V_p$ = plate current
$k_o$ = constant for particular situation

**Keywords:** anode, cathode, charge, diode, plate, voltage
LANGMUIR, Irving, 1881-1957, American chemist; Nobel prize, 1932, chemistry
CHILD, Clement Dexter, 1868-1933, American physicist
**Sources:** Parker, S. P. 1989. 1992; Thewlis, J. 1961-1964; Whitaker, J. C. 1996.
*See also* CHILD-LANGMUIR

## LAPICQUE LAW (MEDICAL)

The chronaxia (time interval necessary to stimulate a nerve or muscle) is inversely proportional to the diameter of the nerve fiber.

**Keywords:** chronaxia, nerve, physiology
LAPICQUE, Louis, 1866-1952, French physiologist
**Sources:** Friel, J. P. 1974; Landau, S. I. 1986; Stedman, T. L. 1976.
*See also* ISOCHRONISM

## LAPLACE FIRST LAW (1774)—SEE LAPLACE DISTRIBUTION

## LAPLACE SECOND LAW OF ERROR—SEE NORMAL DISTRIBUTION

## LAPLACE LAW—SEE AMPERE; AREAS; KEPLERIAN MOTION

**Source:** Freiberger, W. F. 1960. For Laplace law.

## LAPLACE LAW OF FLUID FLOW (BLOOD)

A law of fluid mechanics, as applied to blood flow, states that the wall tension of a blood vessel is directly proportional to its radius and inversely proportional to its thickness.

**Keywords:** blood, fluid, radius, tension, thickness, wall
LAPLACE, Pierre Simon, 1749-1827, French astronomer and mathematician
**Sources:** Friel, J. P. 1974; Gennaro, A. R. 1979; Randall, J. E. 1958.

## LAPLACE LAW OF MAGNETIC FIELD

The strength of the magnetic field at any given point due to any element of a current-carrying conductor is directly proportional to the strength of the current and the projected length of the element, and inversely proportional to the square of the distance of the element from the point in question.

**Keywords:** conductor, current, electricity, magnetic
LAPLACE, Pierre Simon, 1749-1827, French astronomer and mathematician
**Sources:** Gillispie, C. C. 1981 (supp.); Thewlis, J. 1961-1964.
*See also* BIOT-SAVART

## LAPLACE LAW OF SUCCESSION; SOMETIMES CALLED BAYES-LAPLACE THEORY

If H is the hypothesis that a population possesses property $\phi$, and if, in a sample of size n (from sampling from an infinite or large population), all members are found to possess property $\phi$, the probability that the next number will possess property $\phi$ is:

$$(n + 1)/(n + 2)$$

and the next m members that possess property $\phi$ are:

$$(n + 1)/(n + m + 1)$$

**Keywords:** population, property, sample

BAYES, Thomas, 1702-1761, British mathematician

LAPLACE, Pierre Simon, 1749-1827, French astronomer and mathematician

**Sources:** Kotz, S. and Johnson, N. L. 1983. 1988.

## LAPLACE-POISSON LAW OF ADIABATIC CHANGE; POISSON LAW

The adiabatic process of an ideal gas having a constant ratio of specific heats:

$$p\rho^{-\gamma} = \text{constant}$$

$$\frac{p\theta^{\gamma}}{1-\gamma} = \text{constant}$$

$$\theta p^{1-\gamma} = \text{constant}$$

where $\gamma$ = ratio of specific heats, $c_p/c_v$
p = pressure
$\rho$ = density
$\theta$ = temperature

**Keywords:** adiabatic, gas, specific heat

LAPLACE, Pierre Simon, 1749-1827, French astronomer and mathematician

POISSON, Simeon Denis, 1781-1840, French mathematician

**Source:** Truesdell, C. A. 1980.

## LARGE NUMBERS, LAW OF (1689) (1713); BERNOULLI THEOREM (1713)

Various statements are used to represent the law of large numbers, but the idea is the same for each. If the size of a sample of statistically independent variables is increased indefinitely, good sample estimates of population parameters will tend to concentrate more and more closely about the true value. There are strong laws and weak laws of large numbers. Strong laws are concerned with showing that a variable, x, converges to a value $\mu$ with a probability of one. The strong law of large numbers is represented by the Borel theorem.

Weak laws consider conditions under which the probability that $|x - \mu|$ is greater than some given epsilon, $\varepsilon$, tends to zero. The weak law of large numbers is represented by the Bernoulli theorem. For the Bernoulli theorem, we have the following relationship:

$$\lim P\,(|x - m| > \varepsilon) = 0$$

$$\text{as } N \to \infty$$

where $\bar{x}$ = sample means
m = population
N = number of trials

S. Poisson gave the name of the law of large numbers to J. Bernoulli.

**Keywords:** probability, statistics, strong, weak
BERNOULLI, Jacques (Jakob), 1654-1705, Swiss mathematician
BOREL, Felix Edouard Emile, 1871-1956, French mathematician
POISSON, Simeon Denis, 1781-1840, French mathematician
**Sources:** Bothamley, J. 1993; Considine, D. M. 1995; Gillispie, C. C. 1991; Morris, C. G. 1992; Newman, J. R. 1956; Parker, S. P. 1989; Van Nostrand. 1947.
*See also* BERNOULLI; SMALL NUMBERS; STANDARD ERROR; STRONG LAW; WEAK LAW

## LARGE NUMBERS (ECONOMICS), LAW OF

As applied to economics, this law holds that, if an action is repeated many times, the percentage occurrence of each possible result will tend toward the statistical probability of each of those results, with less and less variability.

**Keywords:** econometrics, economics, probability, variability
**Source:** Moffat, D. W. 1976.

## LASÈGUE LAW

Now an obsolete expression, previously used to represent that tendon reflexes are enhanced in the presence of functional disorders but more often diminished in the presence of organic lesions or organic nervous disease.

**Keywords:** diminished, disorders, enhanced, lesions, reflex
LASÈGUE, Ernest Charles, 1816-1883, French physician
**Source:** Landau, S. I. 1986.

## LASKY LAW (MATERIALS)

A rule for determining the depletion of mineral resources. It states that, as the cumulative tonnage of mineralized rock forms at a constant geometric rate, the average grade of the accumulated tonnage decreases at an arithmetic rate.

**Keywords:** accumulated tonnage, grade, mineral resources
LASKY, Samuel Grossman, twentieth century (b. 1901), American geologist
**Source:** Morris, C. G. 1992.

## LAST THEOREM—SEE FERMAT

## LATENT HEAT—SEE TROUTON

## LATERAL PRESSURE—SEE JANSSEN; KETCHUM; RANKINE

## LAVAL NUMBER, La OR $N_{La}$

A dimensionless group in compressible fluid flow that relates the linear velocity of flow with the critical velocity of sound:

$$La = V/[(\gamma^2/\gamma + 1)RT]^{1/2}$$

where V = velocity
$\gamma$ = ratio of specific heats, $c_p/c_v$
R = gas constant
T = temperature

**Keywords:** compressible, fluid flow, velocity, sound
LAVAL, Carl Gustaf Patrick de, 1845-1913, Swedish engineer
**Sources:** Bolz, R. E. and Tuve, G. L. 1970; Encyclopaedia Britannica. 1961; Parker, S. P. 1992; Potter, J. H. 1967.
*See also* MACH

## LAVOISIER—SEE COMBUSTION; CHEMICAL COMBINATION; CONSERVATION OF MASS

## LBB LAW—SEE LAMBERT-BEER-BOUGUER

## LEARNING, LAWS OF

The basic laws of learning include:

1. The contribution of practice
2. Development of reinforcement and the internal stimulus
3. Extent of reward
4. Role of motivation

**Keywords:** motivation, practice, reinforcement, reward, stimulus
**Sources:** Corcini, R. J. 1987; Encyclopedia America. 1993.
*See also* ASSOCIATIVE SHIFTING; BELONGINGNESS; EFFECT; EXERCISE; READINESS; REWARD; THORNDIKE

## LEAST ACTION, PRINCIPLE OF

Dynamic motion always precedes in such a way that action is minimized with the least expenditure of energy. This was originally based on theology and then was placed on a mathematical basis by Lagrange.

**Keywords:** action, minimized, motion, theology
MAUPERTUIS, Pierre Louis Moreau de, 1698-1759, French mathematician
**Sources:** Barrow, J. D. 1991; Encyclopedia America. 1993; Good, C. V. 1973; Webster's Biographical Dictionary. 1959.
*See also* CASTIGLIANO; D'ALEMBERT; KÖNIG; LAGRANGE; LE CHATELIER; MAUPERTUIS

## LEAST EFFORT—SEE ZIPF

## LEAST TIME—SEE FERMAT

## LEAVITT—SEE PERIOD-LUMINOSITY

## LEBEDEV (LEBEDOV) NUMBER, Le OR $N_{Le}$

A dimensionless group used for drying porous materials, it relates the molar expansion flux to the molar vapor transfer flux:

$$Le = eb_t(t_s - t_o)/c_b p \rho_s$$

where $e$ = voidage; porosity
$b_t$ = vapor expansion in capillaries
$t_s$ = temperature of surrounding medium
$t_o$ = temperature of original material

$c_b$ = specific vapor capacity
$p$ = pressure
$\rho_s$ = density of porous material

**Keywords:** drying, expansion, molar, porous, vapor
LEBEDEV, P. D., twentieth century, Russian engineer;* or LEBVEDEV, Sergei Vassilievich, 1874-1934, Russian chemist (other references)
**Sources:** Gillispie, C. C. 1981; Hall, C. W. 1979; *Lykov, A. V. and Mikhaylov, Y. A. 1961; Parker, S. P. 1992.
*See also* LUIKOV; MINIOVICH

## LE CHATELIER LAW OF RADIATION

The intensity of radiation of red light is represented by:

$$I = 10^{6.7}\, T^{-3210/T}$$

where $I$ = intensity
$T$ = absolute temperature

**Keywords:** intensity, radiation, red light
LE CHATELIER, Henry Louis, 1850-1936, French chemist
**Source:** Rothman, M. A. 1963.
*See also* STEFAN-BOLTZMANN

## LE CHATELIER PRINCIPLE OR LAW (1884); PRINCIPLE OF LEAST ACTION; LE CHATELIER-BRAUN PRINCIPLE; LAW OF MOBILE EQUILIBRIUM

In a system initially at equilibrium, if any stress or force is exerted on the system, the equilibrium will shift in a direction that tends to decrease the intensity of the stress or force. The factors of equilibrium are temperature, pressure, electromotive force, corresponding to three forms of energy, heat, electricity, and mechanical energy.

**Keywords:** equilibrium, force, shift, stress
LE CHATELIER, Henri Louis, 1850-1936, French chemist
**Sources:** Considine, D. M. 1955; Honig, J. M. 1953: Lotka, A. J. 1956; Mandel, S. 1972; Parker, S. P. 1989; Stedman, T. L. 1976.
*See also* ROBIN; VAN'T HOFF

## LEDUC LAW OR EFFECT; ALSO RIGHI-LEDUC EFFECT

The volume of a gas mixture is equal to the sum of the volumes that would be occupied by each of the components of the mixture if at the temperature and pressure of the mixture.

**Keywords:** components, gas, mixture, pressure, temperature, volume
LEDUC, Stephane Armand Nicolas, 1853-1939, French physicist
RIGHI, Augusto, 1850-1920, Italian physicist
**Sources:** Ballentyne, D. W. G. and Lovett, D. R. 1972; Freiberger, W. F. 1960; Hunt, V. D. 1979. Also see AMAGAT-LEDUC; LE CHATELIER

## LEIBNIZ (LEIBNITZ) LAW

This law includes two different principles: (1) the principle of the indiscernibility of identicals, and (2) the identity of indiscernibles. The first asserts that if a is identical to b, whatever is true of a is true of b, and vice versa. The second asserts that if whatever is true of a is true of b, and vice versa, then a is identical to b.

**Keywords:** identity, indiscernibles, true
LEIBNIZ, Gottfried Wilhelm von 1646-1710, German philosopher and mathematician
**Source:** Flew, A. 1984.
*See also* CONTINUITY

## LENARD MASS ABSORPTION LAW

The absorption of electrons moving with a velocity of at least one-fifth of that of light is determined only by the mass of the absorbing matter traversed and not by its chemical nature.

**Keywords:** absorption, electrons, light, physics
LENARD, Philipp Eduard Anton von, 1862-1947, Czechoslovakian German physicist
**Source:** Gillispie, C. C. 1973.

## LENS LAW (OPTICS)

This law (Fig. L.2) is expressed in several ways, one of which is as follows. If $s_o$ and $s_i$ are the distance from the object to the principal point and the distance from the second principal point to the image, then the relationship between the object and the image is given by:

$$\phi = n_i/s_i + n_o/s_o$$

where $s_o$ = distance from object to principal point
$s_i$ = distance from second principal point to image

When the distances are measured from the focal points, the image relationship is known as the *Newton imaging equation.*

**Keywords:** distances, image, optics
**Source:** Bass, M. 1995.

## LENZ LAW (1833)

When a current is established by a change of flux through a circuit, the direction of the flux will be so as to oppose the act that caused it.

**Keywords:** current, electricity, flux
LENZ, Heinrich Friedrich Emil, 1804-1865, Russian German physicist
**Sources:** Bothamley, J. 1993; Daintith, J. 1981; Hodgman, C. D. 1952; Lide, D. R. 1990; Mandel, S. 1972; Parker, S. P. 1989. 1992; Thewlis, J. 1961-1964.
*See also* NEUMANN

## LEONARDO DA VINCI—SEE COMPOSITION OF FORCES, LAW OF

## LEOPOLD LAW

When the placenta is inserted upon the posterior (rear) wall of the uterus, the oviducts assume directions converging on the anterior (front) wall, but if inserted on the anterior wall during

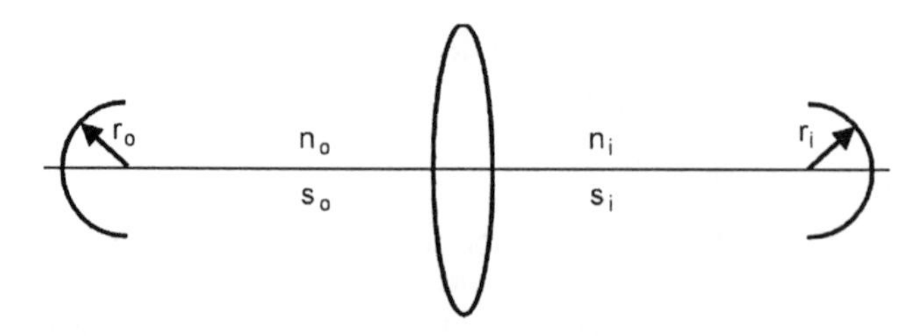


**Figure L.2** Lens Law.

recumbency (resting on surface), the tubes turn backward and become parallel to the axis of the body.

**Keywords:** medical, placenta, uterus
LEOPOLD, Christian Gerhard, 1846-1911, German physician
**Sources:** Friel, J. P. 1974; Landau, S. I. 1986; Stedman, T. L. 1976.

## LEROUX NUMBER, Ler OR $N_{Ler}$

A dimensionless group applicable to cavitation:

$$Ler = 2(p - p_c)/\rho V^2$$

where $p$ = local static pressure
$p_c$ = fluid vapor pressure
$\rho$ = mass density
$V$ = velocity of flow

Also equals the Cavitation number.

**Keywords:** cavitation, flow, pressure
LEROUX, Francois Pierre, French chemist (possible person, not confirmed)
**Sources:** Bolz, R. E. and Tuve, G. L. 1970; NUC; Parker, S. P. 1992; Potter, J. H. 1967.
*See also* CAVITATION NUMBER

## LEVER—SEE MACHINE

## LEVER LAW

This law describes a technique for determining the amount of each phase in a two-phase region on a phase diagram.

**Keywords:** diagram, phase
**Source:** Morris, C. G. 1992.

## LEVERETT NUMBER, J

A dimensionless group on two phase flow in porous materials relating the characteristic dimension of interfacial curvature and the characteristic dimension of pores:

$$J = (k/\varepsilon)^{1/2} p_c/\sigma$$

where $k$ = permeability (1/L2)
$\varepsilon$ = porosity, volume of voids per total volume
$p_c$ = capillary pressure
$\sigma$ = surface tension

**Keywords:** flow, interfacial, porous materials
LEVERETT, Frank, 1859-1943, American geologist
**Sources:** Gillispie, C. C. 1981; Land, N. S. 1972; Potter, J. H. 1967; Webster's Biographical Dictionary. 1959. 1995.

## LEVRET LAW (MEDICAL)

The insertion of the cord is marginal in placenta praevia (previa in Stedman).

**Keywords:** medical, placenta
LEVRET, Andre, 1703-1780, French obstetrician
**Sources:** Friel, J. P. 1974; Landau, S. I. 1986; Stedman, T. L. 1976.

## LÉVY STABLE LAWS (c. 1937)

A probability distribution characterized by a single parameter D, which ranges between 0 and 2, similar to the Gaussian but with infinite variance rather than finite variance. Levy developed the notion of stable, semistable, and quasistable laws. The laws were used by Benoit Mandelbrot for financial market studies. Levy discovered a class of random walks in which the steps vary in size.

**Keywords:** distribution, probability, variance
LEVY, Paul-Pierre, 1886-1971, French mathematician who was a mining engineer
**Sources:** Casti, J. L. 1990; Science News 150(2):104, August 17, 1996.

## LEWIS NUMBER, Le OR $N_{Le}$ (1939); OR LEWIS RELATION

A dimensionless group covering heat and mass transfer that relates mass diffusivity and thermal diffusivity:

$$Le = D\ \rho c_p/k$$

where D = mass diffusivity
$\rho$ = mass density
$c_p$ = specific heat at constant pressure
k = thermal conductivity

The Lewis number is also written as the inverse of this form.

The Lewis number equals the Prandtl number divided by the Schmidt number. The Lewis-Semenov number is the reciprocal of the Lewis number.

**Keywords:** diffusivity, heat, mass transfer
LEWIS, George William,* 1882-1948, American mechanical engineer
**Sources:** Bolz, R. E. and Tuve, G. L. 1970; Jerrard, H. J. and McNeill, D. B. 1992;* Land, N. S. 1972; Parker, S. P. 1992; Perry, R. H. 1967; *Wolko, H. S. 1981.
Also see PRANDTL; SCHMIDT; SEMENOV

## LEWIS-RANDALL RULE OR LAW (1916; 1923)

In dilute solutions, the activity coefficient of a given strong electrolyte is the same in all solutions of the same ionic strength.

**Keywords:** activity, electrolyte, ionic, solution
LEWIS, Gilbert Newton, 1875-1946, American physical chemist
RANDALL, Merle, 188-1950, American chemist and chemical engineer
**Sources:** Abbott, D. 1983; Merriam-Webster's Biographical Dictionary. 1995; National Cyclopaedia, v. 39; Pelletier, R. A. 1994.

## LEWIS RULE (1922)

A procedure for approximating the relationship between heat transfer and mass transfer:

$$a = h/\rho\ c_p = b$$

where a = for heat transfer
b = for mass transfer
h = film heat conductance
$\rho$ = density of fluid
$c_p$ = specific heat at constant pressure

**Keywords:** heat, mass, transfer
LEWIS, George William,* 1882-1948, American engineer
**Source:** National Cyclopedia, v. 39; *Wolko, H. S. 1981.

## LEWIS THEORY (1923)

A substance that in solution can bind and remove hydrogen ions or protons, it is a substance that acts as an electron-pair donor. A base reacts with an acid to give a salt and water (neutralization) and has a pH greater than 7.0.

**Keywords:** base, acid, electron-pair, neutralization, solution
LEWIS, Gilbert Newton, 1875-1946, American physical chemist
**Sources:** Allaby, M. 1994; Chemical Heritage 16(1):24, 1998.

## LICHTENECKER LAW

Permittivity is the effective dielectric constant in ferroelectric mixtures. The law is represented by:

$$\log K = X_1 \log K_1 + X_2 \log K_2$$

where K = permittivity
X = proportion of components, percent volume

**Keywords:** dielectric constant, ferroelectric, permittivity
**Source:** Hix, C. F. and Alley, R. P. 1958.

## LIEBIG LAW OR LAW OF LIMITING FACTORS; LAW OF MINIMA AND MAXIMA AND LAW OF OPTIMA; BLACKMAN PRINCIPLE OF LIMITING FACTORS (MINIMA, LAW OF) (1841)

The several components that promote the growth of the subject are presented in varied abundance. The most important of these are oxygen, nitrogen, hydrogen, and carbon. If one essential component is presented in limited amount, any moderate change in the sample supply of the other components will have little or no observable influence upon the rate of growth of the subject, as well as environmental conditions such as temperature, humidity, sunshine, water, etc. An essential component added in limited supply acts as a check or brake or as a limiting factor. The principle is often applied to nutrients in the soil or ocean and in the distribution or survival of a species. There is also a certain maximum that cannot be exceeded. There is also an optimum of environmental conditions, or combination of conditions, that is particularly favorable for life to flourish.

**Keywords:** component, growth, limiting, minimum, nutrients, species
LIEBIG, Justus von, 1803-1873, German chemist
BLACKMAN, Frederick F., 1866-1947, English plant pathologist
**Sources:** Brown, S. B. and L. B. 1972; Chemical Heritage 16(1): 25, 1998; Gray, P. 1967; Landau, S. I. 1986; Lapedes, D. N. 1974.
*See also* MINIMUM

## LIGHT DISPERSION—SEE RAYLEIGH (RALEIGH)

## LIGHT EMISSION, COSINE LAW OF

A law derived by Lambert relating the emission of radiation in different direction from a radiator surface. The greatest radiation is at a perpendicular to the surface, and intensity falls off as one moves from perpendicular to almost zero. However, the apparent brightness is the same from whichever direction it is viewed. The energy emitted in any direction is proportional to the cosine of the angle from the normal and is known as *Lambert cosine emission law.*

**Keywords:** brightness, cosine, direction, emission
LAMBERT, Johann Heinrich, 1728-1777, German mathematician
**Source:** Thewlis, J. 1961-1964.
*See also* BEER; LAMBERT

### LIGHT SCATTERING—SEE RAYLEIGH

### LIGHT TRANSMISSION, LAWS OF

*See* BOUGUER-LAMBERT
*See* BEER
*See* LAMBERT

### LIKING, LAW OF; LAW OF ATTRACTION

There are fundamental differences between liking and loving. There are two types of love—passionate, which is associated with sexual attraction, and compassionate, which is an extension of liking.

**Keywords:** compassionate; loving; passionate
BERSCHIED, Ellen S., twentieth century, American psychologist
**Sources:** Bothamley, J. 1993; Sheehy, N. 1997.

### LIMITING FACTORS—SEE LIEBIG

### LIMITING LAW—SEE LIEBIG (VON LIEBIG); MINIMA, LAW OF

### LIMITING LAW FOR STRONG ELECTROLYTES—SEE HÜCKEL

### LIMITS LAW

As applied to combustion, the presence of combustible gases affect the limits of flammability. Limits can be calculated from the limits law where L equals the flammability of the mixture; where x, y, and z equal the percent of each gas present in the mixture; and $L_x$, $L_y$, and $L_z$ equal the limits of flammability of the gases x, y, and z, respectively.

**Keywords:** combustion, gases, mixture, percent
**Source:** Thrush, P. W. 1968.
*See also* COMBUSTION

### LINDEMANN LAW (1910)

Solids melt at a temperature at which they are about 30 percent expanded from their volume at absolute zero.

**Keywords:** expand, melt, solids, volume
LINDEMANN, Frederick Alexander, 1886-1957, British scientist
**Sources:** Besancon, R. M. 1974; Thewlis, J. 1961-1964.

### LINDGREN VOLUME LAW

In geological formations, replacement of a material occurs on approximately a volume-by-volume basis.

**Keywords:** geological, material, replacements, volume
LINDGREN, Waldemar, 1860-1939, Swedish American geologist
**Sources:** Bates, R. L. and Jackson, J. A. 1980; Gary, M. et al. 1972.

## LINEAR CREEP, LAW OF

A linear relationship exists between deformation by creep and corresponding stresses, and the law of superposition applies to creep deformation for a homogeneous isotropic body. For a number of materials, the linear relationship between creep deformation and corresponding stress applies only for stresses not exceeding approximately one-half the ultimate strength, sometimes identified as R.

**Keywords:** creep, deformation, linear, stresses, superposition
**Source:** Scott Blair, G. W. 1949.
*See also* NONLINEAR CREEP; SUPERPOSITION

## LINEAR DAMAGE LAW; MINER RULE OF CUMULATIVE FATIGUE

A means of estimating the fatigue resistance of a material under conditions of stress cycles of varying magnitude such that, if the number of cycles endured at each stress range is n, and those required to cause failure is N, the failure will occur when:

$$\Sigma\ (n/N) = 1$$

**Keywords:** cycling, failure, fatigue, materials, stress
**Source:** John, V. 1992.
*See also* LOMNITZ

## LINEAR EXPANSION, LAW OF

The expansion of a solid material when heated as related to the temperature increase is represented by the equation:

$$L_t = L_0\ (1 + \alpha t)$$

where $L_t$ = length after expansion
$L_0$ = original length of the material
$\alpha$ = coefficient of linear expansion when heated
t = final temperature

**Keywords:** expansion, length, temperature
**Source:** Achinstein, P. 1971.

## LINEAR GROWTH CURVE OR LAW

Based on work done at the University of Missouri, a linear growth curve, during the self-inhibiting phase of growth, representing W the weight is:

$$W = A - B\ e^{-kt}$$

where W = weight at age t
t = age
A = weight at maturity
e = natural base of logarithm
k = fractional decline in the monthly gain
B = constant

The instantaneous velocity (rate) of growth is:

$$dW/dt = k (A - W)$$

This equation shows that the velocity (rate) of growth is not dependent on age but on the growth remaining to reach the mature weight.

**Keywords:** animal, growth, rate
**Source:** Robbins, W. J., et al. 1928.
*See also* GOMPERTZ

### LINEAR MOMENTUM—SEE NEWTON

### LINKAGE LAW (GENES)

There is a tendency of different gene pairs to segregate together if they are located on the same pair of homologous chromosomes; linked gene pairs may be separated with a frequency that varies from 0 to 50 percent, depending on the distance between them.

**Keywords:** chromosomes, gene, heredity
**Source:** Critchley, M. 1978.
*See also* BOBILLIER; MENDEL

### LIPSCHÜTZ LAW OF CONSTANT NUMBERS IN OVULATION

The number of ova discharged at each ovulation of a mammal is nearly constant for any given species.

**Keywords:** eggs, ova, reproduction
LIPSCHÜTZ, Benjamin, 1878-1931, Austrian dermatologist
**Source:** Stedman, T. L. 1976.

### LIPSCHÜTZ LAW OF FOLLICULAR CONSTANCY

If one ovary is removed, the follicles ripening at each cycle from the remaining ovary is the same as reached in both ovaries with the female bearing the same number of offspring at each pregnancy.

**Keywords:** birth, follicles, offspring, ovary, pregnancy
LIPSCHÜTZ, Benjamin, 1878-1931, Austrian dermatologist
**Source:** Stedman, T. L. 1976.

### LIPSCHÜTZ LAW OF PUBERTY

The development of changes from child to adult at puberty in the testis of the male and the ovary of the female is not determined by their sex organs but by the maturation of the axial part of the body, such as the head, neck, and trunk.

**Keywords:** adult, maturation, ovary, puberty, testis
LIPSCHÜTZ, Benjamin, 1878-1931, Austrian dermatologist
**Source:** Stedman, T. L. 1976.

### LISTER PRINCIPLE OR LAW (1867)

The principle of antiseptic surgery using chemicals (Lister used carbolic acid, also known as phenol) to destroy germs that made clean operations practical and led the way to modern surgery procedures.

**Keywords:** antiseptic, clean, surgery
LISTER, Baron Joseph, 1827-1912, English surgeon
**Source:** Asimov, I. 1976.

## LISTING LAW

If the eyeball is moved from a resting position, the rotational angle in the second position is the same as if the eye were turned about a fixed axis perpendicular to the first and second positions of the visual line.

**Keywords:** eye, medicine, physiology, position, rotational
LISTING, Johann Benedict, 1808-1882, German physiologist
**Sources:** Friel, J. P. 1974; Landau, S. I. 1986; Stedman, T. L. 1976.

## LLOYD MORGAN CANON—SEE MORGAN

## LOCALIZATION, LAW OF AVERAGE—SEE AVERAGE LOCALIZATION

## LOEB PHENOMENON OR PRINCIPLE (1880S)

Although Fabre originally first demonstrated taxis in animals, the principle was named after Loeb. A taxis is a reflex in which an organism orients its body toward or away from a source of stimulation. Loeb developed it as a counterpart to tropism in plants, the tendency for plants to grow toward light against gravity.

**Keywords:** gravity, light, organism, orientation, stimulation
FABRE, Jean Henri, 1823-1915, French entomologist
LOEB, Jacques, 1859-1924, German American physiologist
**Source:** Scientific American 271(1):74, July 1994

## LOGARITHMIC CREEP LAW (MATERIALS)

In transient creep, some materials (such as rubber, glass, and some metals) obey a law of the form:

$$\text{creep strain} = K \log t + e$$

where K = a constant related to the material and conditions
t = time
e = instantaneous strain when t is one

**Keywords:** creep, materials, mechanics, strain transient
**Source:** Scott Blair, G. W. 1949.

## LOGARITHMS, FUNDAMENTAL LAWS OF (1614)

As developed by J. Napier,

1. The logarithm of the product of two numbers is the sum of the logarithm of each number.
2. The logarithm of the quotient of two numbers is equal to the logarithm of the dividend minus the logarithm of the divisor.
3. The logarithm of a power of a number is equal to the product of the exponent and the logarithm of the number.
4. The logarithm of a root of a number is equal to the quotient of the logarithm of the number and the index of the root.

**Keywords:** index, mathematics, power, product, quotient, root
NAPIER, John, 1550-1617, Scottish mathematician
**Sources:** Asimov, I. 1972; James, R. C. and James, G. 1968.

## LOGIC, LAWS OF—SEE SYLLOGISM

## LOGISTIC LAW—SEE POPULATION GROWTH

## LOMMEL-SEELIGER LAW

A law of diffuse reflection of light, in which a surface will receive a quantity of light, which under any angle of incidence, the element will receive the amount:

$$L \cos i \, d\sigma$$

where L = quantity of light received on unit area under normal incidence
i = angle of incidence
$d\sigma$ =an element of surface in the light

The intensity of the diffusely reflected light at a unit distance in general will depend on both the angle of emanation, $\varepsilon$, and the azimuth of the plane of reflection. The law of diffuse radiation is a function of intensity and emanation:

$$f(i, \varepsilon) = \frac{\cos i \cos \varepsilon}{\cos i + \cos \varepsilon}$$

where $\varepsilon$ = angle of emanation

The mean intensity for all azimuths for a given emanation is proportional to L and $d\sigma$ and may be expressed as $\gamma L \, d\sigma \, f(i, \varepsilon)$, where $\gamma$ is a constant.

**Keywords:** azimuth, diffuse, incidence, light, reflection
LOMMEL, E. von, 1837-1899, German physicist
SEELIGER, H. von, 1849-1924, German astronomer
**Sources:** Barr, E. S. 1973; Bynum, W. F. et al. 1981; Thewlis, J. 1961-1964.
*See also* DIFFUSE REFLECTION

## LOMNITZ LAW

The strain in a solid when a small constant traverse stress is applied at time zero is:

$$\varepsilon = P/\mu[1 + q/a \log (1 + at)]$$

where $\varepsilon$ = strain
P = transverse stress
$\mu$ = rigidity
q, a = constants; a is such that it is large if t = 1

The fraction of the energy dissipated per period under harmonic stress is nearly independent of the period.

**Keywords:** energy, strain, stress, solid, transverse
LOMNITZ, C. Biographical information not located.
**Source:** Runcorn, S. K. et al. 1967.
*See also* LINEAR DAMAGE

## LONDON THEORY—SEE MAGNETIC FIELD DEPTH PENETRATION

## LORENTZ FORCE LAW (1895)

In classical radiation theory:

$$F = \rho(E + j\ \beta)$$

where F = force
ρ = density of charges
β = field of induction
j = current density, which is ρv
v = mean velocity
E = field strength, –grad ϕ – A

**Keywords:** charges, current, physics, plasma, radiation
LORENTZ, Hendrik Antoon, 1853-1928, Dutch physicist; Nobel prize, 1902, physics
**Sources:** Bothamley, J. 1993; Griem, H. R. 1964; Thewlis, J. 1961-1964.

## LORENTZ-LORENZ FORMULA (1869; 1870; 1880*)

A formula that provides that the refractive index of a gas is constant, that leads to the Gladstone-Dale law:

$$\text{constant} = (N^2 - 1)/(N^2 + 2)v$$

where N = refractive index
v = specific volume

**Keywords:** gas, refractive index
LORENTZ, Hendrik Antoon, 1853-1928, Dutch physicist; Nobel prize, 1902, physics
LORENZ, Ludwig Valentin, 1829-1891, Danish physicist
**Sources:** Besancon, R. M. 1974; *Gillispie, C. C. 1981; Isaacs, A. 1996; Parker, S. P. 1983; Thewlis, J. 1961-1964.
*See also* GLADSTONE-DALE

## LORENTZ-LORENZ LAW (1870,* 1878); LORENZ-LORENTZ LAW**

This law is based on the Clausius-Mossotti equation that provides that the refractive index of a gas is constant. When the sum of polarizing abilities of all particles present in an unit volume, α, is defined by:

$$p = \alpha x$$

where p = electric dipole induced by force, x
α = polarizing ability

$$|r| = (n^2 - 1)/(n^2 + 2) = 4\ \pi/3$$

where r = specific refraction
n = refractive index

L. Lorenz originally deduced this relationship. There is some confusion in the literature between the formula and law named after Lorentz and Lorenz.

**Keywords:** dielectric, dipole, electric, physics, polarizing
LORENTZ, Hendrik Antoon, 1853-1928, Dutch physicist; Nobel prize, 1902, physics

LORENZ, Ludwig Valentin, 1829-1891, Danish physicist
**Sources:** **Besancon, R. M. 1974; *Gillispie, C. C. 1981; Honig, J. M. et al. 1953.
*See also* CLAUSIUS-MOSSOTTI; *GLADSTONE AND DALE; REFRACTION; WIEDEMANN-FRANZ-LORENZ

## LORENTZ NUMBER, Lo OR $N_{Lo}$

A dimensionless group in magnetofluid hydrodynamics that relates the fluid velocity and velocity of light:

$$Lo = V/c$$

where V = fluid velocity
c = velocity of light

Also known as the *Einstein number.*

**Keywords:** fluid, light, velocity
LORENTZ, Hendrik Antoon, 1853-1928, Dutch physicist; Nobel prize, 1902, physics
**Sources:** Bolz, R. E. and Tuve, G. L. 1970; Land, N. S. 1972; Parker, S. P. 1992; Potter, J. H. 1967.
*See also* EINSTEIN NUMBER

## LORENZ CURVE—SEE PARETO

## LORENZ HYDRAULIC MODEL (1950) (BIOLOGY); FLUSH-TOILET MODEL

A representation of animal behavior similar to a "flush toilet" in which the contents accumulate over time and increase the animal's drive to perform a particular behavior. When sufficient stimulus occurs, the intensity of animal responses is proportional to the power of the stimulus.

**Keywords:** animal, behavior, intensity, stimulus
LORENZ, Konrad Zacharias, 1903-1989, Austrian zoologist
**Source:** Bothamley, J. 1993.

## LORENZ LAW; ALSO KNOWN AS WIEDEMANN-FRANZ RATIO

Describes the relationship of thermal conductivity ÷ electrical conductivity × the absolute temperature, and is equal to a constant × the Boltzmann gas constant ÷ the charge of an electron:

$$k/\sigma T = C (K/e)$$

where k = thermal conductivity
σ = electrical conductivity
T = absolute temperature
C = constant
K = Boltzmann gas constant
e = charge of electron

**Keywords:** conductivity, electricity, electron, gas, thermal
LORENZ, Ludwig Valentin, 1829-1891, Danish physicist
WIEDEMANN, Gustav Heinrich, 1826-1899, German physicist and chemist
FRANZ, Rudolf, twentieth century (b. 1903), German physicist
**Source:** Menzel, D. H. 1960.
*See also* WIEDEMANN-FRANZ

## LORENZ RULE (1916)

A relationship relating critical, melting, and boiling points of a substance derived from the Guldberg rule by K. Lorenz:*

$$T_c = 0.9(T_m + T_b)$$

where $T_c$ = critical temperature
$T_m$ = melting point
$T_b$ = boiling point

**Keywords:** boiling, critical, melting, temperature
**Sources:** *Fisher, D. J. 1988. 1991; Gillispie, C. C. 1981 (has Hans Lorenz, 1865-1940, German engineer).
*See also* GULDBERG

## LOSCHMIDT—SEE AVOGADRO

LOSCHMIDT, Joseph, 1821-1895, Austrian chemist

## LÖSSEN LAW OR RULE

Hemophilia is not transmitted from males having the disease to their immediate offspring.

**Keywords:** hemophilia, heredity, offspring, physiology
LÖSSEN, Herman Friedrich, 1842-1909, German surgeon
**Source:** Friel, J. P. 1974.

## LOTKA LAW (1956)

An empirical rule relating the number of occurrences of authors' names in bibliographic databases to the number of different authors in that database. The law was originally expressed as the number of people producing at least n papers decreases rapidly approximately as $1/n^2$. The following model can be used to predict the number of author names, $A_n$, that occurs n times:

$$A_n = A_1/n^2$$

**Keywords:** authors, bibliographic, papers
LOTKA, Alfred Jam, 1880-1949, American biophysicist
**Source:** Watter, C. 1992.
*See also* ZIPF

## LOUIS LAW

Pulmonary tuberculosis generally begins in the left lung, and tuberculosis of any part is localized in the lungs.

**Keywords:** lung, medical, tuberculosis
LOUIS, Pierre Charles Alexandre, 1787-1872, French physician
**Sources:** Friel, J. P. 1974; Landau, S. I. 1986; Stedman, T. L. 1976.

## LOVELOCK—SEE GAIA

## LOWRY—SEE BRÖNSTED-LOWRY

## LOW MAGNETIC FIELDS—SEE RAYLEIGH

## LUCCA LAW (TERRITORIES)

Other things being equal, territories will be richer when small and independent than when they are large and dependent.

**Keywords:** independent, size, territories
LUCCA is a small town in Italy.
**Source:** Sale, K. 1980.

## LUIKOV (LYKOV) NUMBER, Lu OR $N_{Lu}$

A dimensionless group used in drying and wetting for diffusivity in moist porous bodies that relates the moisture diffusivity to the thermal diffusivity in a porous body:

$$Lu = D/\alpha = D\, c\rho/k_c$$

where $k_c$ = mass transfer coefficient
D = mass diffusivity
$\alpha$ = thermal diffusivity, k/c
k = thermal conductivity
c = specific heat
$\rho$ = mass density

The Luikov number equals the Prandtl number divided by the Schmidt number.

**Keywords:** diffusivity, drying, moist, porous, wetting
LUIKOV, Aleksei Vassilievich, 1910-1974, Soviet thermophysicist
**Sources:** Begell, W. 1983; Bolz, R. E. and Tuve, G. L. 1970; Hall, C. W. 1979; Land, N. S. 1972; Parker, S. P. 1992; Potter, J. H. 1967.
*See also* LEBEDOV; LEWIS; PRANDTL; SCHMIDT

## LUKOMSKII NUMBER, Luk OR $N_{Luk}$

A dimensionless group that represents the ratio of the thermal diffusivity to the mass transfer of a product where combined heat and mass transfer occurs:

$$Luk = \alpha/a_m$$

where $\alpha$ = thermal diffusivity
$a_m$ = potential conductivity of mass transfer

**Keywords:** diffusivity, mass, thermal, transfer
**Sources:** Bolz, R. E. and Tuve, G. L. 1970; Hall, C. W. 1979; Lykov, A. V. and Mikhaylev, Y. A. 1961; Parker, S. P. 1992; Potter, J. H. 1967.
*See also* LEWIS; LUIKOV (LYKOV)

## LUMINESCENCE EQUIVALENCE LAW

The absorption of radiation by a luminescent system is a quantum process involving one quantum per absorbing particle, for example, an atom.

**Keywords:** absorption, luminescent, quantum
**Sources:** Considine, D. M. 1976 (partial); Parker, S. P. 1983.
*See also* FRANCK-CONDON; KOSCHMEIDER

### LUMINOSITY PERIOD FOR CEPHEIDS, LAW OF

Cepheid variables (stars that have a regular period of light pulsation) have an absolute brightness or luminosity, which is connected by a fixed law to the period of their light variation, with the relative distance related to their apparent brightness, m:

$$M = m + 5 - 5 \log d$$

where M = brightness or luminosity
m = apparent brightness
d = relative distance

**Keywords:** brightness, cepheid, light, luminosity
**Source:** Thewlis, J. 1961-1964.

### LUNDQUIST NUMBER, Lu OR $N_{Lu}$

A dimensionless group in magnetohydrodynamics used to characterize unidirectional Alfvén waves:

$$Lu = \beta \, \sigma \, L \, (\mu/\rho)^{1/2}$$

where $\beta$ = magnetic flux density
$\sigma$ = electrical conductivity
L = fluid layer thickness
$\mu$ = magnetic permeability
$\rho$ = mass density

**Keyword:** electrical, magnetic, magnetohydrodynamics
**Sources:** Bolz, R. E. and Tuve, G. L. 1970; Chemical Engineering Progress 59(8):75, 1963; Jerrard, H. G. and McNeill, D. B. 1992; Land, N. S. 1972; Lerner, R. G. and Trigg, G. L. 1991; Parker, S. P. 1992; Potter, J. H. 1967.
*See also* **ALFVÉN**

### LUSSER PRODUCT LAW OF RELIABILITIES

When the reliability of components is $R_n$, the reliability of the system, $R_{system}$, is the product of the reliabilities of the components, which points to the need for high reliabilities of components if the system reliability is to be high:

$$R_{system} = R_1 \times R_2 \times R_3 \ldots R_n$$

**Keywords:** components, reliability, system
LUSSER, Robert, German space scientist during WWII
**Source:** Bazovsky, I. 1961;
*See also* EXPONENTIAL FAILURE; PRODUCT LAW; RELIABILITY

### LYASHCHENKO NUMBER, Ly OR $N_{Ly}$ (c. 1964)

A dimensionless group in fluidization:

$$Ly = N_{Re}^3/N_{Ar}$$

where $N_{Re}$ = Reynolds number
$N_{Ar}$ = Archimedes number

**Keywords:** dimensionless group, fluidization
**Sources:** Bolz, R. E. and Tuve, G. L. 1970; Parker, S. P. 1992; Potter, J. H. 1967.
*See also* ARCHIMEDES; REYNOLDS

## LYELL—SEE UNIFORM CHANGE, LAW OF

## LYKOUDIS NUMBER, Ly OR $N_{Ly}$

A dimensionless group in heat transfer and magnetohydrodynamics:

$$Ly = (\sigma/\rho)(\mu H)^2(L/g\beta\Delta T)^{1/2}$$

where $\mu$ = magnetic permeability
$H$ = magnetizing force
$\sigma$ = electrical conductivity
$\rho$ = mass density
$L$ = characteristic length
$g$ = gravitational acceleration
$\beta$ = temperature coefficient of volume expansion
$\Delta T$ = temperature differential

The Lykoudis number is equal to the Hartmann number squared divided by the Grashof number.

**Keywords:** electrical, magnetic, magnetohydrodynamics
LYKOUDIS, Paul S., twentieth century (b. 1926), American aeronautical engineer
**Sources:** American Men of Science; Bolz, R. E. and Tuve, G. L. 1970; Kakac, S. et al. 1987; Land, N. S. 1972; Parker, S. P. 1994; Potter, J. H. 1967; Rohsenow, W. M. and Hartnett, J. P. 1973.

## LYKOV NUMBER—SEE LUIKOV NUMBER

## MCADAM RULE (1951)

The relationship between the Young modulus and the density of iron and steel is:

$$E = \rho^{3.4}$$

where E = Young modulus (elasticity)
ρ = density

**Keywords:** density, elasticity, modulus
McADAM, G. D., twentieth century, American engineer
**Sources:** Fisher, D. J. 1988; Iron and Steel Inst. J. 16B, 348, 1951.
*See also* YOUNG MODULUS

## MCADAMS GROUP OR NUMBER, Ma OR $N_{Ma}$ (1933)

A model for a dimensionless group or number that represents condensation for a given surface orientation:

$$Ma = h^4\ L\ \mu\ \Delta t/k^3\ \rho^2\ g\ r\ LH_c$$

where h = heat transfer coefficient
L = characteristic length dimension
μ = absolute viscosity
Δt = temperature, T (T is absolute temperature)
k = thermal conductivity
ρ = mass density
g = acceleration due to gravity
$LH_c$ = latent heat of condensation

**Keywords:** condensation, dimensionless, group
McADAMS, William Henry, 1892-1975, American chemical engineer
**Sources:** Bolz, R. E. and Tuve, G. L. 1970; Land, N. S. 1972; Layton, E. T. and Lienhard, J. H. 1988; Parker, S. P. 1992; Potter, J. H. 1967.
*See also* DITTUS-BOELTER

## MCCABE—SEE DELTA LAW

## MACHINES, LAW OF

A machine is a device in which the work output is always less than the work input, but in which either a force or speed advantage can be obtained. The theoretical mechanical advantage (TMA) is the force output divided by the force input. The actual mechanical advantage is the theoretical mechanical advantage times the efficiency, or output force divided by input force. The simple machines (Fig. M.1) are represented by:

1. Lever — TMA = $L_1/L_2$
2. Inclined plane — TMA = $D_I$ (incline)/$D_0$ (height)
3. Wheel and axle — TMA = R (wheel)/r (axial)

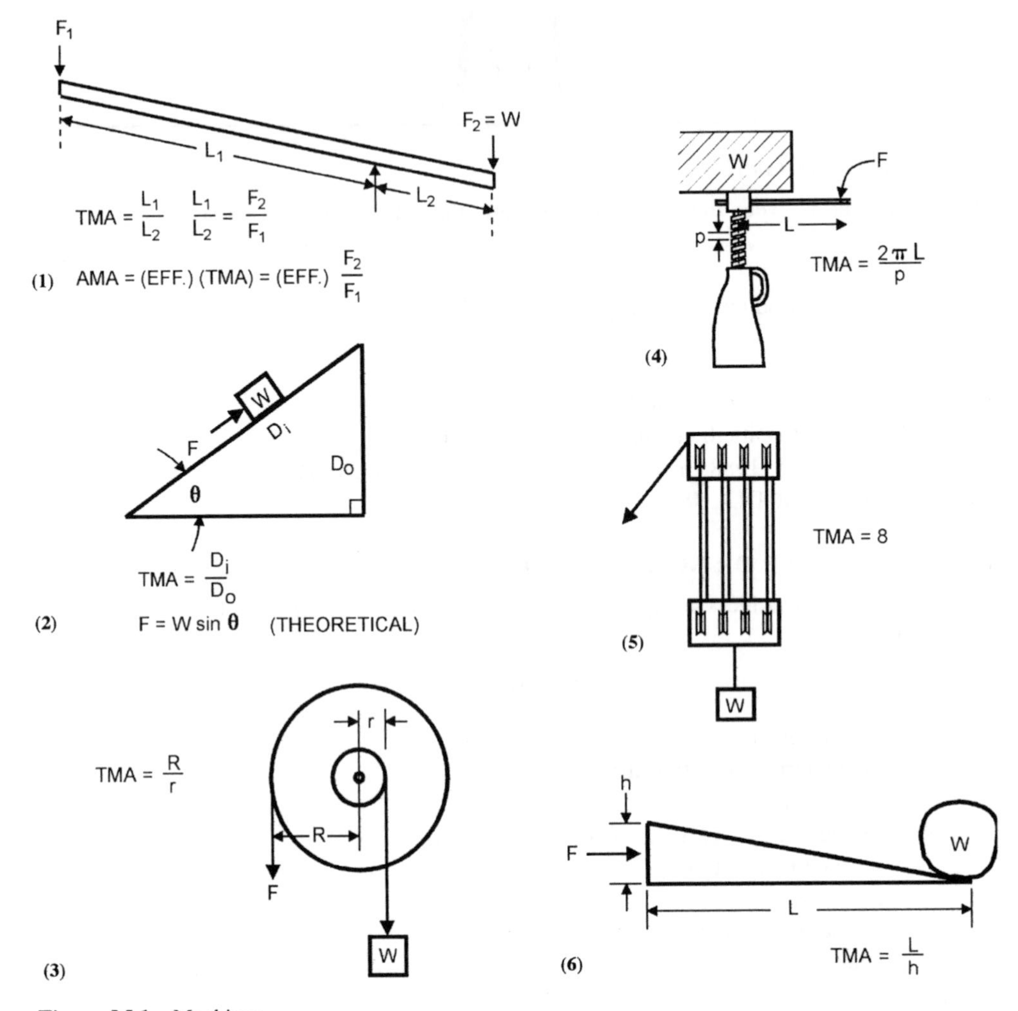


**Figure M.1** Machines.

4. Screw — TMA = $2\pi L/p$ where L = the length of the handle and p = the pitch of the screw
5. Pulley — TMA = $F_o/F_i$, which is equal to the number of ropes supporting the load
6. Wedge — TMA = L (length)/h (height)

**Keywords:** lever, plane, pulley, screw, theoretical mechanical advantage, wedge, wheel
**Sources:** Gillispie, C. C. 1991; Northrup, E. T. 1917; Turner, R. et al. 1981.

## MACH MAGNETIC NUMBER, $Ma_m$ OR $N_{Ma(m)}$

A dimensionless group in magnetohydrodynamics relating the flow speed to the Alfvén wave speed:

$$Ma_m = V\,(\rho\mu)^{1/2}/\beta$$

where V = flow velocity
$\rho$ = mass density
$\mu$ = magnetic permeability
$\beta$ = magnetic induction

The Mach magnetic number equals the Alfvén number and also the reciprocal of the Cowling number. The Mach magnetic number is also the same as the Sarrau number.

**Keywords:** flow, magnetohydrodynamics, wave
MACH, Ernst, 1838-1916, Austrian physicist
**Sources:** Bolz, R. E. and Tuve, G. L. 1970; Land, N. S. 1972; Potter, J. H. 1967.
*See also* ALFVÉN NUMBER; COWLING NUMBER; SARRAU NUMBER

### MACH NUMBER, Ma OR $N_{Ma}$ (1887)

A dimensionless group representing relationship of linear velocity to sonic velocity in compressible gas flow. The Mach 2 represents a speed at twice the speed of sound. The Mach number equals (Cauchy number)$^{1/2}$.

$$Ma = V/V_s \text{ or } V/a$$

where $V_s$ or a = velocity of sound in the fluid (sonic velocity)
V = velocity

Also known as the Maievskii number; previously used for the Bairstow number.

**Keywords:** compressible, dimensionless, sound, velocity
MACH, Ernst, 1838-1916, Austrian physicist and philosopher
BAIRSTOW, Leonard, 1880-1963, engineer
**Sources:** Bolz, R. E. and Tuve, G. L. 1973; Gillispie, C. C. 1981; Grigull, U., et al. 1982; Land, N. S. 1972; Parker, S. P. 1989. 1992; Perry, R. H. 1967; *Potter, J. H. 1967.
*See also* CAUCHY; MACH MAGNETIC; NEWTON

### MACH STATEMENT ON PRINCIPLE OF THE CONSERVATION OF AREAS (1883)

If, from any point in space, radii can be drawn to several masses, and projections be made upon any plane of the areas that the several radii describe, the sum of the products of these areas into the respective masses will be independent of the action of the internal forces.

**Keywords:** areas, forces, mass, mechanics
MACH, Ernst, 1838-1916, Austrian physicist and philosopher
**Source:** Mitton, S. 1973.

### MAGENDIE—ALSO KNOWN AS BELL LAW; SEE BELL-MAGENDIE LAW

### MAGNETIC AND ELECTRIC FIELDS, BASIC LAWS OF

- LORENTZ FORCE EQUATION
- AMPERE CIRCUITAL LAW
- MAGNETIC INTENSITY
- FARADAY LAW OF ELECTROMAGNETIC INDUCTION
- LENZ LAW
- MAXWELL LAWS

**Source:** Potter, J. H. 1967.

### MAGNETIC-DYNAMIC NUMBER, MagDyn

A dimensionless group in magnetohydrodynamics that relates the magnetic pressure and the dynamic pressure:

$$MagDyn = \sigma V \beta^2 L/\rho V^2$$

where $\sigma$ = electrical conductivity
$V$ = velocity
$\beta$ = magnetic induction
$L$ = characteristic length
$\rho$ = mass density

**Keywords:** dynamic, magnetic, magnetohydrodynamics
**Source:** Land, N. S. 1972.
*See also* MAGNETIC PRESSURE NUMBER

## MAGNETIC FIELD—SEE RAYLEIGH LAW FOR LOW MAGNETIC FIELDS

## MAGNETIC FIELD DEPTH PENETRATION, A SEMIEMPIRICAL LAW

The average depth of penetration of a magnetic field into the surface of a bulk specimen is such that there is an increase according to:

$$[\lambda(T)/\lambda(0)]^2 = 1/[1 - (T/T_c)^4]$$

where $\lambda$ = average penetration depth
$T$ = absolute temperature

**Keywords:** depth, magnetic field, penetration, temperature
**Source:** Thewlis, J. 1961-1964.
*See also* MEISSNER

## MAGNETIC FLUX—SEE FARADAY

## MAGNETIC FORCE NUMBER OR PARAMETER

A dimensionless group that represents the ratio between the magnetic body force and the inertia force applicable to studies in magneto-fluid dynamics, represented by:

$$N = \mu^2 H^2 \sigma L/\rho V$$

where $N$ = magnetic force parameter
$\mu$ = magnetic permeability
$H$ = field strength
$\sigma$ = electrical conductivity[sigma]
$L$ = characteristic field dimension
$\rho$ = density
$V$ = fluid velocity

**Keywords:** dimensionless, fluid, inertia force, magnetic
**Sources:** Bolz, R. E. and Tuve, G. L. 1970; Land, N. S. 1972; Parker, S. P. 1992; Potter, J. H. 1967.
*See also* MAGNETIC PRESSURE

## MAGNETIC INTENSITY AND INDUCTION, LAW OF

In a material, the vector quantities of magnetic field intensity and magnetic induction are intensive, as each has a definite value at a given point. Otherwise, for a simple system in a magnetic field, the pressure, temperature, and magnetic field intensity should fix the stable equilibrium state.

**Keywords:** electricity, equilibrium, induction, magnetic, stable
**Source:** Honig, J. M. 1953.

## MAGNETIC INTENSITY PRODUCED BY AN ELECTRIC CURRENT—SEE BIOT-SAVART

## MAGNETIC INTERACTION NUMBER, MagInter

A dimensionless group used to represent dynamics of ferro fluids:

$$\text{MagInter} = \mu H^2 R/2\sigma$$

where $\mu$ = free space permeability
$H$ = magnetizing force
$R$ = tank radius
$\sigma$ = surface tension

**Keywords:** dynamics, ferro, fluids
**Source:** Land, N. S. 1972.

## MAGNETIC (OHM)—SEE ROWLAND

## MAGNETIC PRANDTL NUMBER, $Pr_M$

A dimensionless group in magnetohydrodynamics:

$$Pr_M = \mu\sigma\upsilon$$

where $\mu$ = magnetic permeability
$\sigma$ = electrical conductivity
$\upsilon$ = kinematic viscosity

**Keywords:** electrical, magnetic, magnetohydrodynamics
PRANDTL, Ludwig, 1875-1953, German engineer and physicist
**Sources:** Bolz, R. E. and Tuve, G. L. 1970; Land, N. S. 1972; Parker, S. P. 1992.

## MAGNETIC PRESSURE NUMBER, S

A dimensionless group in magnetohydrodynamics relating the magnetic pressure and the dynamic pressure:

$$S = \mu H^2/\rho V^2$$

where $\mu$ = magnetic permeability
$H$ = magnetizing force
$\rho$ = mass density
$V$ = velocity

**Keywords:** dynamic, magnetic, magnetohydrodynamics
**Sources:** Bolz, R. E. and Tuve, G. L. 1992; Land, N. S. 1972; Parker, S. P. 1992; Potter, J. H. 1967.
*See also* MAGNETIC DYNAMIC

## MAGNETIC PULL, LAWS OF

1. The magnetic force on iron or other magnetic material brought into the field tends to move the material in such a direction to reduce to a minimum the reluctance of the magnetic circuit.
2. The magnitude of the force at any point is proportional to the space rate of change of flux as the iron passes the point or is proportional to the rate of change of the reluctance.

**Keywords:** electricity, flux, magnetic, reluctance
**Source:** Daintith, J. 1981.

## MAGNETIC REYNOLDS NUMBER, $Re_m$

A dimensionless group in magneto fluid dynamics that relates motion induced magnetic field to applied magnetic field:

$$Re_m = \sigma V L \mu$$

where $\sigma$ = electrical conductivity
$1/\sigma\mu$ = magnetic kinematic viscosity
$V$ = velocity
$L$ = characteristic length
$\mu$ = magnetic permeability

**Keywords:** dynamics, fluid, electrical, magnetic
REYNOLDS, Osborne, 1842-1912, English engineer
**Sources:** Bolz, R. E. and Tuve, G. L. 1970; Kakac, S. 1987; Land, N. S. 1972; Oxford English Dictionary. 1989; Potter, J. H. 1967; Rohsenow, W. M. and Hartnett, J. P. 1973.

## MAGNETIC SUSCEPTIBILITY—SEE LANGEVIN-DEBYE

## MAGNETIC $T^{3/2}$ LAW

The saturation intensity of magnetization follows the form:

$$M_s = M_o (1 - CT^{3/2})$$

where $M_s$ = saturation intensity of magnetization
$M_o$ = original state
$T$ = absolute temperature
$C$ = a constant for the material

**Keywords:** electricity, magnetism, saturation
**Source:** Thewlis, J. 1961-1964.

## MAGNETISM, LAW OF

Like charged magnetic poles repel; unlike charged magnetic poles attract.

**Keywords:** charge, electricity, magnetic, poles
**Sources:** Daintith, J. 1981; Morris, C. G. 1992.
*See also* BLOCH; MAGNETIC PULL

## MAGNUS EFFECT (1860)

Refers to the sideways thrust on the axis of a rotating cylinder when placed in a flow of a fluid at right angles to the axis. In the1840s, H. Magnus discovered heat conduction in gases.

**Keywords:** air, cylinder, rotating, thrust
MAGNUS, Heinrich Gustav, 1802-1870, German chemist and physicist
**Sources:** Flugge, W. 1962; Layton, E. T. and Lienhard, J. H. 1988; Parker, S. P. 1989; Webster's Biographical Dictionary. 1959. 1995.
*See also* STROUHAL

## MAIEVSKII NUMBER—SEE MACH

## MAJORANA EFFECT

A magneto-optic effect that deals with the optical anisotropy of colloidal solutions, probably caused by the orientation of particles in the magnetic field.

**Keywords:** anisotropy, colloidal, magnetic, optical, particles
MAJORANA, Ettore, 1906-1938, Italian physicist
**Source:** Parker, S. 1987. 1989.

## MAKEHAM LAW (1860); OR LAW OF HUMAN MORTALITY

The risk of dying is equal to the sum of a constant and a multiple of a constant raised to a power to the age, x, of the life:

$$M = A + B\ e^{x}$$

where $M$ = mortality
$A, B$ = constants
$x$ = age

which is a closer approximation to statistical findings than Gompertz law. Makeham assumed that death was the result of (1) chance and (2) an increased inability to withstand deterioration, as represented by A and B.

**Keywords:** age, death, dying, life
MAKEHAM, William Matthew, ?–1892, British mathematician
GOMPERTZ, Benjamin, 1779-1865, British actuarist
**Source:** Stedman, T. L. 1976.
*See also* GOMPERTZ

## MALLARD LAW

In crystallography, planes or axes of pseudosymmetry in a space lattice may become twinning planes or axes. The formula is:

$$D = K \sin E$$

where $D$ = half the distance between the points of emergence of the optic axes
$K$ = constant for any combination of lenses on a given microscope
$E$ = one half the optic axial angle in air

**Keywords:** axis, crystallography, lattice, pseudosymmetry, space, twinning
MALLARD, Francis Ernest, 1833-1894, French crystallographer and mineralogist
**Sources:** Barr, E. S. 1973; Gary, M. et al. 1972.
*See also* MANEBACH; TWIN

### MALTHUS, LAW OF; MALTHUSIAN LAWS; OR MALTHUSIAN DOCTRINE

Living organisms reproduce at such a high rate that, if the young should survive and reproduce at the same rate, any one species would soon fill the Earth, but since there are not enough necessities of life for all, there is a great struggle for existence in each succeeding generation.

**Keywords:** growth, living, population, reproduce, survival
MALTHUS, Thomas Robert, 1766-1834, English economist
**Sources:** Friel, J. P. 1974; Landau, S. I. 1986.

### MALUS LAW (1808); MALUS COSINE-SQUARED LAW; THEORY OF MALUS AND DUPIN

For a beam of light incident on a polarizer such that the angle between the plane of vibrations of the beam and the plane of vibrations that are transmitted by the polarizer, the intensity of the transmitted beam is:

$$I = I_0 \cos^2\theta$$

where $I$ = intensity of transmitted light
$I_0$ = intensity of light incident on a polarizer
$\theta$ = angle between beam of light and polarizer

**Keywords:** incident; light, optics, polarized
MALUS, Etienne Louis, 1775-1812, French engineer and physicist
DUPIN, Pierre Charles Francois, 1784-1873, French mathematician
**Sources:** Besancon, R. M. 1974; Gillispie, C. C. 1981; Menzel, D. H. 1960; Morris, C. G. 1992; Parker, S. P. 1983. 1992; Thewlis, J. 1961-1964.

### MANEBACH LAW

A type of twinning in the monoclinic crystal system. The basal pinacoid is the "twinning plane."

**Keywords:** crystal, monoclinic, plane, twinning
MANEBACH. Biographical information not located.
**Source:** Thrush, P. W. 1968.
*See also* MALLARD; TWIN (CRYSTALLOGRAPHY)

### MARANGONI NUMBER, Ma OR $N_{Ma}$

A dimensionless group used in cellular convection due to surface tension gradients:

$$Ma = (\Delta\sigma/\Delta T)\ (\Delta T/\Delta L)\ (d^2\ \mu\ D)$$

where $\Delta\sigma/\Delta T$ = surface tension–temperature coefficient
$\Delta T/\Delta L$ = temperature-length gradient
$d$ = fluid depth
$\mu$ = absolute viscosity
$D$ = thermal diffusivity
$L$ = layer thickness

**Keywords:** cellular, convection, gradient, surface, tension
MARANGONI, C. G., 1840-1926, Italian mathematician
**Sources:** Bolz, R. E. and Tuve, G. L. 1970; Land, N. S. 1972; Parker, S. P. 1992; Potter, J. H. 1967.
*See also* THOMPSON

### MARBE LAW

The more common a particular response in word association tasks, the quicker that response is likely to appear, in comparison to a less common response.

**Keywords:** response, word association
MARBE, C., nineteenth century, German psychologist
**Source:** Bothamley, J. 1993.
*See also* THORNDIKE

### MAREY LAW

The heart rate varies inversely with arterial blood pressure. A rise or fall in arterial blood pressure brings about a slowing down or speeding up of the heart rate.

**Keywords:** arterial, blood, heart rate, pressure
MAREY, Etienne Jules, 1830-1904, French physiologist
**Sources:** Friel, J. P. 1074; Landau, S. I. 1986; Stedman, T. L. 1976; Thomas, C. L. 1989.

### MARFAN LAW

The healing of localized tuberculosis protects against subsequent development of pulmonary tuberculosis.*

**Keywords:** healing, pulmonary, tuberculosis
MARFAN, Antoine Bernard-Jean, 1858-1942, French pediatrician
**Sources:** Dorland, W. A. 1980; *Stedman, T. L. 1976.

### MARGULES LAW (ALSO SPELLED MARGOULIS AND MARGULIS)

As the Poiseuille law applies to a capillary instrument, the Margules law applies to a rotation instrument for determining the fluidity of a fluid.

**Keyword:** fluidity, instrument, rotation
MARGULES, Max,1856-1920, Austrian (b. in Ukraine) physicist/meteorologist
**Sources:** Gillispie, C. C. 1981; Pelletier, R. A. 1994; Reiner, M. 1960.
*See also* POISEUILLE

### MARGULES NUMBER, Mr OR $N_{Mr}$ (ALSO SPELLED MARGOULIS AND MARGULIS)

A dimensionless group relating mass transfer in forced convection:

$$Mr = \text{Nusselt number/Péclet number}$$

**Keywords:** fluids, mass transfer
MARGULES, Max, 1856-1920, Austrian (b. in Ukraine) physicist/meteorologist
**Sources:** Bolz, R. E. and Tuve, G. L. 1970; Gillispie, C. C. 1981; Jerrard, H. G. and McNeill, D. B. 1992; Parker, S. P. 1992; Potter, J. H. 1967.
*See also* NUSSELT; PÈCLET

### MARIOTTE LAW (c. 1650) (MARRIOTTE* LAW); OR BOYLE LAW

The product of absolute pressure of a gas multiplied by the volume remains constant by a first approximation:

$$p\,v = p'\,v'$$

where p = absolute pressure
v = volume

E. Mariotte also measured the pressure of water exiting from a pipe.

**Keywords:** gas, pressure, volume

MARIOTTE, Edme, 1620-1684, French physicist

BOYLE, Robert, 1627-1691, British physicist and chemist

**Sources:** *Ballentyne, D. W. G. and Lovett, D. R. 1980; Friel, J. P. 1974; *Glasstone, S. 1946; Landau, S. I. 1986; Stedman, T. L. 1976.

*See also* BERNOULLI; BOYLE; CHARLES; HOOKE

## MARKET, LAW OF—SEE DEMAND, LAW OF

## MARKOV—SEE WEAK LAW OF LARGE NUMBERS

## MASS ABSORPTION—SEE LENARD

## MASS ACTION, LAW OF (1863); LAW OF CONCENTRATION EFFECT

The product of the active masses on one side of a chemical equation, when divided by the product of the active masses on the opposite side of the chemical equation, is constant, with a constant temperature, regardless of the amounts of substances present at the beginning of the action. In addition, the rate of a chemical reaction is directly proportional to the molecular concentrations of the reacting compounds.

**Keywords:** active masses, amounts, chemistry, equation, molecular concentrations, rate

**Sources:** Considine, D. M. 1976; Daintith, J. 1988; Freiberger, W. F. 1960; Honig, J. M 1953; Morris, C. G. 1992; Perry, J. H. 1950; Stedman, T. L. 1976; Thewlis, J. 1961-1964; Walker, P. M. B. 1988.

*See also* EQUILIBRIUM(CHEMICAL); GULDBERG AND WAAGE; OSTWALD DILUTION; REACTION, FIRST ORDER; RICHTER; SECOND ORDER; THIRD ORDER; WENZEL

## MASS LAW (SOUND); NORMAL MASS LAW; RANDOM INCIDENCE MASS LAW

An empirical equation for sound transmission loss is represented by a straight line on the y-axis and the log of the weight on the x-axis, representing a relation between average sound transmission loss and mass per unit area of wall:

$$TL = 23 + 14.5 \log_{10} m$$

where TL = transmission loss, decibels

m = weight, lb per $ft^2$, of the absorbing wall

**Keywords:** building, noise, sound, transmission loss

**Source:** Ballou, G. M. 1987.

*See also* MAYER; SABINE; STIFFNESS

## MASS-LAW EFFECT; COMMON ION EFFECT

The reversal of ionization occurs where a compound is added to a solution of a second compound having a common ion while the volume is kept constant.

**Keywords:** chemistry, compound, ion, ionization

**Source:** Bothamley, J. 1993.

## MASS-LUMINOSITY LAW (1926)

The law, discovered by A. S. Eddington, relates the brightness of normal stars such as the sun to their masses and is expressed by the empirical relation:

$$\log M = 0.26L + 0.06$$

where L = luminosity, expressed in terms of corresponding quantities for the sun
M = mass

The luminosity of a star is approximately proportional to the cube of its mass.

**Keywords:** brightness, luminosity, stars, sun
EDDINGTON, Sir Arthur Stanley, 1882-1944, British mathematician and physicist
**Sources:** Mitton, S. 1973; Scientific American 264(6):95, June 1991.

## MASS RATIO GROUP OR MASS RATIO

A dimensionless group the gives the relationship between the mass of an immersed body to the mass of the surrounding fluid, particularly applicable to airplane flutter and stability:

$$M.R. = m/\pi \rho L^3$$

where m = mass of body
$\rho$ = mass density
L = characteristic body length

**Keywords:** body, fluid, immersed body, mass
**Sources:** Land, N. S. 1972; Potter, J. H. 1967.
*See also* ARCHIMEDES

## MATHIAS—SEE CAILLETET-MATHIAS

## MATTAUCH ISOBAR LAWS OR RULES

**First Law**
There is only one stable nucleus for each odd mass number, with all nuclei to the left of the odd mass number transforming to the stable nucleus by $B^+$ or $K^-$ emission, and all of those to the right by $B^-$ emission.

**Second Law**
There can be several stable doubly even nuclei but these always differ by two units of nuclear charge.

**Keywords:** beta, emission, mass number, nuclei
MATTAUCH, Josef, twentieth century, published in German in 1940s, nuclear physicist
**Sources:** Lerner, R. G. and Trigg, G. L. 1981; NUC.

## MATTHIESSEN RULE FOR RESISTIVITY (1864)

The measured resistivity at a given temperature, T, is composed of temperature-dependent ideal resistivity due to electron scattering by lattice vibrations and temperature-independent residual resistivity caused by impurities and imperfections, such that:

$$\rho = \rho_i + \rho_o$$

where $\rho$ = measured resistivity
$\rho_i$ = ideal resistivity
$\rho_o$ = temperature independent residual resistivity

**Keywords:** electron-scattering, impurities, residual, resistivity
MATTHIESSEN, Augustus,* 1831-1870, English physicist
**Sources:** AIP Handbook. 1972; *Bothamley, J. 1993; Fisher, D. J. 1991; Gillispie, C. C. 1981; Parker, S. P. 1983. 1989; Thewlis, J. 1961-1964.

## MAUPERTUIS, PRINCIPLE OF; OR LAW OF LEAST ACTION (1746)

A particle of given initial kinetic energy will select that path among many possible paths with the same terminal points A and B for which the action, or the integral of the momentum over the path, assumes a stationary path. For a system whose total mechanical energy is consumed, the path to be taken for the system from one configuration to another is the one whose action has the least value as compared to all possible paths.

**Keywords:** electron, kinetic energy, momentum, physics
MAUPERTUIS, Pierre Louis Moreau de, 1698-1759, French mathematician
**Sources:** Carter, E. F. 1996; Encyclopaedia Britannica. 1973; Morris, C. G. 1992.
*See also* CASTIGLIANO; KÖNIG; LAGRANGE; LEAST ACTION

## MAXIMA, LAW OF—SEE LIEBIG

## MAXIMUM SHEARING STRESS—SEE GUEST

## MAXIMUM WORK—SEE BERTHELOT

## MAXWELL-BOLTZMANN DISTRIBUTION LAW; MAXWELL BOLTZMANN LAW (1875)

A more general expression than the Maxwell distribution law, it expresses the relationship between the number of molecules in a gas with velocities between c and c + dc, total number of molecules, mass of the molecules, absolute temperature, and the Boltzmann constant:

$$\frac{dn}{n} = 4\pi N\left(\frac{m}{2\pi kT}\right)^{3/2}\left(\frac{c^2 e^{-mc^2}}{2kT}\right)$$

where dn/n = fraction of total of n molecules = number of molecules
c = for velocities between c + dc
N = total number of molecules
m = mass of molecules having velocities between c and c + dc
T = absolute temperature
k = Boltzmann constant

**Keywords:** gas, molecules, temperature, velocities
MAXWELL, James Clerk, 1831-1879, Scottish mathematician and physicist
BOLTZMANN, Ludwig Edward, 1844-1906, Austrian physicist
**Sources:** Layton, E. T. and Lienhard, J. H. 1988; Thewlis, J. 1961-1964.

## MAXWELL DISTRIBUTION LAW (1860)

Refers to the relative numbers of molecules in an ideal gas with various velocities or kinetic energies of thermal agitation at any particular time. With a total number, N, of molecules, the proportion having speeds in the interval, $\Delta v_1$, between v – 1/2 Δv and v + 1/2 Δv is:

$$\Delta N/N = [4h^3\,(\Delta v_1)/(\pi)^{1/2}]\,(v\,c^{-hv})^2$$

where N = total number of molecules
$h = 6.034 \times 10^7 (m/T)^{1/2}$
c = distribution of molecular speeds
T = absolute temperature
m = mass of one molecule, grams

J. Maxwell developed the above law in 1860 before L. Boltzmann developed the more general distribution law in 1871. When the L. Boltzmann law does not hold,* a more precise distribution law, e.g., the Fermi and Dirac law or Bose and Einstein law, is more applicable.

**Keywords:** gas, molecules, thermal, velocity
MAXWELL, James Clerk, 1831-1879, Scottish mathematician and physicist
**Sources:** *Besancon, R. M. 1974; Thewlis, J. 1961-1964.
*See also* MAXWELL-BOLTZMANN; BOSE AND EINSTEIN; FERMI AND DIRAC

## MAXWELL ELECTROMAGNETIC FIELD EQUATIONS (1864)

Maxwell, who was born in 1831, the year that Faraday published his discovery of electromagnetic induction, expressed Faraday's discoveries mathematically in field theory in 1855-1857, followed by his major work in four electromagnetic field equations, expressed in vector form as follows:

1. $\nabla \cdot D = \rho$, where D is electric displacement. The electric flux lines, if they end, will end on electric charges.
2. $\nabla \cdot \beta = 0$, where β is magnetic flux density. Those magnetic flux lines never terminate.
3. $\nabla \cdot E = -\delta\beta/\delta t$, which is a form of the Faraday law of induction, where E is the electric field density.
4. $\nabla \cdot H = i + \delta D/\delta t$, where H is magnetic field density. Based on the work by Ampere on steady currents, it shows that the line integral of magnetic intensity around a closed curve equals the current encircled, i.

**Keywords:** currents, electricity, electromagnetic, flux, intensity, magnetic
MAXWELL, James Clerk, 1831-1879, Scottish mathematician and physicist
**Sources:** Guillen, M. 1995; Lederman, L. 1993; Mandel, S. 1973; Parker, S. P. 1994; Thewlis, J. 1961-1964.
Also see AMPERE; COULOMB; FARADAY

## MAXWELL-FARADAY LAW (1864-1873); FIRST LAW OF ELECTROMAGNETIC RADIATION (1873)

The line integral of the electric field intensity around a closed path is:

$$-\partial\phi/\partial t = \oint E\, dL$$

where φ = magnetic flux
L = length of the path
E = electric field intensity
C = closed path

**Keywords:** electric field, electromagnetic, intensity
MAXWELL, James Clerk, 1831-1979, Scottish mathematician and physicist
FARADAY, Michael, 1791-1867, English physicist and chemist
**Sources:** Besancon, R. M. 1974; Menzel, D. H. 1960.
*See also* ELECTROMAGNETIC THEORY

## MAXWELL LAW; OR MAXWELL LAW OF VISCOSITY (1866)

The viscosity of air is independent of the density of air, except at very high and very low pressures, which is true to a good approximation in the atmosphere for the range of density observed in the troposphere and lower stratosphere.

**Keywords:** atmosphere, density, stratosphere, troposphere, viscosity
MAXWELL, James Clerk, 1831-1879, Scottish mathematician and physicist
**Source:** Thewlis, J. 1961-1964.

## MAXWELL LAW OF RECIPROCAL DEFLECTIONS

The deflection of a loaded structure with a single load (F) first at point A and second at point B, the deflection at B due to the load at A is equal to the deflection at A due to the load at B, thus $\Delta BA = \Delta AB$ (see Fig. M.2).

**Keywords:** beam, deflection, materials, mechanics, structure, truss
MAXWELL, James Clerk, 1831-1879, Scottish mathematician and physicist
**Sources:** Flugge, W. 1992; Meyers, R. A. 1992.
*See also* BETTI

## MAXWELL LAW OF VELOCITIES

The velocity of any individual molecule in a mass of gas at constant temperature changes after each collision with other molecules, although its mean velocity remains constant.

**Keywords:** gas, molecules, physics, velocity
MAXWELL, James Clerk, 1831-1879, Scottish mathematician and physicist
**Source:** Lerner, R. G. and Trigg, G. L. 1981.

## MAXWELL RELAXATION TIME IN SHEAR LAW (VISCOELASTIC)

A model for representing the relaxation time of a viscoelastic material:

$$\tau(k = 0) = n\ m\ \gamma / G_{\infty}(0)$$

where $\tau$ = relaxation time
$\gamma$ = ratio of specific heats
n, m = constants
G = relaxation modulus of linear viscosity

Further assumptions are needed to establish the k-dependence of $\tau$.

**Keywords:** relaxation, specific heat, time
MAXWELL, James Clerk, 1831-1879, Scottish mathematician and physicist
**Sources:** Blatz, P. J. and Tobolsky, A. V., J. Chem. Physics 14:113, 1946; Eirich, F. R. 1956; Meyers, R. A. 1992.

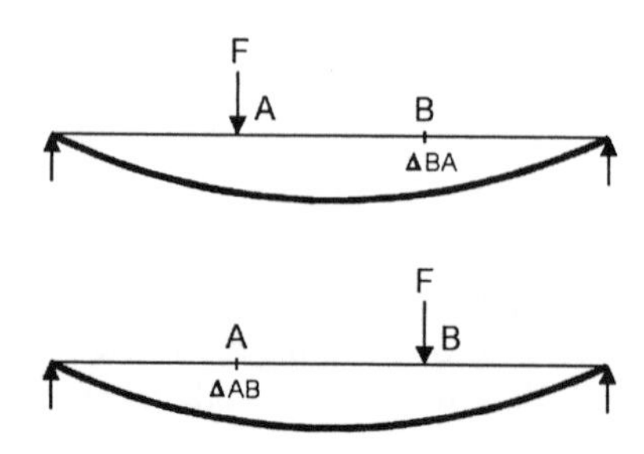


**Figure M.2** Reciprocal deflections.

## MAXWELL RULE (LAW)

Every part of an electric current is acted upon by a force tending to move it in such a direction so as to enclose the maximum amount of magnetic flux.

**Keywords:** electricity, flux, magnetic, physics
MAXWELL, James Clerk, 1831-1879, Scottish mathematician and physicist
**Source:** Hodgman, C. D. 1952.

## MCCABE DELTA LAW—SEE DELTA LAW

## MAYER HYPOTHESIS OR ASSERTION—SEE JOULE LAW

## MAYER LAW

In acoustics, the law gives a quantitative relationship between pitch and the duration of residual auditory sensation.

**Keywords:** acoustics, auditory, pitch, sensation
MAYER, Alfred Marshall, 1836-1897, American physicist
**Source:** Gillispie, C. C. 1981.
*See also* MASS; SABINE; STIFFNESS

## MAYR EFFECT OR FOUNDER EFFECT (1954)

Small population of a species, as they arrive on an island or in isolated area, tend to diverge genetically as a group from the mainland.

**Keywords:** genetic, isolated, population, species
MAYR, Ernst Walter, twentieth century (b. 1904), German American biologist
**Sources:** Bothamley, J. 1993; McMurray, E. J. 1995; Milner, R. 1995; Parker, S. P. 1997.

## MEAN DEVIATION, LAW OF

A measure of the variation of a number from the mean representing a set of numbers:

$$\text{Mean deviation} = \frac{\Sigma|x - \bar{x}|}{n}$$

where $x$ = arithmetic mean of the various values of $\bar{x}$
$n$ = number of numbers

**Keywords:** mean, numbers, set, variation
**Source:** Newman, J. R. 1956.

## MEAN, LAW OF; OR MEAN VALUE THEOREM

A mean value theorem of differential calculus in which where f is continuous for x from $x_1$ to $x_2$, and where f has a derivative for each x between $x_1$ and $x_2$, there is some x for which the following relationship holds true:

$$f'(x) = \frac{f(x_2) - f(x_1)}{x_2 - x_1}$$

This law is credited to J. L. Lagrange.

**Keywords:** continuous, derivative, differential, equation

LAGRANGE, Joseph Louis, 1736-1813, French mathematician
**Sources:** Honig, J. M. 1953; James, R. C. and J. G. 1967; Parker, S. P. 1994.

## MEAN OF DIFFERENTIAL CALCULUS, LAW OF THE

If f(x) is continuous over $a \le x \le b$ and f′(x) exists over $a < x < b$, then there is at least one point $x_0$ such that $a < x_0 < b$ and:

$$\frac{f(b) - f(a)}{b - a} = f'(x_0)$$

**Keywords:** calculus, continuous, differential, exists
**Sources:** James, R. C. and J. G. 1967; Parker, S. P. 1983.

## MEAN OF INTEGRAL CALCULUS, LAW OF

Describes a function that is integratable over an interval with end points a and b, where a may be either less or greater than b, and if m and M are constants such that:

$$m \le f(x) \le M$$

over the interval, then:

$$m \le 1/b - a\int_a^b f(t)dt \le M$$

**Keywords:** function, integratable, interval
**Source:** James, R. C. and J. G. 1967. (doesn't use term law)

## MEAN STRESS, LAW OF—SEE GERBER; GOODMAN; SODERBERG

## MEAN VALUES, LAW OF—SEE MEAN

## MEISSNER EFFECT (1933)

A magnetic field cannot penetrate a superconductor, therefore the magnetic property of a superconductor includes a magnetic field from the interior.

**Keywords:** magnetic, superconductor
MEISSNER, Walther, 1882-1974, German physicist
**Source:** Parker, S. P. 1984. 1989.
*See also* MAGNETIC FIELD

## MELTZER LAW OF CONTRARY INNERVATION

All living functions are continually controlled by two opposing forces, either augmentation or action on one aspect, or inhibition or inaction on the other side.

**Keywords:** biological, forces, functions, life
MELTZER, Samuel James, 1851-1920, American physiologist
**Sources:** Friel, J. P. 1974; Landau, S. I. 1986; Stedman, T. L. 1976.

## MENDEL FIRST LAW OF HEREDITY; OR LAW OF SEGREGATION (1865)

There are genes involved in heredity and development that (1) affect development, (2) retain their individuality from generation to generation, (3) do not become contaminated when they are mixed in a hybrid, and (4) become sorted out from one another when the gametes are formed. The laws of Mendel were published in an obscure journal and remained dormant until 1900, when brought to the attention of the scientific world by H. DeVries.

**Keywords:** gametes, genes, heredity, hybrid, life
MENDEL, Gregor Johann, 1822-1884, Austrian botanist
DEVRIES, Hugo Marie, 1848-1935, Dutch botanist
**Sources:** Bundy Library and Smithsonian Inst. 1980; Friel, J. P. 1974; Gray, P. 1967; Landau, S. I. 1986; Thomas, C. L. 1989.
*See also* MENDEL SECOND; MENDEL THIRD

## MENDEL SECOND LAW OF HEREDITY; OR LAW OF INDEPENDENT ASSORTMENT (1865)

Each pair of genes behaves independently of every other pair of genes.

**Keywords:** genes, heredity, independent
MENDEL, Gregor Johann, 1822-1884, Austrian botanist
**Sources:** Gray, P. 1967; Landau, S. I. 1986; Parker, S. P. 1989; Stedman, T. L. 1976.
*See also* INDEPENDENT ASSORTMENT; MENDEL FIRST; MENDEL THIRD

## MENDEL THIRD LAW OF HEREDITY (1866); OR LAW OF DOMINANCE

This law resulted from the observation that crossing a tall strain of peas with a short strain resulted in the expression of the dominant trait, in this case tallness, or in other cases color of eyes, etc. Thus, some alleles will dominate others in physical expression. Thus, the inheritance of certain traits or characteristics of the progeny are not intermediate in character between the parents but are inherited from one or the other parent.

**Keywords:** alleles, characteristics, dominance, inheritance, parent, progeny, trait
MENDEL, Gregor Johann, 1822-1884, Austrian botanist
**Sources:** Friel, J. P. 1974; Landau, S. I. 1986.
*See also* MENDEL FIRST; MENDEL SECOND

## MENDELEEV (MENDELEEFF) (MENDELEYEV) PERIODIC LAW (1869)

The physical and chemical properties of the elements are periodic functions of their atomic numbers, and the variation of the properties of the elements follows a recurring cycle. Avogadro's work in determining atomic weights helped set the stage to see that valence and other properties were related to atomic weights. It was further noted that, if elements are listed in order of atomic weights, elements with similar properties recur in definite intervals. Sometimes referred to as the *Newlands law.*

**Keywords:** atomic, cycle, elements, properties
MENDELEEV, Dmitri Ivanovich, 1834-1907, Russian chemist
NEWLANDS, John Alexander R., 1838-1898, English chemist
**Sources:** Isaacs, A. et al. 1996; Krauskopf, K. B. 1959; Mandel, S. 1972; Stedman, T. L. 1976.
*See also* AVOGADRO; DALTON; OCTAVES; PAULI; PERIODIC

## MENZEL—SEE KRIGAR-MENZEL

## MERKEL LAW

A psychological law that states that equal sensations of the body above threshold strength correspond to equal stimulus differences.

**Keywords:** sensations, stimulus, threshold

MERKEL, Friedrich Siegismund, 1845-1919, German anatomist
**Sources:** Bothamley, J. 1993; Zusne, L. 1987.
*See also* FECHNER; WEBER

### MERKEL NUMBER, Me OR $N_{Me}$

A dimensionless group applicable to cooling towers that relates the mass of water transferred to the unit humidity difference to the mass of dry gas:

$$Me = MAW/V_m$$

where M = mass transfer rate
A = area of cooling surface per volume
W = total volume
$V_m$ = mass flow rate of gas

**Keywords:** cooling tower, humidity, mass transfer
MERKEL, Friedrich, 1892-1929, German engineer
**Sources:** Bolz, R. E. and Tuve, G. L. 1970; Land, N. S. 1972; Parker, S. P. 1992.

### MERSENNE LAWS

A mathematical expression connecting the frequency, length, diameter, density, and tension of a string:

$$n_p = p/2L(T/M)^{1/2}$$

where $n_p$ = frequency of the harmonic mode
p = number of loops
L = length of string
T = tension
M = mass per unit length of string

Also, for a stringed musical instrument, pitch is inversely proportional to the length, directly proportional to the square root of the tension, and inversely proportional to the square root of the weight of the string:

$$P = k/L(T/W)^{1/2}$$

where P = pitch
L = length of string
T = tension of string
W = weight of string
k = constant based on string and conditions

**Keywords:** frequency, length, mechanics, pitch, string
MERSENNE, Marin, 1588-1648, French mathematician
**Sources:** Thewlis, J. 1961-1964; West, B. H. et al. 1982.
*See also* KRIGAR-MENZEL; PYTHAGOREAN

### MESONS—SEE QUARK

### METABOLIC CONTROL OF ATMOSPHERIC COMPOSITION, LAW OF

The earliest atmosphere of the Earth consisted of gases of the solar system, methane ($CH_4$) and ammonia ($NH_3$). Evaluation of the present atmosphere has occurred in a stepwise fashion by volcanic exhalation from the mantle and by biological, metabolic processes.

**Keywords:** Earth, gases, solar system, volcanic

**Source:** Oliver, J. E. and Fairbridge, R. W. 1987.

## METABOLISM—SEE BRODY; CONSERVATION OF ENERGY; ENZYME; HESS; RESPIRATION QUOTIENT; RUBNER RULE; SURFACE AREA; WILHELMY

## METCHNIKOFF (MECHNIKOV) LAW

When a body is attacked by bacteria, the polymorphonuclear leukocytes and the large mononuclear leukocytes quickly become protective phagocytes.

**Keywords:** bacteria, leukocytes, phagocytes

METCHNIKOFF, Elie Ilich, 1845-1916, Russian zoologist; Nobel prize, 1908, physiology/medicine, shared

**Source:** Friel, J. P. 1974.

## MEYER LAW OF ESTERIFICATION (1894)

When esters are formed from alcohols, the esterification reaction is speeded by the use of acids, but those acids are reduced in their effect on the reaction in the presence of some aromatic acids.

**Keywords:** acids, alcohols, esterification

MEYER, Viktor, 1848-1897, German organic chemist

**Sources:** Chemical Heritage 16(1):25, 1998; Clifford, A. F. 1964; Friel, J. P. 1974; Honig, J. M. et al. 1953; Parker, S. P.1987.

*See also* ESTERIFICATION

## MEYER LAW (MEDICAL)

The internal structure of fully developed normal bone represents the lines of greatest pressure on traction and affords the greatest possible resistance with the least amount of material.

**Keywords:** anatomy, bone, medical, traction

MEYER, George Hermann von, 1815-1892, German anatomist

**Sources:** Landau, S.I. 1986; Merriam-Webster's Biographical Dictionary. 1995.

## MEYER LAW OF INDENTATION (MATERIALS) (1908)

This is an experimental, exponential relation that describes the indentation of a hard metal sphere on a plane metal surface (Fig. M.3) and has been found applicable to plastics, under static conditions. The empirical relation is:

$$W = k\, d^n$$

where $W$ = applied load

$k, n$ = constants depending on materials involved and sphere diameter ($n = 2$ for plastic, $n = 3$ for fully elastic)

$d$ = chordal diameter of remaining indentation

**Keywords:** empirical, exponential, indentation, metal sphere, surface

MEYER, E., twentieth century, German engineer

**Sources:** Gillispie, C. C. 1981; Goldsmith, W. 1960; Tabor, D. 1951.

*See also* HERTZ

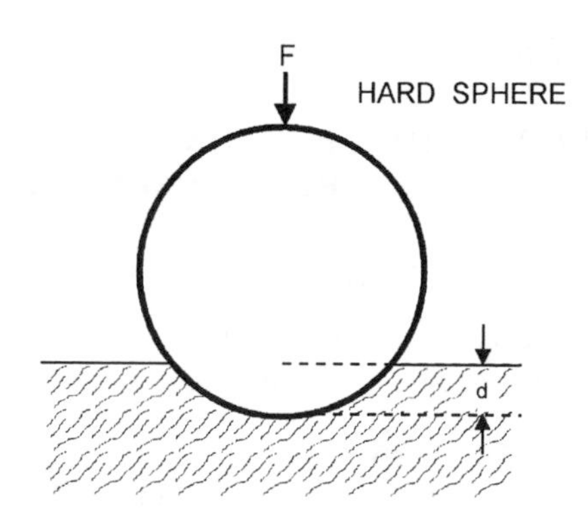


**Figure M.3** Indentation.

## MICHAELIS-MENTEN LAW (1913)

A method developed for determining the rate of enzymatic reaction in relation to the concentration of the substrate. This method evolved from the Michaelis-Menten equation, named after L. Michaelis and his assistant.

**Keywords:** concentration, enzymatic, substrate
MICHAELIS, Leonor, 1875-1949, German American chemist
MENTEN, Maud Lenore, 1879-1960, American physician
**Source:** Gillispie, C. C. 1981.
*See also* ENZYME; FISCHER; NORTHRUP; SCHÜTZ; WILHELMY

## MICROSCOPIC REVERSIBILITY, LAW OF

The pathway and transition states are the same for forward and reverse reactions.

**Keywords:** forward, pathway, reverse, transition
**Source:** Landau, S. I. 1986.

## MILLER EFFECT (ELECTRONICS)

In a vacuum tube, the increase in the effective grid-cathode capacitance is due to the charge induced electrostatically on the grid by the anode through the grid-anode capacitance.

**Keywords:** capacitance, cathode, electrostatically, grid
**Sources:** Michels, W. C. 1956; Morris, C. G. 1992.

## MILLER LAW (CRYSTAL); LAW OF RATIONAL INTERCEPTS (1839)

If the edges formed by the intersections of three faces of a crystal are taken as the three references axes, then the three quantities formed by dividing the intercept of a fourth face with one of these axis by the intercept of a fifth face with the same axis are proportional to small whole number.

**Keywords:** axis, crystal, intersections
MILLER, William Hallowes, 1801-1880, English minerologist
**Sources:** Ballentyne, D. W. G. and Lovett, D. R. 1972; Lapedes, D. 1982; Parker, S. 1989. 1994.
*See also* DE LA RUE AND MILLER; WEISS ZONE

## MINER LAW OR RULE—SEE LINEAR DAMAGE; ODQUIST

## MINIMA, LAW OF; PRINCIPLE OF THE LIMITING FACTOR—SEE LIEBIG; MINIMUM, LAW OF

## MINIMUM, LAW OF; LAW OF LIEBIG

This law, developed by von Liebig, states that plants and animals have minimum requirements of certain nutrients, particularly salts, for growth and development. If a soil does not supply the minimum, the plants requiring those salts (e.g., iron, manganese, zinc, boron, copper, molybdenum, and cobalt) cannot grow, regardless of the abundance of other nutrients.

**Keywords:** growth, nutrients, plants, salts
LIEBIG, Justus von, 1803-1873, German chemist
**Sources:** Allaby, M. 1989; Landau, S. I. 1986; Morris, C. G. 1992; Scott, T. A. 1995; Stedman, T. L. 1976.
*See also* LIEBIG

## MINIMUM LATERAL THRUST, LAW OF (1920)

In geology, the relative displacement of overlying strata to underlying strata surrounding and inclined, the concordant inclusion can be expressed as B-A cotangent θ, in which B is the horizontally measured width of the inclined body, A is the vertically measured width of the horizontal part of the intrusive body, and 0 is the angle of the inclination.

**Keywords:** displacement, geology, inclination, strata
**Sources:** Bates, R. L. and Jackson, J. A. 1980; Gary, M. et al. 1973.

## MINIOVICH (MINYOVICH) NUMBER, Mn OR $N_{Mn}$

A dimensionless group that relates the pore size and porosity of a product being dried:

$$Mn = SR/e$$

where R = pore radius
S = particle area divided by particle volume
e = porosity

**Keywords:** drying, particle, porosity
MINIOVICH, Ya. M., twentieth century, Soviet engineer
**Sources:** Bolz, R. E. and Tuve, G. L. 1970; Hall, C. W. 1979; Land, N. S. 1972; Lykov, A. V. and Mikhaylov, Y. A. 1961; Parker, S. P. 1992; Potter, J. H. 1967.
*See also* LEBEDOV (LEBODEV); LUIKOV (LYKOV)

## MINOT LAW

Organisms age fastest when young.

**Keywords:** age, organisms, young
MINOT, George Richards, 1885-1950, American physician; Nobel prize, 1934, physiology/medicine (shared)
**Sources:** Dorland, W. A. 1974; Friel, J. P. 1974.

## MINUS ONE-QUARTER (ONE-FOURTH) POWER LAW—SEE ALLOMETRIC

## MIRROR SYMMETRY, LAW OF, OR PARITY

As applied to elementary particles, a law that is only partially true: if you look at the subject in a mirror, everything will look and behave exactly the same. There would be no difference between a right-handed and a left-handed universe, and one could not be distinguished from another.

**Keywords:** mirror, particle, physics, symmetry
**Source:** Thewlis, J. 1961-1964.

## MITSCHERLICH LAW OF ISOMORPHISM (1819)

The number of molecules of water of crystallization being the same for any two substances being compared,

1. The form of crystals is the same, as represented by the measurement of their angles.
2. Each crystal deposited from a mixed solution contains both substances and is of the same form as crystals obtained from a solution of either substance.
3. A crystal of one substance, placed in a saturated solution of the other, continues to grow by superposition of the second substance, with the form of the crystal remaining unaltered.

**Keywords:** crystals, materials, solution
MITSCHERLICH, Eilhardt, 1794-1863, German chemist
**Sources:** Isaacs, A. et al. 1996; Van Nostrand. 1947; Walker, P. M. B. 1988.
*See also* BERZELIUS AND MITSCHERLICH; BEUDANT; ISOMORPHISM

## MIXTURES—SEE ARRHENIUS; IDEAL MIXTURES; LE CHATELIER

## MNEMIC CAUSATION, PHENOMENOM, OR LAWS (1921)

The capacity of a living substance or organism to retain after effects of experiences. Laws in which the determined event succeeds the determining event. An example is the Mendelian laws of heredity in which the present characteristics of an organism are determined by the characteristics of its parents. This law was put into words by B. A. W. Russell and associates.

**Keywords:** determined event, determining event, heredity
RUSSELL, Bertrand Arthur William, 1872-1970, British mathematician and philosopher
**Sources:** Audi, R. 1995; Braithwaite, R. B. 1953; Gillispie, C. C. 1981; Oxford English Dictionary. 1989; Simpson, J. A. and Weiner, E. S. C. 1989; Talbott, J. 1970.
*See also* BIOTIC

## MOBILITY OF FREE ELECTRONS—SEE BLANC

## MOBILITY OF IONS—SEE EINSTEIN

## MODEL LAW OF SETTLEMENT—SEE SETTLEMENT

## MODEL LAWS

Laws in which a model represents the relationship of the real world. There are often limits or boundary conditions of applicability of the model. A physical model is the Froude law for drag of a vehicle in a fluid. Differential equations often represent mathematical models of physical phenomena, such as the Newton law. Other mathematical relationships, such as with computer programs, are used to represent events or happenings in the real world.

**Keywords:** computer, mathematical, physical
**Source:** Isaacs, A. et al. 1996.

## MODULUS OF ELASTICITY—SEE HOOKE; YOUNG

## MODULI, LAW OF—SEE VALSON

## MOHR CIRCLE (STRESS, STRAIN)

This is a procedure for graphical solution of structural problems to determine stress based on a circle representation (see Fig. M.4). For the particle out of a loaded member shown in (a), cut across AA, normal and sharing forces are represented in (b), and the biaxial stress field is represented in (c). The approach can be applied to special stress fields and strains. A standard reference should be consulted for examples and applications of graphical solutions of stresses and strains.

**Keywords:** graphical, structural

MOHR, Christian Otto, 1835-1918, German civil engineer

**Sources:** Baumeister, T. 1967; Scott, J. S. 1993.

## MOLAR EXTINCTION COEFFICIENT—SEE LAMBERT-BEER

## MOLECULAR CONDUCTIVITY—SEE OSTWALD

## MOMENTS, LAW OF

For a body to be in equilibrium the sum of the moments around any axis must be zero.

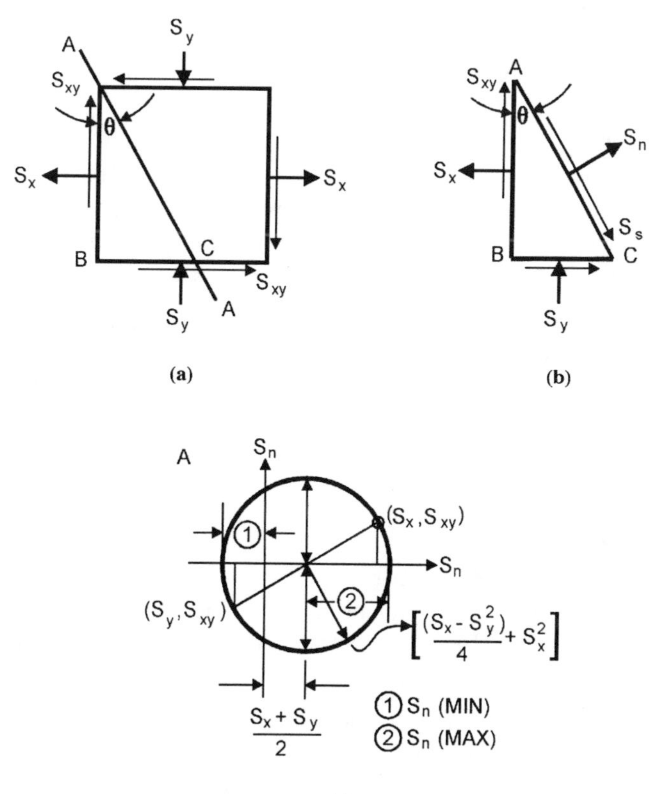


**Figure M.4** Mohr circle.

**Keywords:** equilibrium, mechanics, moments
**Source:** Daintith, J. 1981.

### MOMENTUM—SEE CONSERVATION OF MOMENTUM, LAW OF; NAVIER-STOKES; NEWTON LAW OF MOMENTUM

### MOMENTUM NUMBER, Mo OR $N_{Mo}$

A dimensionless group relating to convection in the ocean and atmosphere:

$$Mo = ML/\gamma \, \Delta V$$

where M = momentum flux
L = thickness of layer
$\gamma$ = kinematic viscosity
$\Delta V$ = velocity difference

**Keywords:** atmosphere, convection, ocean
**Sources:** Land, N. S. 1972; Potter, J. H. 1967.

### MONODROMY, LAW OF; MONODROMY THEOREM; MONODROMY AXIOM; DARBOUX THEOREM

The theorem that a function of a complex variable whose only singularities in a single connected region of the complex plane are poles is one-valued in that region. Characteristic property of the variable is that it returns to original value after it has changed.

**Keywords:** algebra, complex variable, plane, poles
DARBOUX, Jean Gaston, 1842-1917, French mathematician
**Sources:** James, R. C. and James, G. 1968; Oxford English Dictionary. 1989.

### MOORE LAW (1964)

Integrated circuits and microelectronics will double in density every other year (or every one and half years, by some references), according to a binary growth curve, and the design cost and number of functions per circuit will keep pace with complexity (on a 1960 to 1985 time frame). Elements per chip = $2^{(\text{years}/\tau)}$ where $\tau$ is 1, 2, 3, ... years.

**Keywords:** density, design, functions, integrated circuits, microelectronics
MOORE, Gordon, twentieth century (b. 1927), American electrical engineer
**Sources:** High Technology, p. 34, Jan. 1983; Science 274(5294):1834, 13 December 1996; Scientific American 274(1), January 1996.

### MORAT—SEE DASTRE-MORAT

### MORGAN (LLOYD) CANON

An principle that has been accepted as a part of scientific explanation or behavior stating that if there are rival explanations in terms of mental functions for a given phenomenon, the simplest must always be chosen. This canon is fashioned from the same thinking as Occam's Razor.

**Keywords:** explanations, phenomenon, simplest
MORGAN, Conway Lloyd, 1852-1936, British psychologist
OCKHAM, William of, c. 1300-1349, English philosopher
**Sources:** Angeles, P. A. 1981; Eigen, M and Winkler, R. 1981; Walker, P. M. B. 1989.
*See also* OCCAM (OCKHAM); PARSIMONY

## MORLEY REFERRED PAIN LAW

Referred pain arises only from irritation of nerves that are sensitive to those stimuli when applied to the surface of the body that produces pain.

**Keywords:** irritation, nerves, pain, stimuli, surface
**Sources:** Critchley, M. 1978; Landau, S. I. 1986; Stedman, T. L. 1976.

## MORTALITY—SEE GOMPERTZ; MAKEHAM

## MORTON NUMBER, Mo OR $N_{Mo}$

A dimensionless group representing bubble movement:

$$Mo = g\mu^4/\rho\sigma^3$$

where g = gravitational acceleration
$\mu$ = absolute viscosity
$\rho$ = mass density
$\sigma$ = surface tension

**Keywords:** bubble, movement, surface tension, viscosity
MORTON, R. K., twentieth century, hydraulics
**Sources:** Journal Fluid Mechanics 3(1), Oct. 1957; Land, N. S. 1972; Potter, J. H. 1967.

## MOSELEY LAW (1913)

Each chemical element has a simple X-ray spectrum and a characteristic wavelength, with the atomic number as follows:

$$(\upsilon)^{1/2} = a(Z - \sigma)$$

where $\upsilon$ = frequency of given line in spectrum
Z = atomic number of the element emitting the ray

**Keywords:** atomic number, element, wavelength
MOSELEY, Henry Gwyn-Jeffreys, 1887-1915, English physicist
**Sources:** Brown, S. B. and Brown, L. B. 1972; Considine, D. M. 1995; Daintith, J. 1981; Hodgman, C. D. 1952; Isaacs, A. et al. 1996; Parker, S. P. 1987; Thewlis, J. 1961-1964.
*See also* ATOMIC NUMBERS

## MÖSSBAUER EFFECT (1957)

Refers to the decay of a nucleus from an excited state to a ground state of energy with the emission of a gamma ray of energy. These gamma rays are the basis of detectors used to identify a material.

**Keywords:** decay, detector, gamma, nucleus
MÖSSBAUER, Rudolf Ludwig, twentieth century (b. 1929), German American physicist
**Sources:** Considine, D. M. 1989; NYPL Desk Ref.; Parker, S. P. 1989.

## MOSSOTTI-CLAUSIUS—SEE CLAUSIUS-MOSSOTTI

## MOTION (THREE), LAWS OF—SEE NEWTON

## MOTT LAW OF ANTICIPATION

The children of insane parents who also become insane do so at a much earlier age than did their parents.

**Keywords:** children, insane, mental, parents
MOTT, Sir Frederick Walker, 1853-1926, English neurologist
**Sources:** Friel, J. P. 1974; Landau, S. I. 1986.

## MOUTIER LAW

The work done in any isothermal reversible cyclic process is zero, as related to the first law of thermodynamics.

**Keywords:** cyclic, isothermal, reversible, work
MOUTIER, Jules, 1829-1895, French thermodynamicist
**Source:** Clifford, A. F. 1964.
*See also* THERMODYNAMICS

## MOUTON—SEE COTTON-MOUTON

## MÜLLER LAW; LAW OF INSTABILITY

The embryo and the fetus in its development recapitulates that of the ancestral series to which it belongs.

**Keywords:** ancestor, birth, growth, medical
MÜLLER, Johannes Peter, 1801-1858, German physiologist and anatomist
**Sources:** Friel, J. P. 1974; Landau, S.I. 1986.

## MÜLLER LAW OF SPECIFIC ENERGIES (1826); LAW OF SPECIFIC IRRITABILITY (NERVES)

Regardless of the method of excitation, each nerve gives rise to its own specific sensation, regardless of the method excitation. For example, any stimulation to the optic nerve results in a sensation of light. The resonance theory of Helmholtz satisfies the Müller law and provides a physiological explanation of the Ohm law.

**Keyword:** excitation, nerve, response, sensation, stimulus
MÜLLER, Johannes Peter, 1801-1858, German physiologist and anatomist
**Sources:** Landau, S. I. 1986; Michels, W. C. 1961; Stedman, T. L. 1976.
*See also* HELMHOLTZ; OHM

## MÜLLER-HAECKEL LAW—SEE BIOGENETIC LAW; ALSO KNOWN AS HAECKEL LAW; RECAPITULATION THEORY

MÜLLER, Fritz, 1821-1897, German naturalist
HAECKEL, Ernst Heinrich, 1834-1919, German biologist/naturalist
**Sources:** Friel, J. P. 1974; Landau, S. I. 1986.
*See also* BIOGENETIC; RECAPITULATION

## MÜLLERIAN LAW; ICOSACANTHIC LAW

An expression of the regularity in the distribution of 20 radial spines on the shells of Sarcodina or amoeboid protozoa, which are made of silicon compounds.

**Keywords:** amoeboid, silicon, spines
MÜLLER, Johannes Peter, 1801-1858, German physiologist and anatomist
**Source:** Bates, R. L. and Jackson, J. A. 1980.

## MULTINOMIAL PROBABILITY LAW

If the probabilities are $p_1, p_2, \ldots, p_k$ that an event will result in the mutually exclusive categories $A_1, A_2, \ldots$, respectively, the probability that in n trials $A_1$, $x_2$ in $A_2, \ldots, x_k$ in $A_k$ is:

$$m(x_1, x_2, \ldots, x_k) = \{n!/x_1!\ x_2!\ \ldots\ x_k\}\ \{p_1^{x_1}\ p_2^{x_2}\ \ldots\ p_k^{x_k}\}$$

where $x_1 + x_2 \ldots + x_k = n$

$p_1 + p_2 + \ldots + p_k = 1$

**Keywords:** events, mutually exclusive, probabilities
**Sources:** Doty, L. A. 1989; Hazewinkel, M. 1987; Kotz, S. and Johnson, N. L. 1982.

## MULTIPLE PROPORTIONS, LAW OF (1804)

When two elements unite to form more than one compound, and the first element combines with a fixed weight of the second element, the weights of these elements will be in an integer ratio with each other.

**Keywords:** chemistry, combination, compound, elements, weight
**Sources:** Clayton, C. L. 1989; David, I., et al. 1996; Parker, S. P. 1989; Walker, P. M. B. 1988.
*See also* DALTON; DEFINITE COMPOSITION

## MULTIPLE VARIANTS, LAW OF

Any variation from the normal of the hand or foot is always multiple, involving more than one bone.

**Keywords:** anatomy, foot, hand, physiological
**Sources:** Friel, S. P. 1974; Landau, S. I. 1986; Stedman, T. L. 1978.

## MULTIPLICATIVE IDENTITY LAW

One is the multiplicative identity for any real number because a number multiplied by it will give the same number:

$$1 \cdot a = a \cdot 1 \text{ for all values of a}$$

**Keywords:** algebra, mathematics, multiply
**Source:** James, R. C. and James, G. 1968.

## MULTIPLICATIVE INVERSE LAW

For any real numbers, a, for which $a \neq 0$, there is 1/a such that $(a \cdot 1/a = 1)$. For complex numbers, the multiplicative inverse of a + bi is 1/(a + bi), because their product is 1 (one).

**Keywords:** algebra, complex numbers, mathematics, multiplication, real numbers
**Source:** James, R. C. and James, G. 1968.

## MULTIPLICATIVE LAW OF PROBABILITY

The mutual probability of two or more independent events is equal to the product of the probabilities of each event, represented by:

$$P(A \text{ and } B) = P(A) \times P(B)$$

$$P(A \text{ and } B \text{ and } C) = P(A) \times P(B) \times P(C)$$

**Keywords:** events, statistics, probability, mathematics
**Source:** Doty, L. A. 1989.

## MULTIPLICITIES, ALTERNATION—SEE ALTERNATION

## MULTIVARIATE NORMAL PROBABILITY LAW

The probability law of several variables is the multivariate normal probability law of $x_1$, $x_2$, ... $x_k$ as a continuous probability law. The probability density function of this law is:

$$1/\{(2\pi)^{1/2}\,\sigma_1\,\sigma_2 \ldots T_x\,(\Delta)^{1/2}\}\,e^{-1/2}\,Q_k$$

where $Q_k$ = quadratic form in $x_1$, $x_2$ ... $x_k$
$u_i$ and $\sigma$ = mean and variance of $x_i$

where $\Delta$ is the determinant obtained by deleting the ith row and the jth column in $\Delta$.

**Keywords:** mean, mathematics, probability, variance
**Source:** Morris, C. G. 1992.
*See also* MULTINOMIAL PROBABILITY

## MURPHY LAW (ECOLOGY)

Bills of birds are longer that reside on islands than birds of the same species that reside on the mainland.

**Keywords:** bills, birds, islands, mainland
MURPHY, Robert Cushman, 1887-1973, American naturalist (ornithologist and zoologist)
**Sources:** Gray, P. 1967; Sterling, K. B. et al. 1997.

## MURPHY LAW (GENERAL) (1949)

These are general and special laws, of which there are many, delineating the behavior of inanimate objects. One of the most popular is "if anything can go wrong, it will."

Although often portrayed as the probability of failure, the law refers to the certainty of failure and calls for determining the likely causes in advance an acting to prevent failure.*

**Keywords:** behavior, inanimate, failure
MURPHY, Edward A. Jr., 1918-1990, American engineer
**Sources:** *Murphy, Edward A. III, personal correspondence, 1998; Scientific American 276(4):88, April 1997.

## MURPHY LAW (MEDICINE)

Jaundice resulting from an obstruction by a gallstone is preceded by colic, but when caused by a neoplasm or inflammation of the bile ducts, there is no colic.

**Keywords:** bile duct, colic, gallstone, jaundice
MURPHY, John B., 1857-1916, American surgeon (not confirmed)
**Sources:** Critchley, M. 1978; Gray, P. 1967.

## MURRI LAW

This deals with the relationship of the physicopathic or personal disorder compensation in the heart.

**Keywords:** heart, medical, physicopathic
MURRI, Augusto, 1841-1932, Italian clinician
**Source:** Garrison, F. H. 1929.

## MUTUALITY OF PHASES, LAW OF

If two phases, respecting a certain definite reaction, at a certain temperature, are in equilibrium with a third phase, then at the same temperature and respecting the same reaction, they are in equilibrium with each other.

**Keywords:** equilibrium, phase, physical chemistry, reaction
**Sources:** Clifford, A. F. 1964; Northrup, E. T. 1917.

## MYELOGENETIC LAW—SEE FLECHSIG

## MYER—SEE MEYER (ESTERIFICATION)

# N

## NAEGELI (NÄGELI) LAW

A disease in which eosinophilic leucocytes (a type of leukocyte easily stained by eosin dye) are present in numbers greater than half-normal cannot be typhoid fever. The presence of even a few eosinophils may render such a decision as uncertain and demands caution in diagnosis of the patient.

**Keywords:** disease, leucocytes, medical, typhoid fever

NAEGELI, Otto, 1871-1938, Swiss physician

**Sources:** Critchley, M. 1978; Friel, J. P. 1974; Landau, S. I. 1986; Stedman, T. L. 1976.

## NASSE LAW (1820)

Describes the pattern of X-linked recessive inheritance in which hemophilia affects only boys but is transmitted only through females.

**Keywords:** females, hemophilia, inheritance, males

NASSE, Christian Friedrich, 1778-1851, German physicist

**Sources:** Critchley, M. 1978; NUC; Stedman, T. L. 1976.

## NATURAL DECAY—SEE EXPONENTIAL DECREASE

## NATURAL GROWTH, LAW OF; OR LAW OF EXPONENTIAL INCREASE

The differential equation that represents natural growth:

$$dx/dt = kx$$

where $x$ = total number organisms at time, $t$

$k$ = a non-zero constant such that $x = x_0 e^{kt}$ at $t = 0$

**Keyword:** differential equation, growth, organisms

**Sources:** Lapedes, D. N. 1971; Parker, S. P. 1987.

*See also* EXPONENTIAL INCREASE; GOMPERTZ

## NATURE, LAWS OF, OR NATURAL LAW

Refers to an intrinsic orderliness of nature or the necessary conformity of phenomena to reason and understanding. This relationship has been expressed in three laws:

1. The *First Law* of nature is self-preservation.
2. The *Second Law* of nature is reproduction of the species.
3. The *Third Law* of nature is instinct of property rights.

**Keywords:** instinct, preservation, property, reproduction

**Sources:** Fairbridge, R. W. 1967; Flew, A. 1964; Robertson, M. 1898.

*See also* NOMIC

## NAVIER (1823)-STOKES EQUATIONS (1845)

These are three partial differential equations that describe the conservation of momentum for the motion of viscous, incompressible fluids, which for flow in x-direction, $\partial p/\partial x + \mu(\partial^2 u/\partial x^2 + \partial^2 u/\partial y^2 + \partial^2 u/\partial z^2)$, which represents sum of external pressure and viscous forces, $= -\rho(u\partial u/\partial x + v\,\partial u/\partial y + w\partial u/\partial z)$, which represents the change of momentum of the liquid. In the expression:

u, v, w = components of velocity along the x, y, z axes
p = the pressure
ρ = the fluid density
μ = viscosity

The use of these equations, nine to describe the flow of a fluid, is quite complicated and is enhanced greatly with computers.

**Keywords:** conservation, differential, equations, fluids, momentum
NAVIER, Claude Louis Marie Henri, 1785-1836, French physicist and engineer
STOKES, George Gabriel, 1819-1903, British mathematician and physicist
**Sources:** Cheremisinoff, N. P. 1986; Considine, D. M. 1976; Layton, E. T. and Lienhard, J. H. 1982; Parker, S. P. 1989. 1997; Rohsenow, W. M. and Hartnett, J. P. 1973.
*See also* CONSERVATION; NEWTONIAN

## NAZE NUMBER, Na OR $N_{Na}$

A dimensionless group in magneto fluid dynamics that is the velocity of Alfvén wave divided by the velocity of sound.

**Keywords:** magnetohydrodynamcs, sonic
**Sources:** Bolz, R. E. and Tuve, G. L. 1970; Land, N. S. 1972; Potter, J. H. 1967.
*See also* **ALFVÉN**; MACH

## NEGATIVE LAW OF EFFECT (1931)—SEE THORNDIKE

## NERNST DIFFUSIVITY LAW (1893); OR NERNST-THOMSON (THOMPSON*) RULE

At infinite dilution, the diffusivity of an electrolyte is related to the ionic mobilities, which may be obtained from conductance measurements:

$$D_{dil} = R'T\ U + U^-/U^+ + U^-\ (1/Z^+ + 1/Z^-)$$

where $D_{dil}$ = dilute solution
U = absolute velocities of anion and cation
Z = valence of anion and cation
R′ = gas constant
T = absolute temperature, K

W. Nernst is considered to be the founder of modern physical chemistry.

**Keywords:** chemistry, dilution, electrolyte, ions
NERNST, Walther Hermann, 1864-1941, Polish German physical chemist; Nobel prize, 1920, chemistry (Some references use name Hermann Walther** Nernst)
THOMSON, Joseph John, 1856-1940, English physicist; Nobel prize, 1906, physics
**Sources:** **Asimov, I. 1972; *Considine, D. M. 1976; Gillispie, C. C. 1981; Merriam-Webster. 1995; NUC; **Parker, S. P. 1972.
*See also* NERNST-THOMSON

### NERNST DISTRIBUTION LAW

The simple distribution law is modified to use the concentration of the given contents instead of total concentration.

**Keywords:** chemistry, concentration, distribution

NERNST, Walther Hermann, 1864-1941, Polish German physical chemist; Nobel prize, 1920, chemistry

**Source:** Mandel, S. 1972.

### NERNST EFFECT OR LAW (1886)

When heat flows across lines of magnetic force, an electromotive force is produced in the mutually perpendicular direction.

**Keywords:** electromotive, heat magnetic

NERNST, Walther Hermann, 1864-1941, Polish German physical chemist; Nobel prize, 1920, chemistry

**Sources:** Besancon, R. M. 1974; Considine, D. M. 1976; Landau, S. I. 1986.

*See also* ETTINGSHAUSEN; HALL

### NERNST EQUATION

Represents the concentration dependence of the reversible equilibrium potential of a working electrode and electropotential represented by one of many forms of the equation:

$$E_{rev} = E^{O} - \{[0.059/n] \log[A_{acid}/A_{reduc}]\}$$

where $E_{rev}$ = reversible potential

$E^{O}$ = value of amperage when substances in standard state

$n$ = change in valence

$A_{acid}$, $A_{reduc}$ = activities of oxidized and reduced reactants

**Keywords:** electricity, electrode, equilibrium

NERNST, Walther Hermann, 1864-1941, Polish German physical chemist; Nobel prize, 1920, chemistry

**Sources:** Considine, D. M. 1976; Gray, D. E. 1963; Kirk-Othmer. 1984.

Also see BUTLER-VOLLMER

### NERNST HEAT THEOREM (1906); ALSO CALLED THIRD LAW OF THERMODYNAMICS; ALSO CALLED NERNST LAW*

The entropy of a chemically pure substance in a condensed phase vanishes at absolute zero.

**Keywords:** condensed, entropy, heat, phase

NERNST, Walther Hermann, 1864-1941, Polish German physical chemist; Nobel prize, 1920, chemistry

**Sources:** *Asimov, I. 1976; Considine, D. M. 1976.

*See also* THERMODYNAMICS

### NERNST LAW

The amount of electric current needed to stimulate muscle action varies as the square root of the frequency of the current.

**Keywords:** electricity, frequency, muscle, stimulus

NERNST, Walther Hermann, 1854-1941, Polish German physical chemist; Nobel prize, 1920, chemistry
**Source:** Asimov, I. 1972.

### NERNST LAW (1889)

The solubility of a salt is decreased by the presence in solution of another salt with a common ion.

**Keywords:** chemistry, ion, salt, solubility

NERNST, Walther Hermann, 1854-1941, German physical chemist; Nobel prize, 1920, chemistry

**Source:** Gillispie, C. C. 1981.

### NERNST PRINCIPLE OF SUPERPOSITION

The potential difference between junctions in similar pairs of solution that have the same ratio of concentrations are the same, even if the absolute concentrations are different. Thus, the same normal solution of two different chemical solutions gives the same potential difference.

**Keywords:** chemicals, concentrations, junctions, normal, solutions
NERNST, Walther Hermann, 1854-1941, Polish German physical chemist; Nobel prize,1920, chemistry
**Source:** Considine, D. M. 1976.
*See also* SUPERPOSITION

### NERNST-THOMSON RULE (SOME REFERENCES USE NERNST-THOMPSON*); OR THOMSON-NERNST RULE OR LAW

"A solvent of high dielectric constant favors dissociation by reducing the electrostatic attraction between positive and negative ions, and conversely a solvent of low dielectric constant has small dissociating influence on an electrolyte."

**Keywords:** dielectric, dissociation, ions
NERNST, Walther Hermann, 1854-1941, Polish German physical chemist; Nobel prize, 1920, chemistry
THOMSON, Joseph John, 1856-1940, English physicist; Nobel prize, 1906, physics
**Sources:** Asimov, I. 1972; *Considine, D. M. 1976; Travers, B. 1994.

### NERVE CELLS, SIR CHARLES LAW OF

Nerve cells can normally carry impulses in only one direction.

**Keywords:** cells, impulses, nerve
CHARLES, Sir R. Haverlock, twentieth century, English surgeon
**Sources:** Friel, J. P. 1974; Stedman, T. L. 1976.

### NERVE, LAW OF SPECIFIC IRRITABILITY—SEE MÜLLER

### NEUMANN-KOPP LAW (1831)

The molecular weight time the specific heat of a compound is equal to the sum of the atomic heats of its elements.

**Keywords:** atomic weight, compound, elements, molecular weight

NEUMANN, Franz Ernst, 1798-1895, German physicist
KOPP, Hermann, 1817-1892, German chemist
**Sources:** Bothamley, J. 1993; Webster's Biographical Dictionary. 1959.
*See also* DULONG-PETIT

## NEUMANN LAW

In compounds of analogous chemical constitution, the molecular heat, or the product of the specific heat by the atomic heat, is always the same.

**Keywords:** atomic heat, chemical, molecular heat, specific heat
NEUMANN, Franz Ernst, 1798-1895, German physicist
**Sources:** Friel, J. P. 1974; Hensyl, W. R. 1990; Stedman, T. L. 1976; Webster's Biographical Dictionary. 1959.
*See also* DULONG-PETIT

## NEUMANN LAW OF INDUCTANCE (1845); FARADAY-NEUMANN LAW

The inductance of a coil wound on a ferromagnetic core depends on the magnitude of the current. When the flux of magnetic induction enclosed by an electric current is changing, there is an EMF acting in the circuit in addition to the EMF of batteries in the circuit. The amount of this additional EMF is equal to the rate of diminution of the flux of induction. The coefficient of mutual inductance between two circuits is:

$$M_{12} = M_{21} = \mu\mu_0/4\pi \iint_{1\,2} ds_1 ds_2/r_{12}$$

where $ds_1$, $ds_2$ = vector elements of currents in the two circuits
$\mu$ = permeability of core of solenoid
$\mu_0$ = permeability of free space
$r$ = the distance between them

**Keywords:** coil current, electricity, flux, inductance
NEUMANN, Franz Ernst, 1798-1895, German physicist
**Sources:** Ballantyne, D. W. G. and Lovett, D. R. 1980; Thewlis, J. 1961-1964; Webster's Biographical Dictionary. 1959. 1995
*See also* FARADAY; LENZ

## NEUMANN LAW OF CRYSTAL ZONES (1830)

A fundamental law of crystallography that all planes that can occur on a crystal are related to each other in zones or from any four planes, no three of which lie in any one zone. All crystal planes can be devised by means of zones.

**Keywords:** crystallography, planes, zones
NEUMANN, Franz Ernst, 1798-1895, German physicist
**Sources:** Bothamley, J. 1993; Northrup, E. F. 1917.

## NEUMANN LAW ON MOLECULAR SPECIFIC HEAT (1831*)

States that the product of the molecular weight and specific heat remains constant for all compounds belonging to the same general formula(s) and similarly constituted, but that the product varies from one series to another.

**Keywords:** molecular weight, physical chemistry, specific heat
NEUMANN, Franz Ernst, 1798-1895, German physicist

**Sources:** Northrup, E. F. 1917; Preston, T. 1929; *Webster's Biographical Dictionary. 1959. 1995.
*See also* KOPP-NEUMANN

### NEURAL—SEE PSYCHOPHYSICAL

### NEUTRON ABSORPTION LAW

The disappearance of neutrons while passing through matter is neutron absorption. The beam intensity changes as it passes through matter according to an exponential relationship:

$$I = I_0 \exp(-\mu x)$$

where $I_0$ = incident intensity
$I$ = intensity after passing through a material
$\mu$ = neutron absorption coefficient for neutrons of incident energy
$x$ = thickness of material

**Keywords:** absorption, intensity, neutrons
**Source:** Thewlis, J. 1961-1964.

### NEW STATISTICS—SEE DISSOCIATION

### NEWCOMB—SEE FIRST DIGIT

### NEWLANDS LAW (1863)—SEE MENDELEEV; OCTAVES; PERIODIC

Newlands introduced atomic number concept in 1865.

NEWLANDS, John Alexander Reina, 1838-1898, English chemist
**Sources:** Asimov, I. 1972; Isaacs, A. 1996; Jerrard, H. G. and McNeill, D. B. 1992; Stedman, T. L. 1976.

### NEWTON EXPERIMENTAL LAW OF IMPACT

When two bodies meet in impact and their centers of gravity line up on a line through the point of contact, the normal component of relative velocity of their centers of gravity after impact is equal to the relative velocity before impact × the coefficient of resilience, and in the opposite direction:

$$v - v_1 = -e(\mu - \mu_1)$$

where $v$, $v_1$ = velocities of two bodies after impact
$\mu$, $\mu_1$ = velocities of two bodies before impact
$e$ = coefficient of resilience

**Keywords:** impact, mechanics, resilience, velocities
NEWTON, Sir Isaac, 1642-1727, English philosopher and mathematician
**Source:** Northrup, E. T. 1917.
*See also* HERTZ

### NEWTON IMAGING EQUATION—SEE LENS LAW

### NEWTONIAN FRICTION LAW

In the direction of flow of a fluid, the tangential force per unit area acting at an arbitrary position within a fluid between two rigid plates, one stationary and the other moving at a constant velocity, is proportional to the shear of the fluid motion at that position. Mathematically:

$$\tau = \mu\ \partial u/\partial z$$

where $\tau$ = shear
$\mu$ = viscosity, constant of proportionality
u = velocity
$\partial u/\partial z$ = velocity gradient
z = half distance between plates

**Keywords:** dynamics, fluid, friction, shear, velocity
NEWTON, Sir Isaac, 1642-1727, English philosopher and mathematician
**Source:** Gray, G. W. 1972.
*See also* DARCY

### NEWTONIAN PARTICLE LAWS

1. A particle of mass, m, acted on by a resultant force, has an acceleration, F = ma.
2. The idea of action-reaction indicating that, when one particles exerts force on another, the other particle exerts on the one a colinear force equal in magnitude but oppositely directed.
3. A body, a system of particles, acted upon by force is in equilibrium when its constituent particles are in equilibrium. An *internal* force is exerted by one particle on another particle in the body. An *external* force is exerted on a particle or body by a particle not of the body.

**Keywords:** acceleration, action-reaction, body, force, mass, particle
NEWTON, Sir Isaac, 1642-1727, English philosopher and mathematician
**Source:** Parker, S. P. 1992.
*See also* other NEWTON LAWS

### NEWTON INERTIAL FORCE GROUP OR NUMBER, $N_\ell$

A dimensionless group used in agitation relating the imposed force divided by the inertial force:

$$N_\ell = F/\rho\ V^2L^2$$

where F = imposed force
$\rho$ = mass density
V = velocity
L = characteristic length

**Keywords:** agitation, force, inertia
NEWTON, Sir Isaac, 1642-1727, English philosopher and mathematician
**Sources:** Bolz, R. E. and Tuve, G. L. 1970; Parker, S. P. 1992; Potter, J. H. 1967.

### NEWTON-KELVIN MODEL—SEE VISCOELASTIC

### NEWTON LAW FOR MOMENTUM AND MOMENT OF MOMENTUM

Stated in vector representation, the law of momentum is:

$$\Sigma F = D/Dt \int V\ dm$$

where $\Sigma F$ = outside forces on the body, and the moment of momentum is:

$$\Sigma (F \cdot r) = D/Dt \int (V \cdot r)\, dm$$

where r = vector radius from the origin
V = velocity
F = force
t = time
m = mass

Momentum is mass times velocity.

**Keywords:** forces, mechanics, moment, momentum
NEWTON, Sir Isaac, 1642-1727, English philosopher and mathematician
**Source:** Achinstein, P. 1971.

## NEWTON LAW OF CENTRIFUGAL FORCE (1673); NEWTON LAW OF CENTRIPETAL FORCE (1680)

The equal and opposite force, based on the Newton third law, exerted outward (centrifugal) and inward (centripetal) on a rotating on an axis, represented by:

$$F = M\, \omega^2\, r$$

where F = force
M = rotating mass
$\omega$ = angular velocity
r = distance from center of rotation to mass

**Keywords:** force, inward, outward, rotating
NEWTON, Sir Isaac, 1642-1727, English philosopher and mathematician
**Source:** Parker, S. P. 1989.

## NEWTON LAW OF CONSERVATION OF LINEAR MOMENTUM

The quantity of motion, which is obtained by taking the sum of the motions directed toward the same parts, and the difference of those that are directed to the contrary parts, suffers no change from the action of bodies among themselves.

**Keywords:** direction, motion, sum
NEWTON, Sir Isaac, 1642-1727, English philosopher and mathematician
**Source:** Achinstein, P. 1971.

## NEWTON LAW OF COOLING (1701)

The temperature of a heated body surrounded by a medium of lower constant temperature decreases at a rate proportional to the difference in temperature of the body and the medium:

$$q = h\, A\, (T_{sur} - T_{surr})$$

where q = rate of heat transferred, dQ/dt
h = heat transfer coefficient
A = area through which heat is transferred
$T_{sur}$ = surface temperature
$T_{surr}$ = surrounding temperature

**Keywords:** cooling, heat, thermal

NEWTON, Sir Isaac, 1642-1727, English philosopher and mathematician
**Sources:** Clifford, A. F. 1964; Honig, J. M. 1953; Isaacs, I. 1996; Layton, E. T. and Lienhard, J. H. 1988; Mandel, S. 1972; Tapley, B. D. 1990; Thewlis, J. 1961-1964.

## NEWTON LAW OF GRAVITATION (1665)

The essential features of the law of gravitation are as follows:

1. The force depends on the product of the two masses.
2. The force depends on the numerical value of a universal constant that represents the strength of the force.
3. The amount of the force depends on the distance between the two masses.
4. The force is exerted in a straight line between the two masses.
5. The force exerted on the first mass is equal to and opposite to the force on the second mass.

The gravitational force between two masses is represented by:

$$F = G\, m_1\, m_2/d^2$$

where $F$ = force
$G$ = constant of gravitation
$m_1$ and $m_2$ = two masses
$d$ = distance between $m_1$ and $m_2$

**Keywords:** force, gravity, mass
NEWTON, Sir Isaac, 1642-1727, English philosopher and mathematician
**Sources:** Parker, S. P. 1989; Tapley, B. D. 1990.
*See also* NEWTON LAW OF UNIVERSAL GRAVITATION

## NEWTON LAW OF HYDRODYNAMIC RESISTANCE (FLUID)

The force resisting the steady movement of a body through a fluid is proportional to the square of the velocity, the density of the fluid, and the cross-sectional area of the body at moderate velocities. At high velocities, it is nonlinear.

**Keywords:** fluid, hydraulics, mechanics, resistance
NEWTON, Sir Isaac, 1642-1727, English philosopher and mathematician
**Source:** Morris, C. G. 1992.
*See also* FROUDE; REYNOLDS; STRESS-STRAIN

## NEWTON LAW OF IMPACT—SEE HERTZ; INERTIA

## NEWTON LAW OF MOTION—SEE MOTION, LAWS OF

## NEWTON LAW OF OPTICS (OPTICKS) (1704)

Newton expressed the theory that white light is made up of many colors (1672) and postulated a combination of the wave and corpuscular theories that explained light. It is similar to the present view.

**Keywords:** colors, light, white
NEWTON, Sir Isaac, 1642-1727, English philosopher and mathematician
**Sources:** Honig, J. 1953; Landau, S. I. 1986; Mitchell, J. 1983.
*See also* KIRCHHOFF

## NEWTON LAW OF PROPORTIONALITY OF STRESS AND VELOCITY GRADIENT

A proportionality relationship exists between stress and velocity gradient:

$$\tau = \mu\dot{\gamma} = \tau = \mu dv/dx$$

where $\tau$ = yield stress
$\mu$ = a viscosity
$\dot{\gamma}$ = velocity of displacement

**Keywords:** constant, displacement, yield, stress
NEWTON, Sir Isaac, 1642-1727, English philosopher and mathematician
**Source:** Reiner, M. 1960.

## NEWTON LAW OF UNIVERSAL GRAVITATION (1684)

Each particle of matter attracts every other particle with a force directed along a line connecting them and is directly proportional to the product of their masses and inversely proportional to the distances between them:

$$F = G\, m_1\, m_2/d^2$$

where $F$ = force between particles
$G$ = gravitational constant
$m_1$ and $m_2$ = masses of two particles
$d$ = distance between particles

**Keywords:** attraction, force, gravitation, mass
NEWTON, Sir Isaacs, 1642-1727, English philosopher and mathematician
**Sources:** Krauskopf, K. B. 1959; NYPL Desk Ref; Tapley, B. D. 1990.
*See also* NEWTON LAW OF GRAVITATION

## NEWTON LAW OF VELOCITY OF SOUND IN GASES

The velocity of propagation of sound in a gas is directly as the square root of the elasticity of the gas and inversely as the square root of its density:

$$V = (P\, \gamma/\rho)^{1/2}$$

where $V$ = velocity of propagation of sound in a gas
$P$ = elasticity of the gas
$\rho$ = density of gas
$\gamma$ = ratio of specific $\text{heat}_{\text{pressure}}$/specific $\text{heat}_{\text{volume}}$

**Keywords:** density, gas, noise, sound
NEWTON, Sir Isaac, 1642-1727, English philosopher and mathematician
**Source:** Thewlis, J. 1961-1964.

## NEWTON LAW OF VISCOSITY (1687)

The shearing stress in viscous film varies directly with the velocity and viscosity, and inversely with the film thickness, and is constant at a given temperature and pressure:

$$S_s = F/A = \mu V/h$$

The Newton law of viscosity in one dimension is:

$$\tau_{yz} = -\mu \, dv_z/dy$$

where V = velocity
 μ = dynamic viscosity
 τ = shear stress
 h = film thickness

**Keywords:** film, fluid, liquid, shear, stress, velocity, viscosity
NEWTON, Sir Isaac, 1642-1727, English philosopher and mathematician
**Sources:** Greenkorn, R. A. 1983; Kakac. S. et al. 1987; Layton, E. T. and Lienhard, J. H. 1988; Thewlis, J. 1961-1964.

## NEWTON LAWS OF MOTION (THREE LAWS) (1687); LAWS OF DYNAMICS

These three basic laws form the basis of classical mechanics; that is, for mechanical problems not involving atomic particles or smaller, and speeds not involving the speed of light. The first law is a restatement of the discovery by Galileo that no force is required for steady, unchanging motion.

**First Law (Law of Inertia)**
A body at rest remains at rest, a body in motion continues in motion at constant speed along a straight line, unless the body, whether at rest or moving, is acted upon by an unbalanced force.

**Second Law (Law of Constant Acceleration)**
An unbalanced force acting on a body causes the body to accelerate in the direction of the force, with the acceleration directly proportional to the mass (m) of the body:

$$F = m\,a = W/g\,a$$

where F = unbalanced force
 m = mass of the body
 W = weight of the body
 a = acceleration
 g = gravitational constant

**Third Law [Law of Conservation of Momentum (Motion)]**
For every action there is an equal and opposite reaction, applies to all forces—electrical, gravitational, magnetic, etc.

**Keywords:** acceleration, body, force, mechanics, motion
NEWTON, Sir Isaac, 1642-1727, English philosopher and mathematician
**Sources:** Lederman, L. 1993; Mandel, S. 1972; Parker, S. P. 1989; Rouse, H. 1946; Thewlis, J. 1961-1964.
*See also* ANGULAR MOTION; CONSERVATION

## NEWTON NUMBER, $N_c$

A dimensionless group used in friction in fluid flow:

$$N_c = F_R/\rho \, V^2L^2$$

where $F_R$ = resistance force in fluid flow
 ρ = mass density of fluid
 L = characteristic length
 V = velocity

Note the parallel of the Newton number with the Newton inertial force group.

**Keywords:** fluid flow, friction, resistance

NEWTON, Sir Isaac, 1642-1727, English philosopher and mathematician

**Sources:** Bolz, R. E. and Tuve, G. L. 1970; Land, N. S. 1972; Parker, S. P. 1992; Potter, J. H. 1967.

*See also* NEWTON INERTIAL FORCE GROUP

**NEWTON-STOKES FLUID—SEE STRESS-STRAIN**

**NEWTON VELOCITY OF A DISTURBANCE IN AN ELASTIC MEDIUM, LAW OF**

The velocity with which any compressional disturbance is propagated through an elastic medium is given by:

$$V = (E/D)^{1/2}$$

where V = velocity
E = coefficient of volume elasticity
D = density of the medium

**Keywords:** compressional disturbance, elastic, material, volume elasticity

NEWTON, Sir Isaac, 1642-1727, English philosopher and mathematician

Sources (partial): Eirich, F. R. 1956; Reiner, M 1960.

**NIKURADSE SIMILARITY LAW (1942)**

The formation of the boundary is greatly influenced by the shape of the leading edge in a fluid as well as by the pressure gradient which may exist in the internal flow. The dimensionless boundary layer thickness is proportional to the square root of the kinematic viscosity, and is represented by:

$$\delta \sim (U_\infty/\gamma\, x)^{1/2} \text{ and the Reynolds number, for laminar flow}$$

where $\delta$ = a measure of the boundary layer thickness
$\gamma$ = kinematic viscosity
U = velocity of body of fluid
x = length of leading edge of laminar layer

**Keywords:** boundary layer, fluid, leading edge, viscosity

NIKURADSE, J., twentieth century, German scientist/engineer

**Sources:** Int. J. of Heat and Mass Transfer. p. 1131, 1980; Jakob, M. 1949; Schlicting, H. 1960. 1979.

*See also* VELOCITY DEFECT; WALL

**NOISE—SEE FAN; STIFFNESS**

**NOMIC LAW**

A relationship having the general force of a natural law, when the results of the phenomena involved are expected to occur as a part of the system.

**Keywords:** law, natural, phenomena

**Source:** Flew, A. 1979.

*See also* NATURE

**NOMOLOGICAL MODEL—SEE COVERING LAW**

## NONCONTRADICTION, LAW OF

No statement is both true and false. This one of the three laws of logic attributed to Aristotle—the other two being law of identity and the law of excluded middle.

**Keywords:** false, identity, logic, true
ARISTOTLE, 384-322 B.C., Greek philosopher
**Source:** West, B. et al. 1982.
*See also* EXCLUDED MIDDLE

## NONDIMENSIONAL NUMBERS—SEE DIMENSIONLESS GROUPS/NUMBERS

## NONLINEAR CREEP, LAW OF

When the linear relationship between creep deformation and corresponding stresses are disturbed (i.e., where stress, $\sigma > R/2$), and when there are instantaneous deformations proportional to stresses, up to magnitudes that nearly correspond to time of failure, the material creeps in nonlinear manner, there is a modulus of instantaneous deformation that varies with time, and the principle of superposition applies.

**Keywords:** creep, deformation, failure, linear
**Sources:** Eirich, F. R. 1956; Scott Blair, G. W. 1949.
*See also* LINEAR; ODQUIST; SUPERPOSITION

## NON-NEWTONIAN—SEE POWER

## NORMAL LAW. ; NORMAL LAW OF ERROR; NORMAL OR GAUSSIAN DISTRIBUTION LAW; GAUSS ERROR CURVE; PROBABILITY CURVE

The Gaussian distribution and the normal law of error are both often expressed as the same relationship. The Gaussian distribution law is the theoretical frequency distribution for a set of data of any normal, repetitive function, due to chance, represented by a bell-shaped curve (Fig. N.1) symmetrical about a mean. The relationship of the number of events occuring and frequency when the events occur are due to chance only. The probability for distributions that occur due to chance is:

$$f(x) = p = h/(\pi)^{1/2} \exp(-h^2x^2)$$

where $p$ = the probability, often written as $y = p$
$h$ = a constant that depends on spread of the data or is a measure of precision
$x$ = distance, plus or minus, from the center

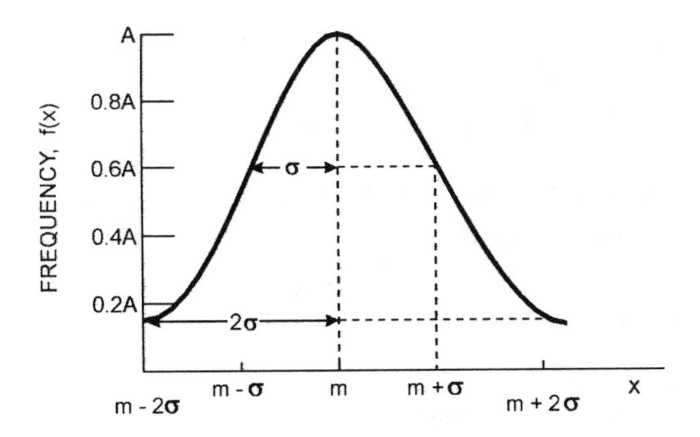


**Figure N.1** Gaussian distribution and normal distribution.

**Keywords:** chance, distribution, error, frequency, statistics
GAUSS, Johann Karl Friedrich, 1777-1855, German mathematician
**Sources:** Considine, D. M. 1961; Freiberger, W. P. 1960; Hall, C. W. 1977; Menzel, D. H. 1960.
*See also* ERROR(S); PASCAL; STANDARD ERROR

## NORMAL INCIDENCE MASS—SEE MASS

## NORMAL STRESS FRACTION—SEE SOHNKE

## NORTHROP LAW

Enzymes are proteins, and pepsin is a crystalline protein.

Northrop, John Howard, 1891-1987, American biochemist; Nobel prize, 1946, chemistry (shared)
**Sources:** Asimov, I. 1972; Gillispie, C. C. 1981;Morris, C. G. 1992.
*See also* ENZYMES; MICHAELIS-MENTEN; SCHÜTZ

## NUCLEAR CONSERVATION LAWS

The following are conserved; that is, in conversion, the property is the same before and after the conversion:

- Angular momentum
- Charge
- Energy
- Linear motion
- Mass
- Parity

**Source:** Parker, S. 1987.

## NUCLEAR DISINTEGRATION—SEE DISPLACEMENT

## NUCLIDE MASS—SEE WEIZSACKER MASS

## NULL FACTORS, LAW OF

A product of numbers can equal zero only if one or more of the factors is zero.

**Keywords:** number, product, zero
**Source:** Mandel, S. 1972.

## NUMBERS, LARGE—SEE LARGE NUMBERS

## NUSSELT BOILING OR BUBBLE NUMBER, $Nu_b$

A dimensionless group for boiling:

$$Nu_b = qD/k\,\Delta T$$

where q = heat flux
D = bubble diameter
k = thermal conductivity of liquid
$\Delta T$ = surface temperature minus saturation temperature

**Keywords:** boiling, bubble, liquid
NUSSELT, Wilhelm Ernst Kraff, 1882-1957, German engineer
**Sources:** Bolz, R. E. and Tuve, G. L. 1970; Land, N. S. 1972; Potter, J. H. 1967.

## NUSSELT (ELECTRICAL) NUMBER—SEE ELECTRICAL NUSSELT

## NUSSELT FILM THICKNESS NUMBER OR GROUP, $Nu_T$ OR $N_{Nu(t)}$

A dimensionless group relating gravitational force and viscous force in falling fluid films:

$$Nu_t = L_f(\rho^2 g/\mu^2)^{1/3}$$

where $\rho$ = mass density
$g$ = gravitational acceleration
$\mu$ = absolute viscosity
$L_f$ = film thickness

**Keywords:** falling film, fluid, gravitational, viscous
NUSSELT, Wilhelm Ernst Kraff, 1882-1957, German physicist
**Sources:** Bolz, R. E. and Tuve, G. L. 1970; Land, N. S. 1972; Parker, S. P. 1992; Potter, J. H. 1967.
*See also* GAY LUSSAC NUMBER

## NUSSELT NUMBER (HEAT TRANSFER) (1915), $Nu_h$ OR $N_{Nu(h)}$

A dimensionless group in forced convection heat transfer that relates the total heat transfer and the conduction heat transfer:

$$Nu_h = hL/k \text{ or } hd/k$$

where h = heat transfer per unit area × time
L = characteristic length
k = heat conductivity of gas

Also defined by the product of the Stanton number, the Prandtl, and the Reynolds number:

$$Nu = St \cdot Pr \cdot Re$$

**Keywords:** convection, forced, heat transfer
NUSSELT, Wilhelm Ernst Kraff, 1882-1957, German engineer
**Sources:** Bolz, R. E. and Tuve, G. L. 1970; Land, N. S. 1972; Layton, E. R. and Lienhard, J. H. 1988; Perry, R. H. 1967; Potter, J. H. 1967; Rohsenow, W. M. and Hartnett, J. P. 1973.
*See also* BIOT; REYNOLDS; STANTON

## NUSSELT NUMBER (MASS TRANSFER), $Nu_m$ OR $N_{Nu(m)}$

A dimensionless group in mass transfer that relates the mass diffusivity and molecular diffusivity or expressed as the intensity of mass flux at interface divided by specific fluid by molecular diffusion in layer of thickness:

$$Nu_m = k_c L/D_m$$

where $k_c$ = mass transfer coefficient, $k_c = \tau/\rho V$
where $\tau$ = fluid shear stress at surface
$\rho$ = mass density of fluid
V = fluid velocity
L = characteristic distance of diffusion
$D_m$ = molecular diffusivity

Same as the Sherwood number.

**Keywords:** diffusivity, flux, mass, molecular
NUSSELT, Wilhelm Ernst Kraff, 1882-1957, German engineer
**Sources:** Bolz, R. E. and Tuve, G. L. 1970; Land, N. S. 1972; Parker, S. P. 1992; Potter, J. H. 1967.
*See also* SHERWOOD

## NUTTING-SCOTT BLAIR EQUATION POWER LAW

This does not represent a physical law in the true sense but is a synthesis of a power formula and quasi-properties, developed for the creep of concrete, expressed in the form:

$$d_c = a\, s^m\, t^n$$

where $d_c$ = deformation due to creep
s = stress
t = time
a, m, n = constants

It was found that the creep increased as the quantity of aggregate in concrete is increased, which implies that the aggregates flow as loose sand may flow, and that it is the cement that bonds the aggregates.

**Keywords:** aggregate, concrete, creep, deformation, sand
NUTTING, Perley Gilman, 1873-1949, American physicist (b. in Germany)
SCOTT BLAIR, George William, twentieth century (b. 1902), English rheologist
**Sources:** Reiner, M. 1960; Who Was Who, v. 2.
*See also* ABRAMS; FERET

## NYQUIST FORMULATION OF SIGNAL FORMULATION

An upper limit on the rate of signal transmission that is related to the bandwidth of the signal used. For binary signals the channel capacity is limited by:

$$C = 2W \text{ bits/sec}$$

where C = rate of signal transmission (channel capacity)
W = bandwidth of signal (frequency)

When signal elements can represent more than two values, the channel capacity increases and for multivalued signals with M values, the channel capacity is limited by:

$$C = 2W\log_2 M \text{ bits/sec}$$

**Keywords:** bandwidth, binary, channel capacity, transmission
NYQUIST, Harry, 1889-1976, Swedish American engineer and physicist
**Sources:** Bothamley, J. 1993; Watters, C. 1992.

## NYSTEN LAW

Rigor mortis first affects the muscles of mastication, next those of the face and neck, then those of the upper trunk and arms, and lastly the muscles of the legs and feet.

**Keywords:** anatomy, muscles, rigor mortis
NYSTEN, Pierre Hubert, 1771-1818, French pediatrician
**Sources:** Friel, J. P. 1974; Landau, S. I. 1986; Stedman, T. L. 1976; Thomas, C. 1989.

## OCCAM (OCKHAM, WILLIAM OF) RAZOR—SEE MORGAN (LLOYD) CANON; PARSIMONY

Ockham, William of, c. 1285-1349, English philosopher.

## OCHOA LAW

The content of the X-chromosome tends to be phylogenetically conserved.

**Keywords:** chromosome, phylogenetically, X-chromosome
OCHOA, Severo, twentieth century (b. 1905), American physician and biochemist; Nobel prize, 1959, physiology/medicine (shared)
**Sources:** Dorland, W. A.; Stedman, T. L. 1976.

## OCKHAM RAZOR—SEE MORGAN; OCCAM

## OCTALS, LAW OF

In the study of semiconductors, the law states that the chemical activity takes place between two atoms lacking eight valence electrons and continues until the requirement of eight electrons is satisfied for all but the first orbit, where only two electrons are required.

**Keywords:** atoms, chemical, semiconductors, electrons
**Sources:** Lerner, R. G. and Trigg, G. L. 1991; Morris, C. G. 1992; Thewlis, J. 1961-1964.
*See also* OCTAVES; OCTET RULE

## OCTAVES, LAW OF (1863); SIMILAR TO PERIODIC LAW

When the chemical elements are arranged in the order of atomic weights, a system of octaves (8) is obtained, the first and last of any consecutive elements possessing similar properties, as observed by Newlands. The name of the octaves was based on a comparison of groups of elements to octaves in music and the piano keyboard.

**Keywords:** atomic weight, eight, elements, properties
NEWLANDS, John Alexander Reina, 1837-1898, English chemist
**Sources:** Asimov, I. 1972; Brown, S. B. and Brown, L. B. 1972; Daintith, J. 1988; Honig, J. M. 1953; Lagowski, J. J. 1997; Mandel, S. 1972; Walker, P. M. B. 1988.
*See also* DALTON; MENDELEEV; NEWLANDS; PERIODIC

## OCTET RULE (1919)

Stable molecular structures are characterized by having eight electrons in the valence shell orbit.

**Keywords:** electrons, orbit, valence
**Sources:** Bothamley, J. 1993; Morris, C. G. 1992.
*See also* OCTALS; OCTAVES

## OCULAR MOVEMENTS—SEE HERING

## OCVIRK NUMBER, Ocv OR $N_{Ocv}$

A dimensionless group applied to bearing lubrication that relates the bearing load force and the viscous force:

$$Ocv = F_L/\mu V(aD/rb)^2$$

where $F_L$ = bearing load/length
$\mu$ = absolute viscosity
$V$ = surface speed
$a$ = clearance width
$r$ = shaft radius
$D$ = shaft diameter
$b$ = bearing length

**Keywords:** bearing, lubrication, viscous
OCVIRK, Fred W., NACA publication, 1940-1950s (bearings)
**Sources:** Bolz, R. E. and Tuve, G. L. 1970; Eirich, F. R. 1960; Land, N. S. 1972; Parker, S. P. 1972; Potter, J. H. 1967.
*See also* HERSEY; SOMMERFELD

## ODQUIST CREEP LAW (1934)

A mathematical expression relating strain rate tensor and the stress deviation tensor on deformation of faces of materials under loading:

$$\dot{\varepsilon}_{ij} = c\, J_2^{\,m}\, s_{ij}$$

where $\varepsilon_{ij}$ = ijth component of the strain rate tensor
$s_{ij}$ = ijth component of the stress deviation tensor
$1/3\ \sigma_{kk} = \sigma_{avg}$ = average normal stress
$c$ = constant, depending on conditions
$J$ = all kinetically admissible strain rate fields
$\varepsilon_{ij} = \mu s_{ij}$ = the strain rate, where $\varepsilon_{ij}$ is strain tensor (Mises law)

These are complicated relationships requiring a study of the sources for use.

**Keywords:** deformation, mechanics, strain rate, stress, tensor
ODQUIST (ODQVIST), Folke Karl Gustav, twentieth century (b. 1899), Swedish engineer and scientist
**Sources:** Eirich, F. R. 1956; NUC; Prager, W. and Hodge, P. G. 1951.
*See also* LINEAR DAMAGE

## OECUMENOGENESIS, LAW OF

"The more organic the subject-matter of a science, the higher the integrative level of the phenomena with which it deals, the longer will be the interval lapsing between the transcurrent point and the fusion point, as between European and Asian civilization."*

**Keywords:** fusion point, integrative, interval, transcurrent
**Source:** *Needham, J. 1970.

## OERSTED ELECTROMAGNETISM LAW (1820)

Oersted discovered that electromagnetism and electricity are connected in that a compass (magnet) placed in the field of an electric current positioned itself with the needle at right angles with the direction of current flow, and that electricity generated magnetism.

**Keywords:** current flow, direction, electricity, magnetism
OERSTED, Hans Christian, 1777-1851, Danish physicist
**Sources:** Bundy Library and Smithsonian Inst. 1985; Layton, E. T. and Lienhard, J. H. 1988; NYPL Desk Ref.
*See also* FARADAY; HENRY

## OHM ACOUSTIC LAW (1843)

The human ear perceives pendular vibrations as simple tones and resolves all other periodic motions of the air into a series of pendular vibrations, hearing the series of simple tones that correspond with these vibrations. This was later disproved by Helmholtz.

**Keywords:** ear, hear, pendular, sounds, tones, vibrations
OHM, Georg Simon, 1787-1854, German physicist
**Source:** Michels, W. C. 1961.
*See also* HELMHOLTZ; OHM (HEARING)

## OHM LAW (1827)

The electric current established in a metallic circuit is directly proportional to the electromotive force (EMF, i.e., voltage) of the source, and inversely proportional to the resistance:

$$I = E/R$$

where I = current or amperage
E = electromotive force (EMF) or voltage
R = resistance, expressed in ohms

The first person to establish the relationship between length and cross-sectional area and resistance to flow of electricity was H. Davy. The power of an electric current is the work per unit time, or EI, volt-amperes, or watts (1 watt is equal to 1 joule per second).

**Keywords:** amperage, current, electromotive force, resistance, voltage
OHM, Georg Simon, 1787-1854, German physicist
DAVY, Sir Humphry, 1778-1829, English chemist
**Sources:** Layton, E. T. and Lienhard, J. H. 1988; Mandel, S. 1972; NYPL Desk Reference; Thewlis, J. 1961-1964.
*See also* JOULE; WATT

## OHM LAW OF HEARING (1843)

The ear perceives only simple harmonic vibrations, even when a complex sound wave strikes the ear.

**Keywords:** ear, harmonic, sound, vibrations
OHM, Georg Simon, 1787-1854, German physicist
**Source:** Thewlis, J. 1961-1964.
*See also* OHM (ACOUSTIC)

## OHNESORGE NUMBER, Z OR $N_Z$

A dimensionless group applicable to low-gravity slosh and capillary jets that relates the viscous force and the surface tension force, partially applicable to atomization:

$$Z = \mu/(L\ \rho\ \sigma)^{1/2}$$

where $\mu$ = absolute viscosity
$L$ = characteristic length
$\rho$ = mass density
$\sigma$ = surface tension

Also represented by the square root of the Weber number (No. 1) × the Reynolds number.

**Keywords:** capillary jets, slosh, surface tension, viscous
**Sources:** Bolz, R. E. and Tuve, G. L. 1970; Land, N. S. 1972; Parker, S. P. 1992; Potter, J. H. 1967.
*See also* REYNOLDS; WEBER

### OHNO LAW (GENETICS)

The x-inactivation mechanism has preserved the ancestral x-chromosomes almost intact during the evolution of mammals. The inactivation of one x-chromosome compensates for different dosages of x-linked alleles in males and females, ensuring genetic balance in both sexes.

**Keywords:** alleles, chromosome, evolution, female, male, mammals
**Source:** Morris, C. G. 1992.

### OKUN LAW (1968)

Relates the loss in aggregate output that is statistically associated with a given short-run increase in the unemployment rate. The law states that the elasticity of the ratio of actual to potential real output with respect to a change in the employment rate is a constant, roughly equal to 3.0.

**Keywords:** aggregate output, economics, elasticity, employment, output
OKUN, Arthur M., 1928-1980, American economist
**Sources:** Pearce, D. W. 1981; Who's Who in America. 1976.

### OLLIER LAW

With two parallel bones that are joined at their extremities by ligaments, the arrest of growth in one of the bones involves the disturbance of growth in the other.

**Keywords:** bones, growth, ligaments
OLLIER, Leopold Louis Xavier Edouard, 1830-1900, French surgeon
**Sources:** Friel, J. P. 1974; Landau, S. I. 1986.

### OLIGOMERIZATION, LAW OF (ECOLOGICAL); ALSO KNOWN AS LAW OF INTEGRATION

An evolutionary trend manifesting a reduction in the number of segments of a body or structure.

**Keywords:** evolutionary, reduction, segments
**Source:** Lincoln, R. 1982.

### ONE-FOURTH POWER LAW—SEE ALLOMETRIC SCALING LAW

### ONE-SEVENTH POWER—SEE PRANDTL

## ONE-THIRD RULE

The fracture of a wire specimen under applied tension tends to occur at a distance of one-third of the length of the specimen from one end or the other.

**Keywords:** fracture, tension, wire
**Source:** Soliman, M. A. et al. J. of Materials Science Letters 5, 329, 1986.

## ONSAGER THEORY OR LAW (1931) (1936*)

The conductance of an electrolyte of dilute solution is given by:

$$\Lambda = \Lambda_0 - (A + B\ \Lambda_0)\ (c)^{1/2}$$

where $\Lambda$ = conductance
$\Lambda_0$ = limiting conductance
A, B = constants, depending on the dielectric constant, viscosity and temperature of the solvent
c = the concentration, a modification of the P. Debye and E. Hückel equation developed in1923

**Keywords:** concentration, conductance, electrolyte, ions
ONSAGER, Lars, 1903-1976, Norwegian American engineer; Nobel prize, 1968, chemistry
**Sources:** Hampel, C. A. and Hawley, G. G. 1973; *Thewlis, J. 1961-1964.
*See also* DEYBE AND HÜCKEL; KOHLRAUSCH

## OPTICAL—SEE KIRCHHOFF; KIRSCHMANN

## OPTICAL DOUBLETS—SEE REGULAR AND IRREGULAR

## OPTICAL ROTATION—SEE OUDEMANS

## OPTIMA, MAXIMA AND MINIMA—SEE (VON) LIEBIG; LIMITING FACTORS, PRINCIPLE OF; MAXIMUM, LAW OF

## ORBITALCANINE LAW

For dogs with a normal denture, the orbital line intersects the cusps of the canine teeth.

**Keywords:** canine, denture, dog, teeth
*See also* HANAU; ORBITALGNATHION

## ORBITALGNATHION LAW

For a person with a normal denture, the lower end of the jaw (gnathion) lies in the orbital line.

**Keywords:** denture, gnathion, jaw, teeth
*See also* HANAU; ORBITALGNATHION

## ORGANIC GROWTH, LAW OF—SEE BACTERIAL GROWTH; GOMPERTZ; MAKEHAM

## ORIGINAL CONTINUITY, LAW OF (GEOLOGY) (1669)

A water-laid stratum, at the time it was formed, must continue laterally in all directions until it thins out as a result of nondeposits or until it abuts the edge of the original basin of deposition, first stated by N. Steno.

**Keywords:** deposition, stratum, water-laid
STENO, Nicholas, 1638-1686, Danish anatomist and geologist
**Sources:** Bates, J. L. and Jackson, J. A. 1980; Gary, M. et al. 1973.
*See also* ORIGINAL HORIZONTALITY; SUPERPOSITION

## ORIGINAL HORIZONTALITY, LAW OF (GEOLOGY) (1669)

Water-laid sediments are deposited in strata that are horizontal or nearly horizontal, and parallel or nearly parallel, to the surface of the Earth, first stated by N. Steno.

**Keywords:** horizontal, parallel, sediments, water-laid
STENO, Nicholas, 1638-1686, Danish anatomist and geologist
**Sources:** Bates, J. L. and Jackson, J. A. 1980; Gary, M. et al. 1973.
*See also* ORIGINAL CONTINUITY

## OSMOTIC PRESSURE LAW, GENERAL

The osmotic pressure is the pressure that must be applied to a solution to prevent the passage of the solut(ion) through an impermeable membrane. The relationship of osmotic pressure for solutions of constant thermodynamic environment, namely:

$$d\,\pi_A = -RT\,dx/V_{oA}\,x_A = -RT/V_{oA}\,d\ln x_A$$

where $\pi$ = osmotic pressure
$x_A$ = mole fraction of solvent in solution
$R$ = gas constant
$T$ = absolute temperature
$V_o$ = molal volume
$\ln = \log_e$

For very dilute solutions, the osmotic pressure law becomes the Van't Hoff law:

$$\pi V_{oA} = x_B\,RT$$

**Keywords:** membrane, physical chemistry, solution, solvent
**Source:** Menzel, D. H. 1960
*See also* VAN'T HOFF; VAN'T HOFF-ARRHENIUS

## OSTROGRADSKY NUMBER, Os OR $N_{Os}$

A dimensionless group that represents the internal heating of a product in relation to the thermal properties of the medium as a solid, liquid, or gas:

$$Os = q_v L^2/k(T_o - T)$$

where $q_v$ = strength of internal heat source
$L$ = characteristic dimension
$k$ = coefficient of thermal conductivity of medium
$(T_o - T)$ = temperature difference

**Keywords:** internal, heating, thermal, properties
OSTROGRADSKY, Mikhail Vasilyevich, 1801-1862, Soviet scientist and mathematician
**Sources:** Bedell, W. 1993; Hall, C. W. 1979; Pelletier, R. A. 1994.

## OSTWALD-DE-WAELE LAW (1923); POWER LAW FLUID

A comparison between Newtonian fluid and power law fluid in free convection systems, the Prandtl number is a dimensionless fluid property, and the Grashof number is generally the dimensionless independent parameter. The Grashof number may be varied by changing the geometry or the temperature difference without affecting the Prandtl number. With the power law, the Prandtl number is not dimensionless but is dependent on the geometry, and the temperature difference is the independent driving variable for the system. This law is the equation of a straight line on a log-log plot of shear stress ($\tau$) versus the strain rate ($\dot{v}$).

$$\tau = m\ \dot{v}^{n}$$

where $\tau$ = shear stress

n = flow index or slope of plots

m = the consistency parameter

$\dot{v}$ = strain rate or shear rate, or velocity gradient, du/dy

**Keywords:** convection, fluid, geometry, Grashof, temperature

OSTWALD, Carl Wilhelm Wolfgang (called Wolfgang by some authors), 1883-1943, German chemist (son of F. W. Ostwald)

de WAELE, A., publications in 1920s and 1930s as referenced by Scott Blair

**Sources:** Cheremisinoff, N. P. 1985; Eirich, F. R. 1960; Holmes, F. L. 1980; Irvine, T., et al. ASME Paper 82-WA/HT-69; Kakac, S. et al. 1987; Scott Blair, G. W. 1949. [Scott Blair has Wo. (Wolfgang)]

*See also* GIBBS-THOMSON; GRASHOF; PRANDTL; WAELE-OSTWALD

## OSTWALD DILUTION OR MOLECULAR CONDUCTIVITY LAW (1888)

For a sufficiently dilute solution of univalent electrolyte, the dissociation or ionization constant approximates for weak organic acids and bases:

$$K = \alpha^2\ c/(1 - \alpha)$$

or more accurately:

$$K = \{\alpha^2\ c/(1 - \alpha)\}\ \{(f_+ + f_-)/f_\pm\}$$

where K = ionization or dissociation constant

c = concentration

$\alpha$ = degree of dissociation

$f_+$ and $f_-$ = the activity coefficients of the anions and cations, respectively

$f_\pm$ = undissociated electrolyte

The law expresses the reversible ionization of weak electrolytes according to the law of mass action.

**Keywords:** acids, bases, dilute, dissociation, electrolyte, organic

OSTWALD, Friedrich Wilhelm, 1853-1932, Russian German Latvian physical chemist; Nobel prize, 1909, chemistry

**Sources:** Ballentyne, D. W. G. and Lovett, D. R. 1980; Holmes, F. L. 1980; Honig, J. M. 1953; Northrup, E. T. 1917; Porter, R. 1994; Thewlis, J. 1961-1964; Thorpe, J. F. et al.1947-1954. 1961-1964; Ulicky, L. and Kemp, T. J. 1992.

*See also* DEBYE-HÜCKEL; KOHLRAUSCH; MASS ACTION; WULLNER

## OSTWALD-FREUNDLICH RELATIONSHIP; ALSO CALLED GIBBS-THOMSON EQUATION

The difference in solubility among (between) crystals of different sizes, is:

$$\ln[c(r)/c^*] = 2\,\gamma\,U/\nu\,k\,T\,r$$

where c(r) = solubility of particle of radius r
r = radius of particle
c* = normal equilibrium solubility
$\gamma$ = surface energy of the solid in contact with its solution
U = molecular volume of solid
k = Boltzmann constant
T = absolute temperature
$\nu$ = number of ions in the solute if it is an electrolyte; if not an electrolyte, $\nu = 1$

This represents the fact that some very small crystals dissolve, even though the solution is supersaturated, showing that there is a difference in solubility of crystals of various sizes.

**Keywords:** Boltzmann, equilibrium, ions, solubility, volume
OSTWALD, Friedrich Wilhelm, 1853-1932, Russian German Latvian physical chemist; Nobel prize, 1909, chemistry
FREUNDLICH, Herbert Max Finlay, 1880-1941, German American chemist
**Sources:** Holmes, F. L. 1980 (supp.); Kirk-Othmer. 1978.

## OSTWALD RULE (1897)

If several forms of a substance can be produced in a chemical reaction, the most labile form is produced first. Following this, successively more stable forms develop. This is a restatement of the Gay-Lussac relationship for chemical reaction.

**Keywords:** chemical, labile, reaction, stable
OSTWALD, Friedrich Wilhelm, 1853-1932, Russian German Latvian physical chemist; Nobel prize, 1909, chemistry
**Source:** Fisher, D. J. 1988.
*See also* GAY-LUSSAC

## OUDEMANS LAW OF OPTICAL ROTATION; LAW OF LANDOLT-OUDEMANS

The molecular rotation of plane polarized light by salts of an optically active base or acid always tends to a definite limiting value as the concentration of the solution diminishes.

**Keywords:** acid, base, concentration, light, optic, polarized, solution
OUDEMANS, Jean Abraham Chretien, 1827-1906, Dutch astronomer
LANDOLT, Hans, 1831-1910, German Swiss chemist
**Sources:** Clifford, A. F. 1964; Michels, W. 1956; Webster's Biographical Dictionary. 1959.

## OUTER LAW—SEE VELOCITY DEFECT; WALL, LAW OF

## OVULATION, LAW OF CONSTANT NUMBERS IN—SEE LIPSCHULTZ

## OWEN FIFTH POWER LAW

The penetrating ability of radiation is roughly proportional to the fifth power of the atomic weight of the metal from which the rays are generated.

**Keywords:** atomic weight, penetration, radiation

## OXIDATION BY GREEN PLANTS, LAW OF

Green plants convert carbon dioxide during photosynthesis during the day to produce oxygen, and at night produce carbon dioxide while consuming oxygen. J. Priestley termed this process as one of *revivifying* air. He wrote about the purification of air by plants and the role of light in that process. Some refer to these new-found relationships as a law.

**Keywords:** carbon dioxide, oxygen, photosynthesis,
PRIESTLEY, Joseph, 1733-1804, English chemist
**Source:** Asimov, I. 1972; Encyclopaedia Britannica. 1961.

# P

## PAJOT LAW

A solid body contained within another body having smooth walls will tend to conform to the shape of those walls, such as the movements of a child before birth while the mother is in labor.

**Keywords:** baby, birth, physiology, shape
PAJOT, Charles, 1816-1896, French obstetrician
**Sources:** Friel, J. P. 1974; Landau, S. I. 1986.

## PARABOLIC LAW

A power function that is a fundamental relationship of natural science, represented by:

$$y = a\,x^n$$

where $y$ = dependent variable
$a$ = constant
$n$ = constant, either positive or negative
$x$ = independent variable

Many physical phenomena follow this relationship; e.g., the weight of material gained by a material is proportional to the time squared.

**Keywords:** function, mathematics, power
**Source:** Cheremisinoff, N. P. 1985.
*See also* POWER LAWS

## PARABOLIC LAW OF METAL DIFFUSION—SEE WAGNER

## PARABOLIC REACTION RATE LAW

In certain types of chemical reactions, known as *consecutive reactions,* the product of the first reaction (A) produces an intermediate product (B), which yields a final product (C),

$$A \xrightarrow{k_1} B \xrightarrow{k_2} C$$

$$dC_B/dt = k_1 C_A - k_2 C_B$$

where A, B, C = consecutive reactions
$k_1$ and $k_2$ = reaction constants

**Keywords:** chemical, consecutive, intermediate, reaction
**Sources:** Considine, D.M. 1976; Hampel, C. A. and Hawley, G. G. 1973; Thewlis, J. 1961-1964.
*See also* ARRHENIUS; REACTION; VAN'T HOFF

## PARADOXIC LAW (PHYSIOLOGY)—SEE ALLEN

## PARALLEL LAW (PSYCHOLOGICAL)

If two stimuli of different intensities are presented to a receptor at the same time, the absolute sensory intensity decreases, but the rates of intensities remain the same, as discovered by G. T. Fechner.

**Keywords:** intensity, receptor, sensory, stimuli
FECHNER, Gustav Theodor, 1801-1887, German psychologist
**Source:** Bothamley, J. 1993.
*See also* FECHNER

## PARALLELOGRAM LAW OF ADDITION

Quantities that can be represented by directed line segments (vectors) are subject to a parallelogram law of addition, by adding the lines representing quantities that have magnitude and direction.

**Keywords:** addition, direction, magnitude, vectors
**Source:** James, R. C. and James, G. 1968.
*See also* ADDITION; ALGEBRA

## PARALLELOGRAM LAW OF FORCES (1586)

For two vectors, the sum of the vectors A and B is the vector C along the diagonal of the parallelogram determined by A and B, which was enunciated by Sevinus. Geometrically, the sum of the vectors can be determined by representing the vectors as arrows joined in sequence with the terminal point of the preceding one. The sum is then the vector with the initial point of the first vector and the terminal point of the last vector (Fig. P.1).

**Keywords:** arrows, first, terminal, vector
**Sources:** Bothamley, J. 1993; James, R. C. and James, G. 1968.
*See also* ADDITION; ALGEBRA

## PARALLEL RESISTANCES, LAW OF

For a group of electrical resistors connected in parallel, the reciprocal of the resistance of the combined effect is the sum of the reciprocals of the individual resistors, or the equivalent resistance,

$$1/R_{eq} = 1/R_1 + 1/R_2 + 1/R_3 + \ldots$$

where $R_{eq}$ = equivalent resistance of the resistors
$R_1, R_2, R_3$, etc. = resistances of the individual resistors

**Keywords:** equivalent resistance, reciprocal, resistors
**Sources:** Fink, D. G and Beaty, H. W. 1993; Mandel, S. 1972.
*See also* SERIES

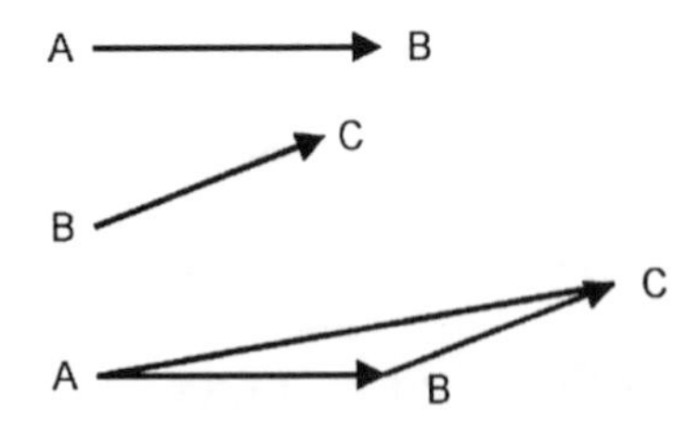


**Figure P.1** Force AC is equivalent to AB + BC.

## PARALLEL SOLENOIDS, LAW OF (METEOROLOGY)

Under stationary conditions, a line on a chart connecting all points of equal or constant pressure and density. The lines at one level must be parallel to those at other levels. Used in meteorological representations for temperature, barometric pressure (isobar), etc., and for physical representations such as temperature, pressure, density, etc.

**Keywords:** chart, lines, meteorological, physical
**Sources:** Lapedes, D. N. 1982; Parker, S. P. 1983 (physics). 1989.

## PARAMAGNETISM—SEE CURIE

## PARAXIAL LAWS (OPTICS)

Quantities expressed in power series, but using the first term only, which relates the object and the image points, as often applied to an optical system.

**Keywords:** image, optical, power, series
**Source:** Meyers, R. A. 1992.

## PARETO LAW (LATE 1800s); PARETO PRINCIPLE; 80/20 RULE

The significant items of a group will normally constitute a relatively small portion of the total (Fig. P.2). First applied to economics, Pareto found that 80 percent of the wealth was owned by 20 percent of the people. The same principle was found to apply to many other phenomena in the physical and biological world. The law can be used as an important management tool. Thus, the law is often referred to as the 80/20 rule. The law is a form of the Lorenz curve.

**Keywords:** 80/20 rule, Lorenz curve, significant, statistics
PARETO, Vilfredo, 1848-1923, Italian sociologist and economist
**Sources:** Parker, S. P. 1989; Today's Chemist at Work 6(10):31, November 1997.

## PARITY—SEE CONSERVATION OF SYMMETRIES; MIRROR SYMMETRY

## PARKINSON LAW (1958)

A statement based on a study by C. N. Parkinson of the British Admiralty, stating that work expands to fill the time available for its completion.

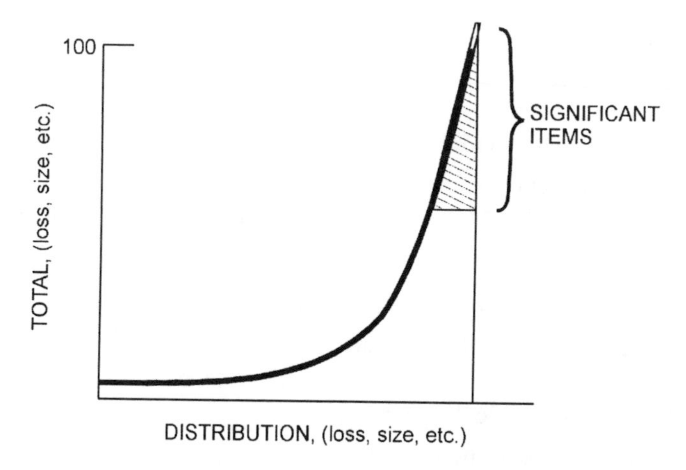


**Figure P.2** Pareto.

**Keywords:** completion, time, work
PARKINSON, Cyril Northcote, 1909-1993, English historian
**Source:** Bothamley, J. 1993.

## PARSIMONY, LAW OF; ALSO KNOWN AS OCCAM RAZOR; MORGAN CANON

If there are competing explanations for a scientific phenomenon, the simplest should be chosen. Animal behavior should be described in the simplest possible terms and should not include the attribution of human mental activities to animals, such as anecdotes, and projection of introspections.

**Keywords:** animal, behavior, mental, simplest
**Sources:** Good, C. V. 1978; Landau, S. I. 1986; Statt, D. 1982.
*See also* OCCAM (OCKHAM) RAZOR; MORGAN CANON

## PARTIAL PRESSURES, LAW OF—SEE AMAGAT; DALTON

## PARTICIPATION, LAW OF

In psychiatry this is the statement, thought, or belief that a person may have the sense that he is two different individuals at the same time, by psychotic patients, particularly in schizophrenia.

**Keywords:** identification, personal, psychiatry, schizophrenia
**Source:** Conger, G. P. 1924.

## PARTICLE NUMBER, Pn OR $N_{Pn}$

A dimensionless group used to represent dust deposit in ducts relating terminal velocity and gravitational acceleration:

$$Pn = UV/gL$$

where U = terminal, free fall velocity of particle
V = fluid velocity
g = gravitational acceleration
L = characteristic length

The particle number is similar to the Froude number.

**Keywords:** deposit, ducts, dust
**Source:** Land, N. S. 1972.
*See also* FROUDE

## PARTITION, LAW OF; OR DISTRIBUTION LAW (1872)

For equilibrium at constant temperature conditions, the ratio of the concentration of the solute in the two phases is a constant, irrespective of the total quantity in which it is present:

$$C_b/C_a = D$$

where D = partition coefficient, or distribution coefficient
$C_a$ and $C_b$ = concentration of the solute

This relationship was expressed by Berthelot.

**Keywords:** concentration, distribution, equilibrium, solute, solution
BERTHELOT, Pierre Eugene, 1827-1907, French chemist

**Sources:** Glasser, O. 1944; Gray, H. J. and Isaacs, A. 1975.
*See also* BERTHELOT; DISTRIBUTION; EQUIPARTITION; HENRY

### PASCAL LAW (c. 1648) (HYDRAULICS)

The fluid pressure due to the reaction of the walls of the containing vessel is the same at all points throughout the fluid. Or, the pressure, force per unit area, exerted anywhere upon a mass of liquid is transmitted undiminished in all directions and acts with the same force on all equal surfaces and in a direction at right angles to those surfaces.

**Keywords:** fluid, hydraulics, mechanics, pressure, surfaces
PASCAL, Blaise, 1623-1662, French mathematician and physicist
**Sources:** Considine, D. M. 1976; Hunt, V. D. 1979; Mandel, S. 1972; Parker, S. P. 1992.
*See also* ARCHIMEDES

### PASCAL LAW OF PROBABILITY

The general relationship for any number of equally probable results of an event is:

$$p = 1/R$$

where p = probability
R = probable results

The probability of getting any particular sequence of results of an event is $1/2^n$ where n is the number of events.

If p is the probability of one event taking place, and q the probability of another independent event, the probability of both events happening together is the product of $p \times q$.

**Keywords:** event, independent, probability
PASCAL, Blaise, 1623-1662, French mathematician and physicist
**Sources:** Asimov, I. 1972; Parker, S. P. 1989.
*See also* NORMAL; PROBABILITY

### PASCHEN LAW

The breakdown voltage or the sparking potential of an electric current between electrodes in a gas is a function only of the product of the pressure × the density of the gas and the distance between the electrodes.

**Keywords:** breakdown, electric, electrodes, gas, pressure, sparking
PASCHEN, Louis Carl Heinrich Frederick, 1865-1947, German physicist
**Sources:** Besancon, R. M. 1974; Considine, D. M. 1976; Honig, J. M. 1953; Mandel, S. 1972; Parker, S. P. 1989; Perry, J. H. 1950; Thewlis, J. 1961-1964; Uvarov, E. and Chapman, I. A. 1964.

### PASCHEN-BACK EFFECT (1921)—SEE ZEEMAN EFFECT

### PASTEUR EFFECT

Fermentation can be inhibited by supplying an abundance of oxygen to replace the anaerobic conditions.

**Keywords:** anaerobic, fermentation, inhibition
PASTEUR, Louis, 1822-1895, French chemist and microbiologist
**Source:** Lapedes, D. N. 1974.

## PASTEUR LAW OF HEMIHEDRAL CORRELATION (c. 1951)

"Any substance that was asymmetric at the molecular level should be asymmetrically hemihedral in the crystalline form and optically active in solution, since optical activity was the external, visible sign of an invisible and otherwise undetectable internal molecular symmetry."

**Keywords:** asymmetric, crystalline, molecular, optically, solution
PASTEUR, Louis, 1822-1895, French chemist and microbiologist
**Sources:** Asimov, I. 1972; Geison, G. L. 1995.

## PAULI EXCLUSION PRINCIPLE (1924, 1925), A FUNDAMENTAL LAW OF NATURE

No two particles (electrons) in the atom can have the same amount of energy and angular momentum; no two electrons in a given system can be in the same quantum state. The fundamental principle explained the Mendeleev periodic table of elements.

**Keywords:** atom, electrons, energy, quantum
PAULI, Wolfgang Ernst, 1900-1958, Austrian Swiss American physicist; Nobel prize, 1945, physics
**Sources:** Lederman, L. 1993; Mandel, S. 1972; Parker, S. P. 1989; Thewlis, J. 1961-1964.
*See also* MENDELEEV

## PAVLOV LAW OF REINFORCEMENT (1889)

Animal response to a stimulus can be strengthened by either unconditioned reflex (bell ringing) as through nerve stimulation, as first proposed by Pavlov, and as latter proved to be of secondary importance, with chemical stimulation being more important, as shown by Bayliss in Pavlov's laboratory.

**Keywords:** conditioned, reinforcement, response, stimulus
PAVLOV, Ivan Petrovich, 1849-1936, Russian physiologist; Nobel prize, 1904, physiology/medicine
BAYLISS, Sir William Maddock, 1860-1924, English physiologist
**Sources:** Corsini, R. J. 1987; David, I. et al. 1996.
*See also* SKINNER; THORNDIKE

## PÉCLET NUMBER (HEAT TRANSFER), $Pe_h$ OR $N_{Pe(h)}$

A dimensionless group used in forced heat transfer by convection that relates bulk heat transfer and conduction heat transfer:

$$Pe_h = \rho\, c_p VL/k$$

where $\rho$ = mass density
$c_p$ = specific heat at constant pressure
V = velocity
L = characteristic length
k = thermal conductivity

This Péclet number is equal to the Reynolds number times the Prandtl number (heat transfer).

**Keywords:** convection, forced, heat transfer
PÉCLET, Jean Claude Eugene (usually referred to as Eugene), 1793-1857, French physicist
**Sources:** Bolz, R. E. and Tuve, G. L. 1970; Grigull, U. et al. 1982; Kakac, S. et al. 1987; Land, N. S. 1972; Lykov, A.and Mikhaylov, Y. A. 1961; NUC; Parker, S. P. 1992; Perry, R. H. 1967; Rohsenow, W. M. and Hartnett, J. P. 1973;
*See also* BIOT; GRAETZ; PRANDTL; REYNOLDS

## PÉCLET NUMBER (MASS TRANSFER), $Pe_m$ OR $N_{Pe(m)}$ (1921)

A dimensionless group used in mass transfer based in the relationship of bulk mass transfer and diffusive mass transfer:

$$Pe_m = LV/D$$

where L = characteristic length
V = velocity
D = mass diffusivity

This Péclet number was first proposed by Grober in 1921. The Péclet number is the Reynolds number times the Schmidt number (mass transfer).

**Keywords:** bulk, diffusive, mass, transfer
PÉCLET, Jean Claude Eugene (usually referred to as Eugene), 1793-1857, French physicist
**Sources:** Bolz, R. E. and Tuve, G. L. 1970; Grigull, U. et al. 1982; Jerrard, H. G. and McNeill, D. B. 1992; Kakac, S.1987; Land, N. S., 1972; Lykov, A V. and Mikhaylov, Y. A. 196l; NUC; Parker, S. P.1992; Perry, R. H. 1967; Potter, J. H. 1967; Wolko, H. S. 1981.
*See also* BIOT; REYNOLDS; SCHMIDT

## PELTIER EFFECT OR LAW (1834)

A current flow across the junction of two unlike metals or semiconductors gives rise to an absorption or liberation of heat due to an electromotive force at the junction. The principle can used for a heating or cooling device.

**Keywords:** cool, current, electricity, heat, metals
PELTIER, Jean Charles Athanase, 1785-1845, French physicist and meteorologist
**Sources:** Besancon, R. M. 1974; Considine, D. M. 1974; Mandel, S. 1972; Thewlis, J. 1961-1964.
*See also* SEEBECK; SUCCESSIVE OR INTERMEDIATE TEMPERATURES; THOMSON EFFECT

## PENDULUM, LAW OF (1581)

The period of vibration of a pendulum varies directly as the square root of the length of the pendulum and inversely as the acceleration due to gravity (isochronicity), as discovered by G. Galileo. The principle was later applied by C. Huygens to make clocks (1685). The time, T, required for a complete swing or period is:

$$T = 2\pi(L/g)^{1/2}$$

where T = period
L = length of pendulum
g = acceleration due to gravity

**Keywords:** acceleration, gravity, length, vibration
GALILEO, Galilei, 1564-1642, Italian scientist and mathematician
HUYGENS, Christiaan, 1629-1695, Dutch physicist and astronomer
**Sources:** Asimov, I. 1972; World Book. 1989.
*See also* COMPOUND PENDULUM

## PEPTIC ACTIVITY, LAW OF

The amount of coagulated protein digested by a peptase is proportional to the time.

**Keywords:** coagulated, digested, peptase, protein
**Source:** Honig, J. M. 1953.
*See also* ENZYME; SCHÜTZ AND BORRISSOW

## PERFECT GAS LAWS

There are seven recognized laws that describe an ideal (perfect) gas:

1. BOYLE
2. CHARLES
3. AVOGADRO
4. DALTON PARTIAL PRESSURE
5. GAY-LUSSAC
6. MAXWELL DISTRIBUTION OF MOLECULES
7. JOULE

**Keywords:** ideal, gas
**Source:** Thewlis, J. 1961-1964.

## PERICLINE LAW

Describes a variety of feldspar (albite) twinned and elongated along the same direction.

**Keywords:** elongated, feldspar, geology, twinned
**Source:** Lapedes, D. N. 1978.
*See also* TWIN

## PERIOD LUMINOSITY LAW (1912); ALSO KNOWN AS LEAVITT "METER STICK"

The longer the period of light of pulsating stars, the higher the absolute magnitude of light by the stars. The distance to any cepheid, a type of star, can be determined by observing its period and apparent brightness. The relationship is:

$$m - M = 5 \log D - 5$$

where $m$ = apparent magnitude of a cepheid
$M$ = absolute magnitude of cepheid
$m - M$ = referred to as the distance modulus
$D$ = parsecs

Pulsating stars with a period, P, are related by a constant formed by P times the square root of the density.

**Keywords:** brightness, distance, light, period, star
LEAVITT, Henrietta Swan, 1868-1921, American astronomer
**Sources:** Asimov, I. 1973; Considine, D. M 1976; Kraus, J. 1995; Mitton, S. 1973.

## PERIODIC LAW OF ELEMENTS (1869); PERIODICITY, LAW OF

Properties of elements are a periodic function of their atomic weights.

**Keywords:** atomic weights, elements, properties
**Sources:** Brown, S. B. and Brown, L. B. 1972; Stedman, T.L 1976.
*See also* DALTON; HARMONIC; MENDELEEV; NEWLAND (OCTAVES)

## PERMISSION, LAW OF—SEE PROHIBITION, LAW OF

## PERMITTIVITY—SEE COULOMB

## PERRIN LAW (1908, 1913)

If particles making up the internal phase of a colloidal system are of low concentration, the particles must arrange themselves as a result of the influence of gravity, as do the molecules of gases under similar conditions.

**Keywords:** colloidal, concentration, gas, gravity, liquid, molecules
PERRIN, Jean Baptista, 1870-1942, French physicist; Nobel prize, 1926, physics
**Sources:** Besancon, R. M. 1974; Parker, S. P. 1989.

## PERRIN VALENCE LAW (RULE) (c. 1895?)

The greater the valence of the cation of a salt that is added to a solution, the greater the decrease in the potential difference between a negative diaphragm and the solvent; the greater the valence of the anion of the added salt, the greater the decrease in potential difference between the positive diaphragm and the solvent.

**Keywords:** anion, cation, diaphragm, potential, solvent
PERRIN, Jean Baptiste, 1870-1942, French physicist; Nobel prize, 1926, physics
**Source:** Besancon, R. M. 1974.

## PETER PRINCIPLE (1969)

In any organization, people are promoted to their level of incompetence and remain there.

**Keywords:** incompetence, organization
PETER, Laurence, 1920-1990, Canadian American sociologist
**Source:** Peter, L. and Hull, R. 1969.

## PETERS LAW (MEDICAL)

Atheroma, a cyst or tumor in which deposits occur in blood vessels and cause sclerosis, most commonly affects the blood vessels at their angles and turns.

**Keywords:** atheroma, blood, vessels
PETERS, Hubert, 1859-1934, Austrian physician
**Source:** Landau, S. I. 1986.

## PETIT—SEE DULONG AND PETIT

## PFIEFFER LAW

The blood serum of an animal immunized against a disease will destroy the bacteria of that disease when introduced into the body of another animal.

**Keywords:** animal, blood, disease, immunity
PFIEFFER, Richard Friedrich Johann, 1858-1945, German bacteriologist
**Sources:** Dorland, W. A. 1980; Friel, J. P. 1974.

## PFLÜGER LAW

A nerve tract is stimulated when catelectrotonus develops or anelectrotonus disappears, which is not a reversible process.

**Keywords:** anelectrotonus, catelectrotonus, nerve, stimulus
PFLÜGER, Edward Friedrich Wilhelm, 1829-1910, German physiologist

**Sources:** Friel, J. P. 1974; Landau, S. I. 1986.
*See also* ALL-OR-NONE

### PFLÜGER LAW OF CONTRACTION; LAW OF MAKE AND BREAK NERVE STIMULATION (1859)

The results of the stimulation of an isolated nerve of a frog by opening and closing electric currents of different intensities and different directions are formulated as follows:

$$\text{CCC} > \text{ACC} > \text{AOC} > \text{COC}$$

Pflüger formulated the laws concerning the behavior of nerves in response to make and break stimulation by an electric current.

**Keywords:** current, frog, make and break, nerves, stimulation
PFLÜGER, Edward Friedrich Wilhelm, 1829-1910, German physiologist
**Sources:** Dorland, W. A. 1980; Schmidt, J. E. 1959; Stedman, T. L. 1976.

### PHASE-CHANGE NUMBER, $K_o$

A dimensionless group defined by relating latent heat and specific heat in phase change:

$$K_o = LH_v/c_p\Delta t$$

where $LH_v$ = latent heat of phase change
$c_p$ = specific heat of liquid (vapor) phase
$\Delta t$ = temperature drop in liquid (vapor) phase

**Keywords:** latent, heat, phase change, specific
**Source:** Bedell, W. 1983.

### pH MONOTONICITY, LAW OF

The principle that, in convection-free electrophoresis, the pH increases monotonically from the anode to the cathode. The principle is of interest for isoelectric focusing.

**Keywords:** anode, cathode, convection, electrophoresis, isoelectric, monotonically
**Source:** Stenesh, J. 1989.

### PHOTOCHEMICAL EQUIVALENCE, LAW OF—SEE STARK-EINSTEIN

### PHOTOCHEMISTRY, FIRST LAW OF—SEE GROTTHUSS-DRAPER LAW

### PHOTOCHEMISTRY, SECOND LAW OF—SEE STARK-EINSTEIN PHOTOCHEMICAL EQUIVALENCE LAW

### PHOTOELECTRIC EFFECT, LAWS OF (1888)(1924); (PUBLISHED IN 1930)

1. For a given frequency of incident light, the kinetic energy of ejected photoelectrons does not change, but their number increases in direct proportion to the light intensity.
2. When the frequency of incident light changes (increases), no electrons are emitted until a certain threshold frequency is reached (depending on the metal). For higher frequencies, the energy of photoelectrons increases in direct proportion to the difference of the frequency used and the threshold frequencies. This is represented by the Einstein equation:

$$W = h\gamma - \phi$$

where W = maximum kinetic energy given off by electrons
h = Planck constant
$\gamma$ = frequency
$\phi$ = minimum energy to remove an electron from a solid (can also be applied to a gas)

**Keywords:** energy, frequency, light, photoelectron
BROGLIE, Louis Victor de, 1892-1987, French physicist; Nobel prize, 1929, physics
DIRAC, Paul Adrien Maurice, 1902-1984, English physicist; Nobel prize, 1933, physics (shared)
**Sources:** Bothamley, J. 1993; Gamow, G. 1961. 1988.
*See also* EINSTEIN

## PHOTOELECTRIC PROPORTIONALITY LAW

As long as there is no change in the "spectral quality of the light causing emission of photoelectrons, the photoelectric current is directly proportional to the illumination on the emitting surface."

**Keywords:** current, emission, illumination, light
**Source:** Thewlis, J. 1961-1964.

## PHOTOGRAPHIC LAW—SEE BUNSEN-ROSCOE

## PHOTOMETRIC LAWS

Many photometric measurements are based on the inverse square law and on Lambert cosine laws of incidence and emission.

**Inverse-Square Law**
Illumination (E) is proportional to intensity of the source (I) divided by the square of the distance from the source perpendicular to the surface ($D^2$):

$$E = I/D^2$$

**Lambert Cosine Law of Incidence**

$$E = 1/D^2 \cos\theta$$

where $\theta$ = the angle of incidence between the normal to the surface and the incident ray

**Lambert Cosine Law of Emission**
The intensity of illumination from a point source in a given direction, radiated or reflected from a perfectly radiating surface, varies as the cosine of the angle between the direction and the normal to the surface.

**Keywords:** angle, illumination, incidence, intensity, radiating
LAMBERT, Johann Heinrich, 1728-1777, German mathematician
**Sources:** Parker, S. 1971. 1987. 1994.
*See also* BEER; LAMBERT

## PHOTOMETRY, LAWS OF

The time rate at which energy is transported in a beam of radiant energy is the ratio of the unabsorbed and absorbed energy passing through a material:

$$T = P/P_0$$

where T = transmittance
P = quantity remaining unabsorbed after passage through a material
$P_0$ = quantity of incident beam

The logarithm (base 10) of the reciprocal of the transmittance is the absorbance,

$$A = -\log T = \log(1/T)$$

where A = absorbance
T = transmittance

**Keywords:** energy, radiant, transported, unabsorbed
**Source:** Dean, J. A. 1992.
Also see BEER

### PHOTOVOLTAIC EFFECT

Refers to the generation of an electromotive force (voltage) by the absorption of electromagnetic waves, which can occur in solids, liquids, and gases.

**Keywords:** absorption, electromagnetic, electromotive, gases, liquids, solids
**Source:** Thewlis, J. 1961-1964.

### PHYSICO-CHEMICAL EVOLUTION, LAW OF; PHYSICOCHEMICAL LAW

The laws of the chemical dynamics of a structured system will be those laws, or at least a very important section of those laws, that govern the evolution of a system comprising living organisms.

**Keywords:** chemical dynamics, evolution, life, organisms
**Source:** Meyers, R. A. 1992.

### PICTET-TROUTON RULE—SEE TROUTON

### PIERCE—SEE BRAGG-PIERCE

### PIEZOELECTRIC EFFECT (LAW)

Certain crystals in an electric field, when expanded along an axis and contracted along another, exhibit changes in voltage. The converse effect, whereby mechanical strain produces opposite charges on different faces of the crystal, also results.

**Keywords:** charge, crystals, electricity, strain
**Source:** Mandel, S. 1972.

### PINCH EFFECT OR LAW

When an electric current, direct current, or alternating current passes through a liquid conductor, the conductor tends to contract in cross-section due to electromagnetic forces.

**Keywords:** conductors, current, electromagnetic
**Source:** Parker, S. P. 1989.

### PIOBERT EFFECT

The front of movement of a plastic wave is visible on the surface while loading a specimen because of change in thickness of the specimen at that point, leaving a roughened surface in the wake.

**Keywords:** front, plastic, wake, wave
**Source:** Parker, S. P. 1987.

### PIPELINE NUMBER OR PARAMETER, $p_n$

A dimensionless group in fluid mechanics used in water hammer that relates the maximum water hammer pressure rise and the static pressure:

$$p_n = aV/2gH$$

where a = pressure wave velocity
V = fluid velocity
g = gravitational acceleration
H = static head

**Keywords:** pressure, static pressure, water hammer
**Sources:** Land, N. S. 1972; Perry, R. H. 1967; Potter, J. H. 1967.
*See also* REYNOLDS NUMBER

### PIPER LAW

For moderate uniform areas of the retina outside the pit of the retina packed with cones, the threshold is inversely proportional to the square root of the stimulated area.

**Keywords:** retina, stimula, threshold
PIPER, Hans Edmund, 1837-1915, German psychologist
**Sources:** Bothamley, J. 1993; Reber, A. S. 1985.
*See also* YOUNG-HELMHOLTZ

### PLANCK DISTRIBUTION LAW

The energy density within the frequency range is the number of oscillations per unit volume × the average energy of an oscillation:

$$e_b = \int_0^\infty (e_b, \lambda)d\lambda$$

where $e_b$ = energy density
$\lambda$ = frequency

**Keywords:** energy, frequency, oscillations
PLANCK, Max Karl Ernst Ludwig, 1858-1947, German physicist; Nobel prize, 1918, physics
**Sources:** Asimov, I. 1972; Thewlis, J. 1961-1964.
*See also* RADIATION

### PLANCK QUANTUM LAW (1900)

Energy radiated from a body occurs in small, separate bundles, called *quanta,* such that the higher the frequency, the larger the quantum of energy:

$$E = h\lambda \quad \text{or} \quad E = hf$$

where E = energy of the quantum
h = Planck constant
$\lambda$ = frequency (or f)

The Planck constant relates wavelength, specific heat of solids, photochemical effects of light, wavelength of spectral lines, Rontgen rays, velocity of rotating gas molecules, and distances between particles in crystals.

**Keywords:** energy, frequency, quantum
PLANCK, Max Karl Ernst Ludwig, 1858-1947, German physicist; Nobel prize, 1918, physics
**Sources:** Bundy Library and Smithsonian Institution. 1980; Lederman, L. 1993; Thewlis, J. 1961-1964.
*See also* X-RAYS

### PLANCK RADIATION LAW (1900)

The spectral distribution of energy of black body radiation is described by:

$$I = hc^{-2v}\ \lambda^3/[\exp\ (h\lambda/kT) - 1]$$

where I = monochromatic specific intensity
h = Planck constant
c = speed of light
λ = frequency per frequency interval
k = Boltzmann constant
T = absolute temperature of black body

This law laid the foundation for the quantum theory, as it was the first physical law to postulate that electromagnetic energy exists in discrete bundles called *quanta*.

**Keywords:** black body, radiation, spectral
PLANCK, Max Karl Ernst Ludwig, 1858-1947, German physicist; Nobel prize, 1918, physics
**Sources:** Isaacs, A. 1996; Parker, S. P. 1992; Thewlis, J. 1961-1964.
*See also* RAYLEIGH-JEANS; WIEN

### PLANETARY—SEE KEPLER

### PLANNING—SEE GRESHAM

### PLASTIC FLOW, LAWS OF

Refer to the deformation of a cell or body where L is the length of the body. Assuming S is the surface; V is the volume with n viscosity; X, Y, Z are the components of the volume force per unit volume in the direction of corresponding coordinate axes; and $X_u$, $Y_v$, and $Z_w$ are the components of pressure at the surface according to Betti's theorem, the average relative rate of change, (1/L) (dL/dt), of L is given for a body of any shape by:

$$1/L\ dL/dt = 1/3nV\ \{\int\int\int [zZ - 1/2(yY + xX)]\ dV + \int\int [zZ_v - 1/2(yY_v + xX_v)]\ dS\}$$

with the first integral over the volume and the second integral over the surface of the body.

**Keywords:** deformation, force, materials
BETTI, Enrico, 1823-1892, Italian mathematician
**Sources:** Gillispie, C. C. 1981; Morris, C. G. 1992.
*See also* BETTI

### PLASTIC STRAIN-RATE LAW

Consists of a series of equations that have been developed by various investigators to represent the strain rate upon loading a body, with longitudinal plastic waves:

$$=\phi < \varepsilon_p \dot{\varepsilon}_p >$$

where $\varepsilon_p$ = plastic component of strain
$\sigma$ = stress
$\varepsilon$ = strain

A general law is:

$$E\dot{\varepsilon}_p = g < \sigma, \varepsilon >$$

where E = Young modulus

Another representation of the plastic strain-rate law is:

$$E\dot{\varepsilon}_p = k^*[\sigma - \sigma_s] \text{ and } G\dot{v}_{0,p} = k^*[\tau - \tau_s]$$

where k* = constant
G = modulus of shear rigidity, G = 1/3 E
$\tau_s$ = shear strain
$\sigma_s$ = shear stress

**Keywords:** plastic, rate, rigidity, strain, stress
**Sources:** Flugge, W. 1962; Goldsmith, W. 1960.

## PLASTICITY NUMBER, P OR $N_P$

A dimensionless group, used in the flow of plastics, that relates the yield stress and viscous stress:

$$P = \sigma L/\mu V$$

where $\sigma$ = stress at elastic yield
L = characteristic length
$\mu$ = absolute viscosity in plastic state
V = velocity

This number is also known as the Bingham number.

**Keywords:** flow, plastics, viscous, yield
**Sources:** Bolz, R. E. and Tuve, G. L. 1970; Land, N. S. 1972; Parker, S. P. Potter, J. H. 1967.
*See also* BINGHAM

## PLATEAU-TALBOT LAW

When the eye is subjected to light stimuli that follow each other rapidly enough to become fused, their apparent brightness is diminished.

**Keywords:** brightness, eye, light, stimuli
PLATEAU, James Antoine Ferdinand, 1801-1883, Belgian physicist
TALBOT, William Henry Fox, 1800-1877, British scientist
**Source:** Stedman, T. L. 1976.

## PLAYFAIR AXIOM

Two intersecting straight lines cannot be parallel to a third straight line.

**Keywords:** lines, parallel, straight
PLAYFAIR, John, 1748-1819, Scottish mathematician and geologist
**Source:** Ballentyne, D. W. G. and Walker, L. E. Q. 1959. 1961.

### PLAYFAIR LAW (1802); LAW OF ACCORDANT TRIBUTARY JUNCTIONS

Every river appears to consist of a main trunk, fed from a variety of branches, each running in a valley proportional to its size, and all of them together forming a system of valleys that are in communication with one another and having an adjustment of their fall such that none of them joins the principal valley at the appropriate level. This is a circumstance that would be infinitely improbable if each of these valleys were not the work of the stream that flows into it.

**Keywords:** geology, river, stream, valley
PLAYFAIR, John, 1748-1819, Scottish mathematician and geologist
**Sources:** Bates, R. and Jackson, J. 1980; Gary, M. et al. 1973; Goudie, A., et al. 1985; Scott, T. A. 1975. 1995.

### PLUNKETT LAW (1975)

From the standpoint of the federal government, given two equally valid technical responses to a national problem, the technology that is larger in scale will invariably be preferred to the smaller, more decentralized technology.

**Keywords:** decentralized, federal, larger, national, technology
PLUNKETT, Jerry, twentieth century, American businessman
**Source:** Sale, K. 1980.

### POISEUILLE LAW (1840) OR EQUATION; POISEUILLE-HAGEN LAW

The rate of flow of a fluid is proportional to the fourth power of the radius of the tube, and the first power of the pressure, and inversely to the length of the tube and viscosity of the fluid:

$$Q = \frac{\pi P r^4}{8 \mu L}$$

where Q = rate of flow
r = radius of tube
P = pressure, difference between two points on axis of tube
L = length of tube
μ = absolute viscosity

**Keywords:** flow, fluid, tube, viscosity
POISEUILLE, Jean Louis Marie, 1799-1869, French physician and physiologist
HAGEN, Gotthilf Heinrich Ludwig, 1797-1884, German hydraulic engineer
**Sources:** Layton, E. T. and Lienhard, J. H. 1988; Mandel, S. 1972; Rashevsky, N. 1960; Reiner, M. 1960; Scott Blair, G. W. 1949; Webster's Biographical Dictionary. 1959.
*See also* CAPILLARY; HAGEN-POISEUILLE; MARGULES; TATE

### POISEUILLE NUMBER, Pois OR $N_{Pois}$

A dimensionless group used to represent laminar fluid friction that relates pressure force and viscous force:

$$Pois = D^2/\mu V \, (dp/dL)$$

where D = pipe diameter

$\mu$ = absolute viscosity
V = velocity
dp/dL = pressure gradient

**Keywords:** fluid friction, laminar, pressure, viscous
POISEUILLE, Jean Louis Marie, 1799-1869, French physician and physiologist
**Sources:** Bolz, R. E. and Tuve, G. L. 1970; Land, N. S. 1972; Parker, S. P. 1992; Perry, R. H. 1962; Potter, J. H. 1967.

## POISSON DISTRIBUTION LAW (1898); LAW OF SMALL NUMBERS

The frequency distribution for the number of random events in a constant time interval is given by:

$$P(r) = m^r e^{-m}/r!$$

where P = probability of the event
m = mean number of events in time interval
r = number of random events in time interval

The name *law of small numbers* was coined by L. von Bortkiewicz in 1898.

**Keywords:** distribution, events, frequency, random, time
POISSON, Simeon Denis, 1781-1840, French mathematician
BORTKIEWICZ, Ladislaus von, 1868-1931, German mathematician
**Sources:** Ballentyne, D. W. G. and Lovett, D. R. 1970. 1980; James, R. C. and James, G. 1968; Kotz, S. and Johnson, N. L. 1983. 1988; NUC.
*See also* POISSON LAW OF LARGE NUMBERS

## POISSON LAW

When a gas expands adiabatically:

$$p V^n = \text{constant}$$

where p = pressure
V = volume
n = ratio of specific heats of the gas, $c_p/c_v$, pressure and volume, respectively

**Keywords:** adiabatically, gas, pressure, volume
POISSON, Simeon Denis, 1781-1840, French mathematician
**Sources:** Ballentyne, D. W. G. and Lovett, D. R. 1970. 1980.
*See also* BOYLE; CHARLES

## POISSON LAW OF LARGE NUMBERS

In a sequence of independent trials, the probability (Pr) that the event E occurs at the jth trial (j = 1, 2, 3, ...), and letting $X_n$ denote the number of times that E occurs in the first n trials, then for any E > 0:

$$\lim_{\to\infty} \Pr[\left|n - 1X_n - n^{-1}\Sigma\rho_j\right| > E] = 0$$

which states that the difference between the relative frequency of occurrence of E and the arithmetic mean of the probabilities $p_j$ (average probability of E) tends to zero as n tends to infinity.

**Keywords:** events, mean, probability, trials
POISSON, Simeon Denis, 1781-1840, French mathematician
**Sources:** Kotz, S. and Johnson, N. L. 1983. 1988.
*See also* BERNOUILLI; LARGE NUMBERS; SMALL NUMBERS

## POISSON LAW OF SMALL NUMBERS—SEE POISSON DISTRIBUTION LAW

## POISSON NUMBER, Po OR $N_{Po}$

A dimensionless group applicable to elasticity and deformation of solid bodies that relates lateral contraction and longitudinal extension:

$$Po = (E/2G) - 1$$

where E = tension modulus of elasticity
G = torsion modulus of elasticity

**Keywords:** contraction, deformation, elasticity, extension, tension, torsion
POISSON, Simeon Denis, 1781-1840, French mathematician
**Sources:** Land, N. S. 1972; Potter, J. H. 1967.
*See also* POISSON RATIO

## POISSON RATIO, σ OR γ (1830)

The ratio of lateral strain to the longitudinal (axial) strain when a body is in tension or compression is $\gamma$ = lateral strain/longitudinal strain (sign depending on direction of loading) (Fig. P.3):

$$\text{Ratio} = d/D \text{ to } \Delta L/L, \text{ in the range of 0.2 to 0.4 for solids}$$

when d = decrease in diameter D (minus for tension)
D = original diameter
ΔL = increase in length
L = original length

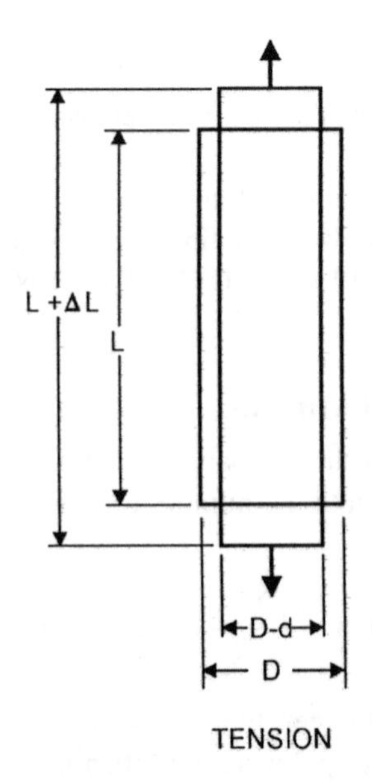


**Figure P.3** Poisson ratio.

**Keywords:** diameter, lateral, longitudinal, strain
POISSON, Simeon Denis, 1781-1840, French mathematician
**Sources:** John, V. 1992; Mandel, S. 1972; Thewlis, J. 1961-1964.
*See also* POISSON NUMBER

## POLAR EXCITATION—SEE PFLUGER; RITTER

## POLARITY, LAW OF (GEOLOGY)

A general observation that the late magnetic processes are characterized by contents that are present in relatively small amounts in the original magma. The polar elements give rise to polar minerals present in the late veins and in contact with the metamorphic zones. In ultrabase magmas, the polar elements are calcium, aluminum, silicon, sodium, and potassium, which produce grossulars, andradite, corundum, albite, etc.

**Keywords:** amounts, magma, magnetic, polar elements, veins
**Source:** Tomkeieff, S. I. 1983.
*See also* DECREASING BASICITY; HILT; SUPERPOSITION

## POLARIZATION—SEE BREWSTER; FARADAY

## POMERANTSEV NUMBER, Pom OR $N_{Pom}$

A dimensionless group in heat transfer with heat source in medium:

$$Pom = H\,L^2/k\,\Delta T$$

where H = heat liberated per volume time
L = characteristic length
k = thermal conductivity
$\Delta T$ = temperature difference of medium and initial temperature of body

**Keywords:** heat, medium, source, transfer
POMERANTSEV. Biographical information not located.
**Sources:** Bolz, R. E. and Tuve, G. L. 1970; Land, N. S. 1972; Parker, S. P. 1992; Potter, J. H. 1967.
*See also* DAMKÖHLER

## PONDEROMOTIVE LAW

This is an electric field equation that represents phenomena independent of the Maxwell equations, such that in "the limiting case where the rest mass of the particle is so small that its contribution to the field is negligible, its motion under the influence of a gravitation field is not distinguishable from a force-free or inertial motion." *Ponderomotive* means mechanical forces of interaction between electric currents and magnetic fields.

**Keywords:** electric, field, gravitation, Maxwell, rest mass
**Source:** Mendel, D. H. 1960.
*See also* LORENTZ

## POPULATION GROWTH, LAW OF; POPULATIONS, LAW OF

For the general character of the population laws, the law of growth is the same for molecules, cells, organisms, chemical kinetics, etc., which involves number of units, total masses, etc., as a property of the species, in terms of time, thus represented by the general equation:

$$dX/dt = F(X)$$

and

$$dX_i/dt = F_i\ (X_i \ldots X_n; P)$$

**Keywords:** chemistry, growth, kinetics, life
**Source:** Lotka, A. J. 1956.
*See also* POPULATION GROWTH, LOGISTIC LAW; VERHULST-PEARL

### POPULATION GROWTH, LOGISTIC LAW (1850s); VERHULST LOGISTIC EQUATION OR MODEL (1838*)

Developed by P. F. Verhulst, this is a mathematical description based on the assumption that, as a population begins to grow, its rate of growth will be constantly diminished by a factor that depends linearly on the population already living and will ultimately tend asymptotically to zero, forming an S-shaped curve. The Verhulst logistic equation is:

$$dN/dt = r\ N[(K - N)/K]$$

where N = population growth from a small population of individuals
r = rate of growth ($r > 0$) or decay ($r < 0$) of population
K = an upper limiting population in a finite environment

**Keywords:** growth, population, rate, upper limit, zero
Verhulst, Pierre F., 1804-1849, Belgian mathematician
**Sources:** *Bio Science 48(7): 540. July 1998; Encyclopedia Britannica. 1961; Lapedes, D. N. 1974; Lincoln, R. J. 1982; Pelletier, P. A. 1994.
*See also* VERHULST-PEARL

### PORTER—SEE FERRY-PORTER

### POSITION EFFECT (LAW) OF GENES

The effect of a gene in heredity depends on both its presence and on its position with respect to its neighboring genes.

**Keywords:** gene, heredity, position, reproduction
**Source:** Lincoln, R. J. 1982.

### POSNOV NUMBER, Pn OR $N_{Pn}$

A dimensionless group for heat and mass transfer in capillary-porous, moist bodies:

$$Pn = \sigma\ \Delta T/\Delta M$$

where $\sigma$ = thermal gradient
$\Delta T$ = temperature difference
$\Delta M$ = difference in moisture content of body (decimal)

The Posnov number is the analog of the Biot mass transfer number, but it characterizes the internal mass transfer without regard to the external process.

**Keywords:** capillary-porous material, heat and mass transfer
POSNOV, published in Russian. Biographical information not located.
**Sources:** Bolz, R. E. and Tuve, G. L. 1970; Land, N. S. 1972; Lykov, A. V. and Mikhaylov, Y. A. 1961.
*See also* BIOT

## POTENTIAL DROP, LAW OF; POTENTIAL, FALL OF—SEE FALL OF POTENTIAL

## POWER FUNCTION LAW (MEDICAL)

A psychophysical hypothesis for relating the amount of physical energy required to evoke a sensory experience, in certain systems, to the subjective intensity of the sense impression. The intensity of a subjective sensory experience is equal to a constant multiplied by the magnitude of the physical stimulus energy raised to a power.

**Keywords:** intensity, stimulus, psychophysical, response
**Source:** Landau, S. I. 1986.

## POWER LAW FLOW—SEE FERRY; HERSCHEL-BULKLEY; DE WAELE-OSTWALD; STEVENS

## –3/2 POWER LAW

Mean plant weight and density in competing populations can be related using the competition-density effect and the reciprocal equation, and the –3/2 power law:

$$w = K\, d^{-3/2}$$

where w = mean plant weight
K = intercept value
d = plant density

**Keywords:** density, plant populations, weight
**Source:** BioScience 31(9), 640. October 1981.
*See also* YODA

## POWER-LAW FLUID; POWER LAW OF FLOW OF NON-NEWTONIAN FLUIDS

A fluid without a yield stress in which the shear stress at any point is proportional to the rate of shear at that location raised to some power:

$$\tau = (\text{constant}) \cdot (\text{shear rate}, \dot{\gamma})^{n}$$

$$\tau = B(\dot{\gamma})^{m}$$

where $\tau$ = shear stress
$\partial u/\partial y = \dot{\gamma}$ = shear rate
B = constant
n = constant:
$n < 1$ = pseudoplastic fluid
$n > 1$ = dilatant fluid (increases in viscosity when worked)
$n = 1$ = Newtonian fluid

**Keywords:** loading, shear rate, shear stress
**Sources:** J. Heat Transfer 94:64-72, 1972; Kakac, S. et al. 1987.
*See also* OSTWALD-DE WAELE

## POWER LAW (ONE-SEVENTH)

A velocity distribution of fluid flow:

$$u/v_x = 8.74(yv_x/\upsilon)^{1/7}$$

where $u$ = velocity of flow
$v_x$ = friction velocity
$y$ = distance from wall
$\upsilon$ = kinematic viscosity

**Keywords:** distribution, flow, fluid
**Source:** Schlicting, H. 1979.
*See also* PRANDTL; WALL

## POWER-LAW PROFILE (METEOROLOGICAL)

The variation of the wind velocity, u, with height, h, in the surface boundary layer is an alternative to the logarithm of the velocity profile:

$$u/u_m = (h/h_m)^p$$

where $p$ = power whose value is 0.02 to 0.87, depending on conditions

**Keywords:** boundary layer, height, logarithm, profile, velocity, wind
**Sources:** Meyers, R. A. 1992; Morris, C. G. 1992.

## POWER LAWS, GENERAL

In general, refers to laws or models (Figs. P.4 through P.6) that follow the form of $y = x^n$. Those with $n > 1$ are parabolas. Those with $n < 0$ are hyperbolas. For example:

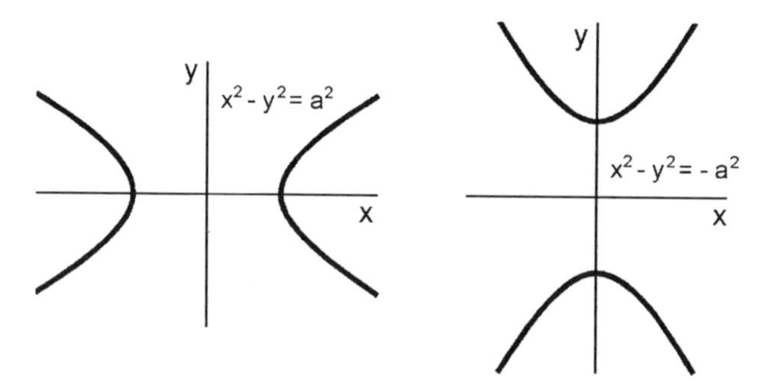


**Figure P.4** Power function: hyperbola.

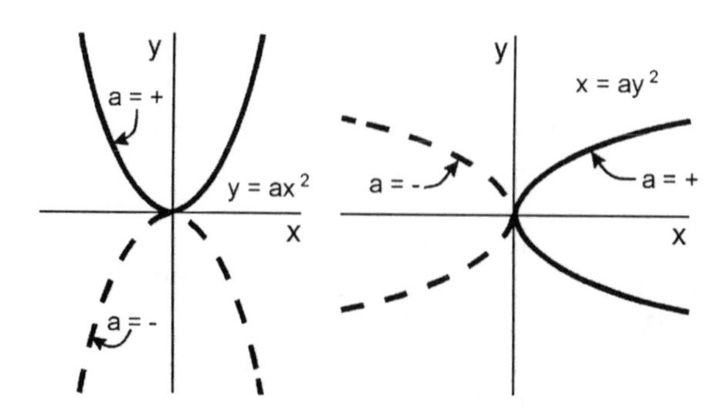


**Figure P.5** Power function: parabola.

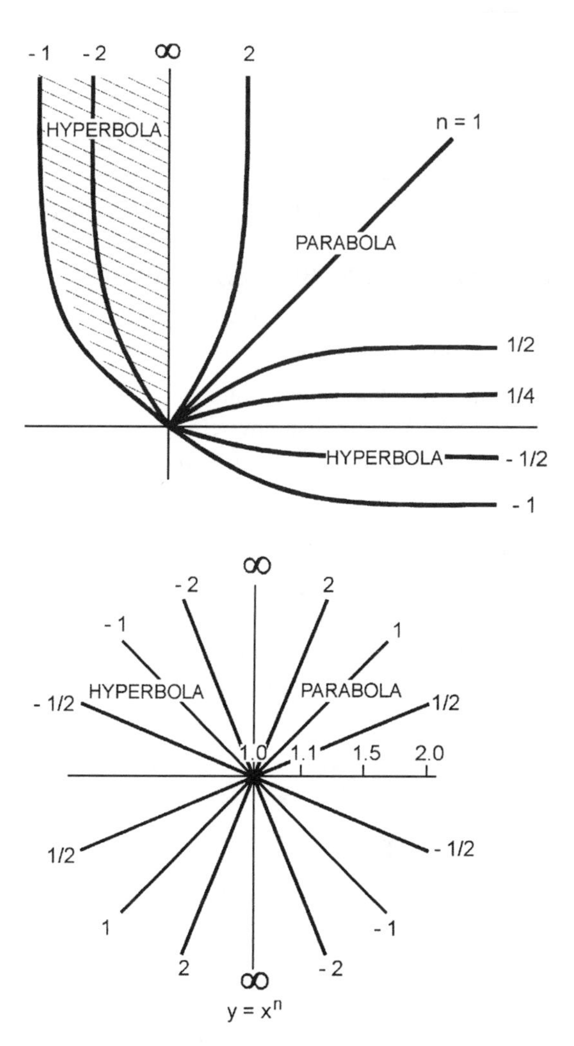


**Figure P.6** Logarithmic power curves.

| | |
|---|---|
| Boyle-Mariotte law | n = –1 |
| Newton law of gravitation | n = –2 |
| Coulomb law of electrostatics | n = –2 |
| Stefan-Boltzmann law for emission of a black body | n = 4 |

**Keywords:** hyperbolic, parabolic
**Sources:** Cheremisinoff, N. P. 1985; Reiner, M. 1960.
*See also* ALLOMETRIC (1/4, 3/4); BOYLE-MARIOTTE; COULOMB; DULONG-PETIT (5/4); IONIZATION (3/2); NEWTON (GRAVITATIONAL); OSTWALD-DE WAELE; OWEN; PRANDTL; STEFAN-BOLTZMANN; YODA (3/2)

## POWER NUMBER, pn

A dimensionless group for agitators and fans that relates the paddle drag and inertia force:

$$pn = P/L^5 \rho n^3$$

where P = agitator power

L = characteristic length
ρ = mass density
n = rotational speed(1/t)

**Keywords:** agitators, drag, fans, inertia
**Sources:** Bolz, R. E. and Tuve, G. L. 1970; Land, N. S. 1972; Parker, S. P. 1992.

## POYNTING LAW; OR POYNTING-ROBERTSON LAW

This is a special case of the Clapeyron equation in which fluid is removed as fast as it forms so that its volume may be ignored, as may happen under a metamorphic stress.

**Keywords:** fluid, geology, metamorphic, stress, volume
POYNTING, John Henry, 1852-1914, English physicist
**Sources:** Morris, C. G. 1992; Northrup, E. F. 1917; Parker, S. P. 1989 (partial).
*See also* CLAPEYRON

## PRAGER WORK HARDENING LAW

The stress of an elastic body is a monotonically increasing function of the strain applicable to work hardening for materials with a conventional limit of elasticity. The elastic limit of such a material will increase with a loading and unloading of the material, establishing a new stress-strain curve, depending on the speed, at a higher elastic limit.

**Keywords:** elasticity, strain, stress
PRAGER, William, 1903-1980, German engineer
**Source:** Prager, W. and Hodge, P. G. 1951.

## PRAGNANZ—SEE WERTHEIMER

## PRANDTL DIMENSIONLESS DISTANCE NUMBER, $y^+$; OR PRANDTL BOUNDARY LAYER (1904)

A dimensionless group used in turbulence studies:

$$y^+ = L(\rho \ \tau \ \omega)^{1/2}$$

where L = distance from wall
ρ = mass density
τ = shear stress
ω = rotational velocity

**Keywords:** shear, stress, turbulence, wall
PRANDTL, Ludwig, 1875-1953, German physicist (considered father of aeronautics)
**Sources:** Layton, E. T. and Lienhart, J. H. 1989; Parker, S. P. 1992; Potter, J. H. 1967.

## PRANDTL-GLAUERT LAW

The influence of compressibility of a fluid on the local pressures associated with a flow past an obstacle, which is negligible at low speeds, is represented on rectangular coordinates (x, y) by:

$$(1 - M_1^2)\ \partial^2\phi/\partial x^2 + \partial^2\phi/\partial y^2 = 0$$

where ϕ = the potential
$M_1$ = the free stream Mach number

**Keywords:** compressibility, fluid, gas, pressure, velocity

PRANDTL, Ludwig, 1875-1953, German physicist (considered father of aeronautics)
GLAUERT, Hermann, 1892-1934, German aeronautics
**Sources:** Pelletier, P. A. 1994; Thewlis, J. 1961-1964; Wolko, H. S. 1981.
*See also* MACH

## PRANDTL NUMBER (HEAT TRANSFER), $Pr_h$ OR $N_{Pr(h)}$ (1910; 1922)

A dimensionless group representing a fluid property in convection heat transfer relating momentum diffusivity and thermal diffusivity:

$$Pr_h = c_p\mu/k$$

where $c_p$ = specific heat at constant pressure
$\mu$ = absolute viscosity
k = thermal conductivity

The number was deduced by E. Nusselt in 1910, then presented by L. Prandtl.

**Keywords:** convection, diffusivity, heat transfer
PRANDTL, Ludwig, 1875-1953, German physicist (considered father of aeronautics)
**Sources:** Bolz, R. E. and Tuve, G. L. 1970; Jerrard, H. G. and McNeill, D. B. 1992; Land, N. S. 1972; Parker, S. P. 1992; Perry, R. H. 1967.
*See also* NUSSELT

## PRANDTL NUMBER (MASS TRANSFER), $Pr_m$ OR $N_{Pr(m)}$

A dimensionless group representing a material property in diffusion of flowing systems relating momentum diffusivity and mass diffusivity:

$$Pr_m = \mu/\rho D$$

where $\mu$ = absolute viscosity
$\rho$ = mass density
D = mass diffusivity

The Prandtl number is equal to the Schmidt number.

**Keywords:** diffusion, diffusivity, mass, momentum
PRANDTL, Ludwig, 1875-1953, German physicist (considered father of aeronautics)
**Sources:** Bolz, R. E. and Tuve, G. L. 1970; Land, N. S. 1972; Parker, S. P. 1992; Potter, J. H. 1967.
*See also* SCHMIDT

## PRANDTL ONE-SEVENTH POWER LAW

The velocity distribution in a pipeline with turbulent flow is represented by:

$$v = v_{max}\,(x/r_o)^{1/7}$$

where v = velocity at x
x = distance from the pipe wall
$v_{max}$ = velocity at the center
$r_o$ = radius of the pipe

**Keywords:** distribution, pipeline, turbulent, velocity
PRANDTL, Ludwig, 1875-1953, German physicist (considered father of aeronautics)
**Source:** Streeter, V. L. 1966.
*See also* VELOCITY-DEFECT; WALL

## PRANDTL VELOCITY RATIO, $u^+$

A dimensionless group used in turbulent flow that relates the inertia force and wall shear force:

$$u^+ = V(\rho/\tau)^{1/2}$$

where V = velocity
ρ = mass density
τ = wall shear stress

**Keywords:** flow, inertia, shear, turbulent
PRANDTL, Ludwig, 1875-1953, German physicist (considered father of aeronautics)
**Sources:** Bolz, R. E. and Tuve, G. L. 1970; Land, N. S. 1972; Parker, S. P. 1992; Potter, J. H. 1967.
*See also* FANNING; WALL

## PREDVODITELEV NUMBER, Pd OR $N_{Pd}$

A dimensionless group used for heat transfer of immersed body that relates the rate of change of temperature of medium to the rate of change of temperature of the body:

$$Pd = \tau L^2/\alpha T_0$$

where τ = rate of change of temperature of medium
L = characteristic length
α = thermal diffusivity
$T_0$ = initial temperature of immersed body

**Keywords:** heat transfer, immersed body, temperature
PREDVODITELEV, Aleksandr Savvich, twentieth century (b. 1891*), Soviet physicist
**Sources:** Bolz, R. E. and Tuve, G. L. 1970; Land, N. S. 1972; Lykov, A. V. and Mikhaylov, Y. A. 1961; Potter, J. H. 1967; *Turkevich, J. 1963.

## PRENTICE LAW

In optics, for each 10 mm of decentralization, a spherical lens displaces an object by the same number of prism dioptres as the dioptric strength of the lens.

**Keywords:** decentralization, dioptric, lens, spherical
PRENTICE, Charles F., 1854-1946, American optician
**Sources:** Critchley, M. 1978; Landau, S. I. 1986.

## PREPOTENCY OF ELEMENTS, LAW OF

A learner tends to select relevant elements in response to a complex situation. The law is attributed to E. L. Thorndike.

**Keywords:** complex situation, elements, learner, select
THORNDIKE, Edward Lee, 1874-1949, American educator
**Source:** Wolman, B. B. 1989.
*See also* THORNDIKE

## PRESSURE (BLOOD)—SEE CIRCULATION

## PRESSURE NUMBER OR PARAMETER, Kp OR $N_{Kp}$

A dimensionless group that relates the absolute pressure of a system and the pressure change across the phase boundary:

$$Kp = p/(\sigma)\ (\gamma_{w'} - \gamma_{w''})^{1/2}$$

where $p$ = absolute pressure
$\sigma$ = surface tension
$\gamma_{w'}$ = the specific weight of liquid
$\gamma_{w''}$ = the specific weight of gas

**Keywords:** absolute pressure, phase boundary, pressure change, system
**Sources:** Bolz, R. E. and Tuve, G. L. 1970; Hall, C. W. 1979; Parker, S. P. 1992; Potter, J. H. 1967.

### PRESSURES OF A GAS, LAW OF (1852)

When the volume of a gas is kept constant, if the temperature is raised by 1° C, the pressure is increased by 1/273 of the pressure at 0° C. This relationship was established by H. Regnault.

**Keywords:** gas, pressure, temperature, volume
REGNAULT, Henri Victor, 1810-1878, German French chemist and physicist
**Source:** Thewlis, J. 1961-1964.
*See also* BOYLE LAW; VAN DE WAAL LAW

### PRESTON LAW IN ZEEMAN EFFECT

The Zeeman pattern of lines from the same series of spectrum lines gives similar patterns in a magnetic field.

**Keywords:** magnetic field, pattern, spectrum
PRESTON, Samuel Tolver, 1844-1917, English physicist
**Source:** Thewlis, J. 1961-1964.
*See also* ZEEMAN

### PRÉVOST LAW (1824) (MEDICAL)

In a lateral cerebral lesion, the head is turned toward the affected side.

**Keywords:** cerebral, head, medical
PRÉVOST, Jean Louis, 1838-1927, Swiss physician
**Sources:** Friel, J. P. 1974; Landau, S. I. 1986.

### PRÉVOST LAW OF EXCHANGES (1791)

In a vacuum chamber, with the walls maintained at constant temperature, contents in the chamber will reach a condition of thermal equilibrium at which they will attain and remain at the temperature of the walls. Each body is constantly exchanging heat energy with its surroundings, the net result of which is that the temperature of the chamber and its contents equalize.

**Keywords:** chamber, contents, equilibrium, exchange, temperature, thermal
PRÉVOST, Pierre, 1751-1839, Swiss physicist
**Sources:** Considine, D. M. 1995; Isaacs, A. 1996; Mandel, S. 1972.

### PRICE SQUARE ROOT LAW (c. 1963)

A relationship postulated that half of the published articles in any given subject area are generated by a group of authors consisting of approximately the square root of the total number of authors in that area. Empirical data have not supported this claim.

**Keywords:** articles, authors, published
PRICE, Derek John de Solle, twentieth century (b. 1922), English science historian
**Sources:** NUC; Watters, C. 1992; Who's Who in America. 1970.
*See also* LOTKA; ZIPF

### PRIMACY, LAW OF (LEARNING)

That which is learned first often has such a strong impression as to be almost unshakable, making it difficult to remove or change what is taught and learned first, as expressed by H. Ebbinghaus.

**Keywords:** learned, taught, remembering
EBBINGHAUS, Hermann, 1850-1909, German psychologist
**Source:** Good, C. V. 1973.
*See also* EBBINGHAUS

### PRINS—SEE DARWIN-EWALT-PRINS

### PRIORITY LAW; LAW OF PRIORITY

A principle in taxonomy is that the proper name of nomenclature of a taxonomic group is based on that under which it was first designated, usually by the priority of publication.

**Keywords:** designated, first, name, taxonomy
**Sources:** Bates, R. L. and Jackson, J. A. 1980. 1984; Gray, P. 1967.

### PROBABILITY—SEE ADDITIVE; COMBINATION, COMPLEMENTARITY; CONDITIONAL; GAUSS; MULTINOMIAL; MULTIPLICATIVE; NORMAL

### PROBABILITY LAW (GENERAL)

The general formula for any number of equally probable results, R, is probability, P, and equals 1/R.

**Keywords:** probable, results
**Sources:** Rothman, M. A. 1963; Thewlis, J. 1961-1964.
*See also* GAUSS; PASCAL

### PROBABILITY, LAW OF (PHYSICAL) (1654)

First introduced by B. Pascal and P. Fermat, there are four different approaches:

1. *Limiting frequency definition by considering an infinite series of trials.* If the event occurs m in the first n trials, the probability (Pr) is defined to be the limit of m/n as n approaches infinity (introduced by L. Ellis).
2. *Considering an infinity of trials, in a certain subset of which the event occurs.* The probability is the ratio of the numbers in the subset and the whole set (introduced by W. Gibbs and R. Fisher).
3. *Considering the set of all possible results and taking the ratio of the number in the subset when the event occurs to the whole number* (attributed to De Moivre).
4. *The relationship is between a set of data and a proposition being considered, given these data, as might be associated with a number between zero and one.* This is expressed as Pr(p/q) = a, which reads, the probability of p given q is a. (This is based on statements by T. Bayes and P. Laplace.)

**Keywords:** data, frequency, proposition, set, subset, trials

PASCAL, Blaise, 1623-1662, French mathematician and physicist
FERMAT, Pierre de, 1601-1665, French mathematician
**Source:** Thewlis, J. 1961-1964.
*See also* ADDITIVE; BAYES; COMBINATION; COMPLEMENTARITY; CONDITIONAL; ERROR; GAUSS; MULTIPLICATIVE; NORMAL; PASCAL

## PRODUCT LAW OF UNRELIABILITY

With n components in parallel, the unreliability of the system of n components, where Q is the unreliability, Q + R (reliability) is one:

$$Q_p(t) = \prod_{i=1}^{n} Q_i(t)$$

$$Q_p(t) = Q_1(t) \cdot Q_2(t) \ldots Q_n(t)$$

**Keywords:** components, reliability, system, unreliability
**Source:** Bazovsky, I. 1960.
*See also* EXPONENTIAL FAILURE; RELIABILITY

## PROFETA LAW; PROFETA IMMUNITY LAW

A nonsyphilitic child born to syphilitic parents is immune to the disease.

**Keywords:** birth, child, immune, syphilitic
PROFETA, Giuseppe, 1840-1910, Italian dermatologist
**Sources:** Friel, J. P. 1974; Landau, S. I. 1986; Stedman, T. L. 1976.

## PROGRAM EVOLUTION, LAWS OF

Repeated observation of phenomenologically similar phenomena based on measurements of a variety of systems has led to a set of five laws. These laws have developed over a period of time and are abstractions based on statistical models.

1. *Continuing change.* A program that is used, and that as an implementation of its specification reflects some other reality, undergoes continual change or becomes progressively less useful. The change of decay process continues until it is judged to be more cost-effective to replace the system with a re-created version.
2. *Increasing complexity.* As an evolving program is continually changed, its complexity, reflecting a deteriorating structure, increases unless work is done to maintain or reduce it.
3. *The fundamental law of program evaluation.* Program evolution is subject to a dynamic that makes the programming process, and hence measures of global project and system attributes, self-regulating with statistically determinable trends and invariances.
4. *Conservation of organizational stability (invariant work rate).* During the active life of a program, the global activity rate in a programming project is statistically invariant.
5. *Conservation of familiarity (perceived complexity).* During the active life of a program, the release content (changes, additions, deletions) of the successive releases of an evolving program is statistically invariant.

**Keywords:** complexity, dynamics, familiarity, invariant, system
**Source:** Proceedings IEEE, 68(9):1068. September, 1980.

## PROGRESS, LAW OF—SEE GASKELL

## PROHIBITION, LAW OF

Classical laws of physics have dealt primarily with change. The older view of a fundamental law of nature was that it is a law of permission that defined what can and must happen in a natural phenomenon. The newer view is that the more fundamental law is a law of prohibition, which defines what cannot happen. A conservation law is, in effect, a law of prohibition that prohibits any phenomenon that would change the conserved quantity. A conservation law is simpler and less restrictive. The seven quantities that are conserved are: energy (including mass), momentum, angular momentum (including spin), charge, electron-family number, muon-family number, and baryon-family number.

**Keywords:** conserved quantities, natural phenomena, permission
**Source:** Brown, S. B. and Brown, L. B. 1972.

## PROJECTION, LAW OF

Simulation of a sensory system at any point central to the sense organ causes a sensation that is projected to the periphery of the sensory system, not at the point of stimulation. An example of this law is the "phantom limb" in which an amputee complains about an itching sensation in the amputated limb.

**Keywords:** amputee, periphery, sensory, stimulation
**Source:** Bronzino, J. D. 1995.

## PROJECTION, LAW OF (MATHEMATICS)

Defines a trigonometric relationship in which:

$$a = b \cos C + c \cos B$$

where a, b, c = the lengths of three sides of a triangle
C = angle opposite side c
B = angle opposite side b

**Keywords:** angle, sides, length, triangle
**Source:** Tuma, J. J. and Walsh, R. A. 1998.
*See also* COSINES

## PROPAGATION OF RADIO WAVES, LAWS FOR GUIDED

The waves are described by solutions of Maxwell equations. Wave propagations along the z-axis are exponential relationships that vary with $\underline{t}$ and $\underline{z}$ according to:

$$\exp(i\omega \underline{t} - \gamma \underline{z})$$

where $i = (-1)^{1/2}$
$\omega$ = frequency of wave
$\underline{t}$ = time
$\underline{z}$ = distance on axis perpendicular to x-y
$\gamma$ = propagation constant with unattenuated wave along z-axis

**Keywords:** communication, exponential, wave propagation, z-axis
**Source:** Thewlis, J. 1961-1964.
*See also* GRICE; PROPAGATION OF WAVES

## PROPAGATION OF WAVES, LAWS FOR

Guided waves are solutions of Maxwell's equations:

$$\nabla \times E = -\mu \, \partial H/\partial t$$

$$\nabla \times H = \varepsilon \, \partial E/\partial t$$

where $E$ = electric field
$H$ = magnetic field
$\mu$ = permeability
$\varepsilon$ = permittivity

**Keywords:** communications, electric field, magnetic field, Maxwell
**Source:** Thewlis, J. 1961-1964.
*See also* MAXWELL; PROPAGATION OF RADIO WAVES

## PROPORTION—SEE ALLEN

## PROUST LAW OF DEFINITE PROPORTIONS; LAW OF FIXED PROPORTIONS (1799)

Any compound always contains the same kind of elements in the same proportions.

**Keywords:** chemistry, compound, elements, proportions
PROUST, Joseph Louis, 1754-1826, French chemist
**Sources:** Asimov, I. 1966; Honig, J. M. 1953; Stedman, T. L. 1974; Van Nostrand. 1947. Webster's Biographical Dictionary. 1973.
*See also* BERTHOLETT; CHEMICAL COMBINATION; FIXED PROPORTIONS

## PSEUDOCONTIGUOUS ACTION, LAWS OF

Laws of a phenomenon at a distance may be written in a form similar to that of formulas for contiguous action. As an example, if we assume that the density of mass in a particular relationship is extremely small, then displacement of the first particle will at the same moment produce a force acting on the last particle, because the inertia of the intervening particles has dropped out. The phenomenon may be represented in the form of a differential equation as a contiguous action, and these pseudo laws prepare the way for the true laws of action. Such laws of pseudocontiguous action are confronted in electricity and magnetism, where they have led to the true laws of contiguous action.

**Keywords:** contiguous, phenomenon at a distance
**Source:** Brown, S. B. and Brown, L B. 1972.

## PSYCHOPHYSICAL LAW (1850)

In order that a sensation may increase in intensity by arithmetic progress, the stimulus must increase by geometrical progression. Equal magnitudes of stimulus ratios in a human produce equal subjective ratios. S. Landau* refers to psychological law as the Weber-Fechner law, an exponential function of the magnitude of the stimulus.

**Keywords:** psychological, response, stimuli
STEVENS, Albert Mason, 1884-1934, American pediatrician
**Sources:** Critchely, M. 1978; Dorland, W. A. 1980; Friel, J. P. 1974; *Landau, S. I. 1986; Stevens, S. S. 1970. Science v. 170, p. 1043-1050. 4 December.
*See also* ALL-OR-NOTHING; STEVENS; WEBER-FECHNER

## PTOLEMY THEORY (c. A.D. 150)

Originally proposed by Apollonius in the third century B.C., it formulated the classical system of astronomy that stated that the Earth was the center of the universe, known as the Ptolemaic system. The Ptolemy system was replaced by the Copernican system.

**Keywords:** astronomy, Earth, universe
APOLLONIUS of Perga, c. 261 B.C., Greek geometer
PTOLEMY (Ptolemaeus Claudius), 100-178, Greek astronomer
**Sources:** Encyclopaedia Britannica. 1961; Isaacs, A. 1996.
*See also* COPERNICUS

## PULLEYS—SEE MACHINE

## PURDY EFFECT

Changes in saturation of light occur with changes in intensity such that, as the brightness of color increases, its saturation increases up to a certain point and then decreases until no color may remain, and only the sensation of brightness is left.

**Keywords:** brightness, color, light, saturation
PURDY, Charles Wesley?, 1846-1901, American physician
**Source:** Physics Today 19(3):34, Mar. 1966.
*See also* ABNEY; BEZOLD-BRUCKE; GRASSMANN

## PURKINJE EFFECT OR LAW (1825)

The human eye is more sensitive to blue light when the illumination is poor and to yellow light when the illumination is good.

**Keywords:** eye, illumination, light, physiological
PURKINJE, Johannes Evangelista, 1787-1869, Czech physiologist
**Sources:** Bothamley, J. 1993; Hodgman, C. D. 1952; McAinsh, T. F. 1986.

## PYTHAGOREAN LAW OF STRINGS (FIFTH CENTURY B.C.)

The strings of a musical instrument deliver sound of a higher pitch if they are shorter, and that pitch can be simply correlated with the length of the strings and as the string tension is increased the pitch is raised. If one string is twice the length of another, the sound emitted is just one octave lower. Harmonic sounds are given by strings whose lengths are in simple numeric ratios, such as 2:1, 3:2, etc. This is perhaps the first physical law expressed and recorded.

**Keywords:** length, music, pitch, string, tension
PYTHAGORAS, c. 582-497 B.C., Greek philosopher
**Sources:** Asimov, I. 1964. 1972. 1976; Gamow, G. 1988.
*See also* KRIGAR-MENZEL; VIBRATING STRINGS

## PYTHAGOREAN THEOREM; PYTHAGORAS THEOREM (SIXTH CENTURY B.C.)

The square of the hypotenuse (the longest side) of a right-angled triangle is equal to the sum of the squares of the sides of that triangle (Fig. P.7):

$$c^2 = a^2 + b^2$$

where c = length of the hypotenuse
a, b = lengths of the other two sides

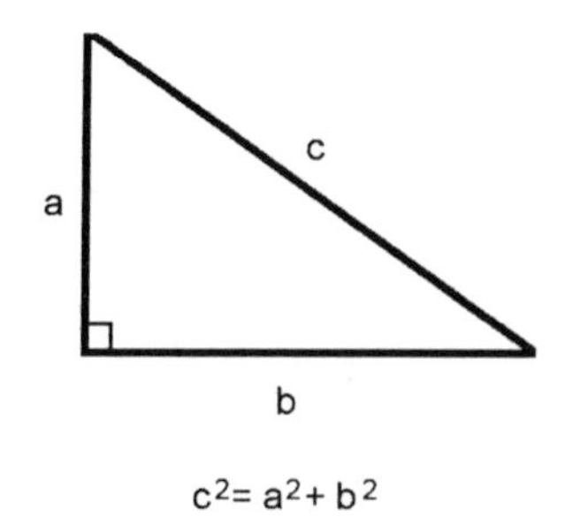


**Figure P.7** Pythagorean.

**Keywords:** hypotenuse, square, triangle
PYTHAGORAS, c. 582-497 B.C., Greek philosopher
**Sources:** Gamow, G. 1961. 1988; Mandel, S. 1972; West, B. H. et al. 1982.
*See also* MERSENNE

## $Q_{10}$ RULE

The rate of response of a process, such as respiration or chemical reaction, in an organism is often doubled or more for each increase of 10° C temperature. This is also defined as the ratio of a chemical reaction to the rate of the reaction at 10° C lower.

**Keywords:** doubled, response, temperature
**Source:** Morris, C. G. 1992.
*See also* METABOLISM; RESPIRATION; VAN'T HOFF; WILHELMY

## QUADRANTS, LAWS OF, FOR A SPHERICAL RIGHT TRIANGLE; OR RULE OF SPECIES

For a spherical right triangle, this defines a relationship concerning the relative sizes of its sides and angles (species). For a spherical right triangle, let C be a right angle, and let a, b, c be the sides opposite vertices A, B, C.

1. Angle A and side a are the same species, and so are B and b.
2. If side c is less than 90°, then a and b are of same species.
3. If side c is greater than 90°, then a and b are of different species.

Any angle and the side opposite it are in the same quadrant, and when two of the sides are in the same quadrant, the third is in the first quadrant, and when two are in different quadrants, the third side is in the second. The quadrants are first, 0° to 90°; second, 90° to 180°; third, 180° to 270°; and fourth, 270° to 360° (Fig. Q.1).

**Keywords:** algebra, angles, quadrant, sides, triangle
**Sources:** James, R. C. and James, G. 1968; Karush, W. 1989.

## QUADRATIC EQUATION OR FORMULA

An equation or formula giving the roots of a quadratic equation in which the highest power of x, a variable, is 2:

$$ax^2 + bx + c = 0, a \neq 0$$

where a, b, and c = real numbers

$$x = [\{-b \pm (b^2 - 4ac)^{1/2}\}/2a]$$

**Keywords:** equation, quadratic, roots
**Sources:** James, R. C. and James, G. 1968; Mandel, S. 1972.

## QUADRATIC RECIPROCITY, LAW OF

When p and q are distinct odd primes, then

$$(p/q)\ (q/p) = (-1)\ [(p - 1)/2][(q - 1)/2]$$

where p/q and g/p = Legendre symbols

J. Gauss gave 6 proofs of the law of quadratic reciprocity, and more than 50 proofs have been devised by others. A number of assertions by P. Fermat can be shown to follow the above law.

QUADRANTS

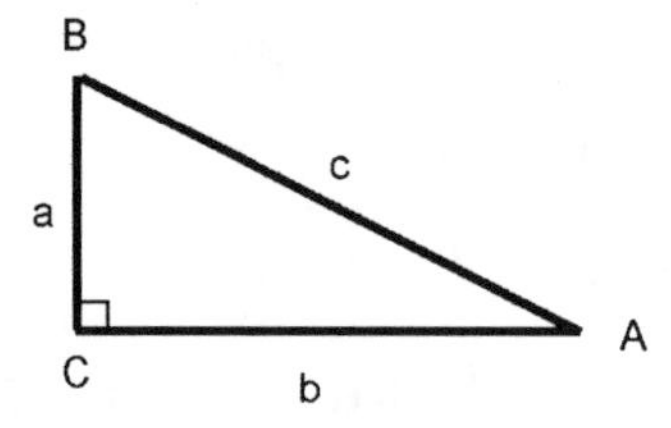


RIGHT TRIANGLE

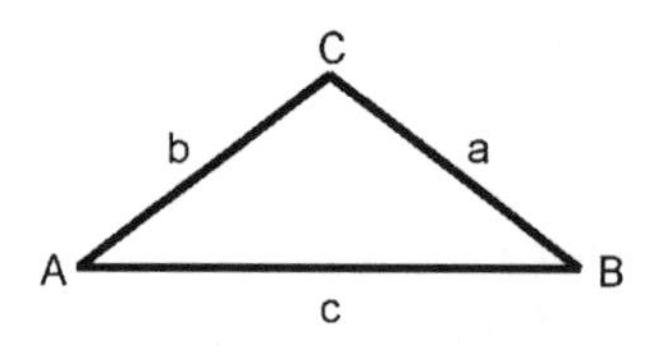


OBLIQUE TRIANGLE

**Figure Q.1** Quadrants.

**Keywords:** assertions, Fermat, Gauss, odd primes, proofs
GAUSS, Johann Karl Friedrich, 1777-1855, German mathematician
FERMAT, Pierre de, 1601-1665, French mathematician
**Sources:** Meyers, R. A. 1987; Morris, C. G. 1992; NYPL Desk Ref.
*See also* GAUSS; FERMAT

## QUANTUM ENERGY LAW—SEE DUANE-HUNT

## QUANTUM LAW OR THEORY (1900)

Radiant energy, such as light, is a continuous stream of tiny packets of energy called *quanta* or *quantum*. The energy in a quantum is the frequency of radiation, $\gamma$, times a universal constant, h, known as Planck's constant. The Planck formula applies to all forms of radiant energy, including ultraviolet light, X-rays, radio waves, microwaves, etc. Forms of radiant energy with high frequencies have higher energy than those with lower frequencies,

$$E = h\,\gamma$$

where E = energy
h = Planck's constant
$\gamma$ = frequency

**Keywords:** energy, frequency, Planck, radiant
PLANCK, Max, Earl Ernst Ludwig,1858-1947, German physicist; Nobel prize, 1918, physics
**Sources:** Parker, S. P. 1989; Speck, G. E. 1965; Uvarov, E. B. et al. 1964.
*See also* BOSE-EINSTEIN; COMPTON; EXCLUSION; HERTZ; PHOTOELECTRIC; PLANCK; THOMSON

## QUANTUM PHYSICS, LAWS OF: HERTZ; EINSTEIN; PLANCK; PHOTOELECTRON; PHOTONS; COMPTON EFFECT; THOMSON; HEISENBERG UNCERTAINTY PRINCIPLE.

**Source:** Rothman, M. A. 1963.

## QUANTUM PHYSICS, TWO BASIC LAWS OF

Small masses at very small distances do not behave as commonly considered. The effects occur in discrete steps based on Planck quantum rather than continuously. The two laws that describe their behavior are:

1. Heisenberg Uncertainty Principle
2. Pauli Exclusion Principle

**Keywords:** discrete, distances, effects
**Source:** Henry Holt Handbook of Current Science and Technology. 1992.
*See also* HEISENBERG; PAULI; PLANCK

## QUANTUM PRINCIPLE, A FUNDAMENTAL LAW

Every particle, radiant or corpuscular, responds as though it possesses both particle properties and wave properties, related as:

$$\lambda = h/u \text{ and } f = E/h$$

where $\lambda$ = wavelength
h = Planck constant
u = velocity
f = frequency
E = energy

The two fundamental principles expressed by N. Bohr are (1) the correspondence principle, that the quantum-mechanical descriptions of the macroscopic world must correspond to the description of classical mechanics, and (2) the complementary principle, that the wave-like and particle-like characteristic radiation are mutually exclusive properties.

**Keywords:** corpuscular, particle, radiant, wave
BOHR, Niels Henrik David, 1885-1962, Danish physicist; Nobel prize, 1922, physics
**Source:** Wasson, T. 1987.
*See also* QUANTUM

## QUARK CONCEPT (1963)

This is the concept that subatomic particles are made up of numerous other smaller particles. Thus, protons and neutrons can be described as being made up of more fundamental objects called *quarks*. Quarks have fractional electric charges. A baryon is composed of three quarks; a meson is composed of a pair of quarks and anti-quarks. Quarks are held together by a very strong force, provided by gluons. The development of the concept paralleled the discovery of atomic nuclei by E. Rutherford in 1910, when he bombarded alpha particles into various materials. The discovery of subatomic particles was initiated by bombarding protons and neutrons with electrons. It is generally agreed that there are six kinds of quarks, of which the "up" and "down" are most prevalent. These may combine in different ways to form 15 possible mesons.

| Name | Electric Charge | |
|---|---|---|
| Up | 2/3 | |
| Down | | −1/3 |
| Charm | 2/3 | |
| Strangeness | | −1/3 |
| Top | 2/3 | |
| Bottom | | −1/3 |

Quarks were discovered by M. Gell-Mann and G. Zweig.

**Keywords:** atomic nuclei, bombarding, electrons, neutrons, protons, subatomic
GELL-MAN, Murray, twentieth century (b. 1929), American physicist; Nobel prize, 1969, physics
ZWEIG, George, twentieth century (b. 1937), Russian American physicist
RUTHERFORD, Ernest Baron, 1871-1937, British physicist
**Sources:** Feynman, R. 1965; Lederman, L. 1993; Mascett, J. A. 1996; NYPL Desk Ref.; Phi Kappi Phi J. Winter,
1983; Science 280(5360):35, 3 April 1989; Thewlis, J. 1961-1964.
*See also* RUTHERFORD AND SODDY

# R

## RADIANT ENERGY—SEE STEFAN; STEFAN-BOLTZMANN

## RADIATION, LAWS OF

The major laws of radiation, as listed elsewhere, are:

- Kirchhoff law
- Plank law
- Wien law
- Rayleigh and Jean law
- Wien Displacement law
- Stefan-Boltzmann law

**Sources:** Morris, C. G.; Thewlis, J. 1961-1964.
*See also* QUANTUM LAW

## RADIATION NUMBER, $\bar{K}_s$

A dimensionless group used in radiant heat transfer:

$$\bar{K}_s = kE/\eta\sigma T^3$$

where k = thermal conductivity
E = modulus of elasticity
η = Stefan-Boltzmann constant
σ = surface tension
T = absolute temperature

The Radiation number equals:

$$\frac{\text{Weber number}}{\text{Hooke number} \times \text{Stefan number}}$$

**Keywords:** heat transfer, radiation
**Sources:** Bolz, R. E. and Tuve, G. L. 1970; Parker, S. P. 1992; Potter, J. H. 1967.
*See also* HOOKE; STEFAN

## RADIATION PRESSURE NUMBER, rp

A dimensionless group used in high-temperature gas flow that relates the radiation pressure to the gas pressure:

$$rp = \eta\, T^4/3p$$

where η = Stefan-Boltzmann constant
T = absolute temperature
p = pressure

**Keywords:** flow, gas, pressure, radiation, temperature
**Sources:** Land, N. S. 1972; Potter, J. H. 1967.

## RADIOACTIVE DECAY—SEE EXPONENTIAL DECAY

## RADIOACTIVE DISPLACEMENT LAW

The nuclear charge of a decaying nucleus is reduced two units by alpha-decay, producing an element two places the left in the periodic table. The positive nuclear charge is increased one unit by beta-decay, so that the resulting nucleus is an isotope of the element on the right of the decaying element in the periodic table. Since alpha-radiation does not imply a change in the charge of the nucleus, transformation of an element does not occur.

**Keywords:** alpha, beta, decay, element, nuclear, periodic
SODDY, Frederick, 1877-1956, English chemist
**Source:** Ballentyne, D. W. G. and Lovett, D. R. 1970.
*See also* SODDY-FAJANS

## RADIOACTIVITY EXPONENTIAL DECAY LAW

The rate of decrease in radioactivity is –dA/dt of the number of atoms present:

$$-dA/dt = A\,\lambda \quad \text{or} \quad \ln A/A_0 = -\lambda\,(t - t_0)$$

where A = the total activity of atoms
–dA/dt = the rate of decrease in number of atoms present
t = time

A generally more useful form for many applications is:

$$A = A_0\, e^{-\lambda t}$$

from which the half-life is calculated.

For radioactive decay:

$$-DI/dt = \lambda I$$

where I = instantaneous value of radiation level
$\lambda$ = rate of decay of substance, radioactive constant
t = time

These relationships can be used for determining the half-life of several different reactions in which there is a decrease in substance.

**Keywords:** activity, atoms, decrease, rate
**Sources:** Honig, J. M. 1953; Thewlis, J. 1961-1964.
*See also* DECAY; DRAPER

## RADIO WAVES (GUIDED)—SEE PROPAGATION OF RADIO WAVES

## RALEIGH—SEE RAYLEIGH

## RAMAN SCATTERING, LAW OF (1928); RAMAN EFFECT; SMEKAL-RAMAN EFFECT

The Raman effect is the phenomenon of light scattering from a material medium whereby the light undergoes a wavelength change in the scattering process. In Raman scattering, there is no change in wavelength. The Raman effect was discovered by C. Raman and K. S. Krishnan in India, following prediction of such by A. Smekal in 1923, inspired by discovery of the Compton effect.

The experimentally confirmed laws of Raman scattering are:

1. The pattern of Raman lines, expressed as frequency shifts, is independent of the exciting frequency.
2. The pattern of Raman frequency shifts in symmetrical about the exciting line.
3. A given Raman line shows a degree of polarization that depends on the origin of the line.
4. The Raman shifts correspond to energy differences between discrete stationary states of the scattering system.

The Raman effect differs from the Tyndall effect, for which there is not a change in frequency.

**Keywords:** light, frequency, scattering, shifts
RAMAN, Chandrasekhara Venkata, 1888-1970, Indian physicist; Nobel prize, 1930, physics
KRISHNAN, K. S., twentieth century, Indian physicist
SMEKAL, Adolf Gustav Stephen, 1895-1959, German Soviet American physicist
TYNDALL, John, 1820-1893, British physicist
**Sources:** Besancon, R. M. 1974. 1990; Holmes, F. L. 1980. 1990; Isaacs, A. 1996; Mandel, S. 1972.
*See also* COMPTON; SMEKAL

## RAMON AND CAJAL LAW OF AVALANCHE (1898)

A single sensation at the periphery of the brain may arouse multiple sensations in the brain.

**Keywords:** brain, medicine, sensations
RAMON Y CAJAL, Santiago, 1852-1934, Spanish neurologist; Nobel Prize, 1906, physiology/medicine (shared)
**Sources:** Asimov, I. 1976; Landau, S. I. 1986.

## RAMSAY AND YOUNG LAW OR RELATIONSHIP (1885); RAMSAY-YOUNG LAW

The relationship of the boiling point at a second temperature can be related to the first temperature by:

$$T_A'/T_B' = T_{Ap}/T_{Bp} + C(T_A' - T_A)$$

where $T_A'$ and $T_B'$ = absolute boiling points of the substances A and B under the same pressure
$T_A$ and $T_B$ = corresponding boiling points under another pressure
C = a constant

**Keywords:** boiling points, liquid, pressure, temperature
RAMSAY, Sir William, 1852-1916, British chemist; Nobel prize, 1904, chemistry
YOUNG, Sidney, 1857-1937, English chemist
**Sources:** Merriam-Webster's Biographical Dictionary. 1995; Perry, J. H. 1950.
*See also* BOILING POINT; CAILLETET-MATHIAS

## RAMZIN NUMBER, Ra

A dimensionless group on molar mass transfer:

$$Ra = c_b\, p\, c_m\, \Delta\Omega$$

where $c_b$ = specific vapor capacity
p = pressure
$c_m$ = mass capacity
$\Delta\Omega$ = difference in mass transfer potential (concentration)

The Ramzin number equals the Bulygin number divided by the Kossovich number.

**Keywords:** concentration, mass, transfer, vapor
RAMZIN, twentieth century, Soviet scientist/engineer
**Sources:** Bolz, R. E. and Tuve, G. L. 1970; Parker, S. P. 1992; Potter, J. H. 1967.
*See also* BULGIN; KOSSOVICH

**RANDOM INCIDENCE MASS LAW—SEE MASS (SOUND)**

**RANKINE-HUGONIOT LAW**

Relates the pressure and density ratios across two sides of a shock wave. This relationship is represented by an equation on the basis of the conservation of mass, momentum, and energy:

$$\frac{p_2}{p_1} = \left(\frac{\nu+1}{\nu-1}\right)\left(\frac{\rho_2}{\rho_1}-1\right)\left(\frac{\nu+1}{\nu-1}\right)-\left(\frac{\rho_2}{\rho_1}\right)$$

where $p$ = pressure
$\rho$ = density
$\nu$ = ratio of specific heats at constant pressure and constant volume (1.6 for air)

The change of entropy through a shock wave is:

$$s_2 - s_1 = c_v \log\left[\frac{p_2}{p_1}\left(\frac{\rho_1}{\rho_2}\right)^r\right]$$

where $s$ = entropy
$c_v$ = volume specific heat
$p$ = pressure of gas on high pressure side
$\rho$ = density

**Keywords:** conservation, entropy, equation, shock, wave
RANKINE, William John Macquorn, 1820-1872, Scottish engineer
HUGONIOT, Pierre Henry, 1851-1887, French physicist
**Sources:** Lerner, R. G. and Trigg, G. L. 1991; Thewlis, J. 1961-1964; Trigg, G. L. 1991. 1996.
Also see MACH

**RANKINE LAW (c. 1860)**

The horizontal or lateral pressure of a granular material against a wall at any point is:

$$L = wh\left(\frac{1-\sin\phi}{1+\sin\phi}\right)$$

where $L$ = lateral pressure at any point, y
$h$ = depth (from top of storage) at which pressure is to be determined
$w$ = weight of granular material per unit volume
$\tan\phi$ = relationship that gives coefficient of friction, $\mu$, and angle of repose, $\phi$, or internal coefficient of friction
$\mu = \tan\phi$

and

$$\text{total } F = \frac{1}{2}wh^2\left(\frac{1-\sin\phi}{1+\sin\phi}\right) = \frac{1}{2}wh^2(K_a)$$

where F = total force on a one foot width of wall
h = depth of material at any location

$$K_a = \frac{1 - \sin\phi}{1 + \sin\phi}$$

The Rankine law is generally used for a shallow storage with a smooth wall. For deep storages, the Janssen law is usually applicable.

**Keywords:** flow, granular material, pressure, repose
RANKINE, William John Macquorn, 1820-1872, Scottish engineer
**Sources:** Ketchum, M. S. 1919; Richey, C. B., Jacobson, P., and Hall, C. W. 1961; Scott, J. S. 1993.
*See also* JANSSEN; KETCHUM; SEMI-FLUIDS

## RAOULT LAW (1886)

This is a general law of freezing in which a molar weight of nonvolatile non-electrolytes, when dissolved in a definite weight of a given solvent under the same conditions, lower the solvent's freezing point, elevate the boiling point, and reduce the vapor pressure equally for all such solutes. For an ideal solution, the volumes of individual liquid- and vapor-phase components in a mixed system are additions.

One molecule of any compound dissolved in 100 molecules of any liquid of a different nature lowers the freezing point of this liquid by a nearly constant value, 0.62° C. The Raoult law is a special case of the Henry law.

**Keywords:** boiling point, freezing point, molecules, solution, solvent
RAOULT, Francois Marie, 1830-1901, French physical chemist
**Sources:** Asimov, I. 1972; Environmental Engineering World 2(3):19, May-June 1996; Hampel, C. A. and Hawley, G. G. 1973; Hunt, V. D. 1979; Isaacs, A. 1996; Mandel, S. 1972; Perry, J. H. 1950; Thewlis, J. 1961-1964.
*See also* BABO; BLADGEN; BOILING POINT; COPPET(DE); FREEZING POINT; HENRY; VAN'T HOFF; WULLNER; RAMSAY AND YOUNG

## RAREFIED GAS LAWS

At a sufficiently low density of a gas, the mean free path, $\lambda$, of the molecule will become commensurate with some linear dimensions of an immersed object or the surrounding enclosure so that deviations from continuum flow theory must be considered, and allowances must be made for the discrete character of the gas.

**Keywords:** continuum, density, enclosure, immersed
**Source:** Thewlis, J. 1961-1964.

## RATE LAW OF A REACTION

The rate of a chemical reaction, as measured by the disappearance of one of the reactants, or the appearance of particular products, as determined initially by experimental means, is generally proportional to the concentration of reactants:

$$-dC_A/dt = k_r\, C_A\, C_B\, C_C \ldots$$

where $k_r$ = rate constant
$C_A$, $C_B$, $C_C$ = chemical reactants
t = time

**Keywords:** chemical, concentration, experimental, reactants

**Source:** Menzel, D. H. 1960.
*See also* MASS ACTION; NEWTON; RADIOACTIVITY EXPONENTIAL

## RATIONAL (OR RATIONALITY) INDICES, LAW OF (1784); RATIONAL (OR RATIONALITY) INTERCEPTS, LAW OF; ALSO KNOWN AS LAW OF DECREMENT; ALSO KNOWN AS THE HAÜY LAW

The intercepts of the faces of a crystal upon the axes of a crystal bear a simple ratio to each other. The axial lengths are characteristic of the mineral.

**Keywords:** axes, crystal, faces, mineral
HAÜY, Rene Just, 1743-1822, French minerologist
WEISS, Christian Samuel, 1780-1856, German chemist
**Sources:** American Geological Institute. 1957; Bates, R. L. and Jackson, J. A. 1980; Gary, M. et al. 1973; Morris, C. G.1992; Sharp, D. W. A. 1981; Thewlis, J. 1961-1964; Walker, P. M. B. 1988.
*See also* CONSTANCY OF INTERFACIAL ANGLES; HAÜY; MILLER

## RAUNKAIER LAW OF FREQUENCY (ECOLOGY)—SEE FREQUENCY, LAW OF

## RAYLEIGH-JEANS LAW (1900)

An equation of radiation energy which is particularly applicable for long wavelengths, or high values of T/ν or λT and approximates Planck law, which is $L\lambda^b \approx c_1/c_2\ \pi\ \lambda^4\ T$, the wavelength for blackbody, and is useful for limited ranges of temperature and wavelength:

$$u(\nu)d = 8\pi\ \nu^2\ k\ T/c^3\ d\nu$$

$$u(\nu) = 8\ \pi\ h\ \nu^3\ n^3/c^3/(e^{h\nu/kt} - 1)$$

where λ = wavelength
T = absolute temperature
h = Planck constant
ν = frequency
k = Boltzmann constant
u = energy density
n = an integer
c = velocity of light

At low frequencies, the Planck law becomes the Rayleigh-Jeans law.

**Keywords:** long, radiation, wavelength
RAYLEIGH, John William Strutt, Third Baron, 1842-1919, English physicist; Nobel prize, 1904, physics
JEANS, James Hopwood, 1877-1946, English physicist and astronomer
**Sources:** Achinstein, P. 1971; Driscoll, W. G. 1978; Thewlis, J. 1961-1964; Ulicky, L. and Kemp, T. J. 1992.
*See also* PLANCK RADIATION; STEFAN-BOLTZMANN; WIEN

## RAYLEIGH LAW FOR DISTRIBUTION OF SPECTRAL INTENSITY—SEE RAYLEIGH-JEANS

## RAYLEIGH LAW FOR LOW MAGNETIC FIELDS

Hysteresis losses per cycle at low inductions, valid only at low frequencies are:

$$W_h = 4\ \pi/3\ d\mu/dH\ H^3$$

where $W_h$ = hysteresis loss per cycle for low inductance
$H$ = maximum field strength during the cycle (Oersteds)
$\mu$ = permeability
$d\mu/dH$ = slope of $\mu/H$ versus H curve near $\mu_o$

**Keywords:** frequencies, hysteresis, induction, magnetic
RAYLEIGH, John William Strutt, Third Baron, 1842-1919, English physicist; Nobel prize, 1904, physics
**Source:** Gray, D. E. 1972.

## RAYLEIGH LAW OF DISTRIBUTION

The Rayleigh distribution* is next to the Gaussian distribution in importance in which:

$$\pi = (x/\sigma^2)\ e^{-x^2}/2\sigma^2$$

where $\pi$ = probability
$\sigma$ = standard distribution
$\sigma^2$ = variance
$x$ = variable

**Keywords:** distribution, Gaussian, probability
RAYLEIGH, John William Strutt, Third Baron, 1842-1919, English physicist; Nobel prize, 1904, physics
**Sources:** Asimov, I. 1972; *Rothbart, H. A. 1996; Thewlis, J. 1961-1964.
*See also* GAUSSIAN; NORMAL; WIEN

## RAYLEIGH LAW OF INDUCTION

The induction for small magnetization may be approximated by:

$$\beta = \mu_o\ H + \nu\ H^2$$

which yields:

$$\mu = \mu_o + \nu\ H$$

where $\beta$ = induction (gauss)
$H$ = field strength (oersted)
$\nu$ = frequency
$\mu$ = normal permeability
$\mu_o$ = initial permeability

**Keywords:** electricity, induction, magnetism, permeability
RAYLEIGH, John William Strutt, Third Baron, 1842-1919, English physicist; Nobel prize, 1904, physics
**Source:** Tapley, B. D. 1990.

## RAYLEIGH LAW OF LIGHT DISPERSION (1871)

With small particles in the atmosphere, which are small as compared to the wavelength of light, $\lambda$, the dispersion of light is proportional to $1/\lambda^4$.

**Keywords:** atmosphere, dispersion, particles, wavelength
RAYLEIGH, John William Strutt, Third Baron, 1842-1919, English physicist; Nobel prize, 1904, physics
**Sources:** Menzel, D. H. 1960; Parker, S. P. 1989.

## RAYLEIGH LAW OF LIGHT SCATTERING (1871)

The intensity of light scattered by a particle through an angle depends on the intensity of incident light, the distance from the scattering volume, and the polarization of the particle:

$$R_0 (1 + \cos^2\theta)$$

which is called the Rayleigh ratio and is $I_0 \, r_s^2/I_o$,

where $I_o$ = the intensity of incident light
$\theta$ = angle of scattering
$r_s$ = the distance from scattering volume

**Keywords:** incidence, light, polarization, scattering
RAYLEIGH, John William Strutt, Third Baron, 1842-1919, English physicist; Nobel prize, 1904, physics
**Sources:** Menzel, D. H. 1960; Parker, S. P. 1989.

## RAYLEIGH NUMBER, Ra OR $N_{Ra}$

A dimensionless group used in free convection that results from density variations in fluid force and that relates gravity and thermal diffusivity:

$$Ra = c_p \, \rho^2 \, g \, L^3 \, \beta \, \Delta T/\mu \, k$$

where $c_p$ = specific heat at constant pressure
$\rho$ = mass density
$g$ = gravitational acceleration
$L$ = characteristic length
$\beta$ = volume expansion coefficient with temperature
$\Delta T$ = temperature difference
$\mu$ = absolute viscosity
$k$ = thermal conductivity

The Rayleigh number equals the Prandtl number times the Grashof number. There are nine different Rayleigh numbers.

**Keywords:** convection, diffusivity, gravity, thermal
RAYLEIGH, John William Strutt, Third Baron, 1842-1919, English physicist; Nobel prize, 1904, physics
**Sources:** Bolz, R. E. and Tuve, G. L. 1970; Kakac, S. et al. 1987; Land, N. S. 1972; Perry, R. H. 1967; Potter, J. H. 1967; Rohsenow, W. M. and Hartnett, J. P. 1973; Tapley, B. D. 1990.
*See also* GRASHOF; PRANDTL

## RAYLEIGH RATIO—SEE RAYLEIGH LAW OF LIGHT SCATTERING

## REACTION, FIRST ORDER LAW OF; OR MASS ACTION, LAW OF

The rate of any unimolecular reaction should at any time be proportional to the concentration in the system at that instance:

$$dC_A/dt = -k_1 \, C_A$$

where $C_A$ = concentration at that time
$t$ = time
$k_1$ = a constant (first order)

**Keywords:** concentration, mass, rate, reaction
**Source:** Perry, J. H. 1950.
*See also* GULDBERG AND WAAGE; REACTION, SECOND ORDER; REACTION, THIRD ORDER; MASS ACTION

### REACTION, SECOND ORDER LAW OF

Any bimolecular reaction can be represented, in general, by:

$$dx/dt = k_2(a - x)(b - x)$$

where x = decrease in concentration of each
t = time
$k_2$ = a constant (second order)
a, b = initial concentrations of A, B respectively
x = decrease in concentration, at time, t

**Keywords:** bimolecular, concentration, molecular, time
**Source:** Perry, J. H. 1950.
*See also* MASS ACTION; REACTION, FIRST ORDER; REACTION, THIRD ORDER

### REACTION, THIRD ORDER LAW OF; TRIMOLECULAR REACTION

The general type of molecular reaction where three different molecules react, with different initial concentrations, and the initial concentrations of each decrease with time as represented by:

$$dx/dt = k_3 (a - x)(b - x)(c - x)$$

where x = decrease in concentration
t = time
$k_3$ = a constant (third order)
a, b, c = initial concentrations of reactants

**Keywords:** concentration, conditions, molecular
**Source:** Perry, J. H. 1950.
*See also* MASS ACTION; REACTION, FIRST ORDER; REACTION, SECOND ORDER

### REACTION RATE, PARABOLIC—SEE PARABOLIC REACTION RATE

### READINESS, LAW OF

A sub-law, attributed to A. H. Thorndike, of the law of effect for learning is the law of readiness, a statement that a learner must work at it to learn. A learner has to be set to respond to specific stimuli of consequence in a situation, and once the learner is ready to learn, learning is satisfying as long as that action is not altered.

**Keywords:** learn, stimuli, readiness, response
THORNDIKE, Ashley Horace, 1871-1933, American educator
**Sources:** Corsini, R. J. 1984. 1994; Deighton, L. C. 1971; Good, C. V. 1973; Wolman, B. B. 1989.
*See also* EDUCATION; EFFECT; THORNDIKE

### REBINDER NUMBER, Rb OR $N_{Rb}$

A dimensionless group used in drying a moist substance to that used in evaporation over a short time increment:

$$Rb = dt^*/dM^*(c/LH_v)$$

where $t^*$ = mean temperature of body
$M^*$ = mean moisture content of body
c = thermal capacity
$LH_v$ = latent heat of vaporization

**Keywords:** drying, evaporation, moist, time
REBINDER, Pyotr Aliksandrovich, twentieth century (b. 1898), Soviet chemist
**Sources:** Hall, C. W. 1979; Morris, C. G. 1992; Parker, S. P. 1992; Pelletier, P. A. 1994; Scott Blair, G. W. 1949; Turkevich, J. 1963.

### REBOUND, LAW OF

The angle of rebound equals the angle of impact for a perfectly elastic ball.

**Keywords:** elastic, impact, mechanics, rebound
**Sources:** Morris, C. G. 1987. 1992 (partial); Thewlis, J. 1961-1964 (partial).

### RECAPITULATION, LAW OF; BAER LAW OR VON BAER LAW

The concept of von Baer's law is a predecessor of the recapitulation theory; that is, an organism in the course of its development goes through the same succession stages as did the species in developing from the lower to the higher forms of animal life.

**Keywords:** animal, development, lower form, higher form, organism, species
BAER, Karl Ernst von, 1792-1876, Estonian naturalist and embryologist
**Sources:** Gray, P. 1967; Stedman, T. L. 1976; Wolman, B. B. 1989.
*See also* BIOGENETIC; HAECKEL

### RECENCY, LAW OF

A specific item will tend to remind a person of a more recent association than one further back in time, and that a more recent item or experience is better remembered.

**Keywords:** association, experience, remember, time
**Sources:** Good, C. V. 1973; Wolman, B. B. 1989.
*See also* THORNDIKE

### RECIPROCAL DEFLECTIONS, LAW OF—SEE MAXWELL

### RECIPROCAL INNERVATION, LAW OF—SEE SHERRINGTON

### RECIPROCAL-LAW FAILURES (LIGHT)

For low-energy quanta as exemplified by light, the Bunsen-Roscoe reciprocity law does not hold, and the reciprocity failure is usually presented in a diagram in which the log of the exposure, $I \times t$, necessary to produce certain exposure is plotted against either log I or log t.

**Keywords:** exposure, light, plot, quanta
**Sources:** Bothamley, J. 1993; Thewlis, J. 1961-1964
*See also* BETTI; BUNSEN-ROSCOE

### RECIPROCAL PROPORTIONS (PROPERTIES), LAW OF (1792)

Two elements that unite with each other will unite singly with a third element in proportions that are the same as, or multiples of, the proportions in which they unite with each other, as expressed by J. Richter.

**Keywords:** chemistry, combine, elements, proportions

RICHTER, Jerimias Benjamin, 1762-1807, German chemist

**Sources:** Hanau, S. I. 1986; Honig, J. M. 1953; Mandel, S. 1972; Walker, P. M. B. 1988; Thomas, C. L. 1989.

*See also* CHEMICAL COMBINATION; COMBINING WEIGHTS; EQUIVALENT PROPORTIONS (RATIOS); RICHTER

## RECIPROCITY, LAW OF—SEE BUNSEN-ROSCOE

## RECTILINEAR DIAMETERS, LAW OF—SEE CAILLETET AND MATHIAS

## REDFIELD—SEE STORMS, LAW OF

## RED QUEEN HYPOTHESIS—SEE CONSTANT EXTINCTION; VAN VALEN

## RED SHIFT, LAW OF—SEE HUBBLE

## REECH NUMBER, Re OR $N_{Re}$

A dimensionless group applicable to surface boats and gravity-affected motions, it relates gravity force and inertia force:

$$Re = gL/V^2$$

where g = gravitational acceleration
L = characteristic length
V = velocity

The Reech number is the reciprocal of the Froude number.

**Keywords:** boats, gravity, inertia force, surface vehicles

REECH, Ferdinand, 1805-1884,* Alsatian scientist

**Sources:** Bolz, R. E. and Tuve, G. L. 1970; Land, N. S. 1972; Parker, S. P. 1992; Potter, J. H. 1967; *Truesdell, C. A. 1960.

*See also* FROUDE NUMBER

## REFERRED PAIN, LAW OF—SEE MORLEY

## REFLECTION (WAVES), LAWS OF; REFLECTION OF LIGHT AT A PLANE SURFACE, LAWS OF

### Law 1

For a ray of light that strikes the surface of the mirror at any point, P, the angle of incidence, i, is equal to the angle of reflection, r (Fig. R.1).

### Law 2

The plane determined by the normal and the reflected ray coincides with the plane determined by the normal and the incident ray.

**Keywords:** incident, light, ray, reflection

**Sources:** Bates, R. L. and Jackson, J. A. 1980. 1984; Daintith, J. 1981; Mandel, S. 1972; Walker, P. M. B. 1988.

*See also* DIFFUSE REFLECTION

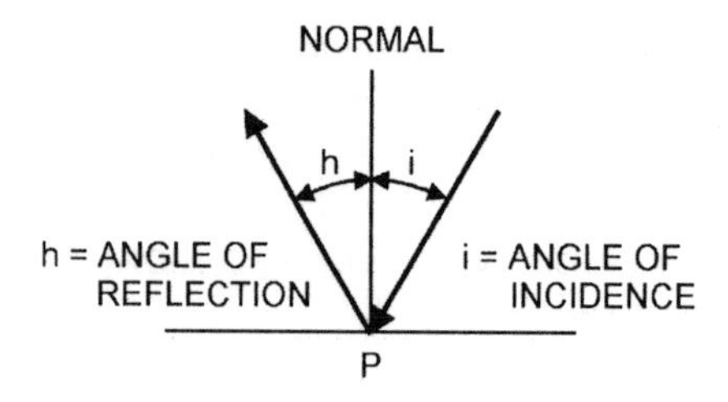


Reflection

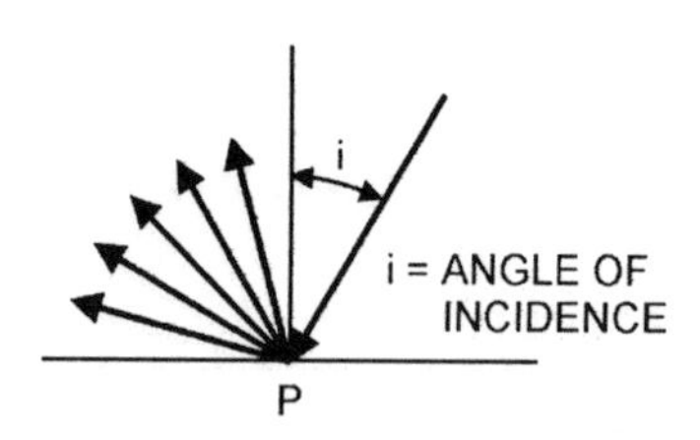


Diffuse Reflection

**Figure R.1** Reflection of light at a plane surface.

## REFLEX IRRADIATION—SEE SHERRINGTON

## REFLEXIVE—SEE BOOLEAN ALGEBRA

## REFRACTION, LAW OF; SNELL LAW (1621); DESCARTES LAW

When a wave crosses a boundary between two isotropic substances, the wave normal changes direction such that the sine of the angle of incidence between wave normal and boundary normal divided by the velocity in the first medium equals the angle of refraction divided by the velocity in the second medium. J. Kepler approximated the law of refraction and is often given credit as being the founder of optics.

**Keywords:** angle, boundary, incidence, normal, velocity

KEPLER, Johannes, 1571-1630, German astronomer

SNELL, Willebrod van Roijen, 1591-1626, Dutch mathematician

**Sources:** Bates, R. L. and Jackson, J. A. 1980; Considine, D. M. 1976; Daintith, J. 1981. 1988; Landau, S. I. 1986; Mandel, S. 1972; Stedman, T. L. 1976; Uvarov, E. G. et al. 1964; Williams, T. 1974.

*See also* DESCARTES; KERR; KIRCHHOFF-CLAUSIS; LAGRANGE; SNELL

## REFRESHMENT, LAW OF

The rate of recovery of a laboring muscle from fatigue depends on, and is proportional to, the rate of supply and flow of arterial blood to the muscle.

**Keywords:** arterial, blood, muscle, physiology, recovery

**Sources:** Critchley, M. 1978; Friel, J. P. 1974; Landau, S. I. 1986.

## REGIER NUMBER, Reg OR $N_{Reg}$

A dimensionless group relating to airfoil and wing flutter:

$$Reg = L\omega\mu^{1/2}/a$$

where L = wing half chord
$\omega$ = angular frequency
a = sonic velocity
$\mu$ = mass ratio

**Keywords:** air foil, chord, sonic velocity
REGIER, Arthur A., twentieth century (b. 1909), American aeronautical engineer
**Sources:** AIAA J. 2(7), July 1964; American Men of Science; Land, N. S. 1972; NUC; Potter, J. H. 1967.
*See also* FRUEH

## REGNAULT, LAW OF—SEE PRESSURES

## REGRESSION, LAW OF—SEE GALTON; RIBOT

## REGULAR AND IRREGULAR LAWS OF OPTICAL DOUBLETS

The regular and irregular doublet laws of optical spectra are relationships adopted from X-ray spectra, which provide predictions in the study of isoelectronic sequence. With the regular double law, the difference in energy levels is:

$$j_1 = L + 1/2 \text{ and } j_2 = L - 1/2$$

which may be expressed approximately by:

$$\Delta\sigma = R_\alpha{}^2(Z - S_1)^4/n^3\ L(L + 1)$$

where $\Delta\sigma$ = separation of the levels
$R = (2\pi^2/ch^3)\ (m\ e^4)$
$\alpha = 2\pi\ e^3/ch$
Z = nuclear charge

The regular doublet law states that the difference between the square roots of the values for such a pair is a constant independent of the atomic number A. Also, the radiated frequency for the transition between such a pair is a linear function of the nuclear charge, Z.

**Keywords:** atomic, doublet, optic, radiated frequency, spectra, X-ray
**Source:** Thewlis, J. 1961-1964.

## REINFORCEMENT, LAW OF CONDITIONING—SEE PAVLOV; SKINNER

## RELATIONS, LAW OF (1859)

This refers to the anticipated law that would explain the relations between various elements, as later assembled in the periodic table, expressed by A. Strecker.

**Keywords:** anticipated, elements, periodic
STRECKER, Adolph Friedrich Ludwig, 1822-1871, German chemist
**Source:** Brown, S. B. and Brown, L. B. 1972.
*See also* MENDELEEV; PERIODIC TABLE

### RELATIVE EFFECT, LAW OF

The delay in behavior of an organism is related to the relative size of the reward in the future. Organisms will distribute their behavior (response) according to the relative gains connected with each action.

**Keywords:** behavior, organisms, response, reward
**Source:** Harre, R. and Lamb, R. 1983.
*See also* EFFECT; THORNDIKE

### RELATIVE INEFFICIENCY, LAW OF (ECONOMICS)

The more we automate the production of goods and lower their pre-unit cost, the more we increase the relative cost of handicrafts and nonautomated services.

**Keywords:** automate, costs, economics, handicrafts
**Source:** Toffler, A. 1980.

### RELATIVE PROPORTIONS, LAW OF CONSTANCY OF—SEE CONSTANCY OF RELATIVE PROPORTIONS

### RELATIVE PROPORTIONS IN EQUILIBRIUM, LAW OF

The condition of physical or chemical equilibrium in a heterogeneous system is independent of the relative mass of each phase present in the system.

**Keywords:** equilibrium, mass, phase, physical chemistry
**Source:** Considine, R. M. 1976 (partial).
*See also* CONSTANCY OF; LE CHATELIER; VAN'T HOFF

### RELATIVITY, GENERAL, LAW OF (1915); RELATIVITY, SPECIAL, LAW OF (1905)—SEE EINSTEIN

The general theory of relativity describes gravity in terms of geometry.

**Source:** Science News, vol. 148 (9):140, August 26, 1995.

### RELATIVITY, PRINCIPLE OF, A FUNDAMENTAL LAW (1905, 1915)

No physical experiment can be performed relative to one inertial system only by means of which the uniform velocity of that system relative to another can be detected. The speed of fight in a vacuum as measured relative to any inertial system is always the same, and the laws of physics are invariant for all inertial systems.

**Keywords:** inertia, light, physics, velocity
**Sources:** Nourse, A. E. 1969; Parker, S. P. 1989; Science News,148(9):140, August 26, 1995.
*See also* EINSTEIN

### RELATIVITY, LAW OF (MEDICINE)

Simultaneous and successive sensations modify each other. The relationship of altering the magnitude of stimuli sensations by comparing simultaneous versus successive stimuli of the same magnitude.

**Keywords:** nerves, sensations, physiology
**Source:** Landau, S. I. 1986.

## RELATIVITY OF ECOLOGICAL VARIETY—SEE ECOLOGICAL VALENCY

**Source:** Lincoln, R. J. 1982.

## RELIABILITY, LAWS OF

Reliability of operation of an item is often expressed in terms of mean time before failure (MTBF) or mean time to failure (MTTF) and is related to the distribution of function of the item. Commonly used distributions are exponential, normal, log normal, Weibull, and gamma functions, of which the first is often used. The product rule for system reliability of components in series is:

$$R_s = \Sigma R_i \text{ from } i = 1 \text{ to } i = n \text{ components}$$

The product rule for system reliability of components in parallel is:

$$R_s = 1 - \prod_{i=1}^{M} (1 - R_i) \text{ from } i = 1 \text{ to } i = n \text{ components}$$

**Keywords:** exponential, failure, functions, gamma, log normal, normal, Weibull
**Sources:** Bazovsky, I. 1960; Rothbart, H. A. 1960. 1985; Ushakov, I. A. and Harrison, R. A. 1994.
*See also* EXPONENTIAL FAILURE; LUSSER; PRODUCT LAW (UNRELIABILITY); WEIBULL

## RENSCH LAWS

**Law 1**
In cold climates, races of mammals have larger litters, and birds have larger clutches of eggs, than do races of the same species in warmer climates.

**Law 2**
Birds have shorter wings, and mammals shorter fur, in warm climates than in colder ones.

**Law 3**
Races of land snails, in colder climates, have brown shells, and those in warmer climates have white shells.

**Law 4**
The thickness of shell is positively associated with strong sunlight and arid conditions.

**Keywords:** birds, climate, eggs, mammals, snail, wings
RENSCH, Bernhard, twentieth century (b. 1900), German biologist
**Sources:** Gray, P. 1967; Lincoln, R. J. 1982; Pelletier, P. A. 1980.
*See also* ECOGEOGRAPHIC RULES

## REPETITION (LEARNING), LAW OF—SEE DISUSE

## REPRODUCTION, LAW OF—SEE BIOGENESIS

## REQUISITE VARIETY, LAW OF (1956)

In cybernetics, the law states that, if $\log V_d$ is the variety in possible ways in which a disturbance D can affect a system E, to be regulated by a regulator R, and if $\log V_r$ is the variety in R's alternatives (optional ways of response to D), the variety in the possible outcomes ($\log V_0$) affecting E cannot be forced by R below the limit $(\log V_d - \log V_r)$or $\log (V_a/V_r) \geq \log V_0$.

This law applies to all forms of regulation and is independent of the field of science or technology or specific mechanism.

**Keywords:** cybernetics, disturbance, regulator, system
**Sources:** Ashby, W. R. 1956. 1968; Gray, P. 1967; Ralston, A. et al. 1993.

## RESEMBLANCE, LAW OF

A thought, idea, or feeling tends to bring to mind another thought that resembles the thought in some respect; formulated by T. Hobbes.

**Keywords:** feeling, idea, thought
HOBBES, Thomas, 1588-1679, English philosopher
**Source:** Bothamley, J. 1993.
*See also* ASSOCIATION

## RESISTANCE OF A SPHERE, LAW OF—SEE STOKES

## RESISTANCE (ELECTRIC) WITH TEMPERATURE, LAW OF; MATTHIESSEN AND SIEMENS LAW (1864)

The increase in electrical resistance of an electrical conductor follows:

$$R_t = R_o (1 + at)$$

where Rt = resistance
$R_o$ = original or starting resistance at time zero
a = coefficient of increase in resistance with and increase in temperature

**Keywords:** conductor, electrical, resistance
MATTHIESSEN, Augustus, 1831-1870, German physicist
SIEMENS, Ernst Werner von, 1816-1892, German electrical engineer
**Sources:** Besancon, R. 1974; Gillispie, C. C. 19811; Parker, S. P. 1989.
*See also* MATTHIESSEN

## RESPIRATION QUOTIENT, LAW OF

A clear, U-tube manometer, with appropriate measures, the number of molecules that an organism respires is:

$$\Delta N = 9.655 \times 10^{16}\ P\ A\ \Delta L/T$$

where $\Delta N$ = number of molecules
P = atmospheric pressure, cm Hg
A = cross-sectional area of capillary tube, $mm^2$
T = temperature of water bath, K

The respiratory quotient, the ratio of carbon dioxide molecules released to the number of oxygen molecules consumed:

$$\frac{\Delta N_{O+NaOH} - \Delta N_O}{\Delta N_{O+NaOH} + \Delta N_O} = \frac{\Delta L_{O+NaOH} - \Delta L_O}{\Delta L_{O+NaOH} + \Delta L_O}$$

where $\Delta N_O$ = the number of molecules removed by the organism
$\Delta N_{O+NaOH}$ = the number of molecules removed when both the organism and the NaOH are in the test tube

If one is looking for the respiratory quotient only, it is not necessary to know the atmospheric pressure, temperature, or area of the capillary tube.

**Keywords:** carbon dioxide, organism, oxygen, respiration
**Source:** Scientific American 273(6):111. December 1995.
*See also* METABOLISM; $Q_{10}$

### RETGERS LAW (RETGER*)

The physical properties of isomorphous mixtures (mixed crystals) are continuous functions of a percentage composition.

**Keywords:** composition, crystals, isomorphous, mixtures, percentage
RETGERS, Jan Willem, 1856-1896, Dutch soil scientist
**Sources:** *Bothamley, J. 1993; Clifford, A. F. 1964; *Honig, J. M. et al. 1953; Morris, C. G. 1992; NUC

### RETURNS, LAW OF—SEE CONSTANT RETURN; INCREASING RETURN

### REY-BLONDEL LAW—SEE BLONDEL-REY

### REYNOLDS BOILING OR BUBBLE NUMBER, $Re_b$ OR $N_{Re(b)}$

A dimensionless group used in boiling that relates bubble diameter and viscosity:

$$Re_b = DG/\mu$$

where D = bubble diameter
G = $\pi/6\ D^3 p_v f\ n$
$\mu$ = absolute viscosity
$p_v$ = mass density of vapor
f = frequency of formation
n = number of nucleation centers per unit area

**Keywords:** boiling, bubble, diameter, frequency of formation
REYNOLDS, Osborne, 1842-1912, English engineer and physicist
**Sources:** Grigull, U. 1982; Land, N. S. 1972; Potter, J. H. 1967.

### REYNOLDS ELECTRIC NUMBER, $Re_e$ OR $N_{Re(e)}$

A dimensionless group in magnetohydrodynamics that relates permittivity and space charge:

$$Re_e = \varepsilon V/qbL$$

where $\varepsilon$ = permittivity
V = velocity
q = space charge density
b = carrier mobility, speed/voltage gradient
L = characteristic dimension

**Keywords:** carrier mobility, magnetohydrodynamics, permittivity, space charge
REYNOLDS, Osborne, 1842-1912, English engineer and physicist
**Sources:** Bolz, R. E. and Tuve, G. L. 1970; Kakac, S. et al. 1987; Land, N. S. 1972.

## REYNOLDS INDEX LAW

If the static head required to maintain a constant velocity in a length of pipe with a constant diameter, the Reynolds index law relates the inertia force and viscous force:

$$h = f\,L\,V^n/d^x$$

where h = static head
f = constant, based on friction
V = constant velocity of flow
L = length of pipe (conduit)
d = diameter of pipe (conduit)
n, x = constants

**Keywords:** fluid, head, pipe, static
REYNOLDS, Osborne, 1842-1912, English engineer and physicist
*See also* REYNOLDS NUMBER

## REYNOLDS NUMBER, Re OR $N_{Re}$ (1879; 1883)

A nondimensional parameter representing the ratio of the momentum force to the viscous force in fluid flow:

$$Re = 2rV\rho\nu \quad \text{or} \quad DV\rho/\mu$$

where Re = Reynolds number
D = diameter of the pipe; r is radius
V = velocity of the fluid
ρ = density of the flowing fluid
μ = absolute viscosity of the fluid
ν = dynamic viscosity

The Reynolds number was named by Sommerfeld (Barenblatt, G. I.). The Reynolds number is also referred to as the *Damköhler fifth law.* For an airfoil, the Reynolds number is air velocity times the chord of the airfoil divided by the kinematic viscosity of the air, expressed as:

$$Re = VL/\nu$$

where V = air velocity
L = chord of airfoil
ν = kinematic viscosity

The lower critical value of the Reynolds is 2000, below which laminar flow exists, and above which, after a transition zone of mixed flow, at a higher critical value above 12,000 to 14,000, turbulent flow exists (Fig. R.2).

**Keywords:** fluid flow, momentum, viscous
REYNOLDS, Osborne, 1842-1912, English engineer
**Sources:** Bolz, R. E. and Tuve, G. L. 1970; Grigull, U. et al. 1982; Land, N. S. 1972; Mandel, S. 1972; Parker, S. P.1992; Perry, R. H. 1967.
*See also* BLAKE; DAMKÖHLER FIFTH LAW; KÁRMÁN (VON); MAGNETIC REYNOLDS

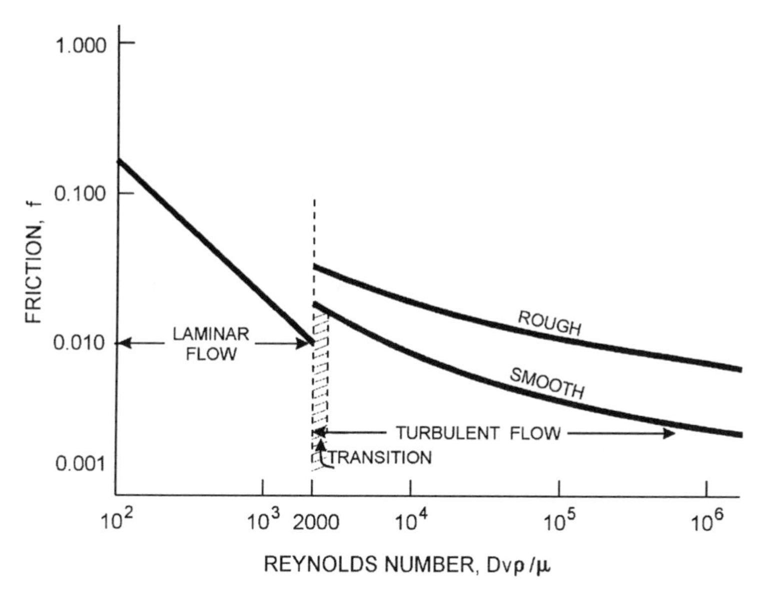


**Figure R.2** Reynolds number.

## RHEOLOGY, AXIOMS OF

### First Axiom

Under isotropic pressure, nearly all materials behave in the same way in that they are purely and simply elastic.

### Second Axiom

In reality, every material possesses all rheological properties, although in varying degrees.

### Third Axiom

There is a hierarchy of ideal bodies such that the rheological equation of the more simple body (lower in the hierarchy) can be derived by putting one or other of the constants of the rheological equation of the less simple body (higher in the hierarchy) equal to zero.

Some of the commonly used rheological models named after people in the field are called Bingham, Burgers, Hooke, Kelvin, and St. Venant. (See illustration under VISCOELASTIC.)

**Keywords:** behavior, hierarchy, simple body, properties
**Source:** Reiner, M. 1960.
*See also* HOOKE; KELVIN; ST. VENANT; VISCOELASTIC

## RIBOT LAW OF REGRESSION (MEMORY)

When mental deterioration occurs, memory for recent events disappears first, and finally only primitive emotions and childhood memories remain.

**Keywords:** memory, mental, thought
RIBOT, Theodule Armand, 1839-1916, French physiologist
**Sources:** Landau, S. I. 1986; Zusne, L. 1987.

## RICCO LAW

For light to be perceived by the eye, the light must be of sufficient intensity and duration. With steady illumination on the macular region of the retina, the condition that a luminous sensation may just barely be perceived is that the total quantity of light, proportional to the product of the source area and brightness, must reach a definite minimum of higher value.

**Keywords:** eve, light, luminous, macular, retina
RICCO, Annibale, 1844-1919, Italian physicist
**Sources:** Friel, J. P. 1974; Stedman, T. L. 1976; Thewlis, J. 1961-1964; Zusne, L. 1987.

### RICHARDSON EQUATION (1901)

The relationship between the current emitted per unit area of a heated metal surface is:

$$I = A\ T^2 \exp - \phi/kT$$

where I = the current emitted per unit area
A = constant
T = absolute temperature on the surface
$\phi$ = constant
k = Boltzmann constant

**Keywords:** current, emitted, heated, metal, surface
RICHARDSON, Sir Owen Willans, 1879-1959, British physicist; Nobel prize, 1928, physics
**Source:** Thewlis, J. 1961-1964.

### RICHARDSON LAW OF EMISSION (1913)

Electrons are emitted from a metal as an exponential function of the temperature and also as a property of the material. The law relates to the saturation current of an emitter of thermionic current, the number of electrons emitted per area, the charge per electron, the absolute temperature, and the electron affinity of the emitting surface, and is represented by:

$$I_s = Ne = AT^{\lambda} e^{-w_0/kT}$$

where $I_s$ = saturation current, per sq cm of surface of emitter
N = number of electrons emitted per sq cm
e = charge per electron
T = absolute temperature
$w_0$ = electron affinity of the emitting surface
k = Boltzmann constant
A = constant
$\lambda$ = number near unity

Also known as the Einstein-de Haas effect.

**Keywords:** affinity, charge, current, electrons, emitted, saturation
RICHARDSON, Sir Owen Willans, 1879-1959, English physicist; Nobel prize, 1928, physics
**Sources:** Gillispie, C. C. 1981; McMurray, E. J. 1995; Thewlis, J. 1961-1964.

### RICHARDSON NUMBER, Ri OR $N_{Ri}$

A dimensionless group in atmospheric shear that relates buoyant force and turbulent force or gravity force and inertial force:

$$Ri = gL\ \Delta\rho/2q$$

where g = gravitation acceleration
L = length (thickness) of liquid layer

$\Delta\rho$ = density difference
q = dynamic pressure, $\rho V^2/2$
V = velocity

The stability of flow is also expressed as the reciprocal of the Froude number squared ($1/Fr^2$).

**Keywords:** atmospheric, buoyant, gravity force, inertial force, shear, turbulent
RICHARDSON, Robert S., twentieth century (b. 1902), American space scientist
**Sources:** Bolz, R. E. and Tuve, G. L. 1970; Jerrard, H., G. and McNeill, D. B. 1992; Land, N. S. 1972; Menzel, D. H. 1960; Potter, J. H. 1967.
*See also* FROUDE

### RICHARDS RULE

The molar heat of fusion for a solid divided by the melting point is approximately two.

**Keywords:** fusion, heat, melting, molar
RICHARDS, Theodore William, 1868-1928, American chemist; Nobel prize, 1914, chemistry
**Sources:** Ballentyne, D. W. G. and Lovett, D. R. 1972; Fisher, D. J. 1991; Gillispie, C. C. 1981; Merriam-Webster. 1995.
Also see TROUTON

### RICHET—SEE RUBNER AND RICHET LAW OF SURFACE AREA

### RICHTER EARTHQUAKE LAW (1935)

The quantity of energy released by an earthquake, or other phenomenon such as nuclear explosions, can be represented on a logarithm scale for comparison of energy involved. Energy is empirically represented by:

$$\log E = 11.4 + 1.5\ M$$

where E = energy in ergs
M = magnitude, with 8.5 used as the largest

An earthquake of an M of 8.0 is 10,000 as powerful as one with an M of 4.0.

**Keywords:** energy, explosions, logarithm
RICHTER, Charles Francis, 1900-1985, American seismologist
**Sources:** Bothamley, J. 1993; Considine, D. M. 1976; Scott, J. S. 1993.
*See also* BIRCH

### RICHTER LAW (1791); LAW OF WENZEL (1777)

When two neutral salts undergo decomposition by interchange of their acid and basic constituents, the two new salts resulting from such interactions are also neutral in character.

**Keywords:** acid, base, chemistry, exchange, salts
RICHTER, Jeremias Benjamin, 1762-1807, German chemist
WENZEL, Karl Friedrich, 1740-1793, German chemist
**Sources:** Honig, J. M. 1953; Morris, C. G. 1992; Van Nostrand. 1947.
*See also* EQUIVALENT RATIOS; PROPORTIONALITY

### RIGHI-LEDUC EFFECT—SEE LEDUC LAW

### RIGID BODY—SEE EQUILIBRIUM FOR A RIGID BODY

### RITTER LAW

Both the opening and the closing of an electric current produce stimulation in a nerve in animals and humans.

**Keywords:** circuit, current, electric, nerve, stimulus
RITTER, Johann Wilhelm, 1776-1810, German physicist
**Sources:** Friel, J. P. 1974; Landau, S. I.1986.
*See also* POLAR EXCITATION

### RITTER-VALLI LAW OR VALLI-RITTER LAW

The primary increase and the secondary loss of irritability in a nerve, produced by a section that separates from the nerve center, travel in a peripheral direction.

**Keywords:** irritability, nerve, peripheral, physiology
RITTER, Johann Wilhelm, 1776-1810, German physicist
VALLI, Eusebo, 1755-1816, Italian physiologist
**Sources:** Friel, J. P. 1974; Landau, S. I. 1986; Stedman, T. L. 1976.

### RITTINGER LAW (1867)

Work done in crushing is proportional to the increased surface produced, which is a function of the square of the diameter of the particle:

$$E = C'(S_2 - S_1) = C''(1/D_2 - 1/D_1)$$

where E = work done or required for the crushing
S = surface area produced
D = diameter of the particle
C = constant of size reduction

**Keywords:** area, crushing, grinding, particle, surface
RITTINGER, Peter Ritter von, 1811-1872, German engineer
**Sources:** Brown, G. G. et al. 1950; Hunt, D. D. 1979; NUC; Perry, J. H. 1950.
*See also* BOND; KICK

### ROBERTS LAW (1875); ROBERTS-CHEBYSHEV (1878) THEOREM

For every mechanical linkage, there are at least two substitutes that will produce the same desired motion, and the two alternative linkages are related to the first by a series of similar triangles. Based on the Roberts-Chebyshev theorem, there are three different but related four-bar mechanisms generating the same coupler curves. For a six-bar cognate linkage, the coupler point curve of a planar four-bar linkage is also described by the joint of the dyad of a proper six-bar linkage.

**Keywords:** design, linkage, mechanical, mechanisms, triangles
ROBERTS, S. nineteenth century, English mathematician
CHEBYSHEV, Pafnuty Lvovich, 1821-1894, Russian mathematician
**Sources:** Machine Design v. 31, n. 8, Apr. 16, 1959; Merriam-Webster Biographical. 1995; Rothbart, H. A. 1996; Shigley, J. E. 1961.

### ROBERTS STRAIGHT LINE SLIDER LINKAGE

For a straight line slider-crank mechanism, there are two different planar mechanisms that will trace identical coupler curves. Roberts straight line mechanisms produce approximately a straight line at B when C is moved.

**Keywords:** coupler, curve, mechanisms, planar
ROBERTS, Richard, early nineteenth century, English engineer
**Source:** Machine Design v. 31, n. 8, Apr. 16, 1959.
*See also* BOBILLIER; ROBERTS-CHEBYSHEV

## ROBIN LAW

When a system is in either or chemical or physical equilibrium, an increase in pressure favors the system formed with a decrease in volume. A reduction in pressure favors the system formed with an increase in volume. A change of pressure has no effect on the system formed without a change in volume.

**Keywords:** chemical, equilibrium, physical, pressure, volume
ROBIN, Charles Phillipe, 1821-1885, French biologist
**Sources:** Gillispie, C. C. 1981; Honig, J. M. et al. 1953.
*See also* LE CHATELIER

## ROBOTICS, THREE LAWS OF

The following were postulated by Asimov in his writing of the future:

### LAW 1

A robot may not injure a human being, or, through inaction, allow a human being to come to harm.

### LAW 2

A robot must obey the orders given to it by human beings except where such orders would conflict with Law 1.

### LAW 3

A robot must protect its own existence as long as such protection does not conflict with Laws 1 and 2.

**Keywords:** existence, humans, orders
ASIMOV, Isaac, 1920-1992, American writer
**Source:** Asimov, I. 1970.

## ROEMER LAW (1977)

Hospitals are always overutilized, as stated by Roemer. He has shown that the supply of beds in a hospital increases the use of beds, and it has been shown that about five percent of all patients admitted to hospitals develop additional infections while staying there.

**Keywords:** beds, deaths, hospitals, infection, patients
ROEMER, Ruth, twentieth century, American writer
**Sources:** Sale, K. 1980; NY Times, 27 September 1977.

## ROME DE LISLE—SEE CONSTANCY OF INTERFACIAL ANGLES

## ROMANKOV NUMBER, Ro´ OR $N_{Ro'}$

A dimensionless group that relates the air dry bulb temperature and the product temperature in drying:

$$Ro' = T_{db}/T_{prod}$$

where $T_{db}$ = the dry bulb temperature (absolute) of the air
$T_{prod}$ = the dry bulb temperature (absolute) of the product

**Keywords:** air, drying, product, temperature
ROMANKOV, Petr Grigorevich, twentieth century (b. 1904), Soviet engineer
**Sources:** Bolz, R. E. and Tuve, G. L. 1970; Hall, C. W. 1979; Lewytzkj, B. 1984; Potter, J. H. 1967.

### ROMMELAERE LAW

In cases of carcinoma (a malignant tumor or a cancerous ulcer), there is a constant diminution of the nitrogen in the urine.

**Keywords:** carcinoma, nitrogen, urine
ROMMELAERE, Guillaume, 1836-1916, Belgian physician
**Source:** Friel, J. P. 1974.

### RÖNTGEN (ROENTGEN)—SEE X-RAY

### ROSA LAW

The possibilities of phyletic (as applied to evolutionary development of plants and animals) variation in an organism decrease in proportion to the extent of its development.

**Keywords:** evolution, organism, phyletic, physiology
ROSA, Danielle, 1857-1944, Italian zoologist
**Source:** Friel, J. P. 1974.

### ROSCOE LAW

The amount of material transformed in a photochemical change is proportional to the product of the light intensity and the time of illumination.

**Keywords:** illumination, light, intensity, photochemical
ROSCOE, Sir Henry E., 1833-1915, British chemist
**Source:** Stedman, T. L. 1976.
*See also* BUNSEN-ROSCOE

### ROSENBACH LAW

In lesions of the nerve centers and nerve trunks, if paralysis occurs, it appears in the extensor muscles before it appears in the flexor muscles.

**Keywords:** extensor, flexor, lesions, muscles, nerves, paralysis
ROSENBACH, Ottomer, 1851-1907, German physician
**Sources:** Landau, S. I. 1986; Stedman, T. L. 1976.
*See also* SEMON (SEMON-ROSENBACH)

### ROSENBUSCH LAW

A description of the sequence and crystallization of minerals from magma (the molten matter under the Earth's crust which, upon cooling, forms igneous rock).

**Keywords:** crystallization, geology, magmas, minerals, rock
ROSENBUSCH, Harry (Karl Heinrich Ferdinand), 1836-1914, German geologist
**Sources:** Gillispie, C. C. 1981; Pelletier, P. A. 1994; Webster's Biographical Dictionary. 1959.

### ROSSBY NUMBER, Ros OR $N_{Ros}$

A dimensionless group that relates to air and ocean currents and fluid spinup, and the effect of Earth's rotation on flow in pipes, based on inertia force and coriolis force:

$$Ros = V/2\ \omega\ L$$

where V = velocity
ω = angular velocity
L = characteristic length

**Keywords:** air, currents, coriolis force, Earth's rotation, fluid spinup, ocean
ROSSBY, Carl Gustav Arvid, 1898-1957, Swedish American meteorologist
**Sources:** Bolz, R. E. and Tuve, G. L. 1970; Jerrard, H. G. and McNeill, D. B. 1992; Land, N. S. 1972; McIntosh, D. H. 1972; NUC; Potter, J. H. 1967; Science 273(5277):942, August 1996.
*See also* CORIOLIS; EKMAN

## ROSSELAND LAW

There is an analogy between continuum conduction and radiation, under some conditions, and the rate of net energy transfer per unit area can be represented by:

$$q = -4/3\ (\text{photon mean path length}) \times (\text{gradient of black body emissivity})$$

where q = net energy transfer per unit area which is analogous to Fourier heat conduction law:

$$q = -k \text{ rad } T$$

where k = Boltzmann constant
T = absolute temperature

**Keywords:** analogy, conduction, energy, Fourier, radiation
ROSSELAND, Svein, twentieth century (b. 1894), Norwegian scientist
**Sources:** Eckert, E. R. G. and Drake, R. M. 1972; NUC.
*See also* BOLTZMANN; FOURIER

## ROSTOW MODEL (1961)

An historical theory of economic growth in which growth is identified as five stages:

1. Traditional
2. Pre-conditions for takeoff
3. Take-off
4. Drive to maturity
5. High mass consumption

**Keywords:** economic, growth, stages
ROSTOW, Walt Whitman, twentieth century (b. 1916), American economist
**Source:** Pearce, D. W. 1986. 1992.

## ROTARY DISPERSION (LIGHT)—SEE BIOT

## ROUSSEL LAW OR SIGN

A sharp pain in the subclavicular region, between the clavicle and fourth rib, which is a sign of incipient tuberculosis.

**Keywords:** clavicle, pain, tuberculosis
ROUSSEL, Theophile, 1816-1903, French physician
**Source:** Friel, J. P. 1974.

## ROUTH RULE OR CRITERIA (1860)

This is a rule used to calculate the moments of inertia about an axis through the center of mass for cuboids, cylinders, and ellipsoids:

$$I = M[(d_1/2)^2 + (d_2/2)^2]/n$$

where $I$ = inertia
$d_1$ = one axis perpendicular to the axis of rotation
$d_2$ = other axis perpendicular to axis of rotation
$M$ = mass
$n$ = 3, 4, or 5 for cuboids, cylinders, or ellipsoids, respectively

**Keywords:** cuboids, cylinders, ellipsoids, inertia, moments
ROUTH, Edward John, 1831-1907, British mathematician
**Sources:** Bothamley, J. 1993; Thewlis, J. 1961-1964.

## ROWLAND LAW

The law for magnetic force is parallel to Ohm's law for resistance, the magnetomotive force (mmf) is proportional to the reluctance and the magnetic g flux; the reluctance is proportional to the length of the flux lines and is inversely proportional to the permeability and area. These relationships are represented by:

$$mmf = \phi R$$

where mmf = the magnetomotive force
$\phi$ = the flux, which is 1.257 NI/R, and NI is for ampere-turns

and

$$R = L/\mu a$$

where $R$ = reluctance
$L$ = length
$I$ = ampere
$\mu$ = permeability
$a$ = area

**Keywords:** flux, magnetomotive, ohms, permeability, reluctance
ROWLAND, Henry Augustus, 1848-1901, American physicist
**Sources:** Perry, J. H. 1950; Perry, R. H. and Green, G. W. 1984.
*See also* BOSANQUET

## RUBNER AND RICHET SURFACE AREA LAW

The basal metabolic rate, heat production, heat loss, and oxygen consumption by an animal are proportional to the free body surface or to the square of a linear dimension, with the temperature remaining constant.

**Keywords:** animal, area, metabolism, surface
RUBNER, Max, 1854-1932, German physiologist
RICHET, Charles Robert, 1850-1935, French physiologist; Nobel prize, physiology/medicine, 1913
**Source:** Gillispie, C. C. 1981.
*See also* BODY; BRODY AND PROCTER; METABOLISM; SURFACE LAW

## RUBNER LAW

The total metabolism of an animal varies approximately as the surface of a body.

**Keywords:** animal, energy, metabolism, surface
RUBNER, Max, 1854-1932, German physiologist
**Sources:** Friel, J. P. 1974; Gillispie, C. C. 1981; Landau, S. I. 1986; Stedman, T. L. 1976; Thomas, C. L. 1989.
*See also* BODY SIZE; BRODY; LAWS OF GROWTH; RUBNER (GROWTH); SURFACE AREA

## RUBNER LAWS OF GROWTH

1. *Constant Energy Consumption, Law of.* The rapidity of growth is proportional to the intensity of the metabolic process.
2. *Constant Growth Quotient, Law of (1902).* The same fractional part of the entire energy consumption is utilized for growth, and this fractional part is called the *growth quotient.* In most young mammals, 24 percent of the entire food energy, or calories, is utilized for growth; in man only 5 percent is utilized.

**Keywords:** consumption, energy, growth, intensity, mammals, man, metabolic
RUBNER, Max,1854-1932, German physiologist
**Sources:** Landau, S. I. 1986; Schmidt, J. E. 1959; Thomas, C. L. 1989.
*See also* BODY SIZE; BRODY

## RUNGE-KUTTA METHOD

An approximate method for solving first-order differential equations. For sufficiently small increments of $x = h$, simple recursion formulas produce stepwise approximations, $y_k$, to the solution of $y = y(x)$. The truncation errors are reduced by higher-order methods as described in the references.

**Keywords:** differential, equations, solving
RUNGE, Carl David Tolmo, 1856-1927, German mathematician
KUTTA, William Martin (or Martin William), 1867-1944, German aeronautics scientist
**Sources:** Bronson, R. 1993; James, R. C. and James, G. 1968; Korn, G. A. and Korn, T. M. 1968; Pelletier, P. A. 1994.
*See also* SIMPSON

## RUNGE LAW OR RULE

The separation of the Zeeman patterns of nonsinglet states, related to the atomic states, are always rational multiples of the normal separation, which is due to the combined effects of the orbit and the spin magnetism and the magnetic anomaly of the spin.

**Keywords:** atomic, magnetic, nonsinglet, orbit, spine, Zeeman
RUNGE, Carl David Tolme, 1856-1927, German mathematician
**Sources:** Ballantyne, D. W. G. and Lovett, D. R. 1972; Fisher, D. J. 1988. 1991.
*See also* ZEEMAN

## RUNNING—SEE HILL POSTULATE

## RUSSELL, FAJANS, AND SODDY DISPLACEMENT LAW (1913)

If radioactive change has its origin in the nucleus, the emission of an alpha particle must cause the positive charge of the nucleus, with the atomic number of the element decreasing by two units. This is illustrated by the element moving two steps to the left in the periodic system if an alpha particle is emitted, or when the nucleus give up a negative electron, the atomic number is increased by one, and the element must advance one place in the periodic system.

**Keywords:** alpha, atomic, electron, element, nucleus, radiation, periodic
RUSSELL, Henry Norris, 1877-1957, American astrophysicist
FAJANS, Kasimir, 1887-1975, German Polish American chemist
SODDY, Frederick, 1877-1956, English chemist; Nobel prize, 1921, chemistry
**Sources:** Gillispie, C. C. 1981; Merriam-Webster's Biographical Dictionary. 1995.
*See also* DISPLACEMENT

## RUSSELL NUMBER, Ru OR $N_{Ru}$

This is a dimensionless group representing waves in stratified flow that is based on the inertia force divided by the buoyancy force:

$$Ru = U/N\ h$$

where U = wind speed
h = height of obstacle
N = natural frequency of an element of fluid about its equilibrium altitude in a density stratified atmosphere

**Keywords:** buoyancy, density, flow, fluid, stratified
RUSSELL, Henry Norris, 1877-1957, American astrophysicist
**Sources:** Land, N. S. 1972; Merriam-Webster's Biographical Dictionary. 1995; Potter, J. H. 1967.

## RUTHERFORD AND SODDY, LAW OF (1902); LAW OF SPONTANEOUS DECAY

These men discovered the law of spontaneous decay, in which an atom can spontaneously decay into another atom—an idea not accepted at that time.

**Keywords:** atom, decay, spontaneous
RUTHERFORD, Ernest, First Baron (Lord), 1871-1937, English chemist and physicist; Nobel prize, 1908, chemistry
SODDY, Frederick, 1877-1956, English chemist; Nobel prize, 1921, chemistry
**Sources:** Gillispie, C. C. 1981; Morris, C. G. 1992.
*See also* RAMSAY

## RYDBERG-SCHUSTER LAW (1900)

In alkali spectra, the difference of the wave numbers of the limit of the principal series and the common limit of the diffuse and sharp series is equal to the wave number of the first line of the principal series. In 1908, W. Ritz proposed this relationship as a fundamental law, which later became known as the *combination principle.*

**Keywords:** alkali, diffuse, principle, spectra, wave
RYDBERG, Johannes Robert, 1854-1919, Swedish physicist
SCHUSTER, Sir Arthur, 1851-1934, British physicist
RITZ, Walter, 1878-1909, Swiss born German physicist
**Sources:** Ballentyne, D. W. G. and Lovett, D. R. 1972. 1980; Lapedes, D. N. 1978.

# S

## SABINE LAW (1910)

The reverberation time of sound in a hall is related to the total absorption of the hall and the volume of the hall:

$$t \propto \frac{V}{A}$$

where $t$ = total reverberation time
$V$ = volume
$A$ = total absorption of hall

The reverberation time is the time it takes for the sound intensity to fall to $10^{-6}$ of its initial value. The unit of sound absorbance is the sabin.

**Keywords:** absorption, reverberation, sound, time
SABINE, Wallace Clement Ware, 1868-1919, American physicist
**Sources:** Gillispie, C. C. 1981; Michels, W. C. 1961; Thewlis, J. 1961-1964.
*See also* MASS; MAYER; STIFFNESS

## SACHS LAW

A cell wall tends to set itself at right angles to another cell wall.

**Keywords:** cell, right angles, wall
SACHS, Julius von, 1832-1897, German botanist. Founder of field of plant pathology
**Sources:** Gray, P. 1967; Webster's Biographical Dictionary. 1959.

## SACHS NUMBER, Sa OR $N_{Sa}$ (1944)

A dimensionless group applied to surface explosions:

$$Sa = Lp_0^{1/3}/E^{1/3}$$

where $L$ = distance from explosive point to reference point
$p_0$ = atmospheric pressure
$E$ = explosive energy

**Keywords:** energy, explosions, surface
SACHS,* Robert Green, twentieth century (b. 1916), American physicist (some references have SACH)
**Sources:** *Land, N. S. 1972; NUC; Potter, J. H. 1967; U. S. Ballistic Research Laboratory, Report 466,1944.

## SAINT-VENANT LAW OR PRINCIPLE (1855); ALSO WRITTEN AS ST. VENANT

Considered to be a simplifying approach to maximum strain theory of an elastic body, it states that, when a test piece is loaded by a complex system of stresses at its ends, the complexity may be disregarded except in their immediate proximity. Forces applied at one part of an elastic body will involve stresses that, except in the region close to that part, will depend almost entirely on their resultant action, and very little on their distribution. The flow condition was postulated independently by C. Coulomb (1801) and H. Tresca (1868).

**Keywords:** distribution, elastic, end, forces
COULOMB, Charles Augustin de, 1736-1806, French physicist
SAINT VENANT, Adhemar Jean Claude Barre de, 1797-1886, French mathematician
TRESCA, Henri Edouard, 1814-1885, French scientist
**Sources:** Barr, E. S. 1973; James, R. C. and James, G. 1968; NUC; Reiner, M. 1960; Scott, J. S. 1993; Scott Blair, G. W. 1949.
*See also* COULOMB

## SARASIN AND DE LA RIVE LAW OF MULTIPLE RESONANCE

Relates to the distance between two nodes on a resonator change, but not in changing the oscillator. The distance between two nodes is the half wavelength of the free oscillations of the resonator only.

**Keywords:** oscillator, physics, resonator, sound, wavelength
SARASIN, Edouard, 1843-1917, Swiss physicist
DE LA RIVE, August Arthur (de), 1801-1873, Swiss physicist
**Sources:** Barr, E. S. 1973; Northrup, E. T. 1917.

## SARRAU NUMBER, Sar OR $N_{Sar}$

A dimensionless group used in compressible flow that relates inertia force and elastic force:

$$Sar = V/a$$

where V = velocity
a = sonic velocity

The Sarrau number equals the Mach number.

**Keywords:** compressible, elastic, flow, inertia
SARRAU, Jacques Rose Ferdinand Emile, 1837-1904, French engineer
**Sources:** Bolz, R. E. and Tuve, G. L. 1970; Land, N. S. 1972; NUC; Potter, J. H. 1967.
*See also* MACH NUMBER

## SATIATION—SEE DIMINISHING UTILITY

## SAVART—SEE BIOT-SAVART

## SAY LAW (ECONOMICS) (1803)

There is an interdependence in an exchange economy such that every act of production incurs costs and an equivalent amount of purchasing power, and every product put on the market creates its own demand, and every demand exerted on the market creates it own supply.

**Keywords:** demand, economy, exchange, production, purchasing power
SAY, Jean Baptiste, 1767-1832, French economist
**Sources:** Pass, C. 1991; Webster's New World Dictionary. 1986.

## SCALING LAWS

A geometrically similar model will behave like its prototype if the model is tested in the same fluid as the prototype.

**Keywords:** fluid, geometrically similar, model, prototype
**Source:** Greenkorn, R. A. 1983.

## SCALING LAWS IN BIOLOGY—SEE ALLOMETRIC SCALING

## SCHILLER NUMBER, Sch OR $N_{Sch}$

A dimensionless group representing flow of fluid over immersed bodies, it involves relationships between the Reynolds number and the Newton number:

$$Sch = V\ [3/4\ \rho\ \nu_m/g\mu\ (\nu_M - \nu_m)]^{1/3}$$

where V = velocity
$\rho$ = mass density
$\nu_m$ = specific gravity of medium
$\nu_M$ = specific gravity of material in bed
g = gravitational acceleration
$\mu$ = viscosity

**Keywords:** flow, fluid, immersed bodies
**Sources:** Bolz, R. E. and Tuve, G. L. 1970; Land, N. S. 1972; Potter, J. H. 1967.
*See also* FROUDE; NEWTON; REYNOLDS

## SCHMIDT NUMBER, Sc OR $N_{Sc}$

A dimensionless group, a fluid property, a transport number, representing diffusion in flowing systems, that relates kinematic viscosity and molecular diffusivity:

$$Sc = \mu/\rho\ D$$

where $\mu$ = absolute viscosity
$\rho$ = mass density
D = mass diffusivity

The Schmidt number equals the Prandtl number divided by the Lewis number, and is also equal to the Colburn number.

**Keywords:** diffusion, flowing system, viscosity
SCHMIDT, Ernst Heinrich Wilhelm, twentieth century (b. 1892), German geophysicist
**Sources:** Bolz, R. R. and Tuve, G. L. 1970; Grigull, U. et al. 1982; Jakob, M. 1957; Land, N. S. 1972; Layton, E. T. and Lienhard, J. H. 1988; Parker, S. P. 1992; Perry, R. H. 1967; Potter, J. H. 1967; Rohsenow, W. M. et al. 1985.
*See also* COLBURN; LEWIS; PRANDTL; SCHMIDT SHEAR

## SCHMIDT NUMBER (2)—SEE SEMENOV NUMBER

## SCHMIDT (OR SCHMID) SHEAR STRESS LAW

Slip in a material takes place along a slip plane and direction, when the shear stress along the plane reaches a critical value. A critical value of the resolved shear stress must be reached or exceeded for the initiation of slip on a substantial scale:

$$\tau_0 = x \cos \phi_x \cos \lambda_x$$

where $\tau_0$ = the shear stress
$\phi$ = angle between slip direction and stress axis

λ = angle between slip plane normal to stress axis
x = yield stress in x-direction

**Keywords:** slip, shear, stress
SCHMIDT, Ernst Heinrich Wilhelm, twentieth century (b. 1892), German engineer
Keyword: mechanics, shear, slip, stress
Sources: Ballentyne, D. W. G. and Lovett, D. R. 1972. 1980; Gillispie, C. C. 1981; Jakob, M. 1957; Thewlis, J. 1966 (Supp. 3).

## SCHRÖDINGER LAW (1925)

Building on the work of N. Bohr, E. Schrödinger postulated that, in the atom, the electron can be in any orbit, around which waves can extend in an exact number of wavelengths, and that any orbit between two permissible orbits where a fractional number of wavelengths would be required is not permissible. A method for obtaining an approximate solution of the one-dimensional, time-dependent Schrödinger equation, when the wavelength varies slowly with position, is known as the Wentzel-Kramers-Brillouin (WKB) method (1962).

**Keywords:** wave mechanics, wavelengths, WKB method
BOHR, Niels Henrik David, 1885-1962, Danish physicist; Nobel prize, 1922, physics
SCHRÖDINGER, Erwin, 1887-1961, Austrian physicist; Nobel prize, 1933, physics
**Sources:** Asimov, I. 1976; Lederman, L. 1993; NYPL Desk Ref.; Parker, S. P. 1987; Peierls, R. E., 1956.
*See also* BOHR; HEISENBERG; WENTZEL-KRAMERS-BRILLOUIN

## SCHROEDER VAN DER KOLK LAW—SEE VAN DER KOLK LAW

## SCHÜLTZ—SEE ARNDT-SCHÜLTZ

## SCHÜLZ(E); SCHÜLTZE—SEE HARDY-SCHÜLZE (OR SCHÜLTZE)

Spelling varies with different authors, i.e., Bothamley, J. 1993; Kirk-Othmer. 1985.

## SCHÜTZ LAW OR RULE; SCHÜTZ-BORRISSOW LAW; LAW OF PEPTIC ACTIVITY

The amount of material digested by an enzyme is directly proportional to the square root of its concentration. The amount of enzyme:

$$x = t\,K(c_f)^{1/2}$$

where x = amount digested
t = time
K = proportionality constant
$c_f$ = amount of coagulated protein digested (peptic activity)

**Keywords:** activity, concentration, enzyme, intensity
SCHÜTZ, Eric, twentieth century (b.1901), German biochemist[umlat]
**Sources:** Friel, J. P. 1974; Honig, J. M. 1953; Stedman, T. L. 1976.
*See also* ENZYME; FISCHER; MICHAELIS-MENTEN; NORTHRUP

## SCREW—SEE MACHINE

## SECANT LAW (EQUIVALENCE THEOREMS)

The relationship between the frequency of the wave incident at an angle on a flat layer of the ionosphere and the equivalent vertical frequency is:

$$f_v = f_{ob} \cos \phi_o$$

or

$$f_{ob} = f_v \sec \phi_o$$

where $f_{ob}$ = the frequency of wave incident at an angle on a flat layer of the ionosphere and the equivalent vertical frequency
$\phi_o$ = the angle between the vertical and the ray at the bottom of the layer of the ionosphere

The relationship shows that a given ionospheric layer can reflect higher frequencies as the angle of the ray paths increases.

**Keywords:** angle, frequency, ionosphere, ray, reflection
**Sources:** Morris, C. G. 1992; Thewlis, J. 1961-1964.

## SECOND LAW OF THERMODYNAMICS—SEE THERMODYNAMICS

## SEEBECK EFFECT OR LAW (1821)

If a circuit consists of two metals with one junction at a higher temperature than the other, a current flows around the circuit, with the direction and amount of current flow dependent on the metals and the temperature of the junctions. This is the conversion of heat into electricity (thermoelectricity). This concept lay in abeyance for over a century before being put to use, and it now is widely used for thermocouples.

**Keywords:** current, electricity, metals, temperature, thermoelectricity
SEEBECK, Thomas Johann, 1770-1831, Russian German physicist
**Sources:** Asimov, I. 1976; Besancon, R. M. 1974; Hodgman, C. D. 1952; NYPL Desk Ref.; Parker, S. P. 1989.
*See also* PELTIER; SUCCESSIVE OR INTERMEDIATE TEMPERATURES; THOMSON

## SEELIGER—SEE LOMMEL-SEELIGER

## SEGREGATION, LAW OF (GENETICS)

"In each generation the ratio of (a) pure dominants, (b) dominants giving descendants in the proportion three dominants to one recessive, and (c) pure recessives is 1:2:1. This ratio follows from the fact that the two alleles of a gene cannot be a part of a single gamete, but must segregate to different gametes."

**Keywords:** alleles, dominants, gametes, recessive
**Sources:** Friel, J. P. 1974; Landau, S. I. 1986; Stedman, T. L. 1976.
*See also* MENDEL; MENDELIAN

## SEMI-FLUIDS PRESSURE AND FLOW, LAW OF

1. The horizontal pressures in very shallow bins vary as the depth.
2. In deep bins, the horizontal pressure is less than the vertical pressure (0.03 to 0.06 of the vertical pressure) and increases very little after a depth of 2.5 to 3 times the width or diameter of the bin is reached.

3. There is not an active upward component of pressure in a granular mass.
4. The flow from an orifice in the side of a deep bin varies approximately as the cube of the orifice and is independent of the head, as long as the orifice is well covered.

**Keywords:** bins, flow, granular materials, orifice, pressure
**Source:** Ketchum, M. S. 1919.
*See also* JANSSEN; KETCHUM; RANKINE

### SEMENOV NUMBER, Sm OR $N_{Sm}$

A dimensionless group representing a fluid property in heat and mass transfer relating mass and thermal diffusivity:

$$Sm = D \rho c_p/k$$

where D = mass diffusivity
ρ = mass density
$c_p$ = specific heat at constant pressure
k = thermal conductivity

This is also known as the Schmidt No. 2. Some references call this the Semenov No. 1, with the Semenov No. 2* being the reciprocal of the Lewis number. In the western countries, the Semenov number is generally called the Lewis number. (Barenblatt, G. I., personal correspondence.)

**Keywords:** diffusivity, fluid, heat, mass, transfer
SEMENOV, Nikolai Nikolaevich, 1896-1986, Soviet chemical physicist; Nobel prize, 1956, chemistry
**Sources:** Bolz, R. E. and Tuve, G. L. 1970; Land, N. S. 1972; Lewytzkj, B. 1984; Morris, C. G. 1992; Pelletier, P. A. 1994; *Parker, S. P. 1992; Turkevich, J. 1963.
*See also* LEWIS; LEWIS-SEMENOV

### SEMON LAW; SEMON-ROSENBACH LAW

In progressive organic diseases of the motor laryngeal nerves, the abductors of the vocal chords (posterior cricoarytenoids) are the first, and occasionally the only, muscles affected. The abductors of the vocal cords succumb much earlier than the adductors.

**Keywords:** laryngeal, muscles, nerves, throat, vocal chords
SEMON, Sir Felix, 1849-1921, German English physician
ROSENBACH, Ottomar, 1851-1907, German physician
**Sources:** Dorland, W. A. 1980; Friel, J. P. 1974; Garrison, F. H. 1929; Landau, S. I. 1986; Smith, G. 1927; Stedman, T. L. 1976.

### SENESCENCE, LAW OF

The law describing the course of declining vitality is that the degree of senescence increases, or the degree of vitality decreases, with age at a constant ratio, indicated by k. The constancy of rate is the fundamental feature characterizing the process. In a man, the constancy of the rate of senescence begins to apply at an age of about 20 yr; of a dairy cow, at about 7 yr; of a domestic fowl, at about 1 yr; and of the fruit fly (Drosophilia) at about 10 days, represented by:

$$V = A e^{-kt}$$

$$S = A e^{kt}$$

where V = index of vitality
S = index of senescence
k = constant representing declining vitality
t = time

**Keywords:** age, declining, senescence, vitality
**Source:** Robbins, W. J. 1928.

## SENIORITY, LAW OF UNIFORM—SEE UNIFORM SENIORITY

## SENSATION, LAW OF—SEE BELL-MAGENDIE; MÜLLER

**Source:** Gillispsie, C. C. 1981.

## SERIES RESISTANCES, LAW OF

For a group of electrical resistances connected in series, the resistance of the combined effect is the sum of the individual resistors:

$$R_{eq} = R_1 + R_2 + R_3 + \ldots$$

where $R_{eq}$ = equivalent resistance
$R_1, R_2, R_3, \ldots$ = individual resistances

For electrical capacitors in series, the capacitance of the combined effect is:

$$1/C_{eq} = 1/C_1 + 1/C_2 + 1/C_3 + \ldots$$

where $C_{eq}$ = the equivalent capacitance
$C_1, C_2, C_3, \ldots$ = individual capacitances

**Keywords:** electrical, capacitors, resistors
**Sources:** Fink, D. G. and Beaty, H. W. 1993; Isaacs, A. 1996; Mandel, S. 1972.
*See also* PARALLEL (RESISTANCES)

## SEWALL WRIGHT LAW OR EFFECT (1931); ALSO KNOWN AS GENETIC DRIFT

A random process as a result of a chance factor in evolution, tied to the size of a breeding population, particularly due to fluctuation of allele frequencies in a small population. If a small population becomes reproductively isolated, its gene pool may no longer represent the range of genetic diversity found in the parent pool. The smaller the breakaway population, the greater the effect of genetic drift. This law was first described by Sewall Wright. Genetic drift is one of the factors that can disturb the Hardy-Weinberg equilibrium.

**Keywords:** evolution, gene, isolated, population
WRIGHT, Sewall, 1889-1988, American geneticist
**Sources:** Barr, E. S. 1973; Gray, P. 1967; Milner, R. 1990; Tootill, E. 1981. 1988.
*See also* EVOLUTION; GENETIC DRIFT; HARDY-WEINBERG

## SHANNON LAW OR FORMULA OR THEOREM (1948)

The information transmitted from a message source over a communications system is represented by:

$$C = W \log_2(1 + P/N)$$

where C = channel capacity in bits per second
W = bandwidth

P = signal power
N = gaussian noise power, N = kTW where k = Boltzmann constant and T = temperature

**Keywords:** communications, information, message, signal
SHANNON, Claude Elwood, twentieth century (b. 1916), American applied mathematician
**Sources:** Reber, A. S. 1985; Science 272(5270);1914. 28 June 1966; Shannon, C. E. and Weaver, W. A. 1949; Weik, M. H. 1996.
*See also* HICK

## SHEAR STRESS—SEE SCHMID (SCHMIDT)

## SHELFORD LAW OF TOLERANCE (1913)

States that any value, quantity, or factor below a critical maximum will exclude certain organisms from that area.

**Keywords:** critical, exclude, factor, organisms
SHELFORD, Victor Ernest, twentieth century (b. 1877), American zoologist
**Sources:** Landau, S. I. 1986; Lincoln, R. J. et al. 1982.
*See also* TOLERATION

## SHERRINGTON LAW; SHERRINGTON LAW OF RECIPROCAL INNERVATION (1906)

Every posterior spinal nerve root supplies a special region of the skin, although fibers from adjacent special segments may move into such a region. When a muscle receives a nerve impulse to contract, its antagonist receives simultaneously an impulse to relax. The inhibition of contracted fibers of the extensor results in resulting relaxation.

**Keywords:** muscles, nerves, posterior, relaxation, response, skin, spinal, stimulus
SHERRINGTON, Sir Charles Scott, 1857-1952, English neurophysiologist; Nobel prize, 1932, physiology/medicine (shared)
**Sources:** Friel, J. P. 1974; Glasser, O. 1944; Landau, S. I. 1986; Stedman, T. L. 1976.

## SHERRINGTON LAWS OF REFLEX IRRADIATION (1895)

The speed of a reflex to an increasing number of motor units takes place according to a certain pattern described by five laws of reflex irradiation:

**Law 1**
The degree of reflex spinal intimacy between the afferent and efferent spinal roots varies directly as their segmental proximity, called the *law of spatial proximity.*

**Law 2**
For each afferent root, there exists in its own segment a reflex motor path of as low a threshold and of as high potency as any open to it anywhere else.

**Law 3**
The motor mechanisms of a segment are unequally accessible to the local different channels.

**Law 4**
The motor neurons simultaneously discharged by a spinal reflex innervate synergic and not antergic muscles.

**Law 5**
The spinal reflex movement elicited in and from one spinal region will exhibit much uniformity, despite considerable variety of locus of incidence of the exciting stimulus.

**Keywords:** muscles, nerves, response, stimulus
SHERRINGTON, Sir Charles Scott, 1857-1952, English neurophysiologist; Nobel prize, 1932, physiology/medicine (shared)
**Sources:** Garrison, F. H. 1929; McMurray, E. J. 1995; Wasson, T. 1987.

## SHERWOOD NUMBER, Sh OR $N_{Sh}$

A dimensionless group used in mass transfer that relates mass diffusivity and molecular diffusivity:

$$Sh = k_c L/D_m$$

where $k_c$ = mass transfer coefficient
$L$ = characteristic length
$D_m$ = molecular diffusivity

$$k_c = \tau/\rho V$$

where $\tau$ = fluid shear stress at surface
$\rho$ = mass density of fluid
$V$ = fluid velocity

The Sherwood number for mass transfer is same as the Taylor number.

**Keywords:** diffusivity, mass, molecular, transfer
SHERWOOD, Thomas K., 1903-1976, American chemical engineer
**Sources:** Bolz, R. E. and Tuve, G. L. 1970; Jerrard, H. G. and McNeill, D. B. 1992; Land, N. S. 1972; Parker, S. P. 1992; Perry, R. H. 1967.
*See also* NUSSELT; TAYLOR

## SIGNS, LAW OF; ALSO KNOWN AS DESCARTES RULE OF THE SIGNS—SEE DESCARTES RULE OF SIGNS; SPECIES, LAW OF

## SIGNS IN DIVISION, LAW OF

When dividing one term by another, the sign of the quotient is positive when the dividend and the divisor have like signs and negative when they have unlike signs.

**Keywords:** algebra, division, quotient, signs
**Sources:** Freiberger, W. F. 1960; James, R. C. and James, G. 1968; Karush, W. 1989; Morris, C. G. 1992; Parker, S. P. 1989.
*See also* SIGNS

## SIGNS IN MULTIPLICATION, LAW OF

When multiplying one term by another, the sign of the product is positive if the multiplier and the multiplicand have like signs and negative if they have unlike signs.

**Keywords:** algebra, multiplication, signs
**Sources:** Freiberger, W. F. 1960; James, R. C. and James, G. 1968; Karush, W. 1989; Morris, C. G. 1992; Parker, S. P. 1989.
*See also* SIGNS

## SIMILARITY OF PHYSICAL AND CHEMICAL CHANGES, NUMBER FOR, K OR $N_K$

A dimensionless group when change of phase relating heat flow for phase change to superheating or supercooling of one of the phases occurs:

$$K = \upsilon/c_p \, \Delta t$$

where $\upsilon$ = specific gravity
$c_p$ = specific heat at constant pressure
$\Delta t$ = change in temperature

**Keywords:** change, heat, phase, supercooling, superheating
**Sources:** Bolz, R. E. and Tuve, G. L. 1970; Potter, J. H. 1967.

## SIMILARS, LAW OF (HEALTH) (1811); HAHNEMANN RULE OR FIRST PRINCIPLE OF HOMEOPATHY (1796)

Expressed by C. Hahnemann, the founder of homeopathy, and one of the tenets of homeopathic medicine in which it is claimed that if a substance produces certain debilitating symptoms in a healthy person, a small dose could be used to treat the same symptoms in an ill person, or expressed as the principle that *like cures like.*

**Keywords:** cure, homeopathy, like
HAHNEMANN, Christian Friedrich Samuel, 1755-1843, German physician
**Sources:** Bynum, W. F. 1981; Friel, J. P. 1974; Stedman, T. L. 1976; Time, Sept. 25, 1995, p. 47.
*See also* HOMEOPATHY; INFINITESIMALS

## SIMILITUDE—SEE BERTRAND; FROUDE; MODEL

## SIMPSON, LAW OF (ECOLOGY); OR SIMPSON INDEX

This is an index of diversity of organisms based on the probability of picking two organisms at:

$$D = 1 - \sum_{I=1}^{I} p_i^2$$

where D = diversity
$p_i$ = the proportion of individual species, i, in a community of s species

**Keywords:** diversity, index, organisms, probability
SIMPSON, George Gaylord, 1902-1984, American paleontologist
**Source:** Lincoln, R. J. 1982.

## SIMPSON RULE—SEE AREA

## SINE LAW; LAW OF SINES

In any triangle the sides are proportional to the sines of the opposite angles, or the sine of an angle is the ratio of the opposite side and hypotenuse of a right angle triangle,

For a plane triangle:

$$\sin A/a = \sin B/b = \sin C/c$$

For a right triangle:

$$a = c \sin A = b \tan A$$

$$b = c \cos A = a \cot A$$

$$c = a \operatorname{cosec} A = b \sec A$$

For a spherical triangle:

$$\sin A/\sin a = \sin B/\sin b = \sin C/\sin c$$

where A, B, and C = angles opposite sides a, b, c.

**Keywords:** algebra, angles, trigonometry
**Sources:** Baumeister, T. 1967; Isaacs, A. 1996; James, R. C. and James, G. 1968; Karush, W. 1989; Morris, C. G. 1992; Parker, S. P. 1992; Tuma, J. J. and Walsh, R. A. 1998.
*See also* ANGLES; COSINE; TANGENTS

## SKIN EFFECT

In any current carrying conductor the current tends to concentrate toward the outer surface as a result of eddy currents. An alternating current flowing through a pipe produces a magnetic field which produces eddy currents. Eddy current losses, the heat generated by eddy currents is given by:

$$P_e = K_e f^2 \beta_m^2 V_{ol}$$

where $P_e$ = power loss by eddy current
$K_e$ = eddy current loss constant
$f$ = frequency of electricity
$\beta_m$ = maximum flux density (webers/m$^2$)
$V_{ol}$ = volume of pipe (conductor)

**Keywords:** conductor, current, eddy, heat
**Sources:** Chemical Engineering 103 (5):112. May 1996; Thewlis, J. 1961-1964.

## SKIN FRICTION, LAW OF—SEE FROUDE

## SKINNER LAWS OF BEHAVIOR

Skinner's philosophy that stimulus-response of an organism, including animals and humans, was more strongly rooted in behaviorism, as contrasted to cognitive psychology. Skinner extended Thorndike's laws. M. Richelle writes that B. Skinner was not a believer in stimulus-response conditioning.

**Keywords:** behavior, cognitive, psychology, stimulus-response
SKINNER, Burrhus Frederic, 1904-1990, American psychologist
**Source:** American Scientist, 1994, v. 82:584. Nov-Dec.
*See also* PAVLOV; THORNDIKE

## SLOSH TIME NUMBER, Sl OR $N_{Sl}$

A dimensionless group applicable to fluid slosh in a low gravity field:

$$Sl = (\sigma/\rho R^3)^{1/2} t$$

where $\sigma$ = surface tension
$\rho$ = mass density
$R$ = tank radius
$t$ = time

**Keywords:** fluid, low gravity, slosh
**Sources:** Land, N. S. 1972; Potter, J. H. 1967.

## SMALL NUMBERS, LAW OF (1898); POISSON DISTRIBUTION

If a variable is distributed according to a a Poisson distribution, the variable is identified as being according to the law of small numbers.

**Keywords:** distribution, Poisson, variable
POISSON, Simeon Denis, 1781-1840, French mathematician
**Source:** Van Nostrand. 1947.
*See also* LARGE NUMBERS; LIMIT THEOREMS; POISSON

## SMEKAL LAW

Ionic activity in a solid can be represented by:

$$1/\rho = A_1 \exp(-B_1/T) + A_2 \exp(-B_2/T)$$

where $\rho$ = the ionic conductivity
$A_1 < A_2 < B_1 < B_2$ = constants

**Keywords:** conductivity, ionic, solid
SMEKAL, Adolf Gustav Stephens, 1895-1959, German Soviet American physicist
**Source:** Gillispie, C. C. 1981.
*See also* RAMAN

## SMITH-HELMHOLTZ LAW (OPTICS)

In spherical lens, the product of the refractive index, off-axis distance, and off-axis angle of the image point is equal to the corresponding product computed at the image plane.

**Keywords:** lens, off-axis, refractive index
SMITH, Robert, 1689-1768, English physicist
HELMHOLTZ, Hermann Ludwig Ferdinand von, 1821-1894, German physicist and physiologist
**Sources:** Gillispie, C. C. 1981; Morris, C. G. 1992.
*See also* LAGRANGE

## SMOLUCKOWSKI NUMBER, Sm OR $N_{Sm}$

A dimensionless group used for rarefied gas flows, relating the characteristic body length to the molecular mean free path:

$$Sm = L/\lambda$$

where L = characteristic length
$\lambda$ = length of mean free path of molecules

The Smoluckowski number is equal to the reciprocal of the Knudsen number.

**Keywords:** flow, gas, molecular mean free path, rarefied
SMOLUCKOWSKI, Marian, 1872-1917, Russian physicist
**Sources:** Gillispie, C. C. 1981; Land, N. S. 1972.
*See also* KNUDSEN

## SNELL LAW (1613*, 1621); SNELL LAW OF REFRACTION; DESCARTES LAW

The relationship of the angle of incidence, the angle of refraction, the velocity light in a first medium, and the velocity in the second medium, gives the index of refraction:

$$n = \sin i/\sin r \qquad i = v/v'$$

where n = index of refraction
i = angle of incidence
r = angle of refraction
v = velocity in first medium
v′ = velocity in second medium

Although the relationship was discovered in 1621, as stated above, and one author claims 1613*, the phenomenon was known at the time of Ptolemy (Claudius) in about the year A.D. 75.

**Keywords:** angle, incidence, light, refraction, velocity
SNELL, Willebrod van Roijen, 1591-1626, Dutch mathematician
PTOLEMY, Claudius, first century, Greek astronomer (possibly born in Egypt)
**Sources:** Asimov, I. 1972; Daintith, J. 1988; Hodgman, C. D. 1952; *Jerrard, H. G. and McNeill, D. B. 1992; Mandel, S. 1972; Parker, S. P. 1989; Stedman, T. L. 1976.
*See also* DESCARTES; FERMAT; REFLECTION; REFRACTION

## SNOEK LAW

Alloys that have an integral number of Bohr magnetrons (the tubes the microwaves come from) per atom exhibit minor or no magnetostriction.

**Keywords:** alloys, Bohr, magnetostriction
SNOEK, Jacob Louis, twentieth century (b. 1902), German physicist
**Sources:** Ballentyne, D. W. G. and Lovett, D. R. 1961.1970. 1972. 1980.

## SODDY-FAJANS DISPLACEMENT LAW (1913)

As an alpha particle is emitted from an atom, the atom moves two places to the left in the periodic table of elements; as a beta particle is emitted from an atom, the atom moves one place to the right in the periodic table of elements. The radioactive series is based on this law. Soddy and Fajans independently discovered these relationships at the same time.

**Keywords:** alpha, atom, beta, elements, periodic table, radioactive
SODDY, Frederick, 1877-1956, British scientist; Nobel prize, 1921, chemistry
FAJANS, Kasimir, 1887-1975, German Polish American chemist.
**Sources:** Ballentyne, D. W. G. and Lovett, D. R. 1970. 1972. 1980; Encyclopaedia Britannica. 1993; Gillispie, C. C. 1981; Science 274(5293):1628, 6 December 1996.
*See also* FAJANS; RUSSELL, FAJANS, AND SODDY.

## SOHNKE LAW (OR SOHNCKE); SOHNKE STRESS LAW FOR FRACTURE (1869)

The stress (force per unit area) normal to a cleavage plane that must be applied to produce a fracture is a characteristic constant for a crystalline substance.

**Keywords:** crystalline, fracture, stress
SOHNKE, Leonhard, 1842-1897, German physicist
**Source:** Thewlis, J, 1961-1964.

## SOIL SETTLEMENT, LAW OF; MODEL LAW FOR CONSOLIDATION PROCESS OF A CLAY (1925)

If two layers of the same clay of different thicknesses have the same number of drainage faces, and attained the same degree of consolidation during time $t_1$ and $t_2$, respectively (generally where $t_1 \neq t_2$), their coefficients of consolidation and their time factors must theoretically be equal. Equating both factors:

$$t_1/t_2 = h_1^2/h_2^2$$

where $t_1$ and $t_2$ = consolidation times
$h_1$ and $h_2$ = thicknesses of two layers

This states that the ratio of the times necessary to attain a certain degree of consolidation in two clay layers of the same material and the same number of drainage faces, but different thicknesses, equals the ratio of the squares of the corresponding thicknesses of the layers.

**Keywords:** clay, consolidation, drainage, settling
**Source:** Scott, J. S. 1993.

## SOLENOIDS (METEOROLOGY)—SEE PARALLEL SOLENOIDS

## SOLUBILITY LAW FOR DILUTE SOLUTIONS

For dilute solutions, the solubility:

$$\log_{10} S'/S = 0.4343\ L_{sl}/R\ [1/T - 1/T']$$

where S = solubility of the solute at T, the temperature
S′ = solubility at temperature T′
$L_{sl}$ = constant related to the particular solution
R = gas constant
T = temperature

**Keywords:** dilute, solubility, solutions
**Source:** Michels, W. C. 1956 (partial).
*See also* OSTWALD; RAOULT; VAN'T HOFF

## SOLUBILITY OF GASES—SEE DALTON

## SOLUBILITY PRODUCT LAW

In sufficiently dilute solutions, saturated with a slightly soluble univalent salt, the product of the concentrations of the ion species of that salt is constant:

$$[C^+][A^-] = \text{constant} = (\alpha_0\ S_0)^2$$

where $\alpha_0$ = degree of ionization
S = solubility of salt in pure water

**Keywords:** concentration, ions, salt, solution, univalent
**Sources:** Considine, D. M. 1976; Morris, C. G. 1992; Thewlis, J. 1961-1964.
*See also* OSTWALD; RAOULT

## SOMMERFELD AND KOSSEL SPECTROSCOPIC DISPLACEMENT LAW

The spectrum of any atom is always similar to that of the singly charged ion that follows it in the periodic table; it is similar to the position of the double-charged positive ion of the element, which is two places to the right in the table.

**Keywords:** atom, ion, periodic, spectrum
SOMMERFELD, Arnold Johannes Wilhelm, 1868-1951, German physicist
KOSSEL, Walter, 1888-1956, German physicist
**Sources:** Besancon, R. M. 1974; Webster's Biographical Dictionary. 1959.
*See also* GRIMM-SOMMERFELD; SPECTROSCOPIC DISPLACEMENT

## SOMMERFELD NUMBER, So OR $N_{So}$

A dimensionless group used in lubrication that relates the viscous force and the load force:

$$So = P\,\Psi^2/\mu\omega$$

where $P$ = the bearing load per unit area
$\Psi$ = the radial clearance/diameter
$\mu$ = absolute viscosity
$\omega$ = angular velocity

According to G. I. Barenblatt, Sommerfeld is credited with naming the Reynolds number. (Personal correspondence. 1999)

**Keywords:** bearing, load, lubrication, viscosity
SOMMERFELD, Arnold Johannes Wilhelm, 1868-1951, German physicist
**Sources:** Bolz, R. E. and Tuve, G. L. 1970; Land, N. S. 1972; Parker, S. P. 1992; Potter, J. H. 1967; Wolko, H. S. 1981.
*See also* HERSEY; OCVIRK

## SORET NUMBER, So OR $N_{So}$

A dimensionless thermodiffusion coefficient coupled heat and mass transfer:

$$So = \theta n'_{20}$$

where $\theta$ = thermodiffusion constant, as in Dufour number
$n'_{20}$ = number of molecules

**Keywords:** coupled, heat, mass, thermodiffusion
SORET, Jacques-Louis, French thermoscientist
**Sources:** Bolz, R. E. and Tuve, G. L. 1970; NUC; Parker, S. P. 1989. 1992.
*See also* DUFOUR

## SOUND LAW—SEE DOPPLER; SABINE; STIFFNESS; MASS SOUND

## SPACE CHARGE, LAW OF—SEE LANGMUIR-CHILD

## SPACIAL PROXIMITY, LAW OF—SEE SHERRINGTON

## SPALDING FUNCTION (NUMBER), Sp OR $N_{Sp}$

A dimensionless group used in combustion and drying that relates the sensible heat and the latent heat of evaporation of a droplet:

$$Sp = c_p \Delta T\ (LH_v - q_r/q)$$

where $c_p$ = the specific heat
$\Delta T$ = the temperature difference
$LH_v$ = latent heat of vaporization
$q_r$ = the radiation flux
$q$ = the rate of heat transfer

**Keywords:** combustion, droplet, drying, evaporation, heat transfer
SPALDING, D. Brian, twentieth century (b. 1923), English engineer
**Sources:** Bolz, R. E. and Tuve, G. L. 1970; Hall, C. W. 1979; Kay, E. 1984; Parker, S. P. 1992. Potter, J. H. 1967.
*See also* SPEEDS OF COMBUSTION

## SPALLANZANI LAW

The regeneration is more complete in younger individuals than in older persons.

**Keywords:** age, physiology, regeneration
SPALLANZANI, Lazaro, 1729-1799, Italian anatomist
**Sources:** Friel, J. P. 1974; Landau, S. I. 1986; Stedman, T. L. 1976.

## SPARKING POTENTIAL, LAW OF—SEE PAULI

## SPATIAL PROXIMITY, LAW OF—SEE SHERRINGTON LAWS OF REFLEX IRRADIATION

## SPECIES, LAW OF (TRIGONOMETRY)

In spherical trigonometry, one-half of the sum of any two sides of a spherical triangle and one-half of the sum of the opposite angles are the same species (type or pattern). Two sides, two angles, or an angle and a side are said to be of the same species (type) if they are both acute or both obtuse, and of different species if one is acute and one is obtuse.

**Keywords:** algebra, angles, spherical, trigonometry
**Source:** James, R. C. and James, G. 1968.
*See also* COSINE; QUADRANTS; SIGNS; TANGENTS

## SPECIES, RULE OF—SEE QUADRANTS

## SPECIFIC BACTERIA, LAW OF—SEE KOCH

## SPECIFIC ENERGIES—SEE MÜLLER LAW

## SPECIFIC HEAT, LAW OF—SEE EINSTEIN; KOPP

## SPECIFIC HEAT RATIO, S.H. RATIO

A dimensionless group representing a material property in gas flow that relates the specific heat at constant pressure to the specific heat at constant volume:

$$\text{S. H. Ratio} = c_p/c_v$$

where $c_p$ = specific heat at constant pressure
$c_v$ = specific heat at constant volume

**Keywords:** flow, gas, specific heat
**Source:** Land, N. S. 1972.

## SPECIFIC INSTABILITY, LAW OF—SEE MÜLLER

## SPECIFIC IRRITABILITY, LAW OF—SEE MÜLLER

## SPECIFICITY OF BACTERIA, LAW OF; KOCH LAW OR POSTULATE

In the case of disease, a microorganism must be present in every case; it must be capable of cultivation in pure culture; when inoculated in pure culture, it reduces the disease in susceptible animals; the organism must be recoverable and again grow in pure culture.

**Keywords:** bacteria, disease, health, medical, microorganism

KOCH, Heinrich Hermann Robert, 1843-1910, German physician and bacteriologist
**Sources:** Landau, S. I. 1986; Lapedes, D. N. 1974; Merriam-Webster's Biographical Dictionary. 1995.
*See also* KOCH

## SPECIFICITY OF NERVOUS ENERGY, LAW OF—SEE SPECIFIC PROPERTIES OF NERVE

## SPECIFIC PROPERTIES OF NERVE, LAW OF THE; OR SPECIFICITY OF NERVOUS ENERGY, LAW OF

Regardless of how excited, each nerve of special sense gives rise to its own peculiar sensation.

**Keywords:** excitation, nerve, response, sensation
**Source:** Thomas, C. L. 1989.

## SPECIFIC SPEED NUMBER, SpSp OR $N_{SpSp}$

A dimensionless group applicable to pumps:

$$SpSp = \omega(Q_v)^{1/2}/(gh)^{3/4}$$

where $\omega$ = rotational speed
$Q_v$ = volume rate of flow
g = gravitational acceleration
h = head produced per stage

**Keywords:** head, pumps, rotational speed
**Sources:** Bolz, R. E. and Tuve, G. L. 1970; Land, N. S. 1972; Potter, J. H. 1967.

## SPECIFICITY OF NERVOUS ENERGY, LAW OF—SEE SPECIFIC PROPERTIES OF NERVE, LAW OF

## SPECTRAL ANALYSIS, LAW OF—SEE KIRCHHOFF

## SPEEDS OF COMBUSTION, LAW OF

The combustion complex of gaseous mixtures with air can be regarded as involving simultaneous, but independent, burning of a number of simple mixtures of the individual gases with air, in which the proportion of inflammable gas and air is such that each mixture, if burning alone, would propagate the flame with the same speed as does the complex mixtures.

**Keywords:** air, combustion, energy, gas, mixtures, thermal
**Source:** Thewlis, J. 1961-1964.
*See also* SPALDING NUMBER

## SPONTANEOUS DECAY, LAW OF—SEE RUTHERFORD AND SODDY

## SPÖRER (SPOERER) LAW OF SUNSPOT LATITUDES (1859)

The interval between successive minima in sunspot activity has varied between 7 and 17 years and averages 11.1 years. The maximum occurrence usually takes place three to five years after the minimum, the spots appearing in latitudes 10° to 20° north or south. For the next six years or so, spots occur less frequently and at still lower latitudes. At the next minimum in sunspot activity, the last few spots may be only 5° to 6° from the equator and may overlap in time with the first high latitude spots of the new cycle.

**Keywords:** astronomy, latitudes, sunspot
SPÖRER, Gustav Friedrich Wilhelm, 1822-1895, German astronomer
**Sources:** Mitton, S. 1973; Thewlis, J. 1961-1964; World Book. 1990.
*See also* HALE

### SQUARE-CUBE LAW

When the dimensions of an object are increased while keeping the shape the same, the area increases as the square of the dimensions, while the volume and mass increase as the cube. Attributed to Galileo Galilei.

**Keywords:** area, dimensions, shape, square, volume, mass
GALILEO GALILEI,1564-1642, Italian scientist
**Sources:** de Camp, S. L. 1962; Thewlis, J. 1961-1964.

### SQUARE LAW DEMODULATOR

An electrical device whose output voltage is proportional to the square of its input voltage. A demodulator is used to obtain information from a modulated waveform from the signal imparted to the wave form by a modulator.

**Keywords:** device, modulator, voltage, waveform
**Source:** Considine, D. M. 1976. 1995.
*See also* SQUARE LAW MODULATOR

### SQUARE LAW MODULATOR

An electrical device whose output is proportional to the square of its input. The carrier and modulating signal are added in the input to produce a modulated carrier in the output.

**Keywords:** device, input, modulated, output
**Source:** Considine, D. M. 1976. 1995.
*See also* SQUARE LAW DEMODULATOR

### SQUARE-ROOT LAW

In statistics, the standard deviation of the number of successes divided by the number of trials is inversely proportional to the square root of the number of trials.

**Keywords:** standard deviation, statistics, trials
**Source:** Lapedes, D. N. 1976. 1982.
*See also* INVERSE SQUARE

### SQUEEZE NUMBER, Sq OR $N_{Sq}$

A dimensionless group describing double-squeeze film damping:

$$Sq = 12\mu\omega/p_a(r/c)^2$$

where $\mu$ = absolute viscosity
$\omega$ = angular frequency
$p_a$ = ambient pressure
r = radius
c = unloaded film thickness

**Keywords:** damping, double, film, squeeze
**Source:** Land, N. S. 1972.

## STABILITY (ELECTRIC FIELD), LAW OF—SEE EARNSHAW

## STABLE EQUILIBRIUM, LAW OF

A system having specified states and an upper bound in volume can reach, from any given state, one and only one stable state and leave no net effect on its environment. This is an application of the first and second laws of thermodynamics.

**Keywords:** environment, stable, state, thermal, volume
**Source:** Morris, C. G. 1992.
*See also* THERMODYNAMICS

## STABILITY LAW OF FLUID FLOW

Deceleration of fluid flow is an inefficient and unstable process resulting in eddy currents and large energy losses, whereas acceleration of flow is an efficient and stable fluid process.

**Keywords:** acceleration, eddy currents, energy, stable, unstable
**Source:** Hix, C. F. and Alley, R. P. 1958.

## STAGES, LAW OF (CRYSTALLIZATION)

In the process of crystallization from the liquid state, the atoms arrive at their final, permanent configuration through successive, temporary arrangements of varying stability, often referred to as the *n/8 law,* in which crystals were arranged in mathematical rather than statistical relationship.

**Keywords:** atoms, crystallization, liquid, stability
**Source:** Gillispie, C. C. 1981.

## STANDARD DEVIATION, LAW OF

The standard deviation from the mean is represented in simple examples, such as coin flipping:

$$\sigma = 1/2(N)^{1/2}$$

where $\sigma$ = standard deviation
N = number of events

There is a probability of about 0.66 that the number of heads of a tossed coin will fall within the ranges of $(1/2N - \sigma)$ to $(1/2 + \sigma)$, and that the total spread is $2\ \sigma$ is $(N)^{1/2}$.

**Keywords:** chance, probability, statistics
**Sources:** Isaacs, A 1996; Hall, C. W. 1977; James, R. C. and James, G. 1968.
*See also* PROBABILITY; STANDARD ERROR

## STANDARD ERROR OF ESTIMATE, LAW OF, S.E.

A relationship for determining the standard error of a number of tests, either estimated or actually performed, often representing a larger population, of data obtained from a number of measurements of calculations:

$$S.E. = [\Sigma(y - y_1)^2/N]^{1/2}$$

where S. E. = standard error of estimate
y = actual or accepted value
$y_1$ = predicted value, such as by equation
N = number of items in sample

**Keywords:** data, population, tests
**Sources:** Considine, D. M. 1976; Hall, C. W. 1977; Newman, J. R. 1956.
*See also* ERRORS; PROBABILITY; STANDARD DEVIATION

## STANTON NUMBER, St OR $N_{St}$

A dimensionless group in forced convection heat transfer relating the heat transferred to fluid and the thermal capacity of the fluid:

$$St = h/\rho\, c_p V$$

where h = heat transfer coefficient
ρ = mass density
$c_p$ = the specific heat at constant pressure
V = velocity

The Stanton number equals the Nusselt number ÷ (Reynolds number × the Prandtl number) and also the Nusselt number ÷ the Péclet number.

**Keywords:** convection, fluid, forced, heat, transfer
STANTON, Sir Thomas Edward, 1865-1931, British engineer (fluid and solid mechanics)
**Sources:** Bolz, R. E. and Tuve, G. L. 1970; Grigull, U. et al. 1982; Land, N. S. 1972; Layton, E. T. and Lienhard, J. H.1988; NUC; Parker, S. P. 1992; Potter, J. H. 1967; Rohsenow, W. M. and Hartnett, J. P. 1973.
*See also* NUSSELT; PRANDTL; PÈCLET; REYNOLDS

## STARK EFFECT OR LAW (1913)

When energy emission takes place in a strong electrical field, a change occurs in the number of spectral lines. The phenomena are dependent on the lines, which differ for different emissions, the geometrical relation between the direction of emission, and the direction of the electric field.

**Keywords:** electrical, emission, energy, physics, spectral
STARK, Johannes, 1874-1957, German physicist; Nobel prize, 1919, physics
**Sources:** Considine, D. M. 1976; Hodgman, C. D. 1952; Mandel, S. 1972; Rigden, J. S. 1996.

## STARK (1908)-EINSTEIN (1912) LAW OF PHOTOCHEMICAL EQUIVALENCE; OR EINSTEIN LAW OF PHOTOCHEMISTRY; FIRST LAW OF PHOTOCHEMISTRY

One quantum of active light is absorbed per molecule of substance that disappears. The primary quantum yield, regardless of frequency (that is, the number of molecules changed per quantum absorbed), is given by the rate of processes divided by the rate of absorption of light, expressed as:

$$E = h\nu$$

where h = Planck constant
ν = frequency (some references use f)

**Keywords:** absorption, molecules, quantum
STARK, Johannes, 1874-1957, German physicist; Nobel Prize, 1919, physics
EINSTEIN, Albert, 1879-1955, German Swiss American physicist; Nobel Prize, 1921, physics
**Sources:** Considine, D. M. 1976; Honig, J. M. 1953; Menzel, W. 1956; Michels, W. 1956; Thewlis, J. 1961-1964.

## STARK NUMBER, Sk OR $N_{Sk}$

A dimensionless group used in radiation heat transfer of a fluid in a duct:

$$Sk = \varepsilon_w \sigma T_e^3 D_h/k$$

where $\varepsilon_w$ = emissivity of direct wall
$\sigma$ = Stefan-Boltzmann constant
$T_e$ = temperature of external environment
$D_h$ = hydraulic diameter of duct
k = thermal conductivity of fluid

**Keywords:** duct, fluid, heat transfer, radiation
STARK, Johannes, 1874-1957, German physicist; Nobel prize, 1919, physics
**Sources:** Bolz, R. E. and Tuve, G. L. 1970; Kakac, S. et al. 1987; Parker, S. P. 1992.

## STARLING LAW (OR HYPOTHESIS) OF HEART (1890S)(1914*)

The heart output per beat is directly proportional to the diastolic (heart cavity) filling. The force of cardiac contraction increases as the heart fills.

**Keywords:** diastolic, heart, physiology
STARLING, Ernest Henry, 1866-1927, English physiologist
**Sources:** Dorland, W. A. 1980. 1988; Friel, J. P. 1974; Landau, S. I. 1986; Science vol. 188, p. 352, 25 April 1975; *Talbott, J. H. 1970.
*See also* HEART, LAW OF

## STATICAL LAW (1924)

Sea level acts like an inverted barometer. This law was expressed by A. Doodson.

**Keywords:** barometer, sea level
DOODSON, Arthur Thomas, 1890-1968, English mathematician and geophysicist
**Sources:** Dictionary of National Biography; Fairbridge, R. W. 1967.

## STATISTICAL LAW OR LAW OF LARGE NUMBERS OR BERNOULLI THEOREM

As physical laws are transformed into statistical laws, the accuracy of the transformation depends on the large number of elements present.

**Keywords:** accuracy, elements, transformation
**Source:** James, R. C. and James, G. 1968.
*See also* BERNOULLI; LARGE NUMBERS; POISSON

## STEFAN LAW (1879; 1886*); STEFAN FOURTH-POWER LAW; STEFAN LAW OF RADIANT ENERGY; STEFAN-BOLTZMANN LAW (1884)

The total amount of radiation by a hot body is proportional to the fourth power of its absolute temperature. If the temperature is doubled, the rate of radiation increases sixteenfold:

$$R = \varepsilon \sigma T^4$$

where R = rate of emission of radiant energy per unit area
$\varepsilon$ = emissivity
$\sigma$ = constant of proportionality, or Stefan-Boltzmann constant
T = absolute temperature

It is also expressed as the energy radiated per unit time by a black body:

$$E = K (T^4 - T_o^4)$$

where E = energy radiated per unit time by a black body
K = constant
T = absolute temperature of body
$T_o$ = absolute temperature of surroundings

The J. Stefan law was stimulated by the work of J. W. Draper. Boltzmann's contribution was to show that the law could be deduced from thermodynamic principles (1844). The Stefan-Boltzmann law can be derived from the Planck law.

**Keywords:** emission, radiation
STEFAN, Josef, 1835-1893, Austrian physicist
BOLTZMANN, Ludwig Edward, 1844-1906, Austrian physicist
**Sources:** Asimov, I. 1976; *Gillispie, C. C. 1981; Hodgman, C. D. 1952; Laowski, J. J. 1997; Lagowski, J. J. 1997; Layton, E. T. and Lienhard, J. H. 1988; Mandel, S. 1972; Parker, S. P. 1989; Thewlis, J. 1961-1964; Ulicky, L. and Kemp, T. J. 1992; Uvarov, E. B. 1964.
*See also* BOLTZMANN; DRAPER; FOURTH POWER LAW; LE CHATELIER; PLANCK; RADIANT ENERGY

### STEFAN NUMBER, Stef OR $N_{Stef}$

A dimensionless group in heat transfer that relates the heat radiated to heat conducted:

$$Stef = \sigma A_1 T^4 / \{k A_2 (\Delta T / \Delta L)\}$$

where σ = Stefan-Boltzmann constant
$A_1$ = radiating area
$A_2$ = conducting area
T = temperature
k = thermal conductivity
$\Delta T/\Delta L$ = temperature gradient

The Stefan number is equivalent to the Margules (Margoulis) number.

**Keywords:** conducted, heat, radiated, transfer
STEFAN, Josef, 1835-1893, Austrian physicist
**Sources:** Bolz, R. E. and Tuve, G. L. 1970; Land, N. S. 1972; Potter, J. H. 1967.
*See also* MARGULES (MARGOULIS)

### STEINMETZ LAW (1916)

An empirical relationship representing the energy lost per unit volume per cycle through hysteresis and the maximum magnetic induction attained during the cycle:

$$W = \eta \, \beta_m^{1.6}$$

where W = energy lost
η = Steinmetz coefficient, a constant for a given material
$\beta_m$ = maximum magnetic induction

**Keywords:** energy, hysteresis, induction, magnetic
STEINMETZ, Charles Proteus, 1865-1923, German American electrical engineer
**Sources:** Campbell, N. R. 1957; Thewlis, J. 1961-1964.

## STELLAR STARS, LAWS OF; OR LAW OF STELLAR STRUCTURE

The laws conform to the following four principles, often represented by differential equations:

1. Total mass interior to the radius is the sum of the mass in each interior shell, which is the product of the shell volume and mass density.
2. A shell's mass produces an increase in gas pressure beneath it equal to its weight per unit area (the product of its density, thickness, and the acceleration of gravity).
3. A temperature difference across a shell results in an energy flow (luminosity per unit area) controlled by the opacity of the material.
4. The luminosity must be the sum of the energy released within the radius by all interior shells.

**Keywords:** energy, flow, mass, opacity, pressure, shell(s)
**Sources:** Brown, S. B. and Brown, L. B. 1972; Maran, S. P. 1992; Travers, B. 1996.

## STENO—SEE SUPERPOSITION, LAW OF (GEOLOGY)

## STERNBERG LAW (1875)

The decline in size of elastic particles transported downstream is proportional to the weight of the particle in the water and to the distance it has traveled, or to the work done against friction along the bed:

$$W = W_0 e^{-as}$$

where $W$ = weight at any distance, s
$W_0$ = initial weight of particle
$a$ = coefficient of size reduction

**Keywords:** friction, size, particles, work
STERNBERG, Hilgard O'Reilly, nineteenth century, German engineer
**Sources:** Gary, M. 1972; NUC.

## STEVENS PSYCHOLOGICAL LAW (1951); STEVENS POWER LAW

A power law stating that equal stimulus ratios produce equal subjective ratios, as represented by:

$$\psi = k\ S^n$$

where $\psi$ = psychologic magnitude
$S$ = stimulus magnitude
$k, n$ = constants on log-log scale, with n the slope of a straight line, and the exponent ranges from 0.3 to 4.0

The sensation magnitude is proportional to the stimulus magnitude raised to an exponent, with the size of the exponent from one sensory continuum to another—from loudness to sweetness to tactual, to vibration to food odor, etc.

**Keywords:** ratio, stimulus, subjective
STEVENS, Stanley Smith, 1906-1973, American psychologist
**Sources:** Bynum, W. F. et al. 1981; Glasser, O. 1944; Science, vol. 188:827-828, 23 May 1995; Wolman, B. B. 1989.
*See also* FECHNER (LOGARITHMIC); PSYCHOPHYSICAL; WEBER

### STEWART AND KIRCHHOFF LAW (1858); KIRCHHOFF OPTICAL LAW

Independent of the nature of the substance, the law is that the emissive power divided by the absorption coefficient for any substance depends only on the frequency and plane of polarization of the radiation and the temperature.

**Keywords:** absorption, emission, polarization
STEWART, Balfour, 1828-1887, Scottish engineer and physicist
KIRCHHOFF, Gustav Robert, 1824-1887, German physicist
**Sources:** Ballentyne, D. W. G. and Lovett, D. R. 1980; Landau, S. I. 1986.
*See also* KIRCHHOFF-CLAUSIUS; KIRCHHOFF OPTICAL

### STEWART LAW

Describes, as related to the flow of a fluid, the smallest shearing force that will cause flow, since there will always be molecules for which cohesion has been broken. (Some say yes, some say no.)

**Keywords:** cohesion, flow, shear
STEWART, Balfour, 1828-1887, Scottish engineer and physicist
**Sources:** Barr, E. S. 1973; Gillispie, C. C. 1981.
*See also* ANDRADE

### STEWART NUMBER, St OR $N_{St}$

A dimensionless group that characterizes magnetic properties and flow of a fluid:

$$St = \beta_o^2 L_o \sigma / \rho \omega_o$$

where $\beta_o$ = magnetic induction
$\omega_o$ = velocity of fluid
$L_o$ = characteristic dimension
$\rho$ = density of fluid
$\sigma$ = electrical conductivity

**Keywords:** conductivity, fluid, induction, magnetic
STEWART, Balfour, 1828-1887, Scottish engineer and physicist
**Sources:** Bolz, R. E. and Tuve, G. L. 1970; Barr, E. S. 1973; Begell, W. 1983; Gillispsie, C. C. 1981; Parker, S. P. 1992; Potter, J. H. 1967.

### STIFFNESS LAW FOR PANELS

This is an empirical law describing noise transmission loss in stiff panels, cylindrical pipes, etc. Stiffness control to reduce noise transmission is a low-frequency characteristic; above a frequency of 20 cps (cycles per second), mass control predominates. The transmission loss increases with decreasing frequency for a very stiff panel (Fig. S.1).

**Keywords:** frequency, noise, panel, sound
**Sources:** Ballou, G. M. 1987. 1991.
*See also* MASS; MAYER; SABINE

### STIMULUS-RESPONSE LAWS—SEE FECHNER; INITIAL VALUE (WILDER); PAVLOV; SKINNER; STEVENS; WEBER

### STOKES-CUNNINGHAM LAW

A relationship that expresses the settling of particles under 3 microns settling in gases, and particles under 0.01 micron for settling in liquids, to refine or modify the Stokes law for resistance to flow, divide by a factor $k_m$:

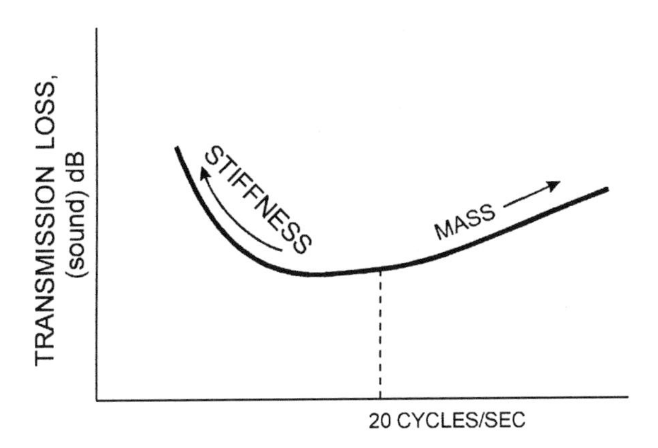


**Figure S.1** Stiffness panels.

$$k_m = 1 + k_{me} (\lambda/D_p)$$

where $\lambda$ = mean path of the fluid molecules
$D_p$ = particle diameter
$k_{me}$ = dimensionless constant (1.3 to 2.3)

**Keywords:** gas, liquids, particle, settling
STOKES, Sir George Gabriel, 1819-1903, British mathematician and physicist
CUNNINGHAM, James, 1800-1878, Scottish scientist
**Sources:** Barr, E. S. 1973; Perry, R. and Green, D. W. 1984.
*See also* BROWNIAN

## STOKES EMISSION LAW—SEE STOKES LAW OF FLUORESCENCE

## STOKES LAW (FLUID FLOW) (1850); STOKES LAW FOR VISCOSITY

The viscous drag on a spherical object in a fluid, where the Reynolds number is less than 0.1, when there is relative motion between them, is given by:

$$F = 6\pi\mu r V$$

where F = viscous drag
$\mu$ = fluid viscosity
r = radius of the sphere
V = relative velocity

**Keywords:** drag, fluid, sphere, viscous
STOKES, Sir George Gabriel, 1819-1903, British mathematician and physicist
**Sources:** Considine, D. M. 1976. 1995.
*See also* DRAG; REYNOLDS NUMBER; STRESS-STRAIN

## STOKES LAW (LIGHT)

The intensity of an elementary wave at an external point varies as cos $\theta$, when $\theta$ is the obliquity, or angle between the wave normal and the line joining the point to the center of the elementary wave. Thus, the effect only vanishes for $\theta = \lambda$ (wavelength), that is, for points directly behind the wave.

**Keywords:** angle, light, optics, physics, wave

STOKES, Sir George Gabriel, 1819-1903, British mathematician and physicist
**Sources:** Ballentyne, D. W. G. and Lovett, D. R. 1980; Clifford, A. F. 1964; Honig, J. H. 1953; Thewlis, J. 1961-1964.
*See also* LAMBERT

## STOKES LAW (MEDICAL)

A muscle located above an inflamed membrane is often affected with paralysis.

**Keywords:** inflamed, membrane, muscle, paralysis
STOKES, William, 1804-1878, Irish physician
**Sources:** Critchley, M. 1978; Friel, J. P. 1974; Glasser, O. 1944; Landau, S. I. 1986; Stedman, T. L. 1976.

## STOKES LAW OF EMULSIONS

The small dispersed globules of an emulsion obey:

$$V = 2\, r^2(d - d') \, g/9\mu$$

where V = velocity of sedimentation of dispersed phase
r = radius of globules
d = density of dispersed phase
d′ = density of the continuous phase
μ = viscosity
g = constant of acceleration of gravity

The separation of an emulsion is termed *creaming* and is usually made visible by the separation of a clear layer of aqueous liquid.

**Keywords:** creaming, emulsion, sedimentation
STOKES, Sir George Gabriel, 1819-1903, British mathematician and physicist
**Sources:** Thorpe, J. 1961-1964; Whiteley, M. 1940.
*See also* STOKES (SETTLING)

## STOKES LAW OF FLUORESCENCE (1852)

If the wavelength of light used to excite a fluorescent material is $\lambda_a$, then the wavelength of light emitted, $\lambda_e$, is such that $\lambda_e > \lambda_a$. The difference in energy $h(\nu_a - \nu_e)$ is dissipated as heat in exciting lattice variations in the phosphors matrix. The emitted photons carry off less energy than applied by exciting photons.

**Keywords:** energy, fluorescence, lattice, phosphors, photons
STOKES, Sir George Gabriel, 1819-1903, British mathematician and physicist
**Sources:** Asimov, I. 1972; NYPL Desk Ref.; Thewlis, J. 1961-1964.
*See also* STARK-EINSTEIN

## STOKES LAW OF RADIATION

In fluorescence, only radiation of wavelengths larger than that of the exciting light can be emitted. Incident radiation is at a higher frequency and shorter wavelength than the re-radiation emitted by an absorber of that incident radiation.

**Keywords:** fluorescence, frequency, incident, wavelength
STOKES, Sir George Gabriel, 1819-1903, British mathematician and physicist
**Sources:** Ballentyne, D. W. G. and Lovett, D. R. 1972. 1980.
*See also* WIEN

## STOKES LAW OF SETTLING (1845–1850)

Describes the rate of fall of a small sphere in a viscous fluid such that a constant velocity, V, the terminal velocity, is reached:

$$V = 2\ g\ r^2\ (d_1 - d_2)/9\mu$$

where V = constant terminal velocity
g = gravitational constant
r = radius of the sphere
$d_1$ and $d_2$ = densities of sphere and medium
$\mu$ = viscosity

**Keywords:** fluid, sphere, terminal, velocity, viscous
STOKES, Sir George Gabriel, 1819-1903, British mathematician and physicist
**Sources:** Hodgman, C. D. 1952; Layton, E. T. and Lienhard, J. H. 1988; Mandel, S. 1972; Northrup, E. T. 1917; Scott Blair, G. W. 1949; Uvarov, E. B. et al. 1964.
*See also* DRAG; STOKES LAW OF TERMINAL VELOCITY

## STOKES LAW OF TERMINAL VELOCITY

A spherical body moves through a viscous fluid in streamline flow with a drag force of:

$$F_d = 6\pi\ \mu\ r\ V$$

where $F_d$ = drag force
$\mu$ = viscosity of fluid
r = radius of spherical body
V = the velocity of movement

**Keywords:** drag, flow, fluid, radius, velocity, viscosity
STOKES, Sir George Gabriel, 1819-1903, British mathematician and physicist
**Source:** Thewlis, J. 1961-1964.
*See also* STOKES LAW OF SETTLING

## STOKES NUMBER, St OR $N_{St}$

A dimensionless group that relates viscous force and gravity force to represent slow flow:

$$St = \mu V/\rho g L^2$$

where $\mu$ = absolute viscosity
V = velocity
$\rho$ = mass density
g = gravitational acceleration
L = characteristic length

The Stokes number is equal to the reciprocal of the Jeffrey number.

**Keywords:** flow, gravity, slow, viscous
STOKES, Sir George Gabriel, 1819-1903, British mathematician and physicist
**Sources:** Bolz, R. E. and Tuve, G. L. 1970; Kakac, S. et al. 1987; Land, N. S. 1972; Parker, S. P. 1992; Potter, J. H. 1967.
*See also* ARCHIMEDES; GALILEO; JEFFREY

### STORMS, LAW OF (1828); OR DOVE LAW OF STORMS

A general statement of (1) the manner in which the winds of a cyclone rotate about the center of a cyclone and (b) the way the entire disturbance moves over the surface of the Earth. The most dangerous half of a typical cyclone is to the right of the path of the cyclone in the northern hemisphere and to the left of the path of the cyclone in the southern hemisphere. A cyclone generally moves from the southwest to the northeast in the northern hemisphere. Three people contributed to the law of storms: H. Brandes, 1826; H. Dove, 1828; and W. Redfield, 1831.*

**Keywords:** atmosphere, cyclone, disturbance, direction, hemisphere (southern and northern)

BRANDES, H. W., 1777-1834, German astronomer

DOVE, Heinrich Wilhelm, 1803-1879, German physicist and meteorologist

REDFIELD, William C., 1789-1857, American meteorologist

**Sources:** Fairbridge, R. W. 1967; *Huschke, R. E. 1959; McIntosh, D. H. 1972; Morris, C. G. 1992; Parker, S. P. 1989.

*See also* BUYS BALLOT; DOVE; EGNELL; FERREL; FAEGRI

### STRAIN-HARDENING—SEE WORK-HARDENING

### STRAIN, LAW OF—SEE LOMNITZ

### STRAIN-RATE LAW

For the plastic potential, the strain rate vector is:

$$\dot{E} = (\dot{k}_r, \dot{k}_e)$$

where $\dot{E}$ = the strain rate vector

$\dot{k}_r$ and $\dot{k}_e$ = vectors at the stress point must be directed along the outward normal to the amount of yield curve at the stress point

**Keywords:** moment, plastic, strain, stress, yield

**Sources:** Goldsmith, W. 1960; Scott Blair, G. W. 1949 (partial).

### STRANGENESS—SEE CONSERVATION OF STRANGENESS, LAW OF

### STREAM GRADIENTS, LAW OF (1945)

A general law that expresses the inverse geometric relationship between the stream order and the mean stream gradient in a given drainage basin. The law is expressed by Horton as a linear regression of the logarithm of mean stream length on stream order, the positive regression coefficient being the logarithm of the stream-length ratio.

**Keywords:** basin, drainage, gradient, logarithm, stream

HORTON, Claude Wendell, twentieth century (b. 1915), American geologist

**Sources:** Bates, R. L. and Jackson, J. A., 1980. 1987.

*See also* STREAM LENGTHS; STREAM NUMBERS

### STREAM LENGTHS, LAW OF (1945); ALSO CHANNEL LENGTH, LAW OF

A general law first stated by Horton that expresses the direct geometric relation between stream order and the main stream lengths of each order in a given drainage basin. The law is expressed as a linear regression of logarithm of mean stream length on stream order, the positive regression coefficient is the logarithm of the stream length ratio.

**Keywords:** drainage, logarithm, numbers, stream
HORTON, Claude Wendell, twentieth century (b. 1915), American geologist
**Sources:** Bates, R. L. and Jackson, J. A. 1980. 1987.
*See also* STREAM GRADIENTS; STREAM NUMBERS

## STREAM NUMBERS, LAW OF (1945)

A general law expressing the inverse geometric relation between stream order and the number of streams of each order, in a given drainage basin. The law is expressed as a linear regression of logarithm of number of streams on stream order, the negative regression coefficient being the logarithm of the bifurcation ratio. The stream order is the designation by a dimensionless integer series the relative position of stream segments in a drainage network: the smallest, unbranched tributary, terminating at the outer point designated as order one, with that stream terminating at a single stage is the highest order and determines the order of the stream bed.

**Keywords:** bifurcation ratio, drainage, order, stream
HORTON, Claude Wendell, twentieth century (b. 1915), American geologist
**Sources:** Bates, R. L. and Jackson, J. A. 1980. 1987.

## STRESS-OPTIC LAW IN TWO DIMENSIONS

In a transparent isotropic plate in which the stresses are two-dimensional and within the elastic limit, the phase difference or relative retardation in wave length between the rectangular wave components traveling through it and produced by temporary double refraction is:

$$R_t = Ct\,(p - q)$$

where $R_t$ = relative retardation in wavelength
C = stress-optic coefficient
t = thickness of the plate
p and q = principle stresses

**Keywords:** optic, photoelastic, refraction, retardation, stresses
**Source:** Frocht, M. M. 1946.

## STRESS-STRAIN LAW OF NEWTONIAN FLUID; OR NEWTON-STOKES FLUID

The viscous force resisting the shearing strain is linearly proportional to the rate of shear, as hypothesized by Newton. Only some fluids have this characteristic, particularly if the deformation rates are not too large. Gases and reasonably simple fluids (chemically) fit the hypothesis (Fig. S.2). The Newton generalization is due to Stokes based on four premises (Stokean):

1. The fluid is isotropic.
2. Translation and rotation do not induce resisting stresses, but any deformation is resisted by viscous stresses.
3. The linear stress-strain law must have the same form for any orientation of the coordinate system.
4. In the absence of deformation, the stress tensor must reduce to the hydraulic pressure.

**Keywords:** deformation, fluid, force, viscous
NEWTON, Sir Isaacs, 1642-1727, English philosopher and mathematician
STOKES, Sir George Gabriel, 1819-1903, British mathematician and physicist
**Source:** Streeter, V. L. 1961.
*See also* NEWTON; STOKES

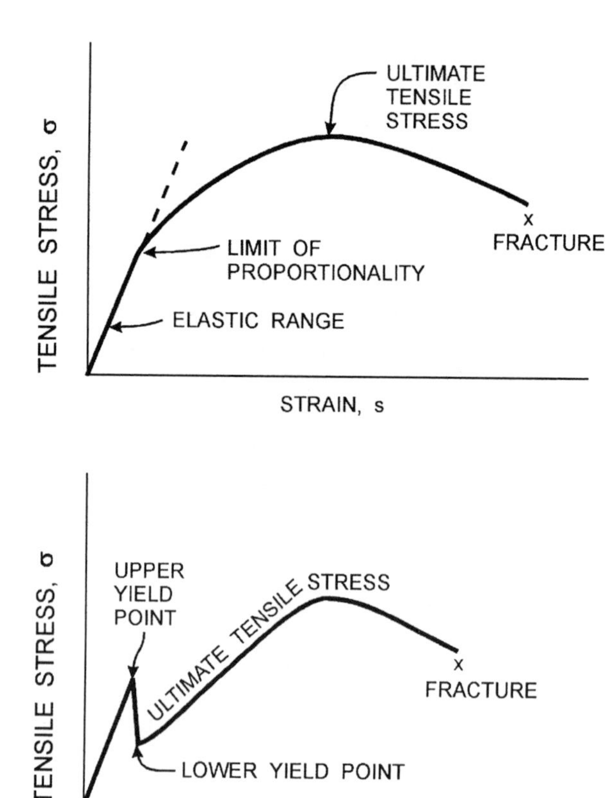


**Figure S.2** Stress law of fracture.

## STRINGS—SEE KRIGAR-MENZEL; MERSENNE; PYTHAGOREAN

## STRONG LAW OF LARGE NUMBERS

The average of a large number of independent, identically distributed random variables (X) is their common mean, such that:

$$\Pr\ [(\lim X_1 + X_2 + \dots + X_n)/n = \mu] = 1$$

as n approaches infinity

where $\Pr$ = probability and lim is limit
$X_1, X_2, \dots X_n$ = independent variables
$\mu$ = mean

**Keywords:** average, distributed, probability, random
**Source:** Downing, D. and Clark, J. 1997.
*See also* LARGE NUMBERS; STRONG LAWS

## STRONG LAWS OF LARGE NUMBERS; OR KOLMOGOROV THEOREM

The only difference between the strong and weak laws of large numbers is the type of convergence. The strong laws employ convergence, and the weak laws use probability of convergence. Since the first implies the second, any sequence satisfying a strong law also satisfies a weak law.

**Keywords:** convergence, numbers, weak
KOLMOGOROV, Andrei Nikolaevich, 1903-1987, Soviet mathematician
**Sources:** Downing, D. and Clark, J. 1997; James, R. C. and James, G. 1968.
*See also* BOREL; KOLMOGOROV; LARGE NUMBERS; WEAK LAW

## STROUHAL LAW

The relationship between the flow of a fluid and the cylinder over which the fluid flows. The shedding frequency depends on, or is related to, the diameter and the flow past the cylinder, and is used for applying to vortex shedding frequency:

$$S = nD/V$$

where S = Strouhal number
D = diameter of cylinder
V = velocity of flow

**Keywords:** cylinder, flow, fluid, frequency, shedding, velocity
STROUHAL, Vincent, 1850-1922, Czech scientist
**Sources:** Journal Fluid Mechanics 10(3):345, 1961; NUC.
*See also* MAGNUS

## STROUHAL NUMBER, Sr OR $N_{Sr}$ (1878)

A dimensionless group used for unsteady flow or streams and vortices and for wind induced vibrations relating vibration speed and translation speed:

$$Sr = L\omega/V$$

where L = characteristic length
$\omega$ = angular frequency of vibration
V = translation speed

**Keywords:** flow, unsteady, speed, translation, vibration, vortices, wind
STROUHAL, Vincent, 1850-1922, Czech scientist
**Sources:** Bolz, R. E. and Tuve, G. L. 1970; Eckert, E. R. G. and Drake, R. M. 1972; Kakac, S. et al. 1987; Jerrard, H. G. and McNeill, D. B. 1992; Land, N. S. 1972; NUC; Parker, S. P. 1992; Potter, J. H. 1967; Rohsenow, W. M. and Hartnett, J. P. 1973.

## STRUCTURAL DEFLECTIONS—SEE MAXWELL RECIPROCAL DEFLECTIONS

## STUFFER LAW

All sulfurs (sulfones) in which sulfone groups are attached to two adjacent carbon atoms can be saponified.

**Keywords:** carbon, chemistry, saponified, sulfurs
**Sources:** Honig, J. M. et al. 1953; Kirk-Othmer. 1985.

## SUBSTITUTION, LAW OF (ECOLOGY)

Any intrinsic factor can be substituted by another as long as its effect on the organism is the same.

**Keywords:** intrinsic, organism, substitution
**Source:** Lincoln, R. J. 1982.

### SUBSTITUTION, LAW OF (ECONOMICS); OR LAW OF DIMINISHING SUBSTITUTION

The principle that the scarcer a good, the greater its substitution value.

**Keywords:** good, substitution, value
**Source:** Ammer, C. and Ammer, D. 1984.
*See also* DIMINISHING RETURNS; DIMINISHING UTILITY

### SUCCESSIVE OR INTERMEDIATE TEMPERATURES, LAW OF

The thermal electromotive force developed by any thermocouple made of homogeneous metals with its junctions at any two temperatures, $T_1$ and $T_3$, is the algebraic sum of the electromotive force of the thermocouple with one junction at $T_1$ and the other at $T_2$ and the electromotive force of the same thermocouple with its junction at $T_2$ and $T_3$.

**Keywords:** algebraic, electromotive, metals, thermocouples
**Source:** Considine, D. M. 1976.
*See also* HOMOGENOUS CIRCUIT; PELTIER; SEEBECK; THERMOCOUPLE

### SUCCESSIVE REACTIONS, LAW OF

As a system passes from a less stable to a more stable state, intermediate conditions of progressively greater stability are covered.

**Keywords:** reaction, stability, state
**Source:** Honig, J. M. 1953.
*See also* OSTWALD

### SUFFICIENT REASON, LAW OR REASON

The principle that there must be a reason why whatever exists or develops does so in the time, place, and manner that it does.

**Keywords:** exists, manner, place, time
LEIBNIZ, Gottfried Wilhelm, 1646-1716, German philosopher
**Source:** Bothamley, J. 1993.
*See also* THOUGHT

### SUNSPOT LATITUDES—SEE SPÖRER

### SUPERCONDUCTIVITY, THEORY OF (1957)—SEE BCS THEORY

### SUPERPOSITION, LAW OF (1669) (GEOLOGY)

In a local sequence of rock layers or undisturbed sedimentary rock, the lower ones are the older, and each successive layer is younger, if there is neither overthrust or inversion, which can be used for evaluation of local geological evolution. This law was first expressed by N. Steno.

**Keywords:** age, geology, layers, rock, stratigraphy
STENO, Nicholaus (Latinized name for STENSEN, Niels), 1638-1686, Danish anatomist and geologist
**Sources:** Bates, R. L. and Jackson, J. L. 1980; Brown, S. B. and Brown, L. B. 1972; Morris, C. G. 1992; Parker, S. P. 1989; Tver, D. F. 1981.
*See also* SURFACE RELATIONSHIP

## SUPERPOSITION, LAW OF (PHYSICAL)

If a physical system is acted upon by a number of independent different influences, the resulting effect of the influence is the sum of the individual influences.

**Keywords:** effect, physical system, sum
**Sources:** Considine, D. M. 1976; Travers, B. 1996.
*See also* NERNST PRINCIPLE OF SUPERPOSITION

## SUPPLY, LAW OF (ECONOMICS)

Producers will be willing to divert more of their productive efforts to a given product when its market price is high than when it is low, other factors being equal. Therefore, there is a direct relationship between price and quantity supplied.

**Keywords:** price, product, quantity
**Source:** Moffat, D. W., 1976.

## SURATMAN NUMBER, Su OR $N_{Su}$

A dimensionless group applicable to particle dynamics:

$$Su = \rho L \sigma / \mu^2$$

where $\rho$ = mass density
$L$ = characteristic length
$\sigma$ = surface tension
$\mu$ = absolute viscosity

The Suratman number is equal to the Reynolds number squared divided by the Weber number.

**Keywords:** dynamics, particle
SURATMAN. Biographical information not located.
**Sources:** Bolz, R. E. and Tuve, G. L. 1970; Land, N. S. 1972; Parker, S. P. 1992; Potter, J. H. 1967.

## SURFACE (AREA) LAW

The metabolic body size of an animal (and human) is that function to which its metabolism rate is proportional. According to the surface law, the metabolic rate is proportional to the surface area. Another approach is:

$$S = k\, W^{2/3}$$

where $S$ = surface area
$k$ = constant determined experimentally
$W$ = body weight

**Keywords:** animal, area, body, metabolic, surface
**Sources:** Friel, J. P. 1974; Glasser, O. 1944; Kleiber, M. 1961. 1975.
*See also* BODY SIZE; BRODY AND PROCTER; DUBOIS; KLEIBER; RUBNER AND RICHET.

## SURFACE RELATIONSHIPS, LAW OF (1964)

The principle developed by E. Wheeler that "time as a stratigraphic dimension has meaning only to the extent that any given moment in the Earth's history may be conceived as precisely coinciding with a corresponding worldwide surfaces of deposition or erosion, and all simultaneous events either occurring thereon or directly related thereto."

**Keywords:** Earth, history, stratigraphic, time, worldwide
WHEELER, Everett Pepperrell II, twentieth century (b. 1900), American geologist
**Sources:** American Men of Science. 1961; Bates, R. L. and Jackson, J. A. 1980. 1987.

## SURFACE VISCOSITY NUMBER, Vi OR $N_{Vi}$

A dimensionless group that describes convection cells in liquid layers with surfactants:

$$Vi = \mu_s/\mu L$$

where $\mu_s$ = surface viscosity
$\mu$ = absolute viscosity
L = depth of liquid layer

**Keywords:** cells, convection, liquid layers, surfactants
**Sources:** Bolz, R. E. and Tuve, G. L. 1970; Land, N. S. 1972; Parker, S. P. 1992; Potter, J. H. 1967.

## SURVIVAL LAW

If one follows the life history of a large group of mammal offspring, or any other kinds of animals born on the same day, one finds that they do not die on the same day, with some living somewhat longer and some living a somewhat shorter time. If one plots the percentage of individuals still alive at a certain time, a typical survival curve is obtained (Fig. S.3a).

The situation for radioactive atoms is different, where a certain percentage disappear each day, represented by a logarithmic curve in which the radioactive atoms survived is based on half-life where 50% of the material disappears, which varies greatly, which for uranium is 4.5 billion years and radium 1,590 years (Fig. S.3b).

**Keywords:** half-life, mammals, organisms, radioactive
**Sources:** Gamow, G. 1961. 1988.
*See also* GOMPERTZ

## SUTHERLAND LAW (1893)

The law expresses the viscosity coefficient of a gas in terms of its absolute temperature:

$$\mu = \mu_o [(S + T_o)/(S + T)] (T/T_o)^{3/2}$$

where $\mu$ = viscosity coefficient
$\mu_o$ = the viscosity
S = Sutherland constant
T = absolute temperature

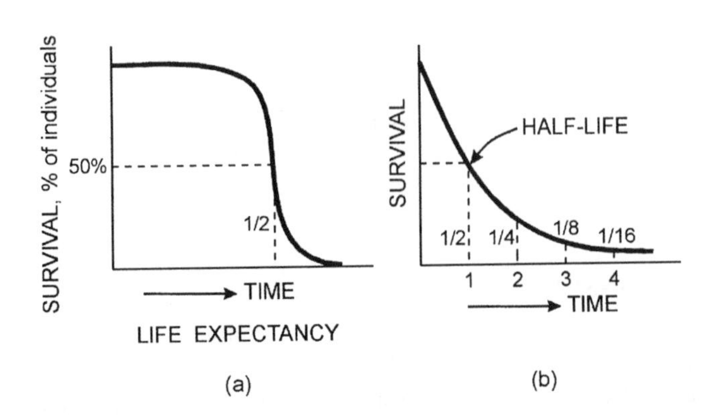


**Figure S.3** Survival.

**Keywords:** gas, temperature, viscosity
SUTHERLAND, William, 1859-1912, Scottish physicist
**Sources:** Barr, E.S. 1973; Bennett, H. 1939; Thewlis, J. 1961-1964; Thorpe, J. and Whiteley, M. 1940.

## SUBSISTENCE THEORY OF WAGES—SEE IRON LAW OF WAGES

## SUFFICIENT REASON, LAW OF OR PRINCIPLE OF

For every fact, there is a reason why it is so and not otherwise, as stated by G. Leibniz.

**Keywords:** fact, reason, why
LEIBNIZ, Gottfried Wilhelm, 1646-1716, German philosopher and mathematician
**Source:** Flew, A. 1979.
*See also* THOUGHT

## SYLLOGISM, LAW OF

The regular logical form of reasoning or argument, consists of three propositions, the first two are called premises and the third the conclusion, such that if p implies q and q implies r, then p implies r.

**Keywords:** argument, conclusion, logic, reasoning
**Sources:** Borchert, D. M. 1996; Runes, D. D. 1983.

## SYLVESTER LAW OF INERTIA; JACOBI-SYLVESTER LAW OF INERTIA (1852)

Two quadratic forms have the same rank and the same index if and only if they can be transformed into each other by an invertible linear transformation. The law was discovered independently by C. Jacobi.

**Keywords:** congruent, index, linear, quadratic, transformation
SYLVESTER, James Joseph, 1814-1897, British mathematician
JACOBI, Carl Gustav Jacob, 1804-1851, German mathematician
**Sources:** Gillispie, C. C. 1991; James, R. C. and James, G. 1968; Morris, C. G. 1992.

## SYLVESTER LAW OF NULLITY

If $r_1$ and $r_2$ are the ranks of two matrices, and if R is the rank of their products:

$$R \leq r_1$$

$$R \leq r_2$$

$$R \geq r_1 + r_2 - n$$

where n = the order of the matrices.

Sylvester defined the nullity of a matrix as the difference between its order and rank.

**Keywords:** matrix, order, rank
SYLVESTER, James Joseph, 1814-1897, British mathematician. He first used words *matrix* and *determinant.*
**Sources:** Gillispie, C. C. 1991; Newman, J. R. 1956.

## SYMMETRY LAWS, PHYSICS

A conservation law exists for each symmetry or invariance as applied to natural laws, such as space-time symmetries and internal symmetries. In general, symmetry means that when one thing is changed, something else remains unchanged. The fact that laws of nature appear to remain unchanged with time is a fundamental symmetry of nature. In a simple case, a circle is most symmetrical if it is rotated about its center it remains indistinguishable from the original circle, or is invariant.

**Keywords:** change, conservation, invariance, time
**Sources:** Brown, S. B. and Brown, L. B. 1972; Parker, S. P. 1989. 1992.

# T

## TAIT LAW

Except when it is known that a disease of the pelvic or abdominal area is malignant, and unless life is otherwise endangered or health ruined, exploration by celiotomy should be made.

**Keywords:** abdominal, celiotomy, health, medical, pelvic
TAIT, Robert L., 1849-1899, British gynecologist
**Source:** Stedman, T. L. 1976.

## TALBOT LAW

A light flashing at a frequency greater than approximately 10 Hz appears steady to the eye. The light has an apparent intensity:

$$I = I_0 \, (t/t_0)$$

where I = actual intensity of light source, for exposure of time, t, during the total time, $t_0$.

The apparent brightness of an object viewed through a slotted disk, rotating over a critical frequency, is proportional to the ratio of the angular aperture of the opening to the opaque sectors.

**Keywords:** apparent, brightness, flashing, frequency (Hz), intensity, light, physiology
TALBOT, William Henry Fox, 1800-1877, English inventor and scientist
**Sources:** Ballentyne, D. W. G. and Lovett, D. R. 1972; Friel, J. P. 1974; Landau, S. I. 1986; Thewlis, J. 1961-1964.
*See also* BLONDEL-REY; TALBOT-PLATEAU

## TALBOT-PLATEAU LAW (1829)

Light waves from two disks rotated at a speed above the flicker point fuse into an intensity the same as that which would be sensed if light waves corresponding to this intensity were directed upon the disks. The total length of a visual impression is approximately one-third of a second. The effect of a color briefly presented to the eye is proportional to the intensity of light and the time of presentation. The brightness is less than the brightness of steady illumination.

**Keywords:** color, disks, flicker, intensity, light waves, visual impression
TALBOT, William Henry Fox, 1800-1877, English inventor and scientist
PLATEAU, Joseph Antoine Ferdinand, 1801-1833, Belgian physicist
**Sources:** Barr, E. S. 1973; Holmes, F. L. 1980. 1990; Wolman, B. B. 1989.
*See also* TALBOT

## TANGENTS, LAW OF

The relation between two sides and the opposite angle of a plane triangle is (Fig. T.1):

$$\frac{a-b}{a+b} = \frac{\tan 1/2(A-B)}{\tan 1/2(A+B)}$$

where a, b, and c = sides of a triangle opposite angles A, B, and C, respectively.

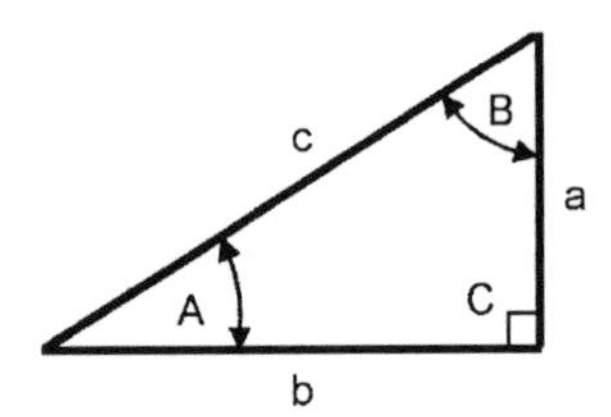


**Figure T.1** Tangents.

**Keywords:** algebra, triangle, trigonometry
**Sources:** Freiberger, W. F. 1960; Isaacs, A. 1996; Karush, W. 1989; James, R. C. and James, G. 1968; Karush, W. 1989; Morris, C. G. 1992; Tuma, J. J. and Walsh, R. A. 1998.
*See also* ANGLES; COSINES; SINES

### TATE LAW

A drop falls through a narrow tube when the weight of the drop is equal to the surface tension that holds the liquid:

$$Mg = 2\pi\gamma r$$

where M = mass of the drop
g = acceleration due to gravity
$\gamma$ = surface tension
r = the radius of the drop and tube

**Keywords:** drop, radius, surface tension, tube, weight
TATE, John Torrence, 1889-1950, American physicist
**Sources:** Ballantyne, D. W. G. and Lovett, D. R. 1980; Gillispie, C. C. 1981; Pelletier, P. A 1994.
*See also* HAGEN-POISEUILLE; MARGULES; POISEUILLE

### TAUTOLOGY, LAWS OF (LOGIC; MATHEMATICS)

For logical multiplication and addition:

$$(p \wedge p) \Leftrightarrow p \quad (p \vee p) \Leftrightarrow p$$

where $\wedge$ = and
$\vee$ = or
$\Leftrightarrow$ = if, and only if

**Keywords:** addition, logic, multiplication
**Source:** Newman, J. R. 1956.
*See also* BOOLEAN; SYLLOGISM; THOUGHT

### TAYLOR NUMBER (1), $Ta_1$ OR $N_{Ta(1)}$

A dimensionless group used to describe the stability of flow pattern between two rotating cylinders based on the relationship of centrifugal force and viscous force:

$$Ta_1 = \omega^2 d^4/\nu^2$$

where $\omega$ = angular velocity
d = clearance between cylinders
$\nu$ = kinematic viscosity

also expressed as:

$$Ta_1 = 2Re^2\frac{(r_1 - r_2)}{(r_1 + r_2)}$$

where Re = Reynolds number
$r_1$ = radius of larger cylinder
$r_2$ = radius of smaller cylinder

**Keywords:** cylinders, flow, rotating
TAYLOR, Geoffrey Ingram, 1886-1975, English engineer and scientist
**Sources:** Abbott, D. 1985; Bolz, R. E. and Tuve, G. L. 1970; Jerrard, H. G. and McNeill, D. B. 1992; Land, N. S. 1972; Parker, S. P. 1992; Potter, J. H. 1967.
*See also* REYNOLDS; SHERWOOD; TAYLOR (2)

## TAYLOR NUMBER (2), $Ta_2$ OR $N_{Ta(2)}$

A dimensionless group that represents the effect of rotation on free convection that relates Coriolis force and viscous force:

$$Ta_2 = (2\omega L^2\rho/\mu)^2$$

where $\omega$ = rate of rotation
L = height of fluid layer
$\rho$ = density
$\mu$ = dynamic viscosity

**Keywords:** convection, Coriolis, rotation, viscous
TAYLOR, Geoffrey Ingram, 1886-1975, English engineer and scientist
**Sources:** Abbott, D. 1985; Bolz, R. E. and Tuve, G. L. 1970; Parker, S. P. 1992; Potter, J. H. 1967.

## TEEVAN LAW

The fractures of bones occur in the line of extension, and not in the line of compression, as the member is stressed.

**Keywords:** anatomy, bones, compression, extension, fractures
TEEVAN, William Frederic, 1834-1887, English surgeon
**Sources:** Friel, J. P. 1974; Landau, S. I. 1986.

## TELEOLOGICAL LAW

A relationship where the determining event succeeds the determined event. The relationship is sometimes stated as *where an event of a certain sort in a system is determined by a later event of a certain sort in the same system.* An event is directed toward an end or shaped by a purpose as applied to natural processes or as by nature as a whole claimed by *devine providence* as opposed to mechanical determinism.

**Keywords:** causes, event, natural
**Sources:** Braithwaite, R. B. 1953; Flew, A. 1984.
*See also* BIOTIC; MNEMIC; SYLLOGISM

## TEMPERATURE OF MIXTURES, LAW OF—SEE BLACK

## TERZAGHI PRINCIPLE (1948)

When a rock is subjected to a stress, it is opposed by the fluid pressure of pores in the rock.

**Keywords:** fluid, pores, rock, stress
TERZAGHI, Karl Anton von, 1883-1961, Austrian civil engineer
**Sources:** Bothamley, J. 1993; Scott, J. S. 1993.
*See also* SOIL SETTLEMENT

## TEULON—SEE GIRAUD-TEULON

## THERMAL RADIATION—SEE STEFAN-BOLTZMANN

## THERMOCHEMICAL LAWS

1. *First Thermochemical Law (1780)*
   "The quantity of heat required to decompose a compound into its elements is equal in magnitude to the quantity of heat evolved when the same compound is formed from its elements." The sign of the heat, of course, will not be the same in each case. This law is a corollary of the first law of thermodynamics, or the law of conservation of energy. Based on work of A. Lavoisier and P. Laplace.

   **Keywords:** decomposition, energy, evolved, formation, heat, thermodynamics
   LAVOISIER, Antoine Laurent, 1743-1794, French chemist
   LAPLACE, Pierre Simon, 1749-1827, French astronomer and mathematician
   **Source:** Thompson, J. J. 1967.
   *See also* CONSERVATION; LAVOISIER; THERMODYNAMICS (FIRST LAW)
2. *Second Thermochemical Law (1840); or Constant Heat Summation, Law of, or Hess Law*
   The heat content change involved in the conversion of one chemical compound to another is independent of the way in which the conversion is carried out. The change in heat content accompanying a chemical reaction is independent of the pathway between the initial and final states.

   **Keywords:** chemical compound, conversion, final, initial, reaction, state
   **Source:** Thompson, J. J. 1967.
   *See also* CONSTANT HEAT SUMMATION; HESS

## THERMOCOUPLES, FUNDAMENTAL LAWS OF

1. *Homogeneous Circuit, Law of.* An electric current cannot be sustained in a circuit of a single homogeneous metal; however, it may vary in sections of the circuit by the application of heat alone.
2. *Intermediate Metals, Law of.* If in any circuit of solid conductors the temperature is uniform from any point P through all conducting matter to a point Q, the algebraic sum of the thermoelectromotive forces in the entire circuit is independent of the intermediate matter and is the same as if P and Q were put in contact.
3. *Successive or Intermediate Temperature, Law of.* The thermal electromotive force developed by any thermocouple of homogeneous metals with its two junctions at any two temperatures is the algebraic sum of the voltage of the thermocouple with one junction at $T_1$ and the other at any other temperature, and the voltage of the same thermocouple with its junctions at other temperatures (Fig. T.2).

**Keywords:** alloys, circuit, current, junction

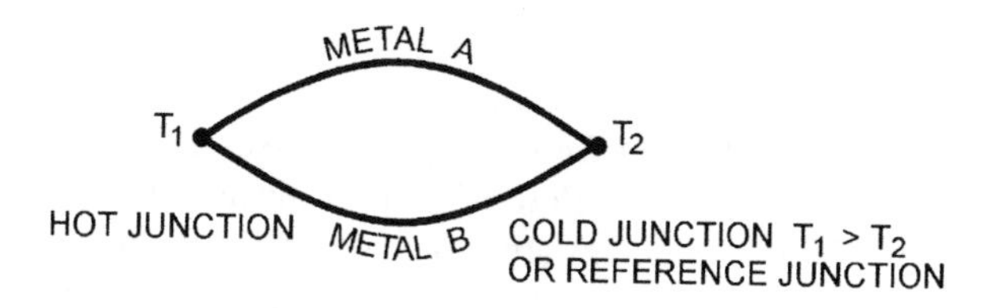


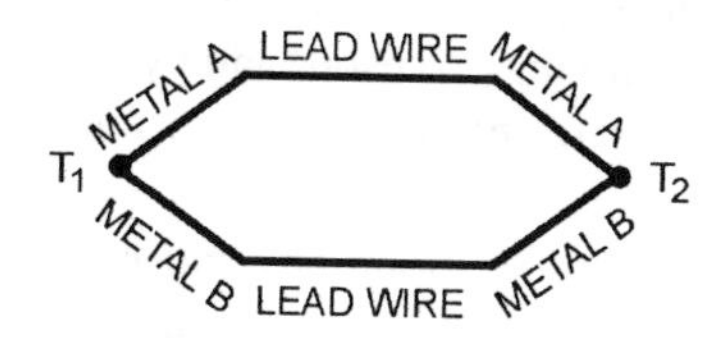


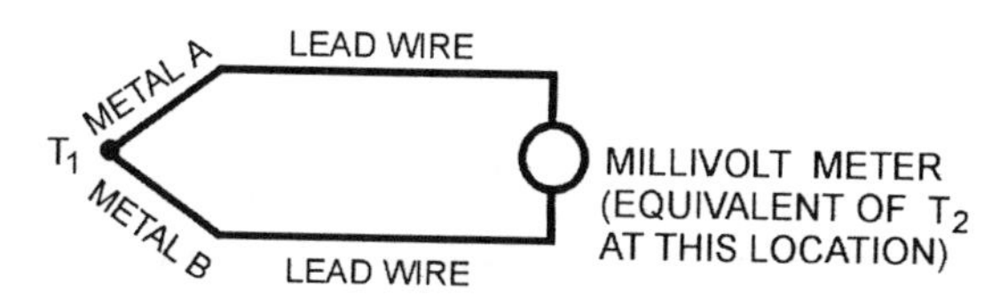


**Figure T.2** Thermocouple.

**Sources:** Considine, D. H. 1995; Parker, S. P. 1989.
*See also* INTERMEDIATE METALS; PELTIER; SEEBECK; THOMSON (KELVIN)

## THERMODYNAMICS, LAWS OF (1847, 1850, 1851, 1906)

The First and Second Laws were stated by R. Clausius in 1850, based on previous work, and were further developed by W. Thomson in 1851.

1. *Zeroth Law of Thermodynamics.* For systems in equilibrium, there is an intrinsic property: internal energy. Any two bodies or systems in equilibrium with a third body are in equilibrium with each other. A function of the state of a substance that takes on the same value for all substances in thermal equilibrium, which is the temperature. For closed systems, changes in the internal energy are:

$$dU = dQ - dW$$

where $dU$ = change internal energy
$dQ$ = heat transferred to system
$W$ = external work

2. *First Law of Thermodynamics (1842*, 1847).* The total energy change of any system together with its surroundings is zero; also called the *Law of Conservation of Energy.* The energy, including that equivalent to mass, of the universe is constant. The First law was expressed by H. Helmholtz and R. Clausius, was based on work by J. von Mayer (1842), and is an extension of the work of J. Joule. The statement of the first law is:

$$\Delta U = Q + W$$

where $\Delta U$ = the change in internal energy of the system
$Q$ = heat absorbed by the system
$W$ = work done on the system

3. *Second Law of Thermodynamics (1850).* A general law of the natural tendency in which the entropy of the universe and of systems in the universe is tending to a maximum, also called *Law of Entropy.* All processes in nature tend to occur with an increase in entropy; the flow of heat is always from higher to lower temperature. Not all forms of energy are equally interchangeable, with other forms of energy tending to go to heat. The Carnot theorem, $\Delta S \geq 0$, published in 1824, provides a working equation embodying the principles of the second law, which was expressed by Lord Kelvin (William Thomson) and by R. Clausius, who coined the word *entropy.* L. Boltzmann provided the statistical foundation of the second law (1877).
4. *Third Law of Thermodynamics (1906).* Solutions and gases are excluded from the third law. The Nernst heat theorem, also identified as the third law of thermodynamics, was extended by Planck by adding the postulate that the absolute entropy of a pure solid or a pure liquid approaches zero at 0 K:

$$\lim_{T \to 0} S \to 0$$

The entropy is related to thermodynamic probability by:

$$S = k \ln W$$

where S = entropy
k = Boltzmann constant
W = statistical probability

Thus, the more random the molecules are arranged, the greater the values of W and S. For a completely ordered system, W = 1 and S = 1. An exception is for a crystalline structure, for which quantum theory shows that the entropy at 0° abs. is not zero, because the crystal may exist in more than one state and have entropy residues from nuclear spin.

**Keywords:** conservation, energy, entropy, equilibrium, heat, temperature, thermal, work

BOLTZMANN, Ludwig, 1844-1906, German physicist

CARNOT, Nicolas Leonard Sadi, 1796-1832, French physicist

CLAUSIUS, Rudolf Julius Emmanuel, 1822-1888, German Polish physicist and engineer

HELMHOLTZ, Hermann, 1821-1894, German scientist, one of the first to state the first Law of the law of the conservation of energy

JOULE, James Prescott, 1818-1889, English physicist

KELVIN, Lord (William Thomson), 1824-1907, Scottish mathematician and physicist who formulated the Second Law

MAYER, Julius Robert von, 1814-1878, German physician and physicist

NERNST, Walter Hermann, 1864-1941, Polish German physical chemist; Nobel prize, 1920, chemistry

**Sources:** Asimov, I. 1972; Barrow, J. D. 1991; Denigh, K. 1961; Harmon, P. M. 1952; James, A. M. 1976; *Layton, E. R. and Lienhard, J. H. 1988; Kirk-Othmer. 1985; Tver, D. F. 1981; Webster's Biographical Dictionary. 1959. 1995.

*See also* CARNOT; CLAUSIUS; CONSERVATION OF ENERGY; JOULE; MAYER; NERNST; ONSAGER

## THERMOELECTRIC EFFECT—SEE SEEBECK; THOMSON EFFECT

## THERMONEUTRALITY, LAW OF

No thermal effect of heat change results when dilute solutions of neutral salts are mixed in the absence of precipitation of the salts.

**Keywords:** heat, precipitation, salts, solutions, thermal
**Source:** Honig, J. M. 1953.
*See also* HESS

## THIELE NUMBER OR MODULUS, $m_T$

A dimensionless group used to represent diffusion in porous catalysts:

$$m_T = Q^{1/2}U^{1/2}L/k^{1/2}t^{1/2}$$

where Q = heat liberated per unit mass
U = reaction rate
L = characteristic length
k = thermal conductivity
t = temperature

The Thiele number is the (Damköhler number II)$^{1/2}$.

**Keywords:** catalysts, diffusion, thermal, porous
THIELE, Fredrich K. Johannes, 1865-1918, German chemist
**Sources:** Bolz, R. E. and Tuve, G. L. 1970; NUC; Pelletier, P. A. 1994; Parker, S. P. 1989; Potter, J. H. 1967.
*See also* DAMKÖHLER NUMBER II

## THINNING LAW—SEE YODA

## THIRD LAW OF THERMODYNAMICS—SEE THERMODYNAMICS

## THOMA LAWS (MEDICAL)

The development of blood vessels is governed by forces acting on the walls of the blood vessels as follows:

1. An increase in velocity of flow causes dilation of the space in the blood vessel.
2. An increase in lateral pressure causes the blood vessel wall to become thicker.
3. An increase in pressure at the end of the blood vessel leads to formation of new capillaries.

**Keywords:** blood, capillaries, dilation, flow, pressure, vessels
THOMA, Richard, 1847-1923, German histologist
**Source:** Stedman, T. L. 1976.

## THOMA NUMBER, $\sigma_T$

A dimensionless group in liquid pump cavitation that relates the pressure margin above cavitation to the pressure rise in the pump (or net positive suction heat to the total head):

$$\sigma_T = (p_1 - p_v)/(p_2 - p_1)$$

where $p_1$ = total pressure at pump inlet
$p_2$ = total pressure at pump outlet
$p_v$ = the vapor pressure

**Keywords:** cavitation, pressure, pump
THOMA, Dietrich, 1881-1943, German engineer
**Sources:** Bolz, R, E. and Tuve, G. L. 1970; Land, N. S. 1972; Parker, S. P. 1992; Potter, J. H. 1967.
*See also* CAVITATION

THOMPSON—some references use William Thompson,* others use William Thomson** for Lord Kelvin (1824-1907).
**Sources:** **Asimov, I. 1972; *Hewitt, G. F. 11th Int. Heat Transfer Conference (Korea), August 1998; **Layton, E. T. and Lienhard, J. H. 1988; *Parker, S. P. 1989.
*See also* KELVIN; JOULE-THOMSON (THOMPSON)

### THOMPSON, LAW OF (1798)

Thompson pursued the concept that the heat produced when boring metal and that heat was a form of motion, and not a fluid as proposed by others. He developed what was referred to as the *mechanical equivalent of heat.*

**Keywords:** boring, fluid, heat, mechanical equivalent, motion
THOMPSON, Benjamin (Count Rumford), 1753-1814, American English physicist
**Sources:** Asimov, I. 1976; Chemical Heritage 16(1):25, 1998.

### THOMPSON NUMBER, Th′ OR $N_{Th'}$—SEE MARANGONI (CELLULAR CONVECTION); THOMSON NUMBER

### THOMSEN—SEE BERTHELOT-THOMSEN

### THOMSON—SEE JOULE-THOMSON (THOMPSON)

### THOMSON (THOMPSON) EFFECT (KELVIN EFFECT) (1854)

When an electric current flows through a conductor, the ends of which are at different temperatures, heat is evolved at a rate approximately proportional to the current and temperature difference. If either the current or the temperature difference is reversed, heat is absorbed.

**Keywords:** conductor, current, heat, rate, temperature
THOMSON, William, 1824-1907, Scottish physicist; Nobel prize, 1906, physics
**Sources:** Besancon, R. M. 1974; Daintith, J. 1988; Isaacs, A. 1966; Mandel, S. 1972.
*See also* KELVIN; PELTIER; SEEBECK; THERMOELECTRIC

### THOMSON NUMBER, Th OR $N_{Th}$

A dimensionless group used for fluid flow that relates characteristic time and characteristic length:

$$Th = tV/L$$

where t = characteristic time
V = velocity
L = characteristic length

Also known as the Strouhal number

**Keywords:** characteristic time, fluid flow
THOMSON, Joseph John, 1856-1940, English physicist; Nobel prize, 1906, physics

**Sources:** Bolz, R. E. and Tuve, G. L. 1970; Land, N. S. 1972; Parker, S. P. 1992; Potter, J. H. 1967.
*See also* STROUHAL

## THOMSON-NERNST RULE (LAW); OR NERNST-THOMSON LAW (NERNST-THOMPSON*)

The electrical forces that hold molecules together will be greatly reduced when the molecule is surrounded by a substance with a large inductive capacity, and by solvents such that those with the greatest dielectric constants exhibit the greatest dissociating powers. Thomson discovered electrolysis in 1897.

**Keywords:** dielectric, dissociating, electricity, inductive, molecules, solvents
THOMSON, Joseph John, 1856-1940, English physicist; Nobel prize, 1906, physics
NERNST, Hermann Walther, 1864-1941, German physical chemist
**Sources:** *Considine, D. M. 1976. 1995; Gray, H. J. and Isaacs, A. 1975.
*See also* NERNST

## THOMSON THERMOELECTRIC EFFECT (LAW)

The magnitude and direction of the potential gradient along an electric conductor varies with the temperature gradient.

**Keywords:** conductor, electrical, gradient, potential, temperature
THOMSON, Joseph John, 1856-1940, English physicist; Nobel prize, 1906, physics
**Source:** Hodgman, C. D. 1952.
Also see THERMOELECTRIC

## THOMSON-WHIDDINGTON LAW

The kinetic energy of a cathode particle at a distance along its path is:

$$e^2V^2 - e^2V_x^2 = ax$$

where V = potential on the tube
$V_x$ = potential that would give the particle the energy at x-distance along plate
a = a constant for the material of the target

**Keywords:** cathode, electronics, kinetic energy, tube
THOMSON, Joseph John, 1856-1940, English physicist; Nobel prize, 1906, physics
WHIDDINGTON, Richard, 1885-1970, British physicist
**Source:** Thewlis, J. 1961-1964.
*See also* WHIDDINGTON

## THORNDIKE LAWS OF LEARNING

1. *Law of Associated Shifting.* Incorporates conditioning into the law of effect. Stimuli associated with the original stimulus-response could come, in time, to participate in the initiation of the response, even in the absence of the original stimulus.
2. *Law of Belongingness.* This law suggests that some kinds of material are more readily learnable than others, in that some things go together more naturally than others.
3. *Law of Effect (1989, 1911).* The stimulus-response connection is strengthened when the response is followed by a reward or satisfier. Elimination of incorrect

reactions to the occurrence of annoyances reinforce desirable responses. One learns or retains responses followed by satisfiers and refrained from responses followed by annoyances. (This is a statement of a hedonic response.) In 1931, he modified the law of effect to drop the negative aspect, such that he believed that punishment had to effect an elimination of incorrect responses.

4. *Law of Exercise.* One needs to attend to the learning process if one is to learn. A learner has to be set to respond to specific stimuli in a situation. Each successful trial adds strength to the connection. The law of exercise adds support to the idea of practices in drill to learn a thing.

**Keywords:** cause, drill, effect, hedonic, stimulus, response
THORNDIKE, Edward Lee, 1874-1949, American educator and psychologist
**Sources:** Corsini, R. J. 1987; Harre, R. and Lamb, R. 1983; Wolman, B. B. 1989.
*See also* EXERCISE; FILIAL REGRESSION; PAVLOV; SKINNER; READINESS; USE

## THOUGHT, LAWS OF

Three laws exemplify the way we think in a way that is not arbitrary.

1. *Law of Contradiction.* A thing cannot be both true and not true:

$$[\sim (p \vee \sim p)]$$

2. *Law of Excluded Middle.* A thing must either be true or not be true:

$$[p \vee (\sim p)]$$

3. *Law of Identity.* If a thing is true, then it is true, symbolically represented by:

$$[p \rightarrow p]$$

In addition, the *Law of Sufficient Reason* is sometimes included as one of the laws of thought.

**Keywords:** logic, not true, think, sufficient reason, true
**Sources:** Angeles, P. A. 1981; Gibson, C. 1981; Honig, J. M. 1953; James, R. C. and James, G. 1968; Karush, W. 1989.
*See also* ARISTOTLE; BOOLEAN; CONTRADICTION; EXCLUDED MIDDLE; IDENTITY; SUFFICIENT REASON; TANGENTS; TAUTOLOGY; TRUTH

## THREE-FOURTH POWER LAW—SEE ALLOMETRIC (BIOLOGY); KLEIBER

## THREE-HALVES POWER LAW—SEE IONIZATION

## THREE-HALVES POWER SCALING LAW—SEE ALLOMETRIC SCALING

## THREE-HALVES THINNING LAW—SEE YODA

## THRING NUMBER, Thr OR $N_{Thr}$

A dimensionless group applicable to radiation heat transfer that relates bulk heat transfer and radiative heat transfer:

$$Thr = \rho\, c_p V / e\, \sigma\, T^3$$

where $\rho$ = mass density
$c_p$ = specific heat at constant pressure
V = velocity
e = surface emissivity

σ = Stefan-Boltzmann constant
T = temperature

The Thring number is also known as the Boltzmann number.

**Keywords:** bulk, heat transfer, radiation
THRING, Meredith Woodbridge, twentieth century (b. 1915), English thermal engineer
**Sources:** Bolz, R. E. and Tuve, G. L. 1970; Land, N. S. 1972; NUC; Parker, S. P. 1992; Potter, J. H. 1967.
*See also* BOLTZMANN

## THROMBOSIS—SEE LANCEREAUX

## THURSTONE LAW (THEORY) OF COMPARATIVE JUDGMENT (1927)

A model based on pair comparisons that assumes that when a subject is asked to express a dominance judgment about two stimuli, the outcome is determined by a comparison of the values for two random variables. If $X_i$ is the random variable associated with stimulus object $o_j$, and $X_j$ is the random variable associated with stimulus object $o_j$, then the subject is said to judge $o_i$ as having more of the attribute of interest than $o_j$ if $X_i$ is greater than $X_j$.

**Keywords:** dominance, judgment, random, stimuli, variable
THURSTONE, Louis Leon, 1887-1955, American educational psychologist
**Sources:** Kotz, S. and Johnson, N. 1982. 1988; Wolman, B. B. 1989.
*See also* CATEGORICAL JUDGMENT; FECHNER; READINESS; THORNDIKE

## TIME REVERSAL, LAW OF

The laws describing the interactions of elementary particles do not depend on the direction of time. Another statement in time reversal is a transformation operating in time, t, that can be replaced with –t.

**Keywords:** interactions, particles, time
**Source:** Morris, C. G. 1992.
*See also* CONSERVATION OF SYMMETRIES

## TIMES, LAW OF (HEAT FLOW)

The times required for any two points to reach the same temperature are proportional to the squares of their distances from the boundary plane.

**Keywords:** boundary, heat, temperature, time

## TITIUS LAW (1772); OR TITIUS-BODE LAW (1772); OR BODE LAW (1772)

This is not a law in the usual sense, but a relation for representing the distances of the planets to the sun. It is based on the series 0, 3, 6, 12, 24, 48, 96, 194..., add 4 to each term and divide by 10, resulting in the series 0.4, 0.7, 1.0, 1.6, 2.8, 5.2, 10, 19.6, in which the distance of the sun to the Earth is 1.0.

**Keywords:** distance, planets, sun
TITIUS, Johann Daniel, 1729-1796, Polish astronomer
BODE, Johann Elert, 1747-1826, German astronomer
**Sources:** American Scientist, vol. 83, p. 116, March-April, 1995; Encyclopaedia Britannica. 1961; Science 183, p. 65, 11 Jan. 1973.
*See also* BODE

## TOLERANCE, LAW OF—SEE SHELFORD; LIEBIG

## TOLERATION, LAW OF; OR TOLERANCE, LAW OF (ECOLOGY)

The distribution of a species of organisms under steady-state conditions is limited by its tolerance to the fluctuation of one or more physical or chemical factors that fall above or below the levels tolerated by the species.

**Keywords:** distribution, fluctuation, tolerance
**Sources:** Art, H. W. 1993; Gray, P. 1967; Lincoln, R. J. 1982.
*See also* SHELFORD

## TOMS NUMBER, To OR $N_{To}$ (1965)

A dimensionless group representing merit of an airplane that relates fuel weight and air drag:

$$To = Q/\rho V^3 L$$

where Q = fuel rate (weight/time)
$\rho$ = mass density
V = velocity
L = characteristic length

**Keywords:** airplane, drag, fuel, merit, weight
TOMS, Charles Frederick, twentieth century, English aerodynamicist
**Sources:** J. Royal Aero. Soc., Feb. 1965; Land, N. S. 1972; NUC; Wolko, H. S. 1981.

## TOOTHED GEARING, FUNDAMENTAL LAW OF

Two gears with centers at $O_1$ and $O_2$, with angular velocities of $\omega_1$ and $\omega_2$, have teeth that are in contact at K, the sides of the teeth must be shaped so that the normal drawn through the point of contact will at all times pass through the pitch point P on the two pitch circles for the two gears (Fig. T.3).

**Keywords:** gears, pitch, teeth
**Sources:** Baumeister, T. 1966; Rothbart, H. A. 1985.

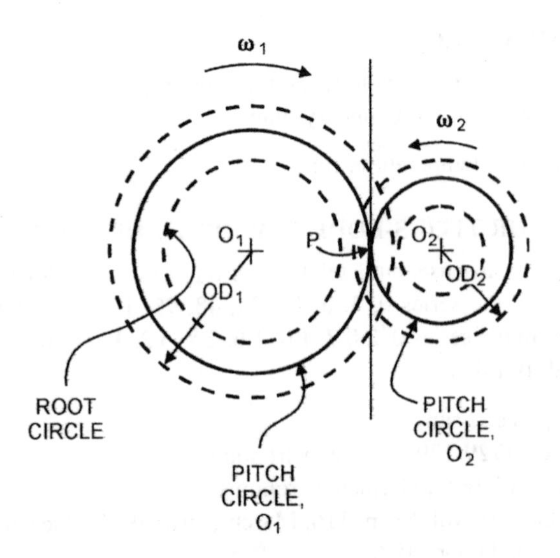


**Figure T.3** Toothed gearing (OD = outer diameter).

## TORRICELLI LAW; TORRICELLI EFFLUX VELOCITY LAW; TORRICELLI LAW OF EFFLUX (1643)

The velocity of a stream coming from a container is the same as a free-falling body would have on reaching the orifice after having started from rest at the surface level, not considering frictional losses, and is represented by:

$$V = (2gh)^{1/2} \text{ or } V^2 = 2gh$$

and the quantity of efflux is theoretically:

$$Q = AV$$

where V = velocity
g = acceleration due to gravity
h = height from surface level to orifice vena contracta
A = cross sectional area of the orifice
Q = quantity of flow from the orifice

In practice, a constant (less than 1.0) is included to account for frictional losses. E. Torricelli invented the barometer in 1643 (1 torr = 1 mm Hg).

**Keywords:** efflux, falling body, fluid, mechanics, velocity
TORRICELLI, Evangelista, 1608-1647, Italian physicist
**Sources:** Asimov, I. 1972; Ballentyne, D. W. G. and Lovett, D. R. 1972. 1980; Mandel, S. 1972; Parker, S. P. 1989.

## TORSION FORCE ON A METAL WIRE—SEE COULOMB

## TOYNBEE LAW (MEDICAL)

Where brain disease occurs otitis (inflammation of the ear), the cerebellum and laterial sinuses are affected from the mastoid and the cerebrum from the tympanic (eardrum) roof.

**Keywords:** brain, cerebellum, medical, sinus
TOYNBEE, Joseph, 1815-1866, British otologist
**Sources:** Friel, J. P. 1974; Landau, S. I. 1986; Stedman, T. L. 1976.
*See also* GULL-TOYNBEE

## TRANSFORMABILITY OF ENERGY, LAW OF (1853)

All different kinds of energy in the universe are mutually convertible. The law of conservation was already known, for example, the sum of the actual (kinetic) and potential energies is unchangeable. The term *energy* at that time was applicable to ordinary motion and mechanical power (work), chemical action, heat, light, electricity and magnetism, as expressed by Rankine.

**Keywords:** conservation, energy, motion, sum, work
RANKINE, William John Macquorn, 1820-1872, English engineer
**Sources:** Harman, P. M. 1982; Tait, P. M. 1899; Thomsen, J. 1908.

## TRANSFORMATION LAWS

Numerous transformation methods exist for a variety of situations. One of the most used transformation is for values in a coordinate system to a vector system, or vice-versa. An excellent summary of these, particularly for tensors and vectors, is provided in the reference.*

**Keywords:** mathematics, tensor, transformed, vector
**Sources:** *James, R. C. and James, G. 1968; Newman, J. R. 1956 (partial)

## TRANSFORMATION LAWS FOR FIELD STRENGTH

The total charge of field strength is an invariant, if the current vector is convective and if the charge has the velocity u in the system:

$$\rho^* u^* = \rho(u_1 - v)/(1 - \beta^2)^{1/2}$$

If the charge is at rest in system $\Sigma$:

$$\rho^* = \rho/(1 - \beta^2)^{1/2}$$

where $\rho^* = u^*$ = velocity of charge in the system $\dot{\Sigma}$
$u$ = velocity of charge in the system $\Sigma$
$v$ = velocity of the system $\dot{\Sigma}$ with respect to system $\Sigma$
$\beta$ = magnetic field

**Keywords:** charge, field strength, velocity
**Sources:** Menzel, D. H. 1960; Thewlis, J. 1961-1964 (partial).

## TRANSIENT CREEP—SEE ANDRADE

## TRANSITIVE—SEE BOOLEAN ALGEBRA

## TRANSONIC SIMILARITY LAW FOR GASES

In many flows of gases where the velocity perturbations are small (it is possible to show that) the pressure, lift, drag, and related factors depend on the various flow parameters in a simple manner, which for two-dimensional flow the pressure coefficient is:

$$c_p = (p - p_\infty)/1/2\ \rho_\infty\ u^2$$

where $c_p$ = pressure coefficient
$p$ = pressure
$\rho$ = density
$u$ = free stream velocity

This holds in general for subsonic, transonic, and supersonic flow, with modifications (Fig. T.4).

**Keywords:** flow, gases, perturbations, simple, two-dimensional
**Sources:** Gray, D. E. 1972; Parker, S. P. 1992.

## TRAPEZOIDAL RULE—SEE AREA

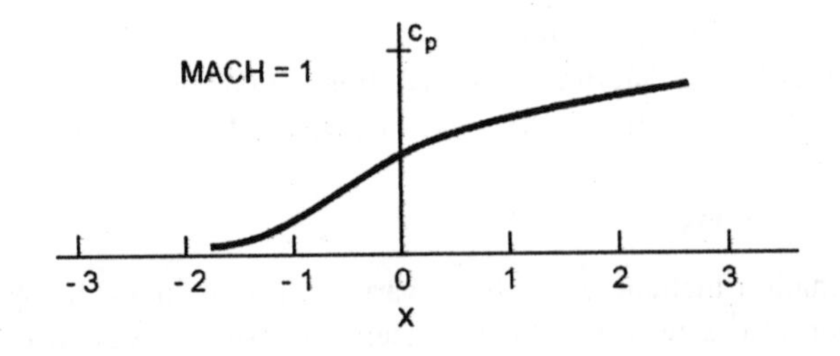


**Figure T.4** Transonic theory.

### TRAUBE RULE (1891)

For a homologous series of fatty acids, concentrated in a dilute water solution, there is a threefold decrease for each methylene group in the series.

**Keywords:** fatty acids, homologous, methylene, solution
TRAUBE, Isidor, 1860-1943, German physical chemist
**Sources:** Bothamley, J. 1993; Glasstone, S. 1946; Webster's Biographical Dictionary. 1959. 1995.

### TRIADS—SEE DÖBEREINER

### TRIBONDEAU—SEE BERGONIÉ-TRIBONDEAU

### TROUTON LAW (1884) OR RULE

The molar heat of vaporization is the molecular weight times the latent heat of vaporization divided by the absolute temperature and is equal to a constant, approximately 23, if the latent heat is expressed in calories:

$$M\ LH_v/T = \text{constant}$$

where $M$ = molecular weight
$LH_v$ = latent heat
$T$ = absolute temperature

This law was generalized by R. Pictet in 1876, rediscovered by W. Ramsay in 1877, and published by and named after F. T. Trouton in 1884 (Fig. T.5).

**Keywords:** latent heat, molecular, temperature, vaporization
TROUTON, Frederick Thomas, 1863-1922, Irish physicist
PICTET, Raoul Pierre, 1846-1929, Swiss chemist (Glasstone has A. Pictet*)
RAMSAY, Sir William, 1852-1916, Scottish chemist
**Sources:** Asimov, I. 1976; Barr, E. S. 1973; Ballentyne, D. W. G. and Walker, L. E. Q. 1959; Fisher, D. J. 1988. 1991; *Glasstone, S. 1946; Mandel, S. 1972; Parker, S. P. 1992; Ulicky, L. and Kemp, T. J. 1992.
*See also* GULDBERG; RICHARDS; WATT

### TRUNCATION NUMBER, $\tau$ OR $N_\tau$

A dimensionless group used in viscous flow that relates the shear stress and normal stress:

$$\tau = \mu\ \alpha/p$$

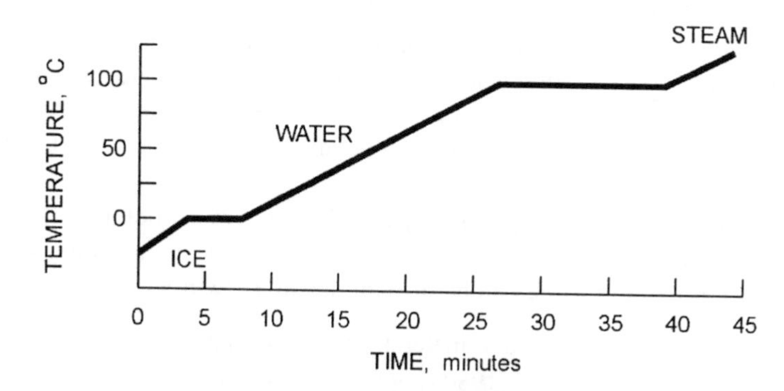


**Figure T.5** Change in temperature when crushed ice is heated.

where $\mu$ = absolute viscosity
$\alpha$ = shear strain rate
p = pressure

The Truncation number is also known as the Hersey number.

**Keywords:** flow, stress, viscous
**Sources:** Bolz, R. E. and Tuve, G. L. 1970; Land, N. S. 1972; Parker, S. P. 1992; Potter, J. H. 1967.
*See also* HERSEY

## TURING MODEL (1936)

"A model used to describe procedures such that any computation that can be described by a Turing machine can also be executed by a computer. The model is analogous to a tape drive with input divided into cells, each of which can contain exactly one symbol, a tape head that scans one cell at a time, and a finite control. The Turing machine can make a move based on the state of the finite control and the symbol in that cell currently scanned by the tape head." The Turing machine allows for unbounded external storage in addition to the finite information represented by the control of the system. There are many modifications of the model.

**Keywords:** cells, computation, computer, control, symbol
TURING, Alan Mathison, 1912-1954, English mathematician
**Sources:** Science vol. 268:545, 28 April 1995; Watters, C. 1992.

## TWIN LAWS IN FELDSPARS

Twinning is frequently present in feldspars, a group of silicate minerals. Twinning can be simple or multiple. The simple twinning is represented by the following laws:

1. *Carlsbad (Karlsbad) Twin Law.* After (100) or (001). A twin law in feldspar, especially orthoclase, that defines a penetration twin in which the twin axis is the crystallographic axis and the composition surface is irregular.
2. *Baveno Twin Law.* After (021). An uncommon twin law in feldspars, in which the twin plane and composition surface are (021). A Baveno twin usually consists of two individuals.
3. *Manebach Twin Law.* After (001). A twin law in the feldspars, both monoclinic and triclinic, usually simple, with the twin plane and composition plane of (001).
4. *Acline Twin Law.* A complex twin law in triclinic feldspar according to which the twin axis is perpendicular to (001) and the composite plane is (001).
5. *Albite (Albitie) Twin Law.* After (010). A twin law in triclinic feldspars in which the twin plane and the composition plane are (010). An albite twin is usually multiple and lamellar, and shows fine striations on the (001) cleavage plane. Albite and Pericline laws cannot occur in monoclinic feldspars and lead to multiple twinning.
6. *Pericline Twin Law.* After (010). A parallel twin law in triclinic feldspars, in which the twin axis is the crystallographic b-axis and the composition surface is the rhombic section. It occurs alone or with the albite twin law.

**Keywords:** complex, crystals, feldspars, material, minerals, multiple, simple
**Sources:** Bates, R. L. and Jackson, J. 1980; Lapedes, D. N. 1978.
*See also* MALLARD

## TWO-DIMENSIONAL NORMAL—SEE BIVARIATE

## TYCHO BRAHE LAW—SEE BRAHE (TYCHO) LAW

## TYNDALL EFFECT

When particles in a colloidal suspension are larger than the wavelength of visible light, there is a visible scattering of light as the particles pass through the medium containing discontinuities, particles.

**Keywords:** colloidal, particles, scattering, suspension, wavelength
TYNDALL, John, 1820-1893, English physicist
**Sources:** Mandel, S. 1972; Parker, S. P. 1989.
*See also* DOPPLER; RAMAN; RAYLEIGH

# U

## UNCERTAINTY PRINCIPLE (1927)—SEE HEISENBERG

## UNEQUAL SLOPES, LAW OF (1877)

As first noted by Gilbert, a stream flowing down the steeper slope of an asymmetric divide or hill erodes its valley more rapidly than one flowing down the gentler slope. The result is that the crest of the divide migrates away from the more actively eroding stream toward the less actively eroding one.

**Keywords:** crest of hill, divide, eroding, slope, stream
GILBERT, Grove Karl, 1843-1918, American geologist
**Source:** Bates, R. L. and Jackson, J. A. 1980.
*See also* STREAM GRADIENTS

## UNIFIED FIELD THEORY (1950)—SEE EINSTEIN; UNIVERSAL UNIFIED FIELD

## UNIFORM ATOMIC PLAN, LAW OF

The pattern of electronic structure of every atom is at first a repetition of the pattern of all atoms having fewer electrons.

**Keywords:** atoms, electronic, electrons, structure
**Source:** Bennett, H. 1962.
*See also* MENDELEEV

## UNIFORM CHANGE, LAW OF; OR LYELL THESIS

The past changes of the Earth's surface are adequately explained by processes now in operation.

**Keywords:** Earth, evolution, geology, processes
LYELL, Sir Charles, 1797-1875, Scottish geologist
**Source:** Krauskopf, K. B. 1959.

## UNIFORMITY, LAW OF

A mathematical approach to the determination of optimum results in fluorograph, where the factors causing a lack of sharpness in the normal roentgenogram of objects in motion and specifically states that the optimum conditions are reached when the following factors are equal:

$$U_m = U_g = U_s$$

where $U_m$ = sharpness resulting from movement
$U_g$ = geometrical lack of sharpness
$U_s$ = lack of sharpness caused by screen factors

**Keywords:** fluorograph, optimum, roentgenogram, sharpness
**Source:** Morris, C. G. 1992.

## UNIFORM SENIORITY, LAW OF

In evaluating joint life insurance policies, the difference between the age that can be used in the computation and the lesser of the unequal two ages is the same for the same difference of the unequal, regardless of the actual ages.

**Keywords:** age, computation, insurance, joint, life, unequal
**Source:** James, R. C. and James, G. 1968.

## UNITY—SEE BOOLEAN

## UNIVERSAL AFFECT,* LAW OF

In psychology, the promise that every idea, thought, or object, no matter how apparently minor or neutral, possesses a distinct quantum of affect.

**Keywords:** idea, thought, effect
**Source:** *Gennaro, A. R. 1979.

## UNIVERSAL CREATION OF MASS-ENERGY, LAW OF

Derived from the mechanics of the no-mass infinite motion that composes the infinity or the absolute space, related to the Unified Field law, where m, n = 1, becomes:

$$\Sigma(V_1 + V_2 + V_3) = 0$$

**Keywords:** energy, field, mass, physics
**Sources:** Thewlis, J. 1961-1964 (partial); World Book. 1990.
*See also* UNIVERSAL UNIFIED FIELD LAW

## UNIVERSAL GAS EQUATION

The combination of the Boyle law, the Charles law, and the pressure law expressed as:

$$pV = nRT$$

where p = pressure
V = volume
n = amount of gas in specimen
R = gas constant
T = absolute temperature

**Keywords:** gas, pressure, temperature, volume
**Source:** Isaacs, A. 1996
*See also* BOYD; CHARLES; GAS

## UNIVERSAL GRAVITATION, LAW OF

Every mass particle in the universe attracts every other mass particle with a force directly proportional to the product of the two masses and inversely proportional to the square of the distance between them, in direction on a line joining the particles. The law applies to particles, not to bodies of finite size.

**Keywords:** attracts, distance, mass, particle, product
**Sources:** Bates, R. L. and Jackson, J. A. 1980; Fairbridge, R. W. 1967; Mandel, S. 1972.
See also NEWTON

## UNIVERSAL LAW OF THE WALL—SEE WALL

## UNIVERSAL UNIFIED FIELD LAW

The existing states and behavior of the structural frames of all mass-energy brings forth a universal, unifying, basic, and prime law that govern all physical creation and all mechanics in motion. All forms of expression of mass-energy, as mass at relative rest or in motion, as energy particle, or as energy dynamics of motion, are regulated by the relationship:

$$\Sigma[n_m(V_1 + V_2 + V_3)] = 0$$

$$m = 1, 2, 3 \ (\rightarrow \infty)$$

$$n = 1, 2, 3 \ (n \neq 0)$$

**Keywords:** energy, mass, mechanics, motion
**Source:** World Book. 1990.

## UNRELIABILITY—SEE EXPONENTIAL FAILURE LAW; PRODUCT LAW OF UNRELIABILITY

## URBACH LAW

Absorption below the absorption edge of spectra of insulation decreases exponentially with energy. One explanation is that there is an exciton interaction with fluctuating electric microfields. This is an empirical law.

**Keywords:** absorption, energy, exciton, exponentially, spectra
URBACH, (possibly URBACH, Erich, 1893-1946, Med. BHMT)
**Source:** Besancon, R. M. 1974.

## URBAN CONCENTRATION, LAW OF (1923)

An empirical law attributed to F. Auerbach, that if the cities of a given country are arranged in order of size, the product of the population and rank is approximately constant, with a hyperbolic curve resulting when the rank of population is plotted against population, or producing a straight line when plotted on log-log coordinates.

**Keywords:** cities, hyperbolic, log-log, population
AUERBACH, F., published in German in 1920s
**Sources:** Lotka, A. J. 1956; NUC.

## USANOVICH THEORY (1939); OR POSITIVE-NEGATIVE THEORY

Acids are substances that form salts with bases, give up cations, and add themselves to anions and to free electrons. Bases give up anions or electrons and add themselves to cations.

**Keywords:** acids, anions, bases, cations, electrons
USANOVICH, M. Biographical information not located.
**Sources:** Parker, S. 1987. 1989. 1992.

## USE, LAW OF (LEARNING)

An association or function in learning is facilitated by use or practice, as stated by E. L. Thorndike.

**Keywords:** association, learning, practice, use
**Source:** Wolman, B. B. 1989.
THORNDIKE, Edward Lee, 1874-1949, American psychologist and educator
*See also* EFFECT; EXERCISE; THORNDIKE; THOUGHT

# V

## VALENSI NUMBER, V OR $N_V$ (c. 1948)

A dimensionless group applicable to oscillation of drops and bubbles:

$$V = \omega L^2 \rho / \mu$$

where $\omega$ = circular oscillation frequency
$L$ = characteristic dimension
$\rho$ = mass density
$\mu$ = dynamic viscosity

**Keywords:** bubbles, drops, oscillation
VALENSI, Jacques, twentieth century, French aerodynamicist
**Sources:** Bolz, R. E. and Tuve, G. L. 1970; NUC; Parker, S. P. 1992; Potter, J. H. 1967.

## VALLI-RITTER—SEE RITTER-VALLI

## VALSON LAW OR MODULI

"For moderate concentrations, the densities of salt solutions show constant differences for substitutes of anion or cation, i.e., the difference in density in normal solutions of potassium and ammonium salts of the same acid is 0.280 to 0.288; and that between chlorides and nitrates of the same base in normal solution is 0.0141 to 0.01015."

**Keywords:** acid, anion, cation, concentrations, densities, salt
VALSON, Claude Alfons, French chemist
**Sources:** Gillispie, C. C. 1981; Honig, J. M. 1953.
*See also* MODULI

## VAN DER HEGGE ZIJNEN COOLING LAW (1956)—SEE COOLING

## VAN DER KOLK LAW; OR SCHROEDER VAN DER KOLK

The sensory fibers of a mixed nerve are distributed to the parts moved by muscles that are stimulated by the motor fibers of the same nerves.

**Keywords:** muscles, nerves, physiology, sensory
VAN DER KOLK, Jacobus L. C. Schroeder, 1979-1862, Dutch physician
**Sources:** Dorland, W. A. 1980. 1988; Friel, J. P. 1974; Landau, S. I. 1986; Stedman, T. L. 1976.

## VAN DER WAALS EQUATION (1873) OR LAW

This equation combines the laws of Boyle and Charles, and Gay Lussac, and represents physical sorption, which consists of weaker forces of adsorption than those of chemical bonds, and includes corrections to provide a greater accuracy in the thermodynamic behavior of real gases and their deviation from the ideal gas laws. For a gram-molecule of a substance in the gaseous and liquid phase, it is represented by:

$$(p + a/v^2)\,(v - b) = R\,T$$

where p = pressure of the gas
v = absolute volume
R = gas constant
T = absolute temperature
a, b = constants

The equation takes into account the volumes and attractions of the particles and is useful at extreme temperatures and pressures. For an ideal gas, pV = RT.

**Keywords:** gases, physical chemistry, thermodynamics
VAN DER WAALS, Johannes Diderik, 1837-1923, Dutch physicist; Nobel prize, 1910, physics
**Sources:** Hampel, C. A. and Hawley, G. G. 1973; Hodgman, C. D. 1952; Mandel, S. 1972; Science vol. 220, pp. 778-794, 20 May 1983; Ulicky, L. and Kemp, T. J. 1992; Uvarov, E. and Chapman, D. 1964.
*See also* BOYLE; CHARLES; GAY LUSSAC; VAPOR PRESSURE

### VAN'T HOFF-ARRHENIUS LAW

The variation of the rate of reaction with temperature can be expressed as a linear decrease of the logarithm of the rate of reaction as the reciprocal of the absolute temperature rises.

**Keywords:** absolute, linear, logarithm, rate, reaction, temperature
VAN'T HOFF, Jacobus Henricus, 1852-1911, Dutch chemist; Nobel prize, 1901, chemistry
ARRHENIUS, Svante August, 1859-1927, Swedish chemist; Nobel prize, 1903, chemistry
**Sources:** Landau, S. 1986; Stedman, T. L. 1976.
*See also* OSMOTIC PRESSURE LAW

### VAN'T HOFF LAWS (1884)

1. In stereochemistry, all optically active substances form an asymmetrical arrangement in space, owing to their having multivalent atoms united to form different atoms or radicals.
2. The osmotic pressure of a substance in a dilute solution is the same that the same substance would exert if present in the state of an ideal gas occupying the same volume as the solution.
3. The velocity of chemical reactions increase between two- and three-fold for each 10° C rise in temperature.

Van't Hoff laws are a special case of Le Chatelier principle.

**Keywords:** atoms, dilute, osmotic, solution, velocity, volume
VAN'T HOFF, Jacobus Henricus, 1852-1911, Dutch physical chemist; Nobel prize, 1901, chemistry
**Sources:** Dox, I. et al. 1979. 1985; Friel, J. P. 1974; Thewlis, J. 1961-1964.
*See also* AVOGADRO; LE CHATELIER; $Q_{10}$ RULE; RAOULT

### VAN'T HOFF-MARIOTTE LAW

Relates the volume of cells to tonicity T, expressed in many different ways, one of which is:

$$V = w\frac{(a - aT)}{(aT + 1)} + 1$$

where V = volume of cells expressed as a fraction of $V_0$ when $V_0$ = 1, originally w = fraction of cell volume occupied by water, and the tonicity is 1 if no shrinking or swelling

a = constant, a ratio of the volume outside the cell to its volume of water in the cell at the beginning of test

T = tonicity, a measure of osmotic tension defined by the osmotic response of cells immersed in a solution

When placed in solution of higher or lower osmotic pressure than its interior, a model such as a red blood cell would shrink or swell by the transfer of water alone and would behave as an "osinometer."

**Keywords:** blood, cells, fraction, osmotic, shrinking, swelling, water

VAN'T HOFF, Jacobus Henricus, 1852-1911, Dutch physical chemist; Nobel prize, 1901, chemistry

MARIOTTE, Edme, 1620-1684, French physicist

**Sources:** Considine, D. M. 1976; Glasser, O. 1944; Lincoln, R. J. 1982.

## VAN'T HOFF PRINCIPLE (LAW) OF MOBILE EQUILIBRIUM (1884); OR RAOULT LAW

If the temperature of interacting substances in equilibrium is raised, the equilibrium concentrations of the reactions are changed so that the products of that reaction that absorb heat are increased in quantity, or if the temperature for such an equilibrium is lowered, the amount of products that evolve heat in their formation are increased.

**Keywords:** equilibrium, heat, reactions, thermal

VAN'T HOFF, Jacobus Henricus, 1852-1911, Dutch physical chemist; Nobel prize, 1901, chemistry

RAOULT, Francois Marie, 1830-1901, French physical chemist

**Sources:** Considine, D. M. 1995; Landau, S. I. 1986; Tver, D. F. 1981.

*See also* OSMOTIC PRESSURE

## VAN VALEN LAW (1974); OR LAW OF CONSTANT EXTINCTION; ALSO CALLED RED QUEEN HYPOTHESIS

For any group of related organisms sharing a common ecology, there is a constant probability of extinction of any taxonomic group per unit time. Since there is not perfect adaptation of organisms, and the environments continually change, natural selection enables organisms to maintain their adaptation.

**Keywords:** ecology, extinction, taxonomic

VAN VALEN, Leigh, twentieth century (b. 1935), American biologist

**Source:** Lincoln, R. J. 1982.

## VAPOR PRESSURE CONDENSATION NUMBER—SEE CONDENSATION NUMBER

## VAPOR PRESSURE LAW, FUNDAMENTAL

If in a solution with two molecular components, A and B, the molecular fraction of A is increased from $x_A$ to $(x_A + dx_A)$, the corresponding increase, $dp_A$, in its partial vapor pressure, $p_A$, is given by:

$$dp_A = f_A\ (T.E.)dx_A$$

and for component B:

$$dp_B = f_B(T.E.)dx_B$$

with the temperature and the total pressure held constant, and where T. E., thermodynamic equilibrium, is held constant. The thermodynamic equilibrium is defined as the heat in body a is equal to the heat in body b, as the exchange between a and b occurs.

**Keywords:** components, fraction, pressure, solution, temperature, thermodynamics
**Sources:** Hampel, C. A. and Hawley, G. G. 1973; Tapley, B. D. 1990 (in part).

## VAPOR PRESSURE LAW, GENERAL

The general equation of a solution in which the thermodynamic environment is constant:

$$[\partial pA/\partial x_A]_{p,T} = k_A$$

where $p_A$ = partial vapor pressure of any molecular species, A, from a solution in which its molecular fraction is $x_A$ and $k_A$ is a constant characteristic of A and the thermodynamics of the solution that surrounds it

**Keywords:** molecular, solution, thermodynamics, vapor pressure
**Sources:** Hampel, C. S. and Hawley, G. G. 1973; Thewlis, J. 1961-1964.
*See also* BOYLE; CHARLES; VAN DER WAALS

## VAPORIZATION—SEE TROUTON

## VARIABLE PROPORTION, LAW OF (ECONOMICS)

There is not a linear relationship between input and output; as the relative amounts of inputs are changed, the output will change, and as the total amount of input is changed, the output will not necessarily change in proportion.

**Keywords:** input, output, relationship
**Source:** Moffat, D. W. 1976.
*See also* DIMINISHING RETURNS

## VARIANCE LAW

For any two, not necessarily statistically independent, random variables $x_1$, $x_2$:

$$E\{x_1 + x_2\} = E\{x_1\} + \{x_2\}$$

$$\text{var}\{x_1 + x_2\} = \text{var}\ \{x_1\} + \text{var}\ \{x_2\} \pm 2\ \text{cov}\{x_1 + x_2\} = \sigma_1^2 + \sigma_2^2 + 2\rho_{12}\sigma_1\sigma_2 \text{ (variance laws)}$$

where $x_1$ and $x_2$ = random variables
$\sigma$ = standard deviation
$\sigma^2$ = variance

**Keywords:** independent, statistics, random, variables
**Sources:** Frieberger, W. F. 1960; Kotz, S. and Johnson, N. 1988; Sneddon, I. N. 1976.

## VARIETY—SEE REQUISITE VARIETY

## VECTOR ADDITION, LAWS OF

The laws of vector addition apply to superposition of forces. Two forces acting at right angles can be resolved into a single force represented by the hypotenuse of the rectangle formed by the forces at right angles.

**Keywords:** force, hypotenuse, rectangle
**Source:** Parker, S. P. 1987.
*See also* ADDITION OF VECTORS; PARALLELOGRAM OF FORCES

### VEGARD LAW (1921)

The average lattice parameter, a, of a substantial solic solution, an alloy system, solid, solution obtained from X-ray or density measurements, is a linear function of the atomic composition of the solution, having values between those for pure elements.

**Keywords:** alloy, atomic, density, lattice, X-ray
VEGARD, Lars, 1880-1963, Norwegian physicist
**Sources:** Ballentyne, D. W. G. and Lovett, D. R. 1972.; Gillispie, C. C. 1981; Hampel, C. A. and Hawley, G. G. 1973; Thewlis, J. 1961-1964; Ulicky, L. and Kemp, T. J. 1992.

### VELOCITIES, LAW OF

Any process changing toward equilibrium becomes slower and slower as the final condition or state is approached.

**Keywords:** cooling, equilibrium, heating, rate
**Source:** Honig, J. M. 1953.
*See also* ADDITION OF VELOCITIES; COOLING; MAXWELL; NEWTON (COOLING)

### VELOCITY-DEFECT LAW; OR OUTER WALL LAW

The relationship:

$$\mu^+ = 1/k \ln y^+ + C$$

where $\mu^+ = \mu/(\tau_w/\rho)^{1/2}$ and $y^+ = y\{(\tau_w/\rho)^{1/2}/\upsilon\}$
$k = 0.4 \qquad C = 5.5$

This describes the velocity profile in the fully turbulent wall region of boundary layer and tube flow with a rough wall if it is used to express the difference between the velocity $\mu$ and the velocity $\mu_0$ at a reference distance $y_0$ (Fig. V.1).

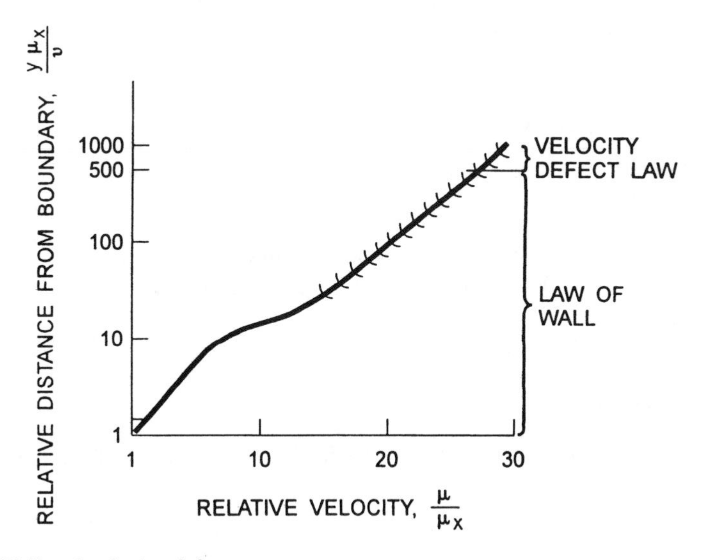


**Figure V.1** Wall and velocity defect.

**Keywords:** boundary, turbulent, velocity, wall
**Sources:** Eckert, E. R. G. and Drake, R. M. Jr. 1972; Roberson, J. and Crowe, C. 1975.
*See also* WALL

## VELOCITY, ELASTIC MEDIUM—SEE NEWTON

## VELOCITY DISTRIBUTION IN A PIPELINE—SEE PRANDTL OR POWER LAW (ONE-SEVENTH)

## VELOCITY HEADS, NUMBER OF, N

A dimensionless group representing friction in conduits that relates the imposed head to the velocity head:

$$N = (F/\rho L^2)/(V^2/2)$$

where F = friction resistance to imposed flow
ρ = mass density
L = length of conduit
V = velocity of flow

**Keywords:** conduits, flow, friction, head, velocity
**Source:** Potter, J. H. 1967.

## VELOCITY NUMBER—SEE MAGNETIC REYNOLDS NUMBER

## VELOCITY OF SOUND—SEE MACH; NEWTON

## VENTURI PRINCIPLE

There is a decrease in pressure of a fluid flowing in a pipe when the diameter has been reduced by a general taper. The principle was used by C. Herschel, who invented the Venturi flowmeter.

**Keywords:** flowmeter, pipe, pressure, taper
VENTURI, Giovanni Battista, 1746-1822, Italian physicist
HERSCHEL, Clemens, 1842-1930, American engineer
**Sources:** Besancom, R. M. 1974; Kirk-Othmer. 1985.

## VERDOORN LAW (1948) (ECONOMICS)

A dynamic relationship exists between the rate of growth in output and growth of productivity due to increasing returns.

**Keywords:** dynamic, growth, productivity, output, returns
VERDOORN, Petrus Johannes, twentieth century, Dutch economist
**Source:** Bothamley, J. 1993.
*See also* INCREASING RETURNS

## VERNER LAW (1875)

Along with the Grimm law, the Verner law provided an explanation of the development of linguistics.

**Keywords:** linguistics
VERNER, Karl, 1846-1896, Danish linguist

**Source:** Bothamley, J. 1993.
*See also* GRIMM

### VERHULST(1838)-PEARL (1920S) LAW OF GROWTH

The growth of organisms can be represented by an S-shaped curve of the form:

$$y = A/\{1 + e^{-B(t - C)}\}$$

where y = number in population
A, B, C = constants
t = time

**Keywords:** curve, growth, organisms, S-shaped
VERHULST, Pierre F., 1804-1849, Belgian mathematician
PEARL, Raymond, 1879-1940, American biologist
**Sources:** BioScience 48(7):541, July 1988; Gillispie, C. C. 1981; Lotka, A. J. 1956; NUC; Pelletier, P. A. 1994; Webster's Biographical Dictionary. 1959.
*See also* POPULATION GROWTH; URBAN CONCENTRATION

### VIBRATING STRINGS, LAWS OF (FREQUENCY)

The number of vibrations of a string is inversely proportional to the length of the string; the vibration frequency of a string is directly proportional to the square root of the tension of the string; and the vibration frequency of a string is inversely proportional to the square of the mass of the string per unit length. Probably the first physical law expressed in mathematical terms, due to Pythagoras. The fundamental frequency of a stretched string is:

$$n = 1/2\ L(T/m)^{1/2}$$

where n = frequency
L = length
T = tension
m = mass per unit length

**Keywords:** frequency, mass, mechanics, string, vibration
PYTHAGORAS, 582-497 B.C., Greek philosopher
**Sources:** Gamow, G. 1961; Hodgman, C. D. 1952.
*See also* KRIGAR-MENZEL; MERSENNE; PYTHAGOREAN (STRINGS); YOUNG-HELMHOLTZ

### VIERDOT LAW (PSYCHOLOGY)

The two-point threshold of a mobile part of the body is directly related to the mobility of that part of the body and the distance of the part of the body from the central axis. The threshold is lower than for a less mobile part of the body.

**Keywords:** body, central axis, distance, mobility, threshold
VIERDOT, Karl von, 1818-1884, German physiologist
**Sources:** Bothamley, J. 1993; Reber, A. S. 1985; Wolman, B. B. 1989.

### VIRCHOW LAW

The cell elements of tumors are derived from normal and preexisting tissue cells.

**Keywords:** cells, physiology, pre-existing, tissue, tumors
VIRCHOW, Rudolf, 1821-1902, German pathologist
**Sources:** Friel, J. P. 1974; Landau, S.I. 1986; Stedman, T. L. 1976.

## VIRIAL—SEE CLAUSIUS

## VISCOELASTIC MATERIAL MODELS

The response of various materials to loading can be represented by various models that represent the material. The basic models (Fig. V.2) are as follows:

**Hooke model**
A viscoelastic material represented by a spring element.

**Maxwell model**
A viscoelastic material represented by a series of a spring (Hookean) element and a viscous (dashpot) element.

**Kelvin model or Voigt-Kelvin model**
A viscoelastic material represented by a parallel combination of a spring (Hookean) and viscous (dashpot) element.

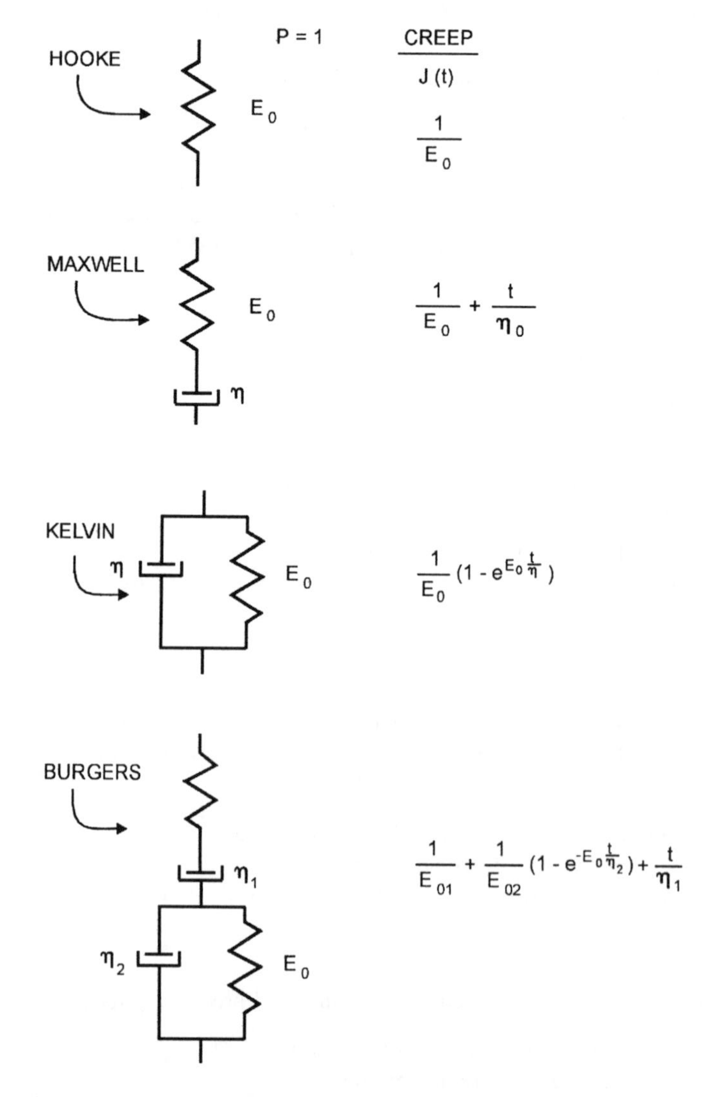


**Figure V.2** Viscoelastic material models.

**Newton-Kelvin Model, Three Element**
A viscous material represented by a Kelvin model or Voigt-Kelvin model in series with a dashpot.

**Newton-Kelvin Model, Four Element**
A material represented by a Kelvin unit (model) in series with a viscous (dashpot) element and a spring (Hookean) model.

**Burgers Model**
A material represented by a Voigt model and Maxwell model in series.

**Keywords:** dashpot, element, Hookean, material, spring, viscous
BURGERS, Johannes Martinus, twentieth century (b. 1895), Dutch rheologist
HOOKE, Robert, 1635-1703, English physicist
KELVIN, Lord (William Thomson), 1824-1907, Scottish mathematician and physicist
MAXWELL, James Clerk, 1831-1879, Scottish mathematician and physicist
NEWTON, Sir Isaac, 1642-1727, English philosopher and mathematician
VOIGT, Woldemar, 1850-1919, German physicist
**Sources:** Flugge, W. 1962; NUC; Scott Blair, G. W. 1949; Vernon, J. 1992.
*See also* BINGHAM; HOOKEAN; KELVIN

## VISCOELASTIC NUMBER, Vis OR $N_{Vis}$

A dimensionless group representing dynamic viscoelasticity that relates the elastic force and viscous force:

$$Vis = G/\omega\mu$$

where G = shear modulus of elasticity
$\omega$ = frequency
$\mu$ = absolute viscosity

**Keywords:** dynamic, elastic, force, frequency
**Sources:** Land, N. S. 1972; Potter, J. H. 1967.

## VISCOSITY—SEE MAXWELL; NEWTON

## VIS VISA CONSERVATION, LAW OF—NOW CALLED LAW OF ENERGY CONSERVATION

**Source:** Guillen, M. 1995
See ENERGY CONSERVATION, LAW OF

## VOIGT EFFECT OR MAGNETIC DOUBLE REFRACTION (1898)

An anisotropic substance placed in a magnetic field becomes birefringent, and its optical properties are similar to a uniaxial crystal. The transverse magneto-optic birefringence is called *Voigt effect.*

**Keywords:** birefringent, crystal, magneto-optic, optical, uniaxial
VOIGT, Woldemar, 1850-1919, German physicist
**Sources:** Parker, S. 1987; Webster's Biographical Dictionary. 1959.

## VOIT LAW OF METABOLISM

The nitrogen in food during metabolism is equivalent to the sum of the nitrogen in the urine and feces. No metabolism nitrogen is in the respiratory gases.

**Keywords:** feces, food, respiratory, urine
VOIT, Carl von, 1831-1908, German (Bavarian) doctor
**Source:** Talbott, J. H. 1970.
*See also* KLEIBER; LIEBIG; RUBNER

## VOLTA LAW (CONDUCTORS) (1796)

When any number of conductors that conduct electricity without electrolytic dissociation are joined together to form a closed chain, and all are at the same temperature, the total electromotive force, or sum of the contact-differences of the potential for the surfaces of union of pairs of elements, is zero. Thus, for three elements a, b, c that form a closed chain:

$$V_{ab} + V_{bc} + V_{ca} = 0$$

In 1796, Volta discovered that an electric current will flow between unlike metals in water and salt. In 1974, Volta proposed that electricity from two metals in contact with a liquid caused twitching of frog legs.

**Keywords:** circuits, conductors, electricity
VOLTA, Alessandro Giuseppe Antonio Anastasio, 1745-1827, Italian physicist
**Sources:** Besancon, R. 1990; Northrup, E. T. 1917.
*See also* ELECTROMOTIVE; GALVANI

## VOLUMES, LAW OF

Volumes of gases in a chemical change have a simple numerical ratio to one another when under the same conditions of temperature and pressure.

**Keywords:** chemical, gases, pressure, temperature
**Source:** Walker, P. M. B. 1988.
*See also* GAY-LUSSAC

## VON BABO—SEE BABO

## VON BAER LAW (BIOLOGY)—SEE BAER

## VON BAER LAW (GEOLOGY)—SEE BAER

## VON EÖTÖS—SEE EÖTÖS

## VON ETTINGSHAUSEN—SEE ETTINGSHAUSEN

## VON KÁRMÁN—SEE KÁRMÁN

## VON LIEBIG—SEE LIEBIG (MINIMA; OPTIMA)

## VON WEIMARN—SEE CORRESPONDING STATES (CRYSTAL)

## VORTEX MOTION LAWS

The following statements cover the vortex laws:

1. A vortex tube always contains the same elements of fluid.
2. The strength of the vortex tube is the same at all parts of the tube and does not change in time, with a perfect fluid.
3. Vortex tubes are either closed surfaces or have their extremities in the surface of the fluid.

A vortex line is a curve whose tangent at every point coincides with the direction of the instantaneous axis of rotation at that point. A space bounded by vortex lines is called a *vortex tube,* and the enclosed fluid is said to have *vortex motion.*

**Keywords:** fluid, mechanics, motion, tube
**Sources:** Considine, D. M. 1976. 1995.
*See also* HELMHOLTZ

## VULPIAN LAWS

When a portion of the brain is destroyed, the functions of the damaged or destroyed functions may be carried on by the remaining parts.

**Keywords:** brain, damaged, medical, physiology
VULPIAN, Edme Felix Alfred, 1826-1887, French physician
**Sources:** Critchley, M. 1978; Friel, J. P. 1974; Landau, S. I. 1986.

## WAAGE-GULDBERG—SEE GULDBERG-WAAGE; MASS ACTION

## WAALS—SEE VAN DER WAALS

## WAELE-OSTWALD LAW (1923); OR DE WAELE-OSTWALD LAW

A power law in rheology in which the relationship for quasi-viscous flow is:

$$f(\tau) = (1/k)\tau^n$$

with

$$D(\tau_r) = \tau_r n/[(n + 3)k]$$

$$Q = \pi_r^3/[(n + 3)k]\ [pr/2L]^n$$

where $\tau$ = orientation time
$r$ = radius
$p$ = pressure
$L$ = length
$n, k$ = constants

which reduces to the Hagen-Poiseuille when $n = 1$; de Waele later refused to recognize the equation as appropriate.

**Keywords:** flow, fluid, quasi-viscous, viscous
WAELE (or de WAELE), A., publications primarily in German and English in 1920s and 1930s
OSTWALD, Friedrich Wilhelm, 1853-1932, Russian German physical chemist; Nobel prize, 1909, chemistry. Scott Blair refers to Carl Wilhelm Wolfgang Ostwald,* 1883-1943, German chemist, son of F. W. Ostwald.
**Sources:** Reiner, M. 1960; *Scott Blair, G. W. 1949; Webster's Biographical Dictionary. 1959.
*See also* DE WAELE; HAGEN-POISEUILLE; OSTWALD-DE WAELE

## WAGES, LAW OF

The real income of workers cannot be improved, as an apparent rise in income will result in a larger family and more rapid population growth. Thus, with an increase in population, there will be an increase in demand for necessities, causing prices to increase and counterbalancing the benefits of a wage increase.

**Keywords:** family size, income, population, prices, wage
Predicted by RICARDO, David, 1772-1823, English economist
**Source:** Moffat. D. W. 1976.

## WAGNER LAW (1876) (ECONOMICS)

The development of an industrialized economy would be accompanied by a rising share of public expenditures in gross national product.

**Keywords:** economy, industrialized, national product, public expenditures

WAGNER, Adolph Heinrich Gotthelf, 1835-1917, German economist
**Sources:** Pass, C. et al. 1991; Pearce, D. W. 1986.

### WAGNER NUMBER, Wa OR $N_{Wa}$

A dimensionless group relating current-density distribution on an electrode:

$$Wa = k(d\eta/di)/L$$

where k = conductivity of electrolyte
$\eta$ = overpotential (v)
i = current
L = characteristic length

**Keywords:** current, density, electrode
WAGNER, Karl Willy, 1883-1953, German electrical engineer
**Sources:** Kirk-Othmer. 1985; NUC; Webster's Biographical Dictionary. 1959.

### WAGNER PARABOLIC LAW OF DIFFUSION

The same as the Fick second law of diffusion, this is a representation of the change of concentration, C, beginning at a distance x as $C_s$, and extending to $C_o$ as time increases, and represented by (Fig. W.1):

$$(C - C_0)/(C_s - C_0) = 1 - \text{erf}\,\{x/(2Dt)^{1/2}\}$$

$$K = 1 - \text{erf}\,\{x/(2Dt)^{1/2}\}$$

where D = diffusivity
x = distance
t = time
erf = error function

$$t = 0 \quad x = 0 \quad C = C_s$$
$$x > 0 \quad C = C_0$$
$$t > 0 \quad x = 0 \quad C = C_s$$
$$x > 0 \quad C_0 < C < C_s$$

**Keywords:** concentration, diffusion, distance, time
WAGNER, Karl Willy, 1883-1953, Rudolf, 1805-1864, German physiologist
**Source:** Gillispie, C. C. 1981.
*See also* ERROR; FICK SECOND LAW; GRAHAM

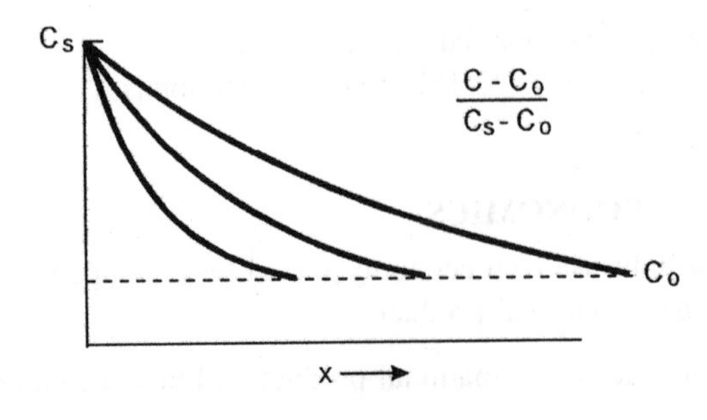


**Figure W.1** Wagner.

## WAKE, LAW OF; FIRST CALLED COLES LAW

The profile shape in the outer layer of an air stream on a flat surface is strongly affected by a pressure variation along the wall, and deviations are especially pronounced where the pressure increases in the flow direction. The relations are similar to the flow in the wake behind a streamlined object, as described by L. Prandtl. The principles and equations describing the velocity field and turbulent diffusivity in this region as the law of the wake. The outer layer for the boundary layer is:

$$\varepsilon_m = 0.0188\ u_s\ \delta^*$$

where $u_s$ = velocity in the mainstream outside the boundary layer
$\delta^*$ = displacement thickness of the boundary layer

**Keywords:** boundary, layer, thickness, turbulent
COLES, Donald Earl, twentieth century (b. 1924), American aeronautical engineer
PRANDTL, Ludwig, 1875-1953, German physicist
**Sources:** Eckert, E. R. G. and Drake, R. M. 1972; Cheremisinoff, N. P. 1986.
*See also* PRANDTL; UNIVERSAL LAW OF THE WALL; VELOCITY-DEFECT

## WALD LAW (1935) (ECONOMICS)

If any good or productive service is in excess supply in equilibrium, its equilibrium price must be zero.

**Keywords:** equilibrium, good, price, service, supply
WALD, Abraham, 1902-1950, Rumanian Austrian economist
**Sources:** Eatwell, J. et al. 1987; Shim, J. K. and Siegel, J. G. 1995.
*See also* WALRAS

## WALDEN LAW (1906); OR WALDEN INVERSION LAW

At infinite dilution, the limiting equivalent conductance of a solution varies inversely with the viscosity coefficient of solvent, the constant being identical for all solvents.

**Keywords:** conductance, dilution, solution, solvent, viscosity
WALDEN, Pavel Ivanovich, 1863-1957, Russian Latvian chemist
**Sources:** Bennett, H. 1962; Bothamley, J. 1993; Hempel, C. G. 1965.

## WALL, LAW OF (1930s)

The velocity distribution in the turbulent boundary layer, is a function of $yu_x/v$ and is much different from that of the laminar sublayer. The relative velocity is $u/u_x$ and the relative distance from the boundary layer is $yu_x/v$. In the range of where the relative distance from the boundary is from 0 to 500, the velocity distribution is called the law of the wall. Above 500, the velocity defect law applies. (See figure with VELOCITY DEFECT LAW.) Some simplifying assumptions in which minuscule viscosity has been overlooked are argued to limit the accuracy of the law. The law was first proposed by T. von Kármán and L. Prandtl.

**Keywords:** distance, experimental, flow, fluid, velocity, wall thickness
**Sources:** Gerhart, P. M., et al. 1992; Roberson, J. and Crowe, C. 1975; Science 272:951, 17 May 1996; Scientific American 275(2):19-20, August 1996.
*See also* KÁRMÁN; PRANDTL; NIKURADSE; VELOCITY-DEFECT LAW; WAKE

### WALL, UNIVERSAL LAW OF

A relationship that describes the velocity profile of air flow of a wall layer, generally 0.2 to 0.3 of the boundary layer thickness.

**Keywords:** boundary, flow, layer, profile
**Source:** Eckert, E. R. G. and Drake, R. M. 1972.
*See also* VELOCITY-DEFECT; PRANDTL; WALL

### WALLER LAW; WALLERIAN LAW OF DEGENERATION (1850)

A nerve fiber can survive only when it maintains continuity with its cell body. If the sensory fibers of the root of the spinal nerve are divided on the central side of the ganglion, the fibers on the peripheral side of the cut do not degenerate, whereas those that are not cut degenerate.

**Keywords:** degenerate, medical, nerve, spinal
WALLER, Augustus Volney, 1816-1870, English physician and physiologist*
**Sources:** Friel, J. P. 1974; Garrison, F. H. 1929; Gray, P. 1967; Landau, S. I. 1986; Morton, L. T. and Moore, R. J. 1987; *Reber, A. S. 1985; Schmidt, J. E. 1959; Stedman, T. L. 1976; Thomas, C. L. 1989.

### WALRAS LAW (1874-1877) (ECONOMICS)

The general form of the law is that, given n markets, if (n – 1) markets are in equilibrium, then the last one must also be in equilibrium, because there cannot be a net excess of demand or supply for goods.

**Keywords:** equilibrium, markets, supply
WALRAS, Leon, 1834-1910, Swiss economist
**Source:** Pearce, D. W. 1981.

### WALTHER LAW OF FACIES (GEOLOGY)

A statement that relates to the manner in which a vertical sedimentary sequence of "facies" develops. The law of facies implies that a vertical sequence of facies will be the product of a series of depositional environments that lie laterally to each other. This law is applicable only to situations where there is no break in the sedimentary sequence.

**Keywords:** depositional, environment, facies, sedimentary
WALTHER, Johannes, 1860-1937, German geologist
**Sources:** Allaby, A. 1990; Bothamley, J. 1993.

### WALTON LAW—SEE RECIPROCAL PROPORTIONS

WALTON, William, 1813-1901, English mathematician

### WARBURG COS$^5\theta$ LAW

The corona discharge depends on localized fields based on geometrical configurations, so it is not possible to specify initial conditions equivalent to those in a uniform field. Also, atmospheric pressure, humidity, and polarity affect the initial values. As voltage is increased, a steady current eventually is established around a stressed electrode. The current density distribution in a point-to-point drift region is given by:

$$j(\theta) = j(0)\cos^m\theta$$

where $\theta$ = angle from the axis of symmetry
m = 4.82 for positive corona
m = 4.65 for negative corona

The distribution is known as the Warburg $\cos^5\theta$ law.

**Keywords:** corona, current, discharge, electric
WARBURG, Emil Gabriel, 1846-1931, German physicist
**Source:** Lerner, R. and Trigg, G. 1991.
Also see WARBURG

## WARBURG LAW

The energy liberated during a complete hysteresis of a magnetic cycle is:

$$E \propto \oint H dM$$

where E = energy liberated during complete hysteresis cycle
H = magnetic field strength
M = magnetization; the relationship between magnetic induction, $\beta$, and magnetization is $\beta = \mu_o H + M$ where $\mu_o$ = permeability in free space

**Keywords:** cycle, energy, hysteresis, magnetic
WARBURG, Emil Gabriel, 1846-1931, German physicist
**Sources:** Ballentyne, D. W. G. and Lovett, D. R. 1972. 1980; Thewlis, J. 1961-1964.
*See also* KNUDSEN; LAMBERT; WARBURG $COS^5\theta$

## WASTAGE OF ENERGY—SEE DISSIPATION OF ENERGY

## WATSON-CRICK MODEL (1953)

A model for the structure of the DNA (deoxyribose nucleic acid) in which sugar phosphate backbones form the twisting shape while providing a rigid framework for pairs of base compounds, in the form of a double helix.

**Keywords:** DNA, base, helix, shape, sugar, twisting
WATSON, James Dewey, twentieth century (b. 1928), American biologist; Nobel prize, 1962, physiology/medicine
CRICK, Francis Harry Compton, twentieth century (b. 1916), English scientist; Nobel prize, 1962, physiology/medicine
**Sources:** Bothamley, J. 1993; Nature 171:737, April 25, 1953.

## WATT LAW

The latent heat of steam at any temperature of generation adds to the sensible heat required to raise the water from 0° C to the temperature is constant. It was found later by H. V. Regnault (1847) that the total heat of steam increases with the temperature and pressure of generation.

**Keywords:** latent heat, sensible, steam, temperature, total
REGNAULT, Henri Victor, 1810-1878, German French chemist
WATT, James, 1736-1819, Scottish engineer
**Sources:** Ballentyne, D. W. G. and Walker, C. E. Q. 1959. 1980.
*See also* OHM; TROUTON

## WAVE MOTION, LAW OF

For waves (water waves and sound waves), the velocity of the wave is equal to the product of the frequency and the wavelength (Fig. W.2):

$$V = f\lambda$$

where V = velocity of the wave
f = frequency
$\lambda$ = wavelength

**Keywords:** frequency, velocity, wavelength
**Source:** Bunch, B. 1992.
*See also* HUYGEN

## WEAK LAWS OF LARGE NUMBERS

These laws are so called because they require convergence in probability.

**Khintchine Theorem or Weak Law of Large Numbers**

**Markov Theorem or Weak Law of Large Numbers**
This is a special case of the Tchebycheff* Theorem.

**Keywords:** convergence, probability
*KHINTCHINE, Aleksandr Iakovlevich (Khinchin, A. Ya), 1894-1959, Russian mathematician
MARKOV, Andrei Andreevich, 1856-1922, Russian mathematician
CHEBYSHEV, Pafnuti Lvovich, 1821-1894, Russian mathematician
**Sources:** Fraser, D. A. S. 1976; Merriam-Webster's Biographical Dictionary. 1995.
*See also* BERNOULLI; LARGE NUMBERS; ROBERTS-CHEBYSHEV

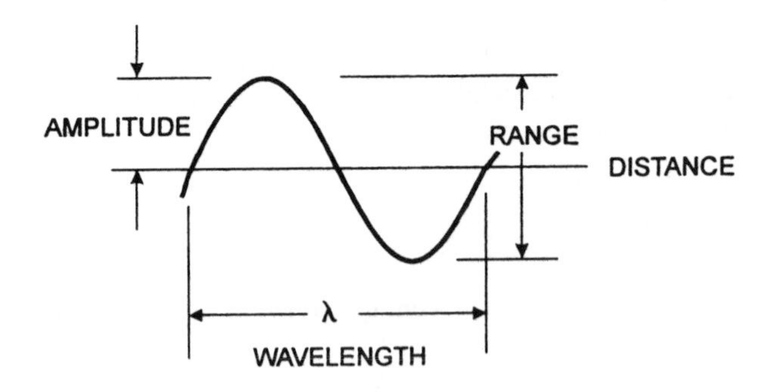


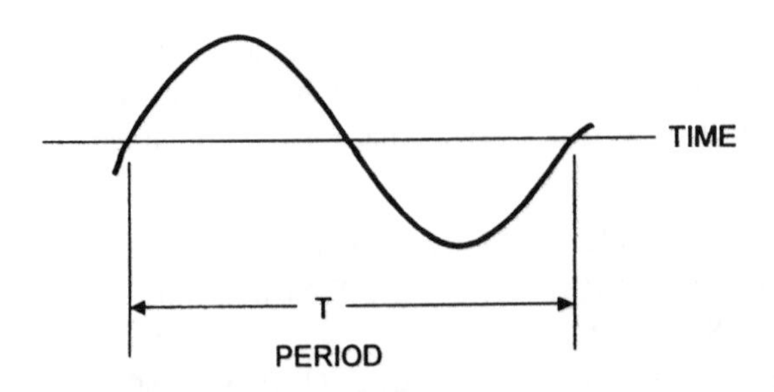


**Figure W.2** Wave motion.

## WEBER-FECHNER LAW (1858)

The stimulus increase required to produce a discernible sensation increase maintains a constant ratio to the total stimulus.* The intensity of sensation is a linear function of the logarithm of the stimulus intensity of a sense organ. E. Weber first stated the law, and G. Fechner popularized it:

$$S = C \log R$$

where R = measured in units of the threshold value
C = constant of proportionality

This law is based on $\Delta S = C\ \Delta R/R$

where $\Delta R$ = the amount of increase in R for a subject to notice any difference at the middle range of behavior response.

**Keywords:** intensity, logarithm, sensation, stimulus
WEBER, Ernst Heinrich, 1795-1878, German physiologist
FECHNER, Gustav Theodor, 1801-1887, German psychologist and physicist
**Sources:** Friel, J. P. 1974; *Glasser, O. 1944; Gray, P. 1967; Stedman, T.L. 1976; Thewlis, J. 1961-1964.
*See also* FECHNER; MÜLLER; PSYCHOPHYSICAL

## WEBER LAW (1834)

The increase in stimulus needed to produce a discernible sensation increase maintains a ratio constant to the total stimulus. Barely noticeable differences in the intensity of a stimulus bear a constant ratio to the intensity. This is found to be limited in its validity.

**Keywords:** response, stimulus, sensation
WEBER, Ernst Heinrich, 1795-1878, German physiologist
**Sources:** Friel, J. P. 1974; Landau, S. I. 1986; McAinsh, T. F. 1986; Thomas, C. L. 1989.
*See also* FECHNER; PSYCHOPHYSICAL; STEVEN

## WEBER LAW OF ELECTRICAL FORCE (1846)

The fundamental law of electrical action expressing the force between moving charge; essential to its derivation were assumptions of central forces and currents as consisting of equal and oppositely directed flow of two kinds of charges, containing:

$$F = e_1 e_2/r^2[1 - 1/c^2(dr/dt)^2 + 2rd^2r/c^2dt^2]$$

where $dr/dt$ = rate at which separation occurs
r = distance of separation between the charges
$e_1$ and $e_2$ = charges
$d^2r/dt^2$ = relative radial acceleration
c = constant of the ratio between electrodynamic and electrostatic units of charges where
c = $(2)^{1/2}$ times the speed of light

**Keywords:** charges, currents, direction, force
WEBER, Wilhelm Eduard, 1804-1891, German physicist
**Sources:** Holmes, F. L. 1980. 1990.
*See also* BIOT-SAVART

### WEBER NUMBER, We OR $N_{We}$ (c. 1919)

A dimensionless group used to represent capillary flow, slosh, bubble formation, and ripples that is the ratio of inertia force and surface tension force:

$$We = \rho V^2 L/\sigma$$

where $\rho$ = mass density
$V$ = velocity
$L$ = characteristic length
$\sigma$ = surface tension

The Weber number 2 is $(We)^{1/2}$. According to Parker*, M. B. Weber first named the Froude, Reynolds, and Cauchy numbers

**Keywords:** bubble formation, capillary, flow, inertia, surface tension
WEBER, M. B.,* twentieth century, German scientist
**Sources:** Bolz, R. E. and Tuve, G. L. 1970; Jerrard, H. G. and McNeill, D. B. 1992; Rohsenow, W. M. and Hartnett, J. P. 1973; *Parker, S. P. 1992; *Potter, J. H. 1967.

### WEBER NUMBER (ROTATING), We(r) OR $N_{We(r)}$

A dimensionless group used in representing agitation of fluid:

$$We(r) = L^3 N^2 \rho/\sigma$$

where $L$ = impeller diameter
$N$ = rate of rotation
$\rho$ = mass density
$\sigma$ = surface tension

**Keywords:** agitation, fluid, rotation
WEBER, W.* possibly, William Eduard, 1804-1891, German physicist
**Sources:** Bolz, R. E. and Tuve, G. L. 1970; Parker, S. P. 1970; *Potter, J. H. 1967.

### WEDDLE RULE—SEE AREA

### WEIBULL DISTRIBUTION OR LAW

A relationship involved in statistical reliability that represents a cumulative distribution function used to estimate the time to failure. The procedure adjusts the distribution parameters. Either equations or Weibull probability graph paper can be used to solve the equations so that one can estimate the age of some item at time of predicted failure. The y-axis of the graph paper is for cumulative failure (percent) and the x-axis is for time.

**Keywords:** failure, reliability, time
WEIBULL, Ernst Hjalmar Waloddi, twentieth century (b. 1887), Swedish materials scientist
**Sources:** Considine, D. M. 1976; Meyers, R. A. 1992; NUC; Rothbart, H. A. 1996.
*See also* ERRORS; PROBABILITY; RELIABILITY

### WEIGERT LAW

The destruction of elements in the organic world is likely to be followed by overproduction of these elements in the reparation process.

**Keywords:** chemistry, organic, production, reparation
WEIGERT, Carl, 1845-1904, German pathologist
**Sources:** Critchley, M. 1978; Friel, J. P. 1974; Landau, S. I. 1986; Stedman, T. L. 1976.

## WEINBURG—SEE HARDY-WEINBURG

## WEIMARN (VON)—SEE CORRESPONDING STATES

## WEISS ZONE LAW (1804); OR ZONE LAW

The face (hkl), called *Miller indices,* is a number of the zone of faces whose axis is [uvw] where u, v, w are integers, if:

$$hU + kV + lW = 0$$

**Keywords:** face, indices, zone
WEISS, Christian Samuel, 1780-1856, German minerologist
**Sources:** Ballentyne, D. W. G. and Lovett, D. R. 1972. 1980; Holmes, F. L. 1980. 1990; Nye, J. F. 1985; Thewlis, J. 1961-1964.
*See also* CURIE-WEISS; ZONE

## WEISSENBERG NUMBER, Wei OR $N_{Wei}$

A dimensionless group used to represent viscoelastic flow through its relaxation time in jets that relates the viscoelastic force and the viscous force:

$$Wei = \{(\lambda_1 - \lambda_2)V\}/D$$

where V = velocity
D = jet diameter
$\lambda_1$, $\lambda_2$ = time constants, t, from $\tau + \lambda_1\dot{\tau} = -\mu_o(\Delta + \lambda_2\dot{\Delta})$
$\tau$ = shear stress
$\mu_o$ = zero shear viscosity
$\Delta$ = rate of deformation
$\cdot$ = d/dt

Different authors express the Weissenberg number in various forms.

**Keywords:** flow, jets, viscoelastic, viscous
WEISSENBERG, Karl von, twentieth century (b. 1893), German chemist and physicist
**Sources:** Bolz, R. E. and Tuve, G. L. 1970; Dorf, R. C. 1995; Land, N. S. 1972; Parker, S. P. 1992; Potter, J. H. 1967; Scott Blair, G. W.1949.
*See also* DEBORAH

## WEIZSACKER MASS LAW (1935)

A complicated semi-empirical expansion relationship in which the binding energy of the nuclei is expressed in terms of the atomic number and the mass number. The mass law is used experimentally to estimate binding energies in the nucleus.

**Keywords:** atomic number, binding energy, mass number, nucleus
WEIZSACKER, Carl Friedrich, twentieth century (b. 1912), German physicist
**Sources:** Ballentyne, D. W. G. and Lovett, D. R. 1980; Gillispie, C. C. 1981; Gray, D. E. 1972; McMurray, E. J. 1995; Menzel, D. H. 1960; Pelletier, P. A. 1994.

## WENTZEL LAW (1927)

Describes the transition of an atom or molecule from a high-energy state to one of lower energy without the emission of radiation, caused by a collision of the second kind and an Auger transition; a collision between an excited atom and a slow particle, such as an electron, whereby the atom undergoes transition to the lower energy state, and the particle is accelerated.

**Keywords:** atom, electron, energy, transition
WENTZEL, Gregor, 1898-1978, German American physicist
**Sources:** Parker, S. 1993; Thewlis, J. 1961-1964.
*See also* AUGER

## WENTZEL-KRAMERS-BRILLOUIN METHOD (1926)

A method of obtaining an approximate solution to the Schrödinger equation developed by G. Wentzel, H. A. Kramers, and M.-L. Brillouin, based on earlier work of several people. These three people contributed to the understanding of the quantum-mechanical relationships of the Schrödinger equation.

**Keywords:** equations, mechanical, quantum
WENTZEL, Gregor, 1898-1978, German American physicist
KRAMERS, Hendrik Anthony, 1894-1952, Dutch physicist
BRILLOUIN, Marcel-Louis, 1854-1948, French physicist
**Sources:** Parker, S. P. 1992; Merriam-Webster's Biographical Dictionary. 1995.
*See also* SCHRÖDINGER

## WENZEL, LAW OF—SEE RICHTER

## WERTHEIMER, LAW OF

Parallel to the second law of thermodynamics, a relationship in Gestalt psychology that the free energy associated with a Gestalt or its complexity tends to decrease to the lowest level consistent with the circumstances. Also, the concept that the whole is something different from the sum of the parts.

**Keywords:** energy, Gestalt, psychology, thermodynamics
WERTHEIMER, Max, 1880-1943, German psychologist and philosopher
**Sources:** Borchert, D. M. 1996; Encyclopaedia Britannica. 1961; Scott Blair, G. W. 1949.
*See also* GOOD CONTINUATION; GOOD SHAPE; THERMODYNAMICS

## WHIDDINGTON LAW (1911)

The energy loss of primary electrons used to bombard a solid to produce secondary electron emission is represented by:

$$-dE/dx = A/E$$

where E = energy of the primary electrons
x = path length
A = constant characteristic of the solid

This leads to the result that the range of the primary electrons is proportional to the square of their energy.

**Keywords:** bombard, electrons, emission, energy
WHIDDINGTON, Richard, 1885-1970, British physicist
**Sources:** Ballentyne, D. W. G. and Lovett, D. R. 1972. 1980.
*See also* THOMSON-WHIDDINGTON

## WHITE LAW

Technological development is expressed in terms of man's control over energy, which underlies certain cultural achievement and social changes.

**Keywords:** development, energy, social
WHITE, Leslie Alvin, 1900-1975, American anthropologist
**Sources:** Steward, J. H. 1955; Who Was Who. 1993.

## WIEDEMANN ADDITIVITY LAW

An empirical law that the mass, molar, or specific magnetic susceptibility of a mixture or solution of components is the sum of the susceptibilities of the portions of each component, and is independent of the concentration of the solute in the solvent. The susceptibility of a solution is represented by:

$$\chi_L = \frac{m\chi_A + M\chi_B}{M + m}$$

where $\chi_L$ = susceptibility of solution
$\chi_A$ = susceptibility of dissolved salt
$\chi_B$ = susceptibility of solvent
M = weight of solvent
m = weight of salt

**Keywords:** concentration, mixture, solute, susceptibility
WIEDEMANN, Gustav Heinrich, 1826-1899, German physicist and chemist
**Sources:** Ballentyne, D. W. G. and Lovett, D. R. 1972; Bennett, H. 1939; Lapedes, D. N. 1978; Thewlis, J. 1961-1964.

## WIEDEMANN-FRANZ LAW (1853); WIEDEMANN-FRANZ-LORENZ LAW (1872)

The ratio of thermal conductivity to electrical conductivity for metals is a constant times the absolute temperature. With the additional work of L. Lorenz, the law was made more general, in that the ratio of the thermal conductivity to the electrical conductivity is the same for metals and alloys at the same absolute temperature, and is directly proportional to the absolute temperature:

$$k_c = L_o \sigma T$$

where $L_o$ = Lorenz ratio, approximately $2.45 \times 10^{-8}$ (volt-kelvin)$^2$
$k_c$ = thermal conductivity
$\sigma$ = electrical conductivity
T = absolute temperature

Another statement of the law is that i*n pure metals, the electronic component accounts for nearly all of the heat conducted, while the lattice component is negligible.* The electronic thermal conductivity is related to the electrical conductivity through the mean free path. Where the mean free path for both thermal and electrical conduction can be assumed to be the same, the Wiedeman-Franz law is applicable.

**Keywords:** alloys, conductivity, electrical, metals
WIEDEMANN, Gustav Heinrich, 1826-1899, German physicist and chemist
FRANZ, Rudolph, nineteenth century (b. 1827), German physicist
LORENZ, Ludwig Valentin, 1829-1891, Danish physicist. Some writers mistakenly use Lorentz.
**Sources:** American Institute of Physics. 1972; Ballentyne, D. W. G. and Lovett, D. R. 1972. 1980; Bothamley, J. 1993; Gray, D. E.1963; Honig, J. M.1953; Mandel, S. 1972; NUC; Parker, S. P. 1992; Thewlis, J. 1961-1964.
*See also* LORENTZ-LORENZ; WIEDEMANN

## WIEDEMANN LAW—SEE WIEDEMANN ADDITIVE LAW

## WIEN DISPLACEMENT LAW; WIEN LAW (1893, 1899)

**First Statement**

The product of the wavelength at which the energy density is a maximum and the absolute temperature of the radiating body is a constant value. Or, the wavelength corresponding to the peak in the radiation spectrum is inversely proportional to the absolute temperature at the wavelength where the temperature lines have their maximum values:

$$\lambda_{max}\ T = \text{constant}$$

where $\lambda$ = wavelength
$T$ = absolute temperature

**Second Statement**

Wavelength of maximum energy:

$$\lambda_v = \text{a constant}$$

This law can be derived from the Planck law.

**Keywords:** energy density, wavelength

WIEN, Wilhelm Carl Werner Otto Fritz Franz, 1864-1928, German physicist; Nobel prize, 1911, physics

**Sources:** Driscoll, W. 1978; Friel, J. P. 1974; Layton, E. T. and Lienhard, J. H. 1988; Menzel, D. H. 1960; Thewlis, J. 1961-1964; Ulicky, L. and Kemp, T. J. 1992.

*See also* PLANCK

## WIEN DISTRIBUTION LAW (1896); OR WIEN LAW OF RADIATION OF WAVELENGTH OF MAXIMUM ENERGY (1896)

The relation between monochromatic emittance of an ideal black body and the body's temperature is:

$$e_\lambda/T^5 = f(\lambda T)$$

where $e_\lambda$ = monochromatic emittance
$\lambda$ = wavelength
$T$ = absolute temperature

**Keywords:** black body, emittance, monochromatic, wavelength

WIEN, Wilhelm Carl Werner Otto Fritz Franz, 1864-1928, German physicist

**Sources:** Considine, D. M. 1976. 1995; Thewlis, J. 1961-1964.

*See also* PLANCK

## WIEN RADIATION LAW (1896)

Often called the Wien third law, the law expresses the spectral energy distribution of the radiation from a black body at temperature, T, black body radiated energy is:

$$e_\lambda = 2\pi\ c_1\ e^{-c_2/\lambda T}/\lambda^5$$

where $e_\lambda$ = black body radiated energy
$c_1$ and $c_2$ = constants
$\lambda$ = wavelength
$T$ = absolute temperature

This law agrees for short wavelength and is today replaced with the Planck radiation law.

**Keywords:** black body, radiation, wavelength
WIEN, Wilhelm Carl Werner Otto Fritz Franz, 1864-1928, German physicist
**Sources:** Considine, D. M. 1976; Driscoll, W. G. 1978; Thewis, J. 1961-1964.
*See also* PLANCK

## WILDER LAW OF INITIAL VALUE

The more intense the function of a vegetative organ, the weaker its capacity for excitation by stimuli and the stronger its reaction to depressing factors, and with extremely high or low initial value, there is a marked tendency to paradoxic reactions, that is, the reversal of direction of reaction.

**Keywords:** excitation, reaction, response, stimuli
WILDER, Joseph, twentieth century (b. 1895), American neuropsychiatrist
**Sources:** Dorland, W. A. 1980. 1988; Friel, J. P. 1974; Landau, S. I. 1986; Stedman, T. L. 1976.

## WILHELMY LAW (1850)

The velocity of chemical reaction at any instant is proportional to the concentration of the reacting substance, which L. Wilhelmy based on the rate of hydrolysis of cane sugar. The variations in acid concentration do not affect the process.

**Keywords:** chemical reaction, concentration, velocity
WILHELMY, Ludwig F., 1812-1864, Polish German chemist
**Sources:** Bennett, H. 1939; Bynum, W. W. et al. 1981; Encyclopaedia Britannica. 1961.
*See also* ENZYME; FISCHER; METABOLISM; $Q_{10}$

## WILLISTON LAW (EVOLUTION)

"Evolution tends to reduce the number of similar parts in organisms and render them more different from each other."

**Keywords:** evolution, organisms, parts
WILLISTON, Samuel Wendell, 1852-1918, American paleontologist
**Sources:** Barr, E. S. 1973; Gary, M. 1972; Stedman, T. L. 1976.
*See also* DARWIN; EVOLUTION

## WIND DIRECTION—SEE BUYS-BALLOT

## WIND LAW; BARIC WIND LAW; BASIC WIND LAW—SEE BUYS BALLOT; DOVE; EGNELL; FAEGRI; FERREL; STORMS

## W-K-B—SEE SCHRÖDINGER; WENTZEL-KRAMERS-BRILLOUIN

## WOESTYN LAW (1848) OR RULE—SEE KOPP

## WÖHLER LAWS (1870)

Wöhler's laws deal primarily with materials elasticity, thermal expansion, and fatigue testing. He distinguished between static, increasing and alternating loads, from which he stated one of four laws: that the failure of a material (he worked primarily with iron and steel) can result from repeated vibrations, no one of which causes the rupture.

**Keywords:** elasticity, loads, materials, rupture, vibrations
WÖHLER, August, 1819-1914, German materials engineer
**Sources:** Engineering 2:160, 1867; Gillispie, C. C. 1981.

## WOLFF LAW (MEDICAL)

All changes that may occur in the form and function of bones are accompanied by definite changes in their internal structure.

**Keywords:** anatomy, bone, changes, structure
WOLFF, Julius, 1836-1902, German anatomist
**Sources:** Friel, J. P. 1974; Landau, S. I. 1986; Rothenberg, R. 1982; Stedman, T. L. 1976; Thomas, C. L. 1989.

## WOOD EFFECT (LAW?)

Alkali metals are transparent to light in the ultraviolet range. According to the free electron theory of metals, the effect should occur for wavelengths of light less than:

$$\lambda_0 = 2\pi(mc^2/4\pi Ne^2)^{1/2}$$

where $\lambda$ = wavelength
m = mass of free electron
c = speed of light
e = charge of free electron
N = number density

**Keywords:** alkali, electron, light, ultraviolet, wavelengths
WOOD, Robert Williams, 1868-1955, American physicist
**Sources:** Michels, W. C. 1961; Merriam-Webster's Dictionary of Biography. 1995.

## WORK HARDENING, GENERAL LAW OF

Four different mechanisms may be responsible for and increase in resistance to plastic deformation of a metal stressed above the yield point:

1. Increase in resistance to slip within a single crystal (dislocation)
2. Increase in resistance to plastic deformation of polycrystalline aggregate produced by fragmentation of crystals and the rotation, elastic deformation, and bending of crystal fragments
3. Stabilization of fragmented and distorted crystal structure by a system of textured stress set up during unloading and fragmentation
4. "Anisotropic change in resistance to plastic deformation produced by practically volume constant deformation associated with rotation, breakup, and rapid reformation of crystal fragments of limiting size"

**Keywords:** anisotropic, crystal, deformation, plastic, polycrystalline
**Source:** Eirich, F. P. 1969.
*See also* PRAGER

## WORK-HARDENING, LAW OF; OR STRAIN-HARDENING, LAW OF

Under tensile test of mild steel, the yield stress increases with increasing deformation. The law of work-hardening relates the square of the tangential strain to the square of the tangential deformation. Thus, $O_t^2$ is related to $e_n^2$. Strain is the recoverable part of the deformation:

$$\vartheta_t^2 = f(\gamma^2)$$

where $\vartheta_t$ = tangential yield stress
$\gamma$ = tangential deformation or deformation gradient

**Keywords:** deformation, strain, tangential
**Sources:** Eirich, F. P. 1969; Reiner, M. 1960.
*See also* PRAGER; WORK HARDENING, GENERAL

### WORK, LAW OF

The work done on the plane by the effort (input) is equal to the work done by the plane (output).

**Keywords:** input, output, plane
**Source:** Thewlis, J. 1961-1964.

### WRIGHT—SEE SEWALL WRIGHT

### WÜLLNER LAW

The modification of the osmotic properties of water by a dissolved substance, notably the reduction of the vapor pressure, is a direct function of the concentration of the solute.

**Keywords:** concentration, osmotic, solute, substance
WÜLLNER, Friedrich Hugo Anton Adolphi, 1835 -1908, German physicist
**Sources:** Considine, D. M. 1976. 1995; Honig, J. M. 1953; Webster's Biographical Dictionary. 1959.
*See also* OSTWALD; RAOLT

### WUNDERLICH LAW OR CURVE

The temperature curve, with ascending oscillations followed by descending oscillations, represents the course of typhoid fever.

**Keywords:** oscillations, temperature, typhoid fever
WUNDERLICH, Carl Reinhold August, 1815-1877, German physician
**Sources:** Critchley, M. 1978; Landau, S. I. 1986.

### WUNDT-LAMANSKY LAW

As the line of vision moves through a vertical plane parallel to the frontal plane, vision moves in straight lines in the vertical and horizontal directions, but in curved paths in other directions.

**Keywords:** medical, planes, sight, vision
WUNDT, Wilhelm Max, 1832-1920, German psychologist
**Source:** Friel, J. P. 1974.

### WYSSAKOVITSCH LAW

The cells covering any part of the body, as long as they preserve their makeup, protect the underlying tissues.

**Keywords:** body, cells, surface, tissues
WYSSAKOVITSCH. Biographical information not located.

## X-RAY PHENOMENON (1895)

Natural fluorescence emanating from a material was discovered by W. Röntgen (Roentgen) and originally called *Röntgen rays*. It was used to see the bones in the hand and soon was applied to many other objects to "see inside." He was working with cathode ray studies at the time, but the phenomenon was entirely different, although with many similarities. W. Crookes observed X-rays earlier but didn't realize what he had seen. Later, in 1896, A. H. Becquerel discovered radioactivity.

**Keywords:** fluorescence, radioactivity, X-rays

RÖNTGEN, Wilhelm Konrad, 1845-1923, German physicist, Nobel prize, 1901, physics

CROOKES, Sir William, 1832-1919, English chemist and physicist

Becquerel, Antoine Henri, 1852-1908, French physicist; Nobel prize, 1903, physics (shared with Madame Curie and Pierre Curie

**Sources:** Asimov, I. 1976; Encyclopaedia Britannica. 1984; Holmes, F. L. 1980. 1990; John, V. 1992.

*See also* BECQUEREL; MOSELEY

# X

# Y

## YERKES-DODSON LAW (1908)

Mice are more able to learn a simple task to avoid a painful electric shock and more able to learn a complex task to avoid a mild electric shock.

**Keywords:** electric, learn, shock
YERKES, Robert Mearns, 1876-1956, American psychologist
DODSON, J. D., twentieth century, American psychologist
**Sources:** Bothamley, J. 1993; Corsini, R. J. 1994; Reber, A. S. 1985.

## YODA LAW; YODA 3/2 POWER LAW; 3/2 THINNING LAW (1963)

A relationship exists between the dry weight of mature surviving plants in a given area and the density of the plant in that area, expressed as (Fig. Y.1):

$$W = C\,P^{-3/2}$$

where W = the dry weight of surviving plants
C = a constant applicable to growth characteristics of the plant species
P = density of plants in the area

**Keywords:** area, density of plants, thickness of plants, weight
YODA, K. Biographical information not located.
**Sources:** Allaby, M. 1964; Art, H. W. 1993; Lincoln, R. J. et al. 1982.
*See also* POWER

## YOUNG-HELMHOLTZ LAW (MECHANICS)

There are two laws that refer to the action of bowed strings.

**First Law**
No overtone with a node at the point of excitation can be present.

**Second Law**
When a string is bowed at a distance 1/n times the length of the string from one of the ends, in which n is an integer, the string moves back and forth with two constant velocities, one

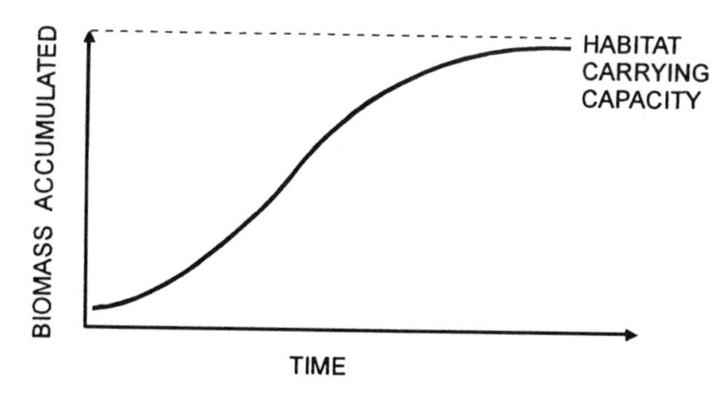


**Figure Y.1** Yoda power law.

of which has the same direction as the bow and equal to it, while the other has the opposite direction and is n – 1 times as large.

**Keywords:** bow, direction, excitation, overtone, strings
YOUNG, Thomas, 1773-1829, English physician and physicist
Helmholtz, Hermann Ludwig Ferdinand von, 1821-1894, German physiologist and physicist
**Source:** Parker, S. P. 1997.
*See also* KRIGAR-MENZEL; PYTHAGOREAN

## YOUNG (1801)-HELMHOLTZ THREE COLOR THEORY OR LAW (1801*, 1805)

Human vision has three separate color sensations, each capable of stimulation in various degrees, which if subjected to separate stimulation will prove to be the sensations produced by red, blue, and green, implying three sites of retinal sensors for color vision. A membrane of rods and cones are mixed together in the retina. Cylinder-shaped rod cells are primarily for dim light, reset slowly, and give black and white responses. Cone cells make color vision possible and provide visual acuity and are densely placed close to the center of the retina and received focused light. Each cone cell synapse has a nerve cell and can send messages via optic nerve fibers to the brain (Fig. Y.2).

**Keywords:** color, retina, sensations, sight, stimulation, vision
YOUNG, Thomas, 1773-1829, English physician and physicist
HELMHOLTZ, Hermann Ludwig Ferdinand von, 1821-1894, German physiologist and physicist
**Sources:** Bothamley, J. 1993; Considine, D. M. 1976; Isaacs, A. 1996; Life Science Library: Light and Vision. 1966; McAinsh, T. F. 1986; Mandel, S. 1972; *NYPL Desk Ref.; Pick, T. P. and Howden, R. 1977; Science News 151: March 29, 1997; Zusne, L. 1987.
*See also* COLOR MIXTURE; COLOR VISION; KIRSCHMANN; PIPER

## YOUNG MODULUS, E

The Young modulus applies to an elastic material and is the ratio of unit stress to elastic strain, produced in tension or compression:

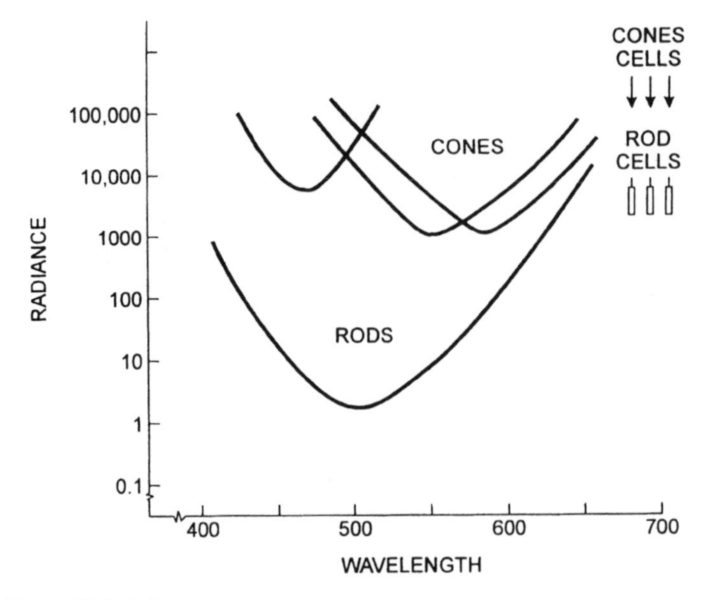


**Figure Y.2** Young-Helmholtz.

$$E = \Delta\sigma/\Delta\varepsilon$$

The Young modulus (Fig. Y.3) for aluminum is $10 \times 10^6$ psi; steel, $30 \times 10^6$ psi; wood, concrete (compression), 5000 psi (34.5 MPa). Elastic materials obey the Hooke law.

The Young modulus by stretching of a wire or rod is:

$$M = mgL/\pi r^2 e$$

where M = modulus
m = mass
L = length
r = radius
e = elongation

The modulus of rigidity for twisting of a bar is:

$$M = CL/\pi r^4 \theta$$

where C = couple, C = mgx
$\theta$ = twist, radians
L = length
r = radius

**Keywords:** elastic, rigidity, shear, stiffness, strain, stress
YOUNG, Thomas, 1773-1829, English physicist and physician
**Sources:** Hodgman, C. D. 1952; John, V. 1992; Mandel, S. 1972; Parker, S. P. 1989.
*See also* ELASTICITY; HOOKE; POISSON

**YOUNG RULE (1813) (MEDICINE)**

A method for calculating the dose of medicine. A child over two years of age should receive a dose based on age. The rule is to divide the age by (age + 12), which represents the fraction of an adult dose. For a person of age four years:

$$4/(4 + 12) = 1/4 \text{ dose}$$

**Keywords:** age, child, dose, medicine
YOUNG, Thomas, 1773-1829, English physicist and physician
**Sources:** Dorland, W. A. 1980; Friel, J. P. 1974; Thomas, C. L. 1989.
*See also* COWLING

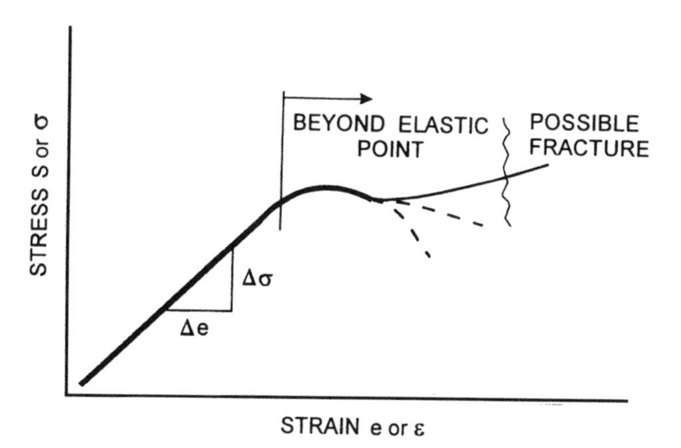


**Figure Y.3** Young modulus.

## YUKAWA INTERACTION (1935)

Yukawa predicted the presence of mesons, because electrons and protons could not otherwise exist in the same atom, and that a strong nuclear binding force exists to hold nucleus together. He first described the meson as pi meson or pions and mumeson or muons. His prediction of the interaction was proven to be true in 1947.

**Keywords:** atom, binding force, electrons, mesons, nuclear

YUKAWA, Hideki, 1907-1981, Japanese physicist; Nobel prize, 1949, physics

**Sources:** Besancon, R. M. 1974; Nourse, A. E. 1969; Thewlis, J. 1961-1964.

# Z

## ZEEMAN EFFECT OR LAW (1885*, 1896)

Discovered by P. Zeeman and his teacher, H. Lorentz, and based on studies of the Kerr effect, when a source of light is placed in a strong magnetic field, the components of the spectrum are polarized, with the directions of polarization and the appearance of the effect depending on the direction from which the source is viewed relative to the lines of force. Zeeman's work concentrated on the light source. In a very strong magnetic field, the result is known as the Paschen-Back effect (1921). The work of P. Zeeman is often considered to be the third magneto-optic effect to be discovered, following M. Faraday (1846) and J. Kerr (1875).

**Keywords:** light, magnetic field, polarization, spectrum
ZEEMAN, Pieter, 1865-1943, Dutch physicist; Nobel Prize, 1902, physics (shared)
LORENTZ, Henrik Antony, 1853-1928, Dutch physicist; Nobel prize, 1902, physics (shared)
**Sources:** *Carter, E. F. 1976; Hodgman, C. D. 1952; Holmes, F. L. 1980. 1990; Mandel, S. 1972; Parker, S. P. 1989; Peierls, R. E. 1956.
*See also* FARADAY; KERR; PRESTON; STARK

## ZERO—SEE BOOLEAN

## ZEROTH LAW (THERMODYNAMICS)

In an isolated system, when each of two bodies are in thermal equilibrium with a third body, the two bodies are also in thermal equilibrium, and the three bodies are at the same temperature. This law is the basis of temperature measurement devices.

**Keywords:** bodies, equilibrium, temperature, thermal
**Source:** Barrow, J. D. 1991.
*See also* THERMODYNAMICS

## ZEUNE LAW

The proportion of people with blindness is less in the temperate zone than in the frigid zone, and increases in the torrid zone as the equator is approached. This law is obsolete.

**Keywords:** blindness, frigid, temperate, torrid
ZEUNE, August, 1778-1853, German scientist
**Sources:** Friel, J. P. 1974; Landau, S. I. 1986; NUC.

## ZHUKOVSKY (ZHUKOVSKI) NUMBER, Zh OR $N_{Zh}$

A dimensionless group relating viscosity and time:

$$Zh = \nu t / L_0^2$$

$\nu$ = kinematic viscosity
t = time
$L_0$ = characteristic length dimension

**Keywords:** length, time, viscosity
ZHUKOVSKY, Nikolai Yegrovich, 1847-1921, Russian physicist/mathematician

**Sources:** Begell, W. 1983; Gillispie, C. C. 1981; Pelletier, A. P. 1994.
*See also* KUTTA-JOUKOWSKY

### ZIPF LAW OF LEAST EFFORT (1935)

An empirical rule that if the words of a sample of spoken language are ranked in terms of the number of times each is used:

$$f(r) = k(1/R)$$

where $f(r)$ = number of times (frequency) a word occurs of rank R
$R$ = rank position of the word
$k$ = constant

As an example, the hundredth most frequent word occurs approximately 1/100th as many times as the most frequent word.

**Keywords:** frequency, language, speech, word
ZIPF, George Kingsley, 1903-1950, American psychologist
**Sources:** Allen, T. J., 1977; Considine, D. M. 1976; Watters, C.1992; Pierce, J. R. 1961; Zusne, L. 1987.
*See also* LOTKA; PRICE SQUARE ROOT

### ZODIACAL LAW

The band of 12 constellations passes through the sun in the course of a year, tracing a path of a circle. The ecliptic is the circle traced out by the sun's "apparent" path over the celestial sphere in the course of a year.

**Keywords:** celestial, constellations, ecliptic, sun
**Source:** World Book Encyclopedia. 1990.

### ZONE LAW OF WEISS

In crystallography, the condition that a crystal face (h k l) shall lie in the zone (U V W) as:

$$U\,h + V\,k + W\,l = 0$$

**Keywords:** crystallography, face, zone
WEISS, Pierre, 1865-1940, French physicist and mechanical engineer
**Sources:** Ballentyne, D. W. G. and Lovett, D. R. 1972; Holmes, F.L. 1980; Thewlis, J. 1961-1964.
*See also* WEISS

# SOURCES AND REFERENCES


Abbott, David. 1983. *The Biographical Dictionary of Scientists: Chemists.* Peter Bedrick Books, New York 203 p. QD21.B4

Abbott, David. 1984. *The Biographical Dictionary of Scientists: Biologists.* Peter Bedrick Books, New York 182 p. QH26.B54

Abbott, David. 1984. *The Biographical Dictionary of Scientists: Physicists.* Peter Bedrick Books, New York 212 p. QC15.B56

Abbott, David. 1985. *The Biographical Dictionary of Scientists: Mathematicians.* Bland Educational, London 175 p. QA28.B565

Abbott, Michael M. and Van Ness, H. C. 1989. *Thermodynamics.* Schaum's Outline Series, McGraw-Hill Book Co., New York, NY 362 p. TJ265.A19

*Academic Press Dictionary of Science and Technology.* See Morris, C. G. 1992. 1995.

Achinstein, Peter. 1971. *Law and Explanation.* Oxford Clarendon Press, Oxford, England 168 p. Q175.A268

Adler, Pierre M. 1992. *Porous Media: Geometry and Transport.* Butterworth-Heinemann, Stoneham, MA 544 p. TA418.9.P6A35

Agnew, Ralph R. 1942. *Differential Equations.* McGraw-Hill Book Co., New York, NY 341 p. QA371.A3

Allaby, Ailsa and Allaby, Michael (ed.). 1990. *The Concise Oxford Dictionary of Earth Sciences.* Oxford University Press, New York, NY and Oxford, UK 410 p. QE5.C66

Allaby, Michael (ed.). 1989 (3rd ed.). *Dictionary of the Environment.* New York University Press, New York, NY 423 p. QH540.4.A44

Allaby, Michael (ed.). 1992. *The Concise Oxford Dictionary of Botany.* Oxford University Press, Oxford, England 442 p. QK9.C67

Allaby, Michael (ed.). 1992. *The Concise Oxford Dictionary of Zoology.* Oxford University Press, New York, NY 508 p. QL9.C66

Allaby, Michael (ed.). 1994. *The Concise Oxford Dictionary of Ecology.* Oxford University Press, New York, NY 415 p. QH540.4.C66

Allen, Thomas J. 1977. *Managing the Flow of Technology.* MIT Press, Cambridge, MA 320 p. T174.3A44

American Council of Learned Societies. *Dictionary of Scientific Biography.* Biographical Dictionary of Mathematics. See Gillispie, C. C. 1981. 1991; Holmes, F. L. 1980. 1990.

American Geological Institute. 1957. 1960. *Glossary of Geology and Related Sciences. NAS-NRC Publication* 501, Washington, DC (Coordinating chairman, J. V. Howell) 325 p. QE5.A48

American Geological Institute. See Gary, M. et al. 1972.

American Institute of Physics. 1972 (3rd ed.). *A. I. P. Handbook.* McGraw-Hill Book Co., New York, NY various paging QC61.A5. See Gray, D. E. 1963. 1972.

Ammer, C. and D. S. (ed.). 1984 (2nd ed.). *Dictionary of Business and Economics.* Free Press, Macmillan, Inc., New York, NY 507 p. HB61.A53

Anderson, K. N., Anderson, L. E. and Glanze, W. D. (ed.). 1986. 1994 (4th ed.). *Mosby's Medical, Nursing, and Allied Health Dictionary.* Mosby Publishing Co., St. Louis, MO 1689 p. 1978 p. R121.M89

Angeles, Peter A. (ed.). 1981. *Dictionary of Philosophy.* Barnes & Noble, New York, NY 320 p. B41.A53

*Applied Science & Technology Index.* 1958 and following, annually. H. W. Wilson Co., New York, NY 7913. I7 (preceded by Industrial Arts Index)

Arey, Leslie B., et al. 1957. *Dorland's Illustrated Medical Dictionary.* W. B. Saunders Co., Philadelphia, PA See Dorland, W. A. 1988. 1989; Newman, W. A. 1985 (26th ed.); Friel, J. P. 1974. 1985.

Art, Henry W. (ed.). 1993. *The Dictionary of Ecology and Environmental Science.* Henry Holt & Co., New York, NY 632 p. GE10.D53

Arutyonyan, Nagush Kh. 1966. (Translated by H. E. Nowattny). *Some Problems in the Theory of Creep.* Pergamon Press, New York, NY 290 p. TA440.A7713

ASCE. 1972. *Biographical Dictionary of American Civil Engineers. ASCE,* Washington, DC 163 p. TA139.A53

Ashby, William R. 1956. *An Introduction to Cybernetics.* J. Wiley, New York and Methuen Press, London 295 p. QA76.15.E48

Asimov, Isaac. 1963. 1970. *I, Robot.* Fawcett. Doubleday, Garden City, NY 218 p. PS3551.S5

Asimov, Isaac. 1966 (Vol. I; Vol. II; Vol. III). *Understanding Physics.* Mentor, New York, NY 248 p.; 249 p.; 268 p. QC23.A8

Asimov, Isaac. 1972 (rev. ed.). *Isaac Asimov's Biographical Encyclopedia of Science and Technology.* Avon Books, New York, NY 805 p. Q141.A74

Asimov, Isaac. 1994. Asimov's *Chronology of Science & Discovery.* HarperCollins Publishers, New York, NY 789 p. Q125.A765

ASME. 1980. *Mechanical Engineers in America Born Prior to 1861 - A Biographical Dictionary.* The American Society of Mechanical Engineers, New York, NY 330 p. TJ139.A47

*(The) Atomic Energy Deskbook.* See Hogerton, John F. 1963.

Audi, Robert (ed.). 1995. *The Cambridge Dictionary of Philosophy.* Cambridge University Press, Cambridge, UK 882 p. B41.C35

Auger, C. P. 1975 (2nd ed.). E*ngineering Eponyms.* The Library Assoc., London, England 122 p. TJ929.A84

Avallone, Eugene and Baumeister, Theodore III. (ed.). 1996 (10th ed.). *Marks' Standard Handbook for Mechanical Engineers.* McGraw-Hill Book Co., New York, NY various paging TJ151.S82

Baker, B. B., et al. (ed.). 1960. 1966. *Glossary of Oceanographic Terms.* U. S. Naval Oceanographic Office, Washington, DC 204 p. GC9.U5

Baker, C. C. T. 1961. 1966. *Dictionary of Mathematics.* Hart Publishing Co., New York, NY 338 p. QA5.B23

Ballentyne, Denis W. G. and Lovett, D. R. 1972. 1980 (4th ed.). *A Dictionary of Named Effects and Laws in Chemistry, Physics, and Mathematics.* Chapman & Hall, Ltd., London, England 335 p. 346 p. QD5.B32

Ballentyne, Denis W. G. and Walker, L. E. Q. 1959. 1961. *A Dictionary of Named Effects and Laws in Chemistry, Physics, and Mathematics.* Macmillan Co., New York, NY 205 p. QD5.B32.

Ballou, Glen M. (ed.). 1987. 1991. *Handbook for Sound Engineering.* Howard W. Sams & Co., Division of Macmillan Computer Publishers, Carmal, IN 1506 p. TK7881.4H37

Barber, J. 1977. *Primary Processes of Photosynthesis.* Elsevier-North Holland, Netherlands 516 p. QK882.P746

Bard, Philip, et al. (ed.). 1941 (9th ed.). *Macleod's Physiology in Modern Medicine.* C. V. Mosby Co., St. Louis, MO 1256 p.

Barr, Ernest Scott. 1973. *An Index to Biographical Fragments in Unpublished Scientific Journals.* The University of Alabama Press, University, Alabama 294 p. Q141.B29

Barrow, John D. 1991. *Theories of Everything.* Fawcett Columbine Book, Ballantine Books, Random House, New York, NY 302 p. Q175.B225

Bass, Michael. 1995 (2nd ed.). *Handbook of Optics.* Optical Society of America. In two volumes QC369.H35.

Bates, R. L., and Jackson, J. A. (ed.). 1980 (2nd ed.). 1987 (3rd ed.). *Glossary of Geology.* American Geological Institute, Alexandria, VA 749 p.; 788 p. QE5.B38

Bates, R. L., and Jackson, J. A. (ed.). 1980. 1984 (3rd ed.). *Dictionary of Geological Terms.* American Geological Institute, Alexandria, VA and Anchor Books, Garden City, NJ 571 p. QE5.D55

Baumeister, Theodore (ed.). 1967 (7th ed.). *Marks' Standard Handbook for Mechanical Engineers.* McGraw-Hill Book Co., New York, NY various paging TJ151.S82

Bazovsky, Igor. 1961. *Reliability Theory and Practice.* Prentice-Hall, Englewood, NJ 292 p. TA168.B33

Begell, William. 1983. *A Glossary of Terms Used in Heat Transfer, Fluid Flow and Related Terms.* Hemisphere Publishing Co., Washington, DC 116 p. QC320.G57

Belzer, Jack, Holtzman, Albert G., and Kent, Allen (ed.). 1975 and following. Beginning in 1986, with vol. 17, editors were Kent, Allen, Williams, James G. and Kent, Rosalind. *Encyclopedia of Computer Science & Technology.* Marcel Dekker, Inc., New York, NY. In 36 volumes QA76.15.E5

Bendiner, Jessica and Bendiner, Elmer. 1990. *Biographical Dictionary of Medicine.* Facts On File, New York, NY 284 p. R134.B455

Benedick, Manson and Pigford, Thomas H. 1957. *Nuclear Chemical Engineering.* McGraw-Hill Book Co., New York, NY 594 p. TK9350.B46

Bennett, Harry. 1939. 1962. 1986 (4th ed.). *Concise Chemical and Technical Dictionary.* Chemical Publishing Co., New York, NY 1271 p. QD5.B4

Bent, Henry A. 1965. *The Second Law.* Oxford University Press, New York, NY 429 p. QC311.B4

Besancon, Robert M. (ed.). 1974 (2nd ed.). 1985 (3rd ed.). *The Encyclopedia of Physics.* Van Nostrand Reinhold, New York, NY 1067 p. 3rd ed. in two volumes QC5.B44

Bever, Michael B. (ed.). 1986. *Encyclopedia of Materials Science and Engineering.* Pergamon Press, Oxford; The MIT Press, Cambridge, MA. In 8 volumes TA402.E53.

*Biographical Dictionary of American Civil Engineers.* See *ASCE.* 1972.

*Biographical Dictionary of Economists.* See Blaug, M. 1986.

*Biographical Dictionary of Mathematics.* See Gillispie, C. C. 1991.

*Biographical Dictionary of Mechanical Engineers Born in America Prior to 1861.* See *ASME.* 1980.

*Biographical Dictionary of Medicine.* See Bendiner, J. and E. 1990; Garrison, F.H. 1967; Talbott, J. H. 1970.

*Biographical Dictionary of Psychology.* See Sheehy, N. et al. 1997.

*Biographical Dictionary of Scientists.* See Abbott, D. 1983. 1984. 1985; Daintith, J., et al 1994; Pelletier, P. A. 1980. 1994; Porter, Roy. 1994; Simmons, J. 1996; Thorne, J. O. 1984; Wasson, T. 1987; Williams, T. 1974. 1994; Vernoff, E. and Shore, R. 1987.

*Biographical Fragments.* See Barr, E. S. 1973.

*Biographical Encyclopedia of Scientists.* See Asimuth, I. 1972; Crystal, D. 1994; Daintith, J., et al. 1994; Spofford, A. R. 1898 ff; Holmes, F. L. 1980. 1990.

*Biography, General.* See Schick, E. A. (Current Biography). 1940 ff; Crystal, D. 1994; Daintith, J. 1981. 1994; Debus, A. G. 1968; Eggenberger, D. I. 1973; Fischer, I. 1962; Greene, J. E. 1966; Lewytzkj, B. 1984; McGraw-Hill *Encyclopedia of World Biography*; Miles, W. D. 1976; Morton, L. T. and Moore, R. S. 1989; *National Cyclopedia of American Biography*; Parker, S. P. 1980; Roysdon, C. and Khatri, L. 1978; Schick, E. A. 1997; Schlessinger, B. H. and J. H. 1991; Schulz, H. 1972; Stankus, T. 1991; Sterling, K. B. et al. 1997; Tobak, A. B. 1990; Turkevich, J. 1963; Vernoff, E. and Shore, R. E. 1987; White, J. T. et al. 1898; *Who Was Who.* 1976. 1991; *Who's Who* (several editions).

Birchon, D. 1965. *Dictionary of Metallurgy.* Philosophical Press, New York, NY 409 p. TN609.B55

Bisio, Attilio, Boots, Sharon and Siegel, Paula. (ed.). 1997. *Wiley Encyclopedia of Energy and the Environment.* John Wiley and Sons, NY 1562 p. In two volumes TJ163.235.W55

Black, Maxine. 1940 ff. *Current Bibliographies.* See Schick, E. A. 1997.

*Black's Medical Dictionary.* See Macpherson, Gordon. 1992; Thomson, W. A. R. 1984.

Blake, Alexander. 1985. *Handbook of Mechanics, Materials, and Structures.* John Wiley & Sons, New York, NY 710 p. TA350.H23

Blake, Leslie S. 1989. 1994 (4th ed.). *Civil Engineering Reference Book.* Butterworth-Heinemann, Oxford, England various paging TA151.C58

*Blakiston's Gould Medical Dictionary.* See Gennaro, Alfonso R. 1979.

Blaug, Mark. 1986. *Who's Who in Economics.* Wheatsheaf Books, Brighton, England 936 p. HB76.W47

*Bobbs-Merrill's Modern Science Dictionary.* See Hechtlinger, A. 1959.

Bockris, J. O'M. 1974. *An Introduction to Electrochemical Science.* Springer Verlag, New York, NY 133 p. QD553.B624

Bolz, R. E. and Tuve, G. S. 1970. 1973 (2nd ed.). *Handbook of Tables for Engineering Science.* CRC Press, Cleveland, OH 975 p.; 1166 p. TA151.C2

Borchert, Daniel M. 1996. *The Encyclopedia of Philosophy.* Simon & Schuster Macmillan, New York, NY. In 8 volumes + supplement B41.85

Bothamley, Jennifer (ed.) 1993. *Dictionary of Theories.* Gale Research, Int., London and Detroit 637 p. AG5.D53

Boyer, Carl B. 1989. *A History of Mathematics.* John Wiley & Sons, New York, NY 715 p. QA21.B767

Braithwaite, R. B. 1953. 1960. *Scientific Explanation, A Study of the Functions of Theory, Probability and Law in Science.* Harpers, New York, NY 374 p. Q175.B7845

Bridgman, P. W. 1961. *The Thermodynamics of Electrical Phenomena in Metals.* Dover Publications, New York, NY 244 p. QC621.B6

Brody, Samuel. 1945. *Bioenergetics and Growth.* Reinhold Publishing Corporation and Hafner, New York, NY 1023 p. SF61.B7B5

Bronson, Richard. 1993 (2nd ed.). *Outline Theory of Proble.s of Differential Equations.* Schaum's Outline Series, McGraw-Hill Book Co., New York, NY 358 p. QA372.B856

Bronzino, J. D. (ed.). 1995. *The Biomaterial Engineering Handbook.* CRC Press and IEEE Press, Boca Raton, FL 2862 p. R856.15B.56

Brown, Laurie M., Pais, Abraham, et al. 1995. *Twentieth Century Physics.* Institute of Physics and American Institute of Physics, Bristol, UK and Philadelphia, PA. In 3 volumes QC7.T84

Brown, Stanley B. and Brown, L. Barbara (ed.). 1972. *The Realm of Science.* Touchstone Publishing Co., Louisville, KY. In 21 volumes Q111.R38

Bunch, Bryan (ed.). 1992. *Henry Holt Handbook of Current Science and Technology.* Henry Holt and Co., New York, NY 689 p. Q158.5B86

Bunch, Bryan and Hellemans, Alexander. 1993. *Timetable™ of Technology.* Simon & Schuster, New York, NY 490 p. T15.B73 See also Hellemans, A. 1988.

*Butterworth's Medical Dictionary.* See Critchley, Macdonald. 1978.

Bynum, W. F., Browne, E. J. and Porter, Roy. 1981. *Dictionary of the History of Science.* Princeton University Press, Princeton, NJ 494 p. Q125.B98

Cahn, Robert W. 1988. *Encyclopedia of Material Science and Engineering,* Supp. vol. 1 & 2. See Bever, M. B. 1986

Cambel, Ali B. 1963. *Plasma Physics and Magnetofluid Mechanics.* McGraw-Hill Book Co., New York, NY 304 p. QC718.C3

Campbell, Norman R. 1957. *Foundations of Science.* Dover Publications, New York, NY 565 p. QC6.C25

*The Cambridge Encyclopedia of Earth Sciences.* See Smith, David G. 1982.

Carey, John (ed.). 1995. *The Faber Book of Science.* Faber & Faber, London and New York 528 p. Q162.F29

Carter, E. F. 1976. *Dictionary of Inventions and Discoveries.* Crane, Russak & Co., New York, NY 208 p. T9.C335

Casti, John L. 1989. *Paradigms Lost.* William Morrow and Company, New York, NY 565 p. Q173.C35

Casti, John L. 1990. *Searching for Certainty.* William Morrow and Company, New York, NY 496 p. Q175.C434

*Chambers Biographical Dictionary.* See Thorne, J. O. 1984.

*Chambers Science and Technology Dictionary.* See Walker, P. M. B. 1988.

Chandrasekhar, S. 1960. *Radiative Transfer.* Dover Publications, New York, NY 393 p. QB461.C46

*Chemical Abstracts.* 1907 and following years. Chemical Abstracts Service, American Chemical Society, Columbus, OH QD1.A51

*Chemical Engineers' Handbook.* See Perry, J. H. 1950; Perry, R. H. 1984.

Cheremisinoff, N. P. (ed.). 1986. *Encyclopedia of Fluid Mechanics.* Gulf Publishing Co., Houston, TX. In 10 volumes TA357.E53

*Churchill's Medical Dictionary.* See Koenigsberg, R. 1989.

Clapper, Russell B. 1961. *Glossary of Genetics and Other Biological Terms.* Vantage Press, New York, NY 200 p. QH13.C53

Clark, George L. and Hawley, Gessner. 1957. 1966. *The Encyclopedia of Chemistry.* Reinhold Publishing Co., New York, NY 1144 p. QD5.E58

Clifford, A. F., et al. 1964. *International Encyclopedia of Chemical Sciences.* D. Van Nostrand, Princeton, NJ 1331 p. QD5.I5 (First listed person, not listed as author)

*Concise Dictionary of Atomics.* See Vecchio, A. del. 1964

*Concise Dictionary of Biology.* See Martin, E. A. 1985. 1990.

*Concise Encyclopedia of Biology.* See Scott, T. A. 1995.

*Concise Encyclopedia of Chemistry.* See Eagleson, M. (trans.)1994; Gruyter, Walter de. 1994.

*Concise Science Dictionary.* See Isaacs, A., et al. 1991. 1996.

*Condensed Chemical Dictionary.* See Lewis, R. J. 1993.

Condon, Edward U. and Odishaw, Hugh. 1958. *Handbook of Physics.* McGraw-Hill Book Co., New York, NY various paging QC21.C7

Conger, George P. 1924. *A Course in Philosophy.* Harcourt, Brace and Co., New York, NY 603 p. BD32.C65

Considine, Douglas M. (ed.). 1957. 1974. *Process Instruments and Controls Handbook.* McGraw-Hill Book Co., New York, NY various paging TA165.C65 TS156.8.C674

Considine, Douglas M. (ed.). 1976. 1995 (8th ed.). *Van Nostrand's Scientific Encyclopedia.* Van Nostrand Reinhold, New York, NY. In two volumes 2370 p.; 3455 p. Q121.V3

Constant, Frank Woodbridge. 1963. *Fundamental Laws of Physics.* Addison-Wesley Publishing Co., Reading, MA 403 p. QC23.C764

Corsini, Raymond J., et al. (ed.). 1984. 1994 (2nd ed.). 1994. *Encyclopedia of Psychology.* J. Wiley & Sons, New York, NY. In 4 volumes BF31.E52

Corsini, Raymond J. (ed.). 1987. 1996. *Concise Encyclopedia of Psychology.* John Wiley & Sons, New York, NY 1242 p.; 1035 p. BF31.E553

*CRC Handbook of Chemistry & Physics.* See Lide, David R. 1990-1991. 1996-1997; Hodgman, C. D. 1952. Weast, R. C. 1988.

Creighton, H. Jermain. 1935. 1943 (4th ed.). *Principles and Applications of Electrochemistry.* John Wiley & Sons, New York, NY. In two volumes QD553.C9

Critchley, MacDonald (ed.). 1978. *Butterworth's Medical Dictionary.* Butterworth, Stoneham, MA 1942 p. R121.B75

Crystal, David. 1994. *The Cambridge Biographical Encyclopedia.* Cambridge University Press, Cambridge, England 1304 p. CT103.C26

*Current Biography.* Annual, beginning in 1940. After 1985 called Current Biography Index. See Schick, E. A. 1997. CT100C8

*Current Technology Index. Annual.* 1981-1996. Bowker Saur, London. Z7913.B7. Replaced by British Technology Index, 1962-1980. In 1997 replaced by Abstracts in New Technology and Engineering.

Daintith, John (ed.). 1981. 1994. *A Biographical Encyclopedia of Scientists.* Facts On File, Inc., New York, NY 1075 p. Q141.B53

Daintith, John (ed.). 1985. *Concise Dictionary of Biology.* Oxford University Press, Oxford, England 261 p. QH302.5C66

Daintith, John (ed.). 1985. *Concise Dictionary of Chemistry.* Oxford University Press, Oxford, England 308 p. QD5.D26

Daintith, John (ed.). 1981. 1988. *Facts On File Dictionary of Chemistry.* Facts On File, Inc., New York, NY 249 p. QD5.F33

Daintith, John (ed.). 1981. 1988. *Facts On File Dictionary of Physics.* Facts On File, Inc., New York, NY 235 p. QC5.F34

Daintith, John, Mitchell, Sarah and Tootill, Elizabeth. 1994 (2nd ed.). *A Biographical Encyclopedia of Scientists.* Bristol Pub. for American Institute of Physics, Philadelphia, PA. In two volumes Q141.B53

Dallavalle, Joseph Marius. 1948 (2nd ed.) *Micromeritics: The Technology of Fine Particles.* Pitman Publishing Corporation, New York, NY and Chicago, IL 555 p. TA407.D3

Darwin, Charles. 1859 (original). 1990. *On The Origin of the Species.* Modern Library, New York, NY various pages for different editions M967.2188 and several other LoC call numbers

Dasch, E. Julius. 1996. *Encyclopedia of Earth Sciences.* Simon & Schuster, Macmillan, New York, NY 1273 p. In two volumes QE5.E5137

Davies, Charles Norman. 1966. *Aerosol Sciences.* Academic Press, New York, NY 468 p. TP244.A3D4

Davies, Kenneth. 1969. *Ionospheric Radio Waves.* Blaisdell Publ. Co., Waltham, MA 460 p. TK6553.D293

Day, Lance and McNeil, Ian. 1996. *Biographical Dictionary of the History of Technology.* Routledge, London and New York 844 p. ISBN 0-415-06042-7

Dean, John A. 1979. 1992. *Lange's Handbook of Chemistry.* MacGraw-Hill, Nw York, NY. In various sections QD65.L362

Debus, A. G. 1968. *World Who's Who in Science, from Antiquity to the Present.* Marquis Who's Who, Chicago, IL 1855 p. Q141.W7

de Camp, L. Sprague. 1962. 1970. *The Ancient Engineers.* Ballantine Books, Random House, New York, NY 450 p. TA16.D4

Deighton, Lee C. (ed.). 1971. T*he Encyclopedia of Education.* Crowell-Collier Educational Corp., Macmillan Co., New York, NY. In 10 volumes LB15.E47

de Lorenzi, Otto. 1951. *Combustion Engineering - A Reference Book on Fuel Burning and Steam Generation.* Combustion Engineering, Inc., New York, NY various paging TJ285.F7 Now authored by Fryling, Glenn R. 1966.

Denbigh, Kenneth. 1961. 1971. *The Principles of Chemical Equilibrium.* Cambridge University Press, London. 494 p. QD501.D365

Denny, R. C. 1973. 1982 (2nd ed.). *A Dictionary of Spectroscopy.* Halsted Press, John Wiley & Sons, New York, NY 205 p. QC450.3D46

*Dictionary.* See McKeever, S. *(Random House)* 1993; Morris, W. *(American Heritage)* 1969; Simpson, J. A. et al. (Oxford) 1987; Webster and Merriam-Webster. 1993. 1986.

*Dictionary of American Biography.* 1937. Charles Scribner's Sons, New York, NY Vol. I - X + Supplement & Index E176.D564; Toback, A B 1990

*Dictionary of Applied Chemistry.* See Thorpe, J., et al. 1940.

*Dictionary of Applied Mathematics.* See Dictionary of Mathematics.

*Dictionary of Astronomy.* See Mitton, J. 1991. 1993; Wallenquist, A. 1966.

*Dictionary of Atomic Energy (Atomics).* See Vecchio, A. del. 1964.

*Dictionary of Behavioral Science.* See Wolman, B. B. 1973. 1989.

*Dictionary of Biochemistry.* See Stenesh, J. 1989.

*Dictionary of Biography.* See Abbott, D. 1983. 1984. 1985; *ASCE.* 1972; *ASME.* 1980. 1981; Blaug, M. 1985; *Dictionary of American Biography.* 1937; Fischer, I. 1962; Gillispie, C. C. 1981; Greene, J. E. 1966; Holmes, F. L. 1990; McMurray, E. J. 1995; Marquis. 1976; Merriam Webster. 1995; Morton, L. T. 1989; Muir, H. 1994; Parker, S. 1980; Pelletier, P. A. 1980; Porter, R. 1994; Sheehy, N. et al. 1997; Smith, G. 1882; Thorne, J. O. and Collocott, T. C. (Chambers) 1962. 1984; Toback, A. B. 1990; Wasson, T. (Nobel)1987; Webster. 1959. 1988. 1995; Williams, T. I. 1974. 1994.

*Dictionary of Biology (Biological Sciences).* See Daintith, J. 1985; Gray, P. 1967; Henderson, I. F. et al. 1963. 1989; Kenneth, J. H. 1963; Landau, S. I. 1986; Lapedes, D. N. 1976; Martin, E. A. 1985. 1990. 1996; Toothill, E. 1981. 1988; Walker, P. M. B. 1989.

*Dictionary of Botany.* See Allaby, M. 1992; Martin, E. A. 1985. 1990. 1996.

*Dictionary of Business.* See Ammer, C. and D. S. 1984; Link, A. N. 1993.

*Dictionary of Chemistry.* See Bennett, H. 1986; Daintith, J. 1981.1985. 1988; Honig, J. M. 1953; Lewis, R. J. 1987. 1993; Miall, L. M. 1961.1968 and Sharp, D. W. A. 1981; Thorpe, J. 1940; Van Nostrand. 1953.

*Dictionary of Civil Engineering.* See Scott, J. S. 1993.

*Dictionary of Dangerous Pollutants, Ecology, and Environment.* See Tver, D. R. 1981.

*Dictionary of Drying.* See Hall, C. W. 1977.

*(Dictionary) Communications Standards.* See Weik, M. H. 1996.

*Dictionary of Earth Sciences.* See Allaby, A. and Allaby, M. 1990.

*Dictionary of Ecology.* See Allaby, M. 1994; Hanson, H. C. 1962; Lincoln, R. J. 1982; Tver, D. F. 1981.

*Dictionary of Ecology and Environmental Science.* See Art, H. W. 1993; Tver, D. F. 1981.

*Dictionary of Economics (and Business).* See Eatwell, J., et al. 1987; Greenwald, D. et al. 1983; Hanson, J. L. 1986; Link, A. N. 1993; Moffat, D. W. 1976. 1983; Pass, C., et al. 1991; Pearce, D. W. 1986. 1992; Shin, J. A. and Siegel, J. G. 1995; Sloan, H. S. and Zurcher, A. 1970.

*Dictionary of Education.* See Good, C. V. 1973.

*Dictionary of Electronics.* See Graf, R. F. 1968; Markus, J. and Cooke, N. M. 1960.

*Dictionary of Energy.* See Hall, C. W. and Hinman, G. W. 1983; Hunt, V. D. 1979.

*Dictionary of English.* See McKeever, S. (Random House) 1993; Morris, W. (American Heritage) 1969; Simpson, J. A. et al. (Oxford) 1989.

*Dictionary of Environment and Environmental Science.* See Allaby, M. 1989; Stevenson, L. H. and Wyman, B. C. 1991.

*Dictionary of Genetics.* See Knight, R. L. et al. 1948; King, R. C. 1990.

*Dictionary of Geological Science (Terms).* See American Geological Institute. 1957. 1960; Bates, R. L. and Jackson, J. A. 1980. 1987; Lapedes, D. N. 1978.

*Dictionary of Geophysics.* See Runcorn, S. K. 1967.

*Dictionary of Health, etc.* See Anderson, R. N. et al. 1994.

*Dictionary of History of Science.* See Bynum, W. F. et al. 1981.

*Dictionary of Information Science and Technology.* See Watters, C. 1992.

*Dictionary of Inventions and Discoveries.* See Carter, E. F. 1976.

*Dictionary of LIfe Sciences.* See Lapedes, D. N. 1976.

*Dictionary of Mathematics and Mathematicians.* See Baker, C. C. T. 1961. 1966; Gibson, C. 1981. 1988; Gillispie, C. C. 1991; Freiberger, W. F. 1960; Ito, K. 1987; James, R. C. 1992; James, R. C. and James, G. 1968; Ito, K. 1987: Karush, W. (Webster) 1989; Lapedes, D. N. 1978; Sneddon, I. N. 1976.

*Dictionary of Medical Ethics.* See Duncan, S. A., et al. 1981.

*Dictionary of Medicine.* See Anderson, K. N. et. al. 1986; 1994; Arey, L. B. 1957; Bendiner, J. & E. 1990; Critchley, M. (Butterworth) 1978; Dorland, W. A. 1980. 1988; Doty, I. 1979. 1985; Dox, I. (Melloni) 1979. 1985; Duncan, S. A. 1981; Friel, J. P. (Dorland) 1974. 1985; Garrison, F. H. (History) 1967; Gennaro, A. R. (Gould) 1974; Hensyl, W. R. (Stedman) 1990; Konigsberg, R. (Churchill) 1989; Landau, S. I. 1986; Macpherson, G. (Black) 1972. 1992; Merriam-Webster. 1993; Miller, B and Keane, C. 1978; Newman, W. A. (Dorland) 1985; O'Toole, M. O. (Miller-Kane) 1992; Rothenberg, R. E. 1982. 1988; Stedman, T. L. 1976; Talbott, J. H. 1970; Taylor, N. B. (Stedman) 1953; Thomas, C. L. (Taber) 1989; Thomson, W. A. R. (Black) 1984; Webster. 1987.

*Dictionary of Metallurgy.* See Birchon, D. 1965; Merriman, A. D. 1958. 1965.

*Dictionary of Meteorology.* See Huschke, R. E. 1959.

*Dictionary of Mineralogy.* See Hurlbut, C. S. 1952.

*Dictionary of Mining, Minerals, and Related Terms.* See Thrush, P. W. 1968.

*Dictionary of Modern Economics.* See Pearce, D. W. (MIT) 1986; Greenwald, D. et al. 1983.

*(Dictionary) Military and Defense.* See Dupuy, T. N. 1993.

*Dictionary of Named Effects.* See Ballentyne, D. W. G. et al. 1959. 1961. 1972. 1980.

*Dictionary of National Biography.* See Nicholls, C. S. 1996; Smith, George. 1920. 1981; Williams, E. T. 1981.

*Dictionary of Petroleum.* See Hyne, N. J. 1991; Tver, D. F. 1980.

*Dictionary of Petrology.* See Tomkeieff, S. I. 1983.

*Dictionary of Philosophy.* See Angeles, P. A. 1981; Audi, R. 1995; Flew, A. 1964. 1984; Honerich, T. 1995; Runes, D. D. 1983.

*Dictionary of Physical Chemistry (Comprehensive).* See Ulicky, L. and Kemp, T. J. 1992.

*Dictionary of Physical Science.* See Emiliani, C. 1987.

*Dictionary of Physics.* See Daintith, J. 1981. 1988; Gray, H. J. 1958. 1962; Gray, H. J. and Isaacs, A. 1975; Isaacs, A. 1996; Lapedes, D. N. 1978; Michels, W. C. 1956. 1961; Parker, S. P. 1984; Rigden, J. S. 1996; Thewlis, J. 1961-1964.

*Dictionary of Physics and Electronics.* See Michels, W. C. 1956. 1961.

*Dictionary of Physics and Mathematics.* See Lapedes, D. N. 1978.

*Dictionary of Pollutants.* See Tver, D. R. 1981.

*Dictionary of Psychology.* See Harre, R. and Lamb, R. 1983; Goldenson, R. M. (Longman) 1984; Harriman, P. L. 1947; Reber, A. S. 1985; Sheehy, N. 1997; Statt, D. 1982; Verplanck, W. S. 1957; Zusne, L. 1987.

*Dictionary of Science.* See Bobbs-Merrill. 1959; Graham, E. C. 1967; Hechtlinger, A. 1959; Isaacs, A. et al. 1991. 1996; Mandel, S. 1969. 1972; McKeever, S. 1993; Morris, C. G. 1987. 1992; Speck, G. and Jaffe, B.1965; Uvarov, E. B. and Chapman, D. R. 1964; Walker, P. M. B. 1988.

*Dictionary of Science and Technology.* See Lapedes, D.N. 1978. 1982; Morris, C. G. 1992; Parker, S. 1960. 1989. 1994; Walker, P. M. B. (Chambers) 1988.

*Dictionary of Scientific Biography.* See Gillispie, C. C. 1981; Holmes, F. L. 1980. 1990; Williams, T. I. 1982. 1984.

*Dictionary of Scientific Terms.* See Kenneth, J. H. 1957. 1963; Lapedes, D. N. 1978. 1982; Parker, S. P. 1989. 1994.

*Dictionary of Scientific Units.* See Jerrard, H. G. and McNeill, D. B. 1992.

*Dictionary of Scientists (and Engineers).* See Abbott, D. 1983. 1985; Greene, J. E. 1966; Millar, D., et al. (Cambridge)1996; Muir, H. (Larousse) 1994; Parker, S. P. 1980; Porter, R. 1994; Simmons, J. 1996.

*Dictionary of Spectroscopy.* See Denny, R. C. 1973. 1980.

*Dictionary of the Life Sciences.* See Lapedes, D. N. 1976.

*Dictionary of Theories.* See Bothamley, J. 1993.

*Dictionary of Thermodynamics.* See James, A. M. 1976.

*Dictionary of 20th Century Scientists (International).* See Vernoff, E. and Shore, R. 1987.

*Dictionary of Zoology.* See Allaby, M. 1992.

Dorf, Richard C. (ed.). 1995. *The Engineering Handbook.* CRC Press and IEEE Press, Roca Raton, FL 2298 p. TA151.E424

Dorland William A. (ed.). 1980. *Dorland's Medical Dictionary.* Saunder's Press, Holt, Rinehart & Winston, Philadelphia, PA 741 p. R121.D74

Dorland, William A. 1988. (ed.). 1994 (28th ed.). *Dorland's Illustrated Medical Dictionary.* W. A. Saunders, a Division of Harcourt, Brace & Co., Philadelphia, PA 1940 p. R121.D73. See also Arey, L. B. 1988; Friel, J. P. 1994. 1985; Newmans, W. A. for early editions.

Doty, Leonard A. 1989. *Reliability for Technology.* Industrial Press, New York, NY 307 p. TA 169.D68

Downing, Douglas and Clark, Jeffrey. 1997 (3rd ed.). *Statistics the Easy Way.* Barron's Educational Series, Hauppauge, NY 394 p. QA276.12F68

Dox, Ida. (ed.). 1979. 1985 (2nd ed.). *Melloni's Illustrated Medical Dictionary.* Williams & Wilkins, Baltimore, MD 533 p. R121.D76

Driscoll, W. G. (ed.). 1978. 1995. *Handbook of Optics.* Optical Society of America. McGraw-Hill Book Co., New York, NY various paging QC399.H35

du Nouy, Pierre L. 1966. *Between Knowing and Believing.* David McKay Co., New York, NY 272 p. Q171.L413

Dulbecco, Renato (ed.). 1991. *Encyclopedia of Human Biology.* Academic Press, HBJ, San Diego, CA. In 8 volumes QP11.E53 or QH302.5E56

Dull, Raymond W. 1926. *Mathematics for Engineers.* McGraw-Hill Book Co., New York, NY 780 p. QA37.D85

Duncan, S. A. et al. (ed.). 1981. *Dictionary of Medical Ethics.* Crossroad Publ., New York, NY 459 p. R724.D53

Dupuy, Trevor N. 1993. *International Military and Defense Dictionary.* Maxwell Macmillan, Washington, DC 3132 p. In 6 volumes U24.I 58

Eagleson, Mary (translator and reviser). 1994. (Jakubke, H. D. and Jeschkeit, H. ed.). *Concise Encyclopedia of Chemistry.* Walter Gruyter Publ., Berlin and New York 1201 p. QD4.A2313

Eatwell, John, et al. 1987. *The New Palgrave Dictionary of Economics.* Macmillan Press, London. In 4 volumes HB61.N49

Eckert, E. R. G. and Drake, R. M., Jr. 1972. *Analysis of Heat and Mass Transfer.* McGraw-Hill Book Co., New York, NY 806 p. QC320.E27

Eckman, Donald P. 1958. *Automatic Process Control.* John Wiley & Sons, New York, NY 368 p. TJ213.E25

Eggenberger, David I. 1973. *Encyclopedia of World Biography.* McGraw-Hill Book Co., New York, NY. In 12 volumes + suppl. 13-18 CT103.M3

Eigen, M. Manfred and Winkler, Ruthild. 1981. *Laws of the Game.* Knopf, distributed by Random House, New York, NY 347 p. Q175.E3713

Eirich, Frederick R. 1956. 1958. 1960. 1967. 1969. *Rheology.* Academic Press, New York, NY. In five volumes QC189.E55

Ellul, Jacques. 1964. *The Technological Society.* Knopf, New York, NY 449 p. T14.E553

Emiliani, Cesare. 1987. *Dictionary of the Physical Sciences.* Oxford University Press, New York and Oxford 365 p. Q123.E46

*Encyclopaedia.* Also see Encyclopedia.

*Encyclopaedia Britannica.* 1961. 1973. 1984. 1993 (15th ed.). Encyclopaedia Britannica, Inc., Chicago, IL. In 29 volumes + Index and Guide AE5.E363

*Encyclopaedia of American Biography.* See Spofford, A. R. 1898 ff.

*Encyclopaedia of Mathematics.* See Hazewinkel, M. 1987; Vinogradov, I. M. (Soviet) 1987.

*Encyclopaedic Dictionary of Mathematics for Engineers and Applied Scientists.* See Sneddon, I. N. 1976.

*Encyclopaedic Dictionary of Physics.* See Thewlis, James. 1961-1964.

*Encyclopedia Americana.* 1997. Grolier, Inc., Danbury, CT. In 30 volumes AE5.E333

*Encyclopedia and Dictionary of Medicine, Nursing, and Allied Health.* See Miller, B. F. and Keene, C. B. 1978; O'Toole, M. O. 1992.

*Encyclopedia of Applied Geology.* See Finkl, C. W. 1984.

*(Encyclopedia) Astronomy and Astrophysics.* See Maran, S. P. 1992; Mitton, S. 1973.

*Encyclopedia of Atmospheric Sciences.* See Fairbridge, R. W. 1967.

*Encyclopedia of Biochemistry.* See Williams, R. J. and Lansford, E. M. 1967.

*Encyclopedia of Biography.* See Asimov, I. 1972; Crystal, D. 1994; Daintith, J. et al. 1981. 1994; Eggenberger, D. I. 1973; Fischer, I. 1962; *National Cyclopaedia. 1898* and following years; Parker, S. P. 1980; White, J. T. 1898 ff.

*Encyclopedia of Biology and Biological Sciences.* See Dulbecco, R. 1991; Gray, P. 1970; Scott, T. A. 1995.

*Encyclopedia of Chemical Process and Design.* See McKetta, J. J. and Cunningham, W. 1976-1996.

*Encyclopedia of Chemical Technology,* etc. See Grayson, M. and Eckroth, D. (Kirk-Othmer) 1978.1984. 1995-1998; Kroschwitz, J. I. (Kirk-Othmer) 1995; McKetta, J. J. and Cunningham, W. 1976-1996.

*Encyclopedia of Chemistry,* etc. See Clark, G. L. and Hawley, G. G. 1957. 1966.; Clifford, A. F. 1964; Eagleson, M. 1994; Grayson, M. 1995; Gruyter, W. de. 1994; Hampel, C. A. and Hawley, G. G. 1966. 1973; Kroschwitz, J. I. 1995-1998; Lagowski, J. J. 1997.

*Encyclopedia of Computer Science & Technology.* See Belzer, J., et al. 1975 ff.; Ralston, A., et al. 1993.

*Encyclopedia of Earth (Geological) Sciences.* See Dasch, E. J. 1996; Finkl, C. W. 1984; Lapedes, D. N. 1978; Smith, D. G. (Cambridge) 1982.

*Encyclopedia of Education.* See Deighton, L. C. 1971.

*Encyclopedia of Electrochemistry.* See Hampel, C. A. 1964.

*Encyclopedia of Energy and Environment.* See Bisio, A., Boots, S., et al. 1997.

*Encyclopedia of Evolution.* See Milner, R. 1990.

*Encyclopedia of Fluid Mechanics.* See Cheremisinoff, N. P. 1986.

*Encyclopedia of Food Science and Technology.* See Hui, Y. H. 1992.

*Encyclopedia of (Applied) Geology.* See Finkl, C. W. 1984; Lapedes, D. N. 1978.

*Encyclopedia of Heat (and Mass Transfer) Transfer.* See Hewitt, G. F., et al. 1997; Kutateladze, S. S. and Borishanskii, V. M. 1966.

*Encyclopedia of Human Biology.* See Dulbecco, R. 1991.

*Encyclopedia of Life Sciences.* See Friday, A. and Ingram, D. S. 1985; O'Daly, A. 1996.

*Encyclopedia of Materials Science and Engineering.* See Bever, M. 1986.

*Encyclopedia of Mathematics.* See Hazewinkel, M. 1987; Ito, K. 1987; Sneddon, I. N. 1976; Vinogradov, I. M. 1987; West, B. H., et al. 1982.

*Encyclopedia of Medicine.* See Glanze, W. O. (Mosby) 1985. 1992; Miller, B. F. and Keane, C. B. 1978; O'Toole, M. O. (Miller-Kane) 1992; Talbott, J. H. 1970.

*Encyclopedia of Metallurgy.* See Merriman, A. D. 1965.

*Encyclopedia of Minerology.* See Frye, K. 1981.

*Encyclopedia of Oceanography.* See Fairbridge, R. W. 1967.

*Encyclopedia of Philosophy.* See Borchert, D. M. 1996.

*Encyclopedia of Physical Sciences and Technology.* See Meyers, R. A. 1992.

*Encyclopedia of Physics.* See Besancon, R. M. 1974. 1985; Lerner, R. G. and Trigg, G. L. 1981. 1991; Parker, S. P. 1985; Rigden, J. S. 1996; Thewlis, J. 1961-1964; Trigg, G. L. 1991. 1996; Truesdell, C. A. 1960.

*Encyclopedia of Physics in Medicine & Biology.* See McAinsh, T. F. 1986.

*Encyclopedia of Psychology.* See Corsini, R. J. 1984. 1987. 1996; Harre, R. and Lamb, R. 1983; Harris, W. and Levey, J. (Columbia) 1974.

*Encyclopedia* (Random House). See Mitchell, J. 1983.

*Encyclopedia of Science and (Technology).* See Asimov, I. 1972; Considine, D. M. 1976. 1995; Lapedes, D. N. 1971. 1974; Parker, S. P. 1989. 1992. 1994; Travers, B. E. 1996; Van Nostrand. 1947.

*Encyclopedia of Scientists.* See Daintith, J. et al. 1981. 1994.

*Encyclopedia of Statistical Sciences.* See Kotz, S. and Johnson, N. L. 1983. 1988.

*Encyclopedia of World Biography.* See Eggenberger, D. I. 1973.

*Encyclopedic Dictionary of Mathematics.* See Ito, K. 1987.

*Encyclopedic Dictionary of Physical Geography.* See Goudie, A., et al. 1985.

*Encyclopedic Dictionary of Psychology.* See Harre, R. and Lamb, R. 1983.

*Encyclopedia Medical Dictionary.* See Thomas, C. 1989

*Encyclopedia of Education.* See Deighton, L. C. 1971.

Eshbach, Ovid W. and Saunders, Matt. 1975 (3rd ed.). *Handbook of Engineering Fundamentals.* John Wiley & Sons, New York, NY 1530 p. TA151.E8 See also Tapley, B. D. 1990.

Facts On File

*Biographical Encyclopedia* (J. Daintith. 1981. 1994)

*Biology* (D. Abbott. 1984; J. Daintith, 1985; E. Toothill. 1981)

*Chemistry* (D. Abbott. 1983; J. Daintith. 1981. 1988)

*Environmental Science* (L. Harold Stevenson and B. C. Wyman, 1991)

*Geology and Geophysics* (L. D. Farris, 1987)

*History of Science and Technology* (E. Marcorini. 1988)

*Mathematics* (D. Abbott. 1985; Carol Gibson. 1981)

*Physics* (J. Daintith. 1981. 1988)

*Science* (E. B. Uvarov. 1986)

Fairbridge, Rhodes W. 1967. *The Encyclopedia of Atmospheric Sciences and Astrogeology.* Reinhold Publishing Co., New York, NY 1200 p. QC854.F34

Fairbridge, Rhodes W. 1966. *The Encyclopedia of Oceanography.* Reinhold Publishing Co., New York, NY 1021 p. GC9.F3

Feinberg, Gerald, and Goldhauser, Maurice. 1963. *The Conservation Laws of Physics.* Scientific American. October.

Feynman, Richard. 1965. 1994. *The Character of Physical Law.* MIT Press, Cambridge, MA and London, England. 173 p. QC28.F4

Fink, Donald G. and Beaty, H. Wayne. 1993 (13th ed.). *Standard Handbook for Electrical Engineers.* McGraw-Hill, Inc., New York, NY variable paging TK151.S8

Finkelnburg, Wolfgang. (translated from German). 1950. *Atomic Physics.* McGraw-Hill Book Co., New York, NY 496 p. QC173.F513

Finkl, Charles W., Jr. 1984. *The Encyclopedia of Applied Geology.* Van Nostrand Reinhold Co., New York, NY 644 p. QE5.E5

Fischer, I. (ed.). 1962. *Biographisches Lexikon der hervorragenden Arzteder letzten (1880-1930).* Urban & Schwartzberg, Berlin (In two volumes)

Fisher, David J. 1988. 1991. *Rules of Thumb for Engineers and Scientists.* Trans Tech Publications, Ltd. Gulf Publishing Co., Houston, TX 242 p. Q199.F57

Flew, Antony. 1964. 1984 (2nd ed.). *Dictionary of Philosophy.* St. Martin's Press, New York, NY 380 p. LCCC No. 78-68699

Flugge, Wilhelm. 1962. 1992. *Handbook of Engineering Mechanics.* McGraw-Hill Book Co., New York, NY various pagings TA350.F58

Folk, George Edgar. 1966. *Introduction to Environmental Physiology.* Lea & Febiger, Philadelphia, PA 308 p. QP82.F62

Fraser, D. A. S. 1976. *Probability and Statistics: Theory and Application.* Duxbury Press, North Scituate, MA 623 p. QA273.F82

Freiberger, W. F. 1960. *International Dictionary of Applied Mathematics.* D. Van Nostrand, New York, NY 1173 p. QA5.I5

Freudenthal, A. M. 1950. *The Inelastic Behavior of Engineering Materials and Structures.* John Wiley and Sons, New York, NY 587 p. TA350.F87

Friday, Adrian and Ingram, David S. 1985. *The Cambridge Encyclopedia of Life Sciences.* Cambridge University Press, Cambridge, England 432 p. QH307.2.C36

Friel, John P. (ed.). 1974 (25th ed.).1985 (26th ed.). *Dorland's Illustrated Medical Dictionary.* W. B. Saunders, Philadelphia, PA 1748 p. R121.D73

Frocht, M. M. 1946. *Photoelasticity.* John Wiley & Sons, New York, NY. In two volumes TA406.F74

Frye, Keith. 1981. The Encyclopedia of Minerology. Hutchinson Press Publishing Co., Stroudsburg, PA 794 p. QE355.E49

Fryling, Glenn R. 1966. See de Lorenzi, O. 1951.

Fuller, Robert W., Brownlee, R. B., and Baker, D. Lee. 1932 and following. Baker became senior author in 1953. *First Principles of Physics.* Allyn and Bacon, New York, NY 799 p. QC23.F84

Fulton, John F., et al. (ed.). 1950. *Howell's Textbook of Physiology.* W. B. Saunders Co., Philadelphia, PA 1304 p. QP31.F8

Gamow, George. 1966. 1972. *Thirty Years That Shook Physics.* Doubleday Anchor, Garden City, NY 224 p. QC173.98.G35

Gamow, George. 1970. *My World Line.* Viking Press, New York, NY 178 p. QC16.G37A3

Gamow, George. 1961. 1988. *The Great Physicists from Galileo to Einstein.* Dover Publications, New York, NY 378 p. QC7.G27

Garrison, Fielding H. 1929 (4th ed.). 1967. *An Introduction to the History of Medicine.* W. B. Saunders Co., Philadelphia, PA 996 p R131.G3

Gartenhaus, Solomon. 1964. *Elements of Plasma Physics.* Holt, Rinehart and Winston, New York, NY 198 p. QC718.G37

Gary, M., McAfee, R. and Wolf, C. L. (ed.). 1972. *Glossary of Geology and Related Sciences.* American Geological Institution, Washington, DC 805 p + app. QE5.G37

Geison, Gerald L. 1995. *The Private Science of Louis Pasteur.* Princeton University Press, Princeton, NJ 378 p. Q143.P2G35

Gennaro, Alfonso R. 1979 (4th ed.). *Blakiston's Gould Medical Dictionary.* McGraw-Hill Book Co., New York, NY 1632 p. R121.B62

Gerhart, Philip M., et al. 1992 (2nd ed.). *Fundamentals of Fluid Mechanics.* Addison-Wesley Publishing Co., Reading, PA 983 p. TA357.G46

Gibson, Carol (ed.). 1981. 1988 (3rd ed.). *Dictionary of Mathematics.* Facts On File, New York, NY 235 p. QA5.G52

Giedt, Warren H. 1957. *Principles of Engineering Heat Transfer.* Van Nostrand, Princeton, NY 372 p. TJ260.G5

Gillispie, Charles C. 1981. *Dictionary of Scientific Biography.* Charles Scribner's Sons, New York, NY Supplement I, vol 15 & 16 818 + 503 p. Q141.D5 (Council of Learned Societies) See also Holmes, F. L. 1980. 1990

Gillispie, Charles C. 1991. *Biographical Dictionary of Mathematics.* Charles Scribner's Sons, Hew York, NY. In four volumes 2696 p. QA28.B534 (Council of Learned Societies)

Glanze, Walter O., Anderson, K. N. and Anderson, Lois E. 1985. 1992. *The Mosby Medical Encyclopedia.* New American Library, New York, NY; Plume, New York, NY 905 p. R125.M63 RC81.A2M67

Glasser, Otto. 1944. *Medical Physics.* The Yearbook Publishers, Chicago, IL 1744 p. QH505.G55

Glasstone, Samuel. 1946 (2nd ed.). Textbook of Physical Chemistry. D. Van Nostrand Co., New York, NY 1320 p. QD453.G55

Glasstone, Samuel. 1958. 1964. *Sourcebook on Atomic Energy.* U. S. Atomic Energy Commission, Washington, DC and Van Nostrand, Princeton, NJ 641 p. QC776.G6

*Glossary of Behavior.* See Verplanck, W. S. 1957.

*Glossary of Chemical Terms.* See Hampel, C. A. and Hawley, G. G. 1982.

*Glossary of Genetics and Other Biological Terms.* See Clapper, R. B. 1961.

*Glossary of Geology and Related Sciences.* See American Geological Institute (A. G. I.) 1957; Bates, R. L. and Jackson, J. A. 1980. 1987; Gary, M., et al. 1972.

*Glossary of Meteorology.* See Huschke, R. (American Meteorology Society) 1959. 1960; McIntosh, D. H. 1972.

*Glossary of Oceanographic Terms.* See Baker, B. B., et al. 1960. 1966.

*Glossary of Organic Chemistry.* See Patai, S. 1962.

*Glossary of Physics.* See Weld, L. D. 1937.

*Glossary of Some Terms Used in the Objective Science of Psychology.* See Verplanck, W. S. 1957.

*Glossary of Terms Used in Heat Transfer, Fluid Flow and Related Terms.* See Begell, W. 1983.

Goldenson, Robert M. 1984. *Longman Dictionary of Psychology and Psychiatry.* Longman, New York and London 816 p. BF31.L66

Golding, Edward W. 1955. *The Generation of Electricity by Wind Power.* Philosophical Library, New York, NY and E. & F. N. Spon, London 318 p. TK1541.G6

Goldsmith, Werner. 1960. *Impact - The Theory and Physical Behavior of Colliding Solids.* Edward Arnold, Ltd., London, England 379 p. TA354.G63

Golob, Richard and Brus, Eric. 1990. *Almanac of Science & Technology - What's New.* Harcourt, Brace, Jovanovich, Boston, MA 530 p. Q158.5.A47

Good, Carter V. 1973. *Dictionary of Education.* McGraw-Hill, Inc., New York, NY 681 p. LB15.G6

Goss, Charles M. 1973 (29th ed.). *Gray's Anatomy.* Lea & Febigner, Philadelphia, PA 1466 p. QM23.2. See also Williams, Peter L. 1989.

Goudie, Andrew, et al. (ed.). 1985. *Encyclopedic Dictionary of Physical Geography.* Blackwell, Oxford 528 p. GB10.E53

Graf, Rudolf F. 1968. *Modern Dictionary of Electronics.* Bobbs-Merrill Co., Inc., New York, NY 593 p. TK7804.H6

Graham, Elsie C. (ed.). 1967. *The Basic Dictionary of Science.* Macmillan Co., New York, NY 568 p. Q123.G7

Gray, Dwight E. (ed.). 1963. 1972. *American Institute of Physics Handbook.* McGraw-Hill Book Co., New York, NY various paging QC61.A5

Gray, Harold J. 1958. 1962. *Dictionary of Physics.* Longman's, Green and Company, London, England and John Wiley, New York, NY 554 p. QC5.G7

Gray, Harold J. and Isaacs, Alan. 1975. *A New Dictionary of Physics.* Longmans, London, England. 619 p. QC5.G7

Gray, Peter. 1967. *The Dictionary of the Biological Sciences.* Reinhold Publishing Corp., New York, NY 602 p. QH13.G68

Gray, Peter. 1970 (2nd ed.). *The Encyclopedia of the Biological Sciences.* Van Nostrand Reinhold Co., New York, NY 1027 p. QH13.G7

Grayson, Martin. 1978. 1984. 1995-1998. *Kirk-Othmer Encyclopedia of Chemical Technology.* John Wiley & Sons, New York, NY. In 24 volumes TP9.E685

Greene, Jay E. (ed.). 1966. 1968. *McGraw-Hill Modern Men of Science.* McGraw-Hill Book Co., New York, NY. In two volumes Followed by a later edition - Parker, S. P. 1980. Modern Scientists and Engineers. In three volumes Q141.M15

Greenkorn, Robert A. 1983. *Flow Phenomena in Porous Media.* Marcel Dekker, Inc., New York, NY 550 p. GB1197.7.G73

Greenwald, Douglas, et al. 1983. *McGraw-Hill Dictionary of Modern Economics.* McGraw-Hill Book Co., New York, NY 632 p. HB61.M3

Griem, Hans R. 1964. *Plasma Spectroscopy.* McGraw-Hill Book Co., New York, NY 580 p. QC718.G7

Grigull, Ulrich, et al. 1982. *Origins of Dimensionless Groups of Heat and Mass Transfer,* 7th International Heat Transfer Conference, Technical University of Munich, Munich, Germany 17 p.

Gruyter, Walter de (publisher). 1994. *Concise Encyclopedia of Chemistry.* Gruyter, Berlin and New York 1201 p. QD4.A2313 Translated and revised from German by Mary Eagleson

Guillen, Michael. 1995. *Five Equations that Changed the World.* Hyperon, New York, NY 277p. QC24.5G85

Hall, Allen S., Jr. 1961. *Kinematics and Linkage Design.* Prentice-Hall, Englewood Cliffs, NJ 162 p. TJ182.H3

Hall, Carl W. 1977. *Errors in Experimentation.* Matrix Publications, Inc., Champaign, IL 170 p. QC39.H284

Hall, Carl W. 1979. *Dictionary of Drying.* Marcel Dekker, Inc., New York, NY 350 p. TP363.H274.

Hampel, Clifford A. (ed.). 1964. T*he Encyclopedia of Electrochemistry.* Reinhold Publishing Co., New York, NY 1206 p. QD553.H3

Hampel, Clifford A. and Hawley, G. G. (ed.). 1966. 1973 (3rd ed.). *The Encyclopedia of Chemistry.* Van Nostrand Reinhold Co., New York, NY 1198 p. QD5.E58

Hampel, Clifford A. and Hawley, G. G. 1982 (2nd ed.). *Glossary of Chemical Terms.* Van Nostrand Reinhold, New York, NY 306 p QD5.H34

*Handbook (Agricultural Engineers).* See Richey, C. B., et al. 1961.

*Handbook for Mechanical Engineers.* See Avallone, E. and Baumeister, T. 1996; Baumeister, T. 1967; Kreith, F. 1997; Tapley, B. D. 1990.

*Handbook for Sound Engineering.* See Ballou, G. M. 1987. 1991.

*Handbook of Biography.* See Holmes, F. L. 1980. 1990.

*Handbook of Biomaterials Engineering.* See Bronzino, J. D. 1995.

*Handbook (Chemical Engineering).* See Perry, J. H. 1950. 1963. 1973; Perry, R. 1984.

*Handbook of Chemistry.* See Dean, J. A. (Lange) 1979. 1992; Lange, N. A. 1952.

*Handbook of Chemistry and Physics* (CRC). See Hodgman, C. D. 1952 ff.; See Lide, D. R. 1980 through 1996; Weast, R. C. through 1997.

*Handbook of Current Science and Technology.* See Bunch, B. 1992.

*Handbook of Electronics.* See Whitaker, J. C. 1996.

*Handbook of Engineering Fundamentals* (Eshbach). See Tapley, B. D. 1990.

*Handbook of Engineering (Sciences).* See Bolz, R. E. and Tuve, G. L. 1970. 1973; Dorf, R. C. 1995; Eshbach, O. W. and Saunders, M. 1975; Potter, James H. 1967; Tapley, B. D. 1990.

*Handbook of Engineering Mathematics.* See Tuma, J. J. and Walsh, R. A. 1970. 1998.

*Handbook of Engineering Mechanics.* See Flugge, W. 1962. 1992.

*Handbook of Fluid Dynamics and Fluid Machinery.* See Schetz, J. A. and Fuhs, A. G. 1996.

*Handbook of Fluid Mechanics.* See Streeter, V. L. 1961. 1966.

*Handbook of Heat Transfer.* See Kakac, S. et al. 1987; Rohsenow, W. M. and Hartnett, J. P. 1973. 1985.

*Handbook of Instruments and Control.* See Considine, D. M. 1957. 1974.

*Handbook of Mathematics for Scientists and Engineers.* See Korn, G. A. and T. M. 1961. 1968.

*Handbook of Mathematics, Scientific, and Engineering Formula.* See Research and Education Association (REA). 1994.

*Handbook (of Mechanical Design).* See Rothbart, M. A. 1985. 1996.

*Handbook of Mechanics, Materials, and Structures.* See Blake, A. 1985.

*Handbook of Noise Control.* See Harris, C. M. 1957.

*Handbook of Optics.* See Bass, M. 1995; Driscoll, W. G. 1978. 1995.

*Handbook of Physics.* See Condon, E.U. and Odishaw, H. 1958; Gray, D. W. (AIP) 1963. 1972; Lide, D. R. 1980-1997; Trigg, George L. 1991.

*Handbook of Reliability Engineering.* See Ushokov, I. A. and Harrison, R. A. 1994.

*Handbook of Science.* See Bunch, B. 1992.

*Handbook of Sound Engineering.* See Ballou, G. M. 1987. 1991.

*Handbook of System Engineering.* See Machol, R. E. 1965.

*Handbook of Tables for Engineering Science.* See Bolz, R. E. and Tuve, G. L. 1970. 1973.

*Handbook of X-rays.* See Kaeble E. F. 1967.

Hanson, Herbert C. 1962. *Dictionary of Ecology.* Philosophical Library, New York, NY 382 p. QH541.H25

Hanson, J. L. 1986 (6th ed). *A Dictionary of Economics and Business.* Pitman Publishing Ltd., London 396 p. HB61.H35

Harmon, Peter M. 1982. *Energy, Force, and Matter.* Cambridge University Press, Cambridge, England 182 p. QC7.H257

Harre, Rom and Lamb, Roger (ed.). 1983. *The Encyclopedic Dictionary of Psychology.* MIT Press, Cambridge, MA 718 p. BF31.E4 (or .E555)

Harriman, Philip Lawrence. 1947. *The New Dictionary of Psychology.* Philosophical Library, New York, NY 364 p. BF31.H34

Harris, Cyril M. 1957. *Handbook of Noise Control.* McGraw-Hill Book Co., New York, NY various paging TA365.H3

Harris, William, and Levey, Judith (ed.). 1974 (4th ed.) *The New Columbia Encyclopedia.* Columbia University Press, New York, NY 3052 p. AG5.C725

Hatsopoulos, George N. and Keenan, Joseph H. 1965. *Principles of General Thermodynamics.* John Wiley & Sons, New York, NY 788 p. QC311.H328

Haus, Hermann and Melcher, James R. 1989. *Electromagnetic Fields and Energy.* Prentice Hall, Englewood Cliffs, NJ 742p. QC665.E4.H38

Hausmann, Erich and Slack, Edgar P. 1946 (2nd ed.). *Physics.* D. Van Nostrand Co., New York, NY 756 p. QC21.H385

Hawkes, Herbert E., Luby, W. A. and Touton, F. C. 1911. *Second Course in Algebra.* Ginn & Co., Boston, MA 264 p. QA152.H44

*Hawley's Condensed Chemical Dictionary.* See Lewis, R. J., Sr. (ed.). 1987. 1993.

Hazewinkel, M., managing editor. 1987; I. M. Vinogradov, editor-in-chief of the original Soviet volumes. *Encyclopedia of Mathematics.* Kluwer Academic Publishers, Dordrecht/Boston/London. In 10 volumes QA5.M3717

Hazelrigg, George A. 1996. *Systems Engineering: An Approach to Information-Based Design.* Prentice-Hall, Upper Saddle River, NJ 469 p. TA168.H39

Hechtlinger, Adelaide. 1959. *Bobbs-Merrill's Modern Science Dictionary.* Franklin Publishing Co., Palisade, NJ 784 p. Q123.H35

Hellemans, Alexander and Bunch, Bryan (ed.). 1988. 1991. *The Timetables™ of Science.* Simon and Schuster, New York, NY 660 p. Q125.H557

Hempel, Carl G. 1965. *Aspects of Scientific Explanations.* The Free Press, New York, NY 505 p. Q175.H4834

Henderson, Isabelle F. and Henderson, W. D. (ed.). 1963. 1989 (10th ed.). *Dictionary of Biological Terms.* John Wiley & Sons, New York, NY 637 p. QH13.H38 See also Kenneth, J. H. 1963.

*Henry Holt Handbook of Current Science and Technology.* 1992. Bryan Bunch, (ed). Henry Holt and Co., New York, NY 689 p. Q158.5B86

Hensyl, William R. 1966. 1982. 1990 (25th ed.). *Stedman's Medical Dictionary* (Thomas L. Stedman). Williams and Wilkins, Baltimore, MD 1678 p. R121.S8

Hewitt, Geoffrey F., Shires, G. L. and Polyzhaev, Y. V. 1997. *International Encyclopedia of Heat and Mass Transfer.* CRC Press, Boca Raton, FL 1824 p. ISBN 0-8493-9356-6

Heyman, Jacques and Leckie, F. A. (ed.). 1968. *Engineering Plasticity.* Cambridge University Press, London, England 706 p. TA1418.14E5

Higdon, A., Ohlsen, E. H., Stiles, W. B., Weese, J. A. and Riley, W. F. 1985. *Mechanics of Materials.* John Wiley & Sons, New York, NY 744 p. TA405.M515

Hix, C. F. and Alley, R. P. 1958. *Physical Laws and Effects.* John Wiley & Sons, New York, NY 291 p. QC28.H53

Hodgman, Charles, D. 1952 (34th ed.) and many succeeding editions. *Handbook of Chemistry and Physics.* Chemical Rubber Co., Cleveland, OH See also Lide, D. R. and Weast, R. C. for more recent editions. QD65.H3

Hogerton, John F. 1963. *The Atomic Energy Deskbook.* Reinhold Publishing Corp., New York, NY 673 p. QC772.H64

Holmes, Frederic L. (ed.). 1980. 1990. *Dictionary of Scientific Biography. American Council of Learned Societies,* Charles Scribner's Sons, New York, NY Vol. 17 and 18 Q141.D5 See Gillispie, C. C. 1981

Honerich, T. 1995. The Oxford Companion to Philosophy. Oxford University Press, Oxford, UK 1009 p. B51.O94

Honig, Jurgen M., et al. 1953. *The Van Nostrand Chemist's Dictionary.* D. Van Nostrand Co., New York, NY 761 p. QD5.V36

Houghton, John T. 1986. *The Physics of Atmospheres.* Cambridge University Press, Cambridge, England 271 p. QC880.H68

Houssay, Bernardo A., et al. 1955 (2nd ed.). *Human Physiology.* McGraw-Hill Book Co., New York, NY 1177 p. QP34.F5314

Howell, J. V. See American Geological Institute (AGI). 1957. 1960.

Houwink, R., and DeDecker, H. K. 1937. 1971 (3rd ed.). *Elasticity, Plasticity, and Structure of Matter.* Dover Publications, New York, NY 470 p. QC189.H68

Hui, Y. H. 1992. *Encyclopedia of Food Science and Technology.* John Wiley & Sons, New York, NY 2972 p. In four volumes TP368.2.E62

Hull, David L. 1974. *Philosophy of Biological Sciences,* Prentice-Hall, Englewood Cliffs, NJ 148 p. QH331.H84

Hunt, V. Daniel. 1979. *Energy Dictionary.* Van Nostrand Reinhold, New York, NY 518 p. TJ163.2S3

Hurlbut, C. S., Jr. 1952 (16th ed.). 1971 (18th ed.). *Dana's Manual of Minerology.* John Wiley & Sons, New York, NY 579 p. QE372.D2

Huschke, Ralph E. (ed.). 1959. *Glossary of Meteorology.* American Meteorology Society, Boston, MA 638 p. QC854.G55

Hyne, Norman J. (ed.). 1991. *Dictionary of Petroleum.* PennWell Books, Tulsa, OK 625p. TN865.H96

*Industrial Arts Index,* 1914-1957. See Applied Science & Technology Index.

Ingersoll, Leonard R., Zobel, Otto J., and Ingersoll, Alfred C. 1954. *Heat Conduction.* University of Wisconsin Press, Madison, WI 325 p. QC321.I68

*International Dictionary of Applied Mathematics.* See Freiberger, W. F. 1960.

*International Dictionary of Geophysics.* See Runcorn, S. K. 1967.

*International Dictionary of Medicine and Biology.* See Landau, S. I. 1986.

*International Dictionary of Physics and Electronics.* See Michels, W. C. 1956. 1961.

*International Dictionary of 20th Century Scientists.* See Vernoff, E. and Shore, R. 1987.

*International Encyclopedia of Chemical Science.* 1964. No editor listed; first contributor is Clifford, A. F. 1964.

*International Encyclopedia of Heat and Mass Transfer.* See Hewitt, G. F. et al. 1997.

Ipsen, C. C. 1960. *Units, Dimensions, and Dimensionless Numbers.* McGraw-Hill Book Co., New York, NY 236 p. QC39.I69

Irodov, Igor E. 1988. *Problems in General Physics.* Mir Publications, Moscow and Chicago 386 p. QC32.I698

Isaacs, Alan, Daintith, J. and Martin, E. 1991. 1996 (3rd ed.). *Concise Science Dictionary.* Oxford University Press, Oxford and New York 794 p. Q123.C68

Isaacs, Alan. 1996 (3rd ed.). *A Dictionary of Physics.* Oxford University Press, Oxford and New York 475 p. QC5.D496

Ito, Kiyosi (ed.). 1987. *Encyclopedic Dictionary of Mathematics,* by the Mathematical Society of Japan, published by MIT Press, Cambridge, MA. In 4 volumes QA5.I8313

Jakob, Max. 1949 (v. 1) and 1957 (v.2). *Heat Transfer.* John Wiley & Sons, New York, NY and Chapman and Hall, London 758 p. (v. 1) and 652 p. (v. 2) QC320.J22

Jakubke, H. D., et al. See Eagleson, M. 1994.

James, A. M. 1976. *A Dictionary of Thermodynamics.* John Wiley & Sons, New York, NY 262 p. QC310.3.J35

James, Robert C. 1992 (5th ed.). *Mathematics Dictionary.* D. Van Nostrand Co., New York, NY 548 p. QA5.J33

James, Robert C. and James, Glenn. 1968 (3rd ed.). *Mathematics Dictionary.* D. Van Nostrand Co., New York, NY 517 p. QA5.J3

Jammer, Max. 1957. *Concepts of Force.* Harvard University Press, Cambridge, MA 269 p. QC73.J3

Jerrard, H. G. and McNeill, D. B. 1992. *A Dictionary of Scientific Units including Dimensionless Numbers and Scales.* Chapman & Hall, London 255 p. QC82.J4

John, Vernon. 1992 (3rd ed.). *Introduction to Engineering Materials.* Industrial Press, New York, NY 536 p. TA403.J54

Kaelble, Emmett F. 1967. *Handbook of X-rays.* McGraw-Hill Book Co., New York, NY various pagings QC481.K15

Kakac, Sadik, Shah, R. K. and Aung, W. 1987. *Handbook of Single-Phase Convective Heat Transfer.* Wiley-Interscience, New York, NY various pagings QC320.4.H37

Karush, William. 1989. *Webster's New World Dictionary of Mathematics.* Macmillan, New York, NY 317 p. QA5.K27

Kay, Ernest. 1984. *International Who's Who in Engineering.* International Biographical Centre, Cambridge, England 589 p. ISBN 0 900332 719

Kaye, G. W. C. and Laby, T. H. 1995 (16th ed.). *Tables of Physical and Chemical Constants.* Longman, London 611 p. QC61.K3

Keenan, Joseph H. 1948. *Thermodynamics.* John Wiley & Sons, New York, NY 499 p. QC311.K35

Kenneth, John H. (ed.). 1957. 1963 (8th ed.). *A Dictionary of Scientific Terms (Biology)* by Henderson, Isabella and Henderson, W. D. Van Nostrand Co., Princeton, NJ 640 p. QH13.H496. See also Henderson, I. F. and Henderson, W. D.

Ketchum, Milo S. 1919 (3rd ed.). *The Design of Walls, Bins, and Grain Elevators.* McGraw-Hill Book Co., New York, NY 536 p. TH4498.K4

King, Robert C. and Stansfield, William D. 1990 (4th ed.). *Dictionary of Genetics.* Oxford University Press, New York and Oxford 406 p. QH427.K55

Kirk-Othmer *Concise Encyclopedia of Chemical Technology.* See Grayson, M. 1995.

Kirk-Othmer *Encyclopedia of Chemical Technology.* See Grayson, M. 1978. 1984; Kroschwitz, J. I. and Howe-Grant, M. 1991-1998.

Kleiber, Max. 1961. 1975. *The Fire of Life.* John Wiley & Sons, New York, NY 454 p. QP141.K56

Koenigsberg, Ruth (managing editor). 1989. *Churchill's Medical Dictionary.* Published by Churchill Livingstone, New York, NY 2120 p + appendix R121.I58

Korn, Granino A. and Korn, Theresa M. 1961. 1968 (2nd ed.). *Mathematical Handbook for Scientists and Engineers.* McGraw-Hill Book Co., New York, NY 1130 p. QA40.K598

Kotz, Samuel and Johnson, Norman L. 1983. 1988. *Encyclopedia of Statistical Sciences.* John Wiley & Sons, New York, NY. In 9 vol. + supp. QA276.14.E5

Kraus, John D. 1995 (2nd ed.). *Big Ear Two.* Cygnus-Quasar Books, Powell, OH 370 p. QB36.K7A33

Krauskopf, Konrad B. and Beiser, Arthur. 1959 (4th ed.). 1971 (6th ed.). *Fundamentals of Physical Science.* McGraw-Hill Book Co., New York, NY 947 p. Q160.2K7

Kreith, Frank. 1997. *Mechanical Engineering Handbook.* CRC Press, Boca Raton, FL c. 2000 p. ISBN 0-8493-9418-X

Kroschwitz, Jacqueline I. and Howe-Grant, Mary (ed.). 1991-1998 (4th ed.). *Kirk-Othmer Encyclopedia of Chemical Technology.* John Wiley & Sons, New York, NY. In 27 volumes TP9.E685

Kutateladze, S. S. and Borishanskii, V. M. 1966. *A Concise Encyclopedia of Heat Transfer.* Pergamon Press, Oxford, New York, NY 489 p. QC320. K83613

Lagowsky, Joseph J. 1997. *Macmillan Encyclopedia of Chemistry.* Simon & Schuster Macmillan, New York, NY 1696 p. In four volumes QD4.M33

Land, Norman S. 1972. *A Compilation of Nondimensional Numbers.* National Aeronautics and Space Administration, NASA SP-274, Washington, DC 123 p. TA347.D52L36

Landau, Sidney I.(ed.). 1986. *International Dictionary of Medicine and Biology.* John Wiley & Sons, New York, NY 3200 p. In 3 volumes R121.I58

Lange, Norbert A. 1952. (8th ed.). *Handbook of Chemistry.* Handbook Publishers, Sandusky, OH. See Dean, John A. 1979 (12th ed.) for more recent editions, published by McGraw-Hill Book Co., New York, NY 1100 p. QD65.L36

Lapedes, Daniel N. (ed.). 1971. 1974. *McGraw-Hill Encyclopedia of Science and Technology.* McGraw-Hill Book Co., New York, NY See Parker, S. P. (ed.) for more recent editions.

Lapedes, Daniel N. (ed.). 1978. *McGraw-Hill Dictionary of Physics and Mathematics.* McGraw-Hill Book Co., New York, NY 1074+A46 p. QC5.M23

Lapedes, Daniel N. (ed.). 1978. *McGraw-Hill Encyclopedia of Geological Sciences.* McGraw-Hill Book Co., New York, NY 915 p. QE5.M29

Lapedes, Daniel N. (ed.). 1978 (2nd ed.). 1982. *McGraw-Hill Dictionary of Scientific and Technical Terms.* McGraw-Hill Book Co., New York, NY 1771+A58 p. Q123.M15

Lapedes, Daniel N. (ed.). 1976. *Dictionary of the Life Sciences.* McGraw-Hill Book Co., New York, NY 907 p.+app. QH302.5.M3

Lapp, Ralph E., et al. 1963. *Matter.* Life Science Library, Time, Inc., New York, NY 200 p. QC171.L33

*Larousse Dictionary of Scientists.* See Muir, Hazel. 1994.

Lay, Joachim Joe E. 1963. *Thermodynamics.* Charles E. Merrill Books, Columbus, OH 814 p. TJ265.L35

Layton, Edwin T. and Lienhard, John H. 1988. *History of Heat Transfer.* American Society of Mechanical Engineers, New York, NY 260 p. TJ260.H57

Lederman, Leon. 1993. *The God Particle.* Dell Publishing, a division of Bantam, New York, NY 434 p. QC793.5B62.L43

Leicester, Henry M. and Klickstein, Herbert S. 1952. *A Source Book in Chemistry,* 1400-1900. McGraw-Hill Book Co., New York, NY 554 p. QD3.L47

Lerner, Rita G. and Trigg, George L. 1981. 1991. *Encyclopedia of Physics.* Addison-Wesley Publishing Co., Reading, MA 1408 p. QC5.E545

Levine, Ira N. 1988. 1995 (4th ed.). *Physical Chemistry.* McGraw-Hill Book Co., New York, NY 901 p. QD453.2L48

Lewis, Richard J. Sr. 1987. 1993. *Hawley's Condensed Chemical Dictionary.* D. Van Nostrand Reinhold, New York, NY 1275 p. QD5.C5

Lewytkj, Borys. 1984. *Who's Who in the Soviet Union.* K. G. Saur, Munich, Germany 428 p. DK37.W4

Lipschutz, Seymour and Schiller, John J. 1966. 1995. *Finite Mathematics.* Schaum Outline Series, McGraw-Hill Book Co., New York, NY 483 p. QA43.L67

Lide, David R. 1990-91. 1996-97 (77th ed.). *CRC Handbook of Chemistry and Physics.* CRC Press, Roca Raton, FL various paging QD65.H3 Previous editions by Hodgman, C. D. and Weast, R. C.

Lincoln, R. J., et al. 1982. *A Dictionary of Ecology, Evolution and Systematics.* Cambridge University Press, Cambridge, England and New York 298 p. QH540.4.L56

Link, Albert N. 1993. *Link's International Dictionary of Business Economics.* Probus Publishing Co., Chicago, IL and Cambridge, England 193 p. RB61.L55

Lorenzi, Otto. 1951. See de Lorenzi.

Lotka, Alfred J. 1956. *Elements of Mathematical Biology.* Dover Publications, New York, NY 306 p. QH307.L75

Lowrie, Robert S. 1965. *Concise Notes on Physical Chemistry.* Oxford and Pergamon Press, New York, NY 142 p. QD453.L6C6

Lykov (Luikov), A. V. and Mikhaylov, Y. A. 1961. Translated from Russian by William Begell. *Theory of Energy and Mass Transfer.* Prentice-Hall, Inc., Englewood Cliffs, NJ 324 p. QC320.L883

McAinsh, T. F. (ed.). 1986. *Physics in Medicine & Biology Encyclopedia.* Pergamon Press, Oxford, England. In two volumes 980 p. R895.A3P47

McCabe, Warren L., Smith, Julian C. and Harriott, Peter. 1956. 1985 (4th ed.). *Unit Operations of Chemical Engineering.* McGraw-Hill Book Co., New York, NY 945 p.; 960 p. TP155.M12U5

*McGraw-Hill Concise Encyclopedia of Science and Technology.* See Parker, Sybil P. 1989 (2nd ed.).

*McGraw-Hill Dictionary of Geological Sciences.* See Lapedes, D. N. 1978.

*McGraw-Hill Dictionary of Science and Technology.* See Parker, Sybil. 1984. 1994.

*McGraw-Hill Dictionary of Scientists and Engineers.* See Greene, J. E. 1966. For more recent editions See Parker, S. P. 1980.

*McGraw-Hill Encyclopedia of Science and Technology.* See Lapedes, D. N. 1971. 1974; Parker, Sybil P. 1988. 1994 (7th ed.).

*McGraw-Hill Encyclopedia of World Biography.* See Eggenberger, D. I. 1973.

*McGraw-Hill Modern Men of Science.* See Greene, Jay E. 1966. Also see Parker, S. P. 1980.

Machol, Robert E. 1965. *System Engineering Handbook.* McGraw-Hill Book Co., New York, NY various paging TA168.M25

McIntosh, D. H. 1972 (5th ed.). *Meteorology Glossary.* Chemical Publishing Co., New York, NY 319 p. QC854.G7

McKeever, Susan (ed.). 1993. *Random House Science Dictionary.* Dorling Kindersley, London 448 p. Q121.R36

McKetta, John J. and Cunningham, William. 1976-1996. *Encyclopedia of Chemical Process Design.* Marcel Dekker, Inc., New York, NY. In 54 volumes TP9E66

*Macleod's Physiology in Modern Medicine.* See Bard, P. 1941.

McMurray, Emily J. (ed.) 1995. *Notable Twentieth Century Scientists*. Gale Research, Inc., New York, NY. In four volumes Q141.N73

Macpherson, Gordon (ed.). 1992 (37th ed.). *Black's Medical Dictionary*. Barnes & Noble, Lanham, MD 645 p. R121.B598

McPherson, W., Henderson, W. E., Fernelius, W. C. and Quill, L. L. 1942. *Introduction to College Chemistry*. Ginn and Company, Boston, MA 606 p. QD31.M315

Maeterlinck, Maurice. 1969. *The Supreme Law*. Kennikat Press, Port Washington, NY 160 p. QC178.M32

Magill, Frank N. (ed.). 1989. *The Great Scientists*. Grolier Educational Corp., Danbury, CT. In 12 volumes Q141.G767

Malinowsky, Harold R. and Richardson, Jeanne. 1980 (3rd ed.). *Science and Engineering Literature*. Libraries Unlimited, Littleton, CO 342 p. Z7401.M28

Mandel, Siegfried. 1969. 1972. *Dictionary of Science*. Dell Publishing Co., New York, NY 407 p. Q123.M25

Maran, Stephen P. 1992. *The Astronomy and Astrophysics Encyclopedia*. Van Nostrand Reinhold, New York and Cambridge University Press, Cambridge, England 1002 p. QB14.A837

Marcorini, Edgardo. 1975. 1988. *The History of Science and Technology*. Facts On File, New York, NY. In two volumes Q125.S43713

*Marks' Standard Handbook for Mechanical Engineers*. See Avallone, E. and Baumeister, T. 1996; Baumeister, T. 1967.

Markus, John and Cooke, N. M. 1960. *Electronics and Nucleonics Dictionary*. McGraw-Hill Book Co., New York, NY 543 p. TK7804.C62

Maron, Samuel H. and Prutton, Carl F. 1965. *Fundamental Principles of Physical Chemistry*. Macmillan Co., New York, NY 836 p. QD453.P78

Mascetta, Joseph A. 1996 (3rd ed.). *Chemistry the Easy Way*. Barron's Educational Series, Hauppauge, NY 448 p. QD41.M38

Marquis Who's Who. 1976. *Who Was Who in American History - Science and Technology*. Chicago, IL 688 p. Q141.W5

Martin, Elizabeth A., et al. 1985. 1990. 1996 (3rd ed.). *(Concise) Dictionary of Biology*. Oxford University Press, New York, NY 553 p. QH302.5D5

*Mechanical Engineers in America Born Prior to 1861 - A Biographical Dictionary*. 1980. American Society of Mechanical Engineers, New York, NY 330 p. TJ139.A47

*Melloni's Illustrated Medical Dictionary*. See Dox, Ida, et al. 1979. 1985.

Mendelsohn, Everett. 1964. *Heat and Life*. Harvard University Press, Cambridge, MA 208 p. Q135.M45

Menzel, Donald H. 1960. *Fundamental Formulas of Physics*, Vol. 1 and Vol. 2. Dover Publications, New York, NY 741 p. In two volumes QA401.M492

*Merriam-Webster Biographical Dictionary*. 1995. Merriam-Webster, Inc., Springfield, MA 1170 p CT103.M41

*Merriam-Webster Medical Desk Dictionary*. 1993. Merriam-Webster, Springfield, MA 790 p. R121.M564

Merriman, Arthur D. 1958. *A Dictionary of Metallurgy*. (In 1965. A Concise Dictionary of Metallurgy) Mcdonald & Evans, London, England 1178 p. TN609.M475

Merriman, Arthur D. 1965. *A Concise Encyclopedia of Metallurgy*. American Elsevier Publishing Co., New York, NY 1178 p. TN609.M475

Mettler, Cecilia C. 1947. *History of Medicine*. The Blakiston Co., Philadelphia and Toronto 1215 p. R131.M59

Meyers, R. A. (ed.) 1992. *Encyclopedia of Physical Sciences and Technology*. Academic Press, San Diego, CA In18 volumes Q123.E497

Miall, L. Mackenzie and Sharp, D. W. A. 1961. 1968 (4th ed.). *A New Dictionary of Chemistry*. Interscience Publishers, New York, NY 638 p. QD5.M48

Michels, Walter C. 1956. 1961. *The International Dictionary of Physics and Electronics*. D. Van Nostrand, Princeton, NJ 1004 p. QC5.I5

Miles, Wyndham D. 1976. *American Chemists and Chemical Engineers*. American Chemical Society, Washington, DC 544 p. QD21.A43

Miles, Wyndham D. and Gould, Robert F. 1994. *American Chemists and Chemical Engineers,* vol. 2. Gould Books, Guilford, CT 365 p. QD21.A43

Millar, David, Ian, John and Margaret. 1996. *Cambridge Dictionary of Scientists.* Cambridge University Press, Cambridge, England 387 p. Q141.C128

Miller, Benjamin F. and Keane, Claire B. 1978 (2nd ed.). *Miller-Kane Encyclopedia and Dictionary of Medicine, Nursing, and Allied Health.* Saunders, Philadelphia, PA 1148 p. R121.M65 See O'Toole, Marie O. (ed.). 1992.

Milner, Richard. 1990. *The Encyclopedia of Evolution.* Facts On File, New York, NY 481 p. GN281.M53

Mitchell, James (ed.). 1983. (rev. ed.). *Random House Encyclopedia, Inc.,* New York, NY 2918 p. AG5.R25

Mitton, Jacqueline. 1991. 1993. *Dictionary of Astronomy. A Concise Dictionary of Astronomy.* Pergamon Press and Penguin Books, New York, NY 423 p. 431 p. QB14.M58

Mitton, Simon (ed.). 1973. *The Cambridge Encyclopaedia of Astronomy.* Cambridge Press, London.

*Modern Scientists and Engineers.* See Parker, Sybil. 1980.

Moffat, Donald W. 1976. 1983 (2nd ed.). *Economics Dictionary.* Elsevier, New York, NY 301 p. 331 p. HB61.M54

Moore, Herbert F. and Kommers, J. B. 1927. *The Fatigue of Metals.* McGraw-Hill Book Co., New York, NY 326 p. TA405.M65

Moore, Walter J. 1962. *Physical Chemistry.* Prentice-Hall, Englewood Cliffs, NJ 844 p. QD453.M65

Morris, Christopher G. (ed.). 1987. 1992. 1995. *Academic Press Dictionary of Science and Technology.* Academic Press, HBJ, San Diego, CA 2432 p. Q123.A33

Morris, William. 1969. 1973. *American Heritage Dictionary of English Language.* American Heritage Publishing Co., New York, NY 1550 p. PE1625.A54

Morton, L. T. and Moore, R. J. 1989. *A Bibliography of Medical and Biomedical Biography.* Scolar Press, Hants, England 208 p. R134.M67

*Mosby Medical Encyclopedia.* See Glanze, W. D. 1994.

*Mosby's Medical, Nursing, and Allied Health Dictionary.* See Anderson, Kenneth. 1994.

Muir, Hazel. 1994. *Larousse Dictionary of Scientists.* Larousse Kingfisher Chambers, Edinburgh 595 p. Q141.L36

National Academy of Sciences. Several years. Biographical Memoirs.

*The National Cyclopaedia of American Biography.* 1898 and following. James T. White Co., New York, NY. First editor was Spofford, A. R. E176.N27 (early); E176.N283 (more recent volumes)

The *National Union Catalogue.* 1968. Pre-1956 Imprint. Mansell Information, London and American Library Association, Chicago, IL 718 vol. Z663.7.L5115

Nawatty, H. E. See Arutyonyan, W. K. 1966.

Needham, Joseph. 1970. *Clerks and Craftsman in China and the West.* Cambridge University Press, Cambridge 470 p. Q127.C5N414

Nernst, Walther. 1895 ff. *Theoretical Chemistry.* Macmillan & Co., New York, NY 697 p. QD453.N44

Newman, Frederick H. and Searle, V. H. L. 1929. 1933. 1949. *The General Properties of Matter.* Macmillan and Company, and E. Arnold, London, England 431 p. QC171.N55G3

Newman, James R. 1956. *The World of Mathematics.* Simon and Schuster, New York, NY. In four volumes QA3.N48

Newman, W. A. 1985 (26th ed.). *Dorland's Illustrated Medical Dictionary.* Dorland. 1485 p. R121.D73

Nicholls, Christine S. (ed.). 1996. *The Dictionary of National Biography.* Oxford University Press, Oxford and New York 607 p. DA28.D525

Nobel Prize Winners: *Biographical Directory.* See Wasson, Taylor. 1987.

Northrup, Edwin F. 1917. *Laws of Physical Science - A Reference Book.* J. B. Lippincott Co., Philadelphia, PA 210 p. QC21.N75

Nourse, Alan E. 1969. *Universe, Earth, and Atom: The Story of Physics.* Harper & Row, Publishers, New York, NY 688 p. Q25.N67. See Miller, B. F. 1973.

Nye, J. F. 1985 (2nd ed.). *Physical Properties of Crystals.* Oxford Science Publications, Clarendon Press, Oxford. 329 p. QD931.N9

NYPL. 1989. *Desk Reference.* New York Public Library, Simon & Schuster, NY 836 p. AG6.N49 and AOL

O'Daly, Anne (ed.). 1996. *Encyclopedia of Life Sciences.* Marshall Cavendish Corp., Tarrytown, NY. In 11 volumes 1584 p. QH302.5.E53

Ochoa, George and Corey, Melinda. 1995. *The Timeline Book of Science.* Ballantine Books, New York, NY 434 p. Q125.O24

Oesper, Ralph E. 1975. *The Human Side of Scientists.* University Publications, University of Cincinnati, Cincinnati, OH 218 p. Q141.O33

Oliver, John E. and Fairbridge, Rhodes W. 1987. *The Encyclopedia of Climatology.* Van Nostrand Reinhold, New York, NY 986 p. QC854.E525

O'Toole, Marie O. (ed.). 1992 (5th ed.). *Miller-Kane Encyclopedia & Dictionary of Medicine, Nursing, & Allied Health.* W. B. Saunders Co., Philadelphia, PA 1792 p. R121.M65

*(The) Oxford English Dictionary.* 1989. See Simpson, J. A. and Weiner, E. S. C. First edition by James A. H. Murray, et al. 1923.

*Oxford Textbook of Medicine.* See Weatherall., D. J., et al. 1987. 1996.

Parker, Sybil P. (ed.). 1980. *McGraw-Hill Modern Scientists and Engineers.* McGraw-Hill Book Co., New York, NY 1366 p. In three volumes Q141.M15. For earlier editions see Greene, J. E. 1966.

Parker, Sybil P. (ed.). 1983. *McGraw-Hill Encyclopedia of Physics.* McGraw-Hill Book Co., New York, NY 1343 p. QC5.M425

Parker, Sybil P. (ed.). 1984. 1997. *McGraw-Hill Dictionary of Physics.* McGraw-Hill Book Co., New York, NY 646 p. 498 p. QC5.M424

Parker, Sybil P. (ed.). 1989. 1994 (5th ed.). *Dictionary of Science and Technology Terms.* McGraw-Hill Book Co., New York, NY 2194 p. + app. Q123. M34

Parker, Sybil P. (ed.). 1989 (2nd ed.). 1994 (3rd ed.). *Concise Encyclopedia of Science and Technology.* McGraw-Hill Book Co., New York, NY 2222 p. Q121.M29

Parker, Sybil P. (ed.). 1987. 1992. 1997 (8th ed.). *McGraw-Hill Encyclopedia of Science and Technology,* New York, NY. In 20 volumes Q121.M3

Parkinson, Claire L. 1985. *Breakthroughs: Chronology of Great Achievements in Science and Mathematics 1200-1900.* G. K. Hall & Co., Boston, MA 576 p. Q125.P327

Partington, James R. 1961. *A History of Chemistry.* Macmillan, London & St. Martin's Press, New York. In four volumes QD11.P28

Pass, Christopher, et al. 1991. *Dictionary of Economics.* HarperCollins, New York 562 p. HB61.P39

Patai, Saul. 1962. *Glossary of Organic Chemistry.* Interscience Publishers, New York, NY 227 p. QD251.P25

Pearce, David W. (ed.). 1986 (3rd ed.). 1992 (4th ed.). *(The MIT) Dictionary of Modern Economics.* The Macmillan Press, New York, NY and MIT Press, Cambridge, MA 464 p. HB61.M49 & HB61.P4

Peierls, Rudolf E. 1956. *The Laws of Nature.* Charles Scribner's Sons, New York, NY 284 p. QC23.P4

Pelletier, Paul A. (ed.). 1980. 1994 (3rd ed.). *Prominent Scientists.* Neal-Schuman Publishers, New York, NY 311 p. 353 p. Q141.P398

Perry, John H. (ed.). 1950. 1963. 1973 (5th ed.). *Chemical Engineers' Handbook.* McGraw-Hill Book Co., New York, NY 1946 p. TP155.C52

Perry, Robert H. (ed.). 1967(2nd ed.). *Engineering Manual.* McGraw-Hill Book Co., New York, NY various sections TA151.P645

Perry, Robert H. and Green, Donald W. (ed.). 1984. (6th ed). *Perry's Chemical Engineers' Handbook.* McGraw-Hill Book Co., New York, NY 1846 p. TP151.P45

Peter, Laurence and Hull, R. 1969. *The Peter Principle.* W. Morrow, New York, NY 170 p. PN6231.M2P4

Pick, T. Pickering and Howden, R. 1977. *Gray's Anatomy.* Gramercy Book, New York, NY See Williams, Peter L. 1989; Goss, Charles M. 1973.

Pierce, John R. 1961. 1965. *Symbols, Signals, and Noise.* Harper & Brothers, New York, NY 305 p. Q360.P5

Porter, Roy. 1994 (2nd ed.). *The Biographical Dictionary of Scientists.* Oxford University Press, New York, NY 891 p. Q141.B528

Porter, Roy (ed.). 1996. *The Cambridge Illustrated History of Medicine.* Cambridge University Press, Cambridge, England 400 p. R131.C232

Potter, James H., editor. 1967. *Handbook of Engineering Sciences.* D. Van Nostrand Co., Princeton, NJ. In two volumes. Vol. I, 1347 p; Vol. II, 1428 p. TA151.P79

Powell, Ralph W. 1940. *Mechanics of Liquids.* Macmillan Co., New York, NY 271 p. TC160.P88

Prager, William and Hodge, Philip G. 1951. *Theory of Perfectly Elastic Solids.* John Wiley & Sons, New York, NY 264 p. QA931.P787

Preston, Thomas. 1929 (4th ed.). *Theory of Heat.* Macmillan Co., London 838 p. QC254.P93

Prutton, Carl E. and Maron, Samuel H. 1944. *Fundamental Principles of Physical Chemistry.* Macmillan Co., New York, NY 780 p. QD453.P78 See Maron, Samuel H. and Prutton, C. F. 1965.

Raistrick, Arthur and Marshall, Charles E. 1939. 1948. *The Nature and Origin of Coal Seams.* English Universities Press, Ltd., London 282 p. TN800.R3

Ralston, Anthony, Reilly, Edwin D. and Dahlin, Caryl D. 1993 (3rd ed.). *Encyclopedia of Computer Science.* Van Nostrand Reinhold, New York, NY 558 p. QA76.15.E48

Randall, James E. 1958. *Elements of Biophysics.* Year Book Publishers, Chicago, IL 333 p. QH505.R26

*Random House Science Dictionary.* See McKeever, S. 1993.

Rashevsky, Nicholas. 1960 (3rd ed.). *Mathematical Biophysics: Physico-Mathematical.* Foundations of Biology. Dover Publications, New York, NY. In two volumes QH505.R3

*Realm of Science.* See Brown, S. B. & L. B. 1972.

Reber, Arthur S. 1985. *The Penguin Dictionary of Psychology.* Viking, New York, NY 848 p. BF31.R42

Research and Education Association (REA). 1994. *Handbook of Mathematics, Scientific, and Engineering Formula...REA,* Piscataway, NJ 1030 p. QA40.H357

Reiner, Markus. 1960 (2nd ed.). *Deformation, Strain and Flow.* H. K. Lewis, London, England 347 p. QC189.R3

Richey, C. B., Jacobson, P., and Hall, Carl W. 1961. *Agricultural Engineers' Handbook.* McGraw-Hill Book Co., New York, NY 880 p. S675.R5

Rigden, John S. (ed.) 1996. *Macmillan Encyclopedia of Physics.* Simon & Schuster, Macmillan, New York, NY 1881 p. In four volumes QC5.M15

Roback, Abraham A. 1961. *History of Psychology and Psychiatry.* Philosophical Library, New York, NY 422p. BF81.R6

Robbins, W. J., Brody, S., Hogan, A. G., Jackson, C. and Greene, C. 1928. *Growth.* Yale University Press, New Haven, CT 189 p. QP84.R63

Roberson, John A. and Crowe, Clayton T. 1975. *Engineering Fluid Mechanics.* Houghton Mifflin, Boston, MA 525 p. TA357.R6

Robertson, Morgan. 1898. *The Three Laws and the Golden Rule.* McKinley, Stone & Mackenzie, New York, NY 249 p. 813R54st

Rohsenow, Warren M., Hartnett, J. P. and Ganic, E. N. (ed.). 1973. 1985. *Handbook of Heat Transfer.* McGraw-Hill Book Co., New York, NY various pagings QC320.R528

Rosenkrantz, Roger D. 1977. *Inference, Method, and Decision.* D. Reidel Publ. Co., Boston, MA 262 p. Q175.R6

Rothbart, Harold A. (ed.). 1964. 1985. 1996. *Mechanical Design Handbook.* McGraw-Hill Book Co., New York, NY various pagings TJ230.M433 (previously titled Machine Design and Systems Handbook)

Rothenberg, Robert E. 1982. 1988. *Medical Dictionary and Health Manual.* New American Library, New York, NY 555 p. R121.R855

Rothman, Milton A. 1963. *The Laws of Physics.* Basic Books, New York, NY; also Fawcett Premier Book, Greenwich, CT 254 p. QC25.R6

Rouse, Hunter. 1946. *Elementary Fluid Mechanics.* John Wiley & Sons, New York, NY 376 p. QA911.R67

Roysdon, Christine and Khatri, Linda. 1978. *American Engineers of the Nineteenth Century - A Biographical Index.* Garland Publishing Inc., New York, NY 247 p. TA139.R7

Ruch, Floyd L. and Zimbardo, Philip G. 1971 (8th ed.). *Psychology and Life.* Scott, Foresman, and Co., Glenview, IL 756 p. BF131.R89

Runcorn, S. K. (ed.). 1967. *International Dictionary of Geophysics.* Pergamon Press, Ltd., Oxford 1728 p. In two volumes QC801.9.I.5

Runes, Dagobert, D. 1983. *Dictionary of Philosophy.* Philosophical Library, New York, NY 360 p. B41D53

Sale, Kirkpatrick. 1980. *Human Scale.* Coward, McCann & Geoghegan, New York, NY 558 p. HC106.7.S24

Sarantites, A. D. 1963. *The Universal Unified Field Law.* Universal Science Foundation, Phoenix, AR 281 p. QC6.5.S25

Schetz, Joseph A. and Fuhs, Allen E. (ed.). 1996. *Handbook of Fluid Dynamics and Fluid Machinery.* John Wiley & Sons, New York, NY. In three volumes 2628 p.+ index TA357.H286

Schick, Elizabeth A. (ed.). 1997. *Current Biography Yearbook.* H. W. Wilson Co., New York, NY First edition in 1940 by Maxine Black. CT100.C8

Schlessinger, B. S. and J. H. 1980. 1991 (2nd ed.). *Who's Who of Nobel Prize Winners 1901-1990.* ORYX Press, Phoenix, AR 234 p. AS911.N9W5

Schlicting, Hermann. 1968. 1979 (7th ed.). (Translated from German by J. Kestin). *Boundary Layer Theory.* McGraw-Hill Book Co., New York, NY 747 p. 817 p. TL574.B6S283

Schmidt, Jacob E. 1959. *Medical Discoveries Who and When.* Charles C. Thomas Publishers, Springfield, IL 555 p. R131.S35

Schneer, Cecil J. 1960. *The Evolution of Physical Science.* Grove Press, Inc., New York, NY 398 p. Q125.S48

Schneider, Walter A. and Ham, Lloyd B. 1943. 1960. *Experimental Physics for Colleges.* Macmillan Co., New York, NY 442 p. QC37.S3

Schulz, Heinrick E. et al. 1972. *Who Was Who in the USSR.* The Scarecrow Press, Inc., Metuchen, NJ 677 p. CT1212.I57

Scott, John S. (ed.). 1993(4th ed.). *Dictionary of Civil Engineering.* Chapman & Hall, New York, NY 534 p. TA9.S35

Scott, Thomas A. 1995. *Concise Encyclopedia of Biology.* Walter der Gruyter, Berlin 1287 p. QA302.5.12313

Scott Blair, George W. 1949 (2nd ed.). *A Survey of General and Applied Rheology.* Sir Isaac Pitman & Sons, London, England 314 p. QC189.S44

Shannon, Claude E. and Weaver, W. A. 1949. *The Mathematical Theory of Communications.* University of Illinois Press, Urbana, IL 125 p. TK5101.S52

Shapiro, Ascher H. 1961. *Shape and Flow, the Fluid Dynamics of Flow.* Doubleday Anchor, Garden City, NY 186 p. TL574.D7S5

Sharp, D. W. A. (ed.). 1981 (5th ed.). *Miall's Dictionary of Chemistry.* Longmans, Harlow, England 501 p. QD5.M618

Sheehy, Noel, et. al. 1997. *Biographical Dictionary of Psychology.* Routledge, London and New York 645 p. BF109.A1.B56

Shigley, Joseph E. 1961. *Theory of Machines.* McGraw-Hill Book Co., New York, NY 657 p. TJ175.S56

Shin, Jae K. and Siegel, Joel G. 1995. *Dictionary of Economics.* John Wiley & Sons, New York, NY 373 p. HB61.S466

Shlain, Leonard. 1991. *Arts & Physics: Parallel Visions in Space, Time, and Light.* Quill, William Morrow, New York, NY 480 p. N70.S48

Simmons, John G. 1996. The Scientific 100. Citadell Press Book, Secaucus, NJ 504 p. Q148.S55

Simon, Herbert A. 1965. *The Shape of Automation.* Harper & Row Publishers, New York, NY 111 p. HD31.S56

Simpson, J. A. and Weiner, E. S. C. 1989 (2nd ed.). *The Oxford English Dictionary.* Clarendon Press, Oxford, England. In 20 volumes PE1625.O87 First edition by James A. H. Murray. 1933. In 12 volumes

Sloan, Harold S. and Zurcher, A. 1970. *A Dictionary of Economics.* Barnes & Noble, New York, NY 520 p. HB61.S54

Smith, David G. (ed.). 1982. *Cambridge Encyclopedia of Earth Sciences.* Cambridge University Press, Cambridge 496 p. QE26.2.C35

Smith, George. 1882 (Founding date) to 1993. *Dictionary of National Biography.* Oxford University Press, Oxford, England. In two volumes: Part I, 1-1900 A.D.; Part II, 1901-1970 A.D. 768 p. in 1993. DA28.D55 (1901-1990). See also Williams, E. T. recent volumes.

Smith, R. R., Lankford, T. G. and Payner, J. N. 1963. *Contemporary Algebra, Book Two.* Harcourt, Brace and World, New York, NY

Sneddon, I. N. 1976. *Encyclopaedic Dictionary of Mathematics for Engineers and Applied Scientists.* Pergamon, New York and Oxford, Cambridge 800 p. TA330.S66

Snyder, L. H. and David, Paul R. 1946 (5th ed.). *The Principles of Heredity.* Heath and Co., Lexington, MA 507 p. QH431.S

Southwell, R. V. 1941. *An Introduction to the Theory of Elasticity for Engineers and Physicists.* Dover Publications, New York, NY 509 p. QA931.S67

Speck, Gerald E. and Jaffe, B. (ed.). 1965. *A Compact Science Dictionary, or A Dictionary of Science Terms.* Fawcett Publications, Inc., Greenwich, CT and Hawthorn Books, New York, NY 272 p. Q123.S6

Spofford, A. R. 1898. See *National Cyclopaedia of American Biography.*

Spotts, M. F. 1948. 1971 (4th ed.). *Design of Machine Elements.* Prentice-Hall, Englewood Cliffs, NJ 620 p. TJ230.S8

Stankus, Tony (ed.). 1991. *Biographies of Scientists for Sci-Tech Libraries.* The Haworth Press, Inc., New York, NY 228 p. Q141.B535

Statt, David. 1982. *Dictionary of Psychology, or Concise Dictionary of Psychology.* Barnes & Noble, New York, NY 132 p. 136 p. BF31.S73

Stedman, Thomas L. 1976 (23 ed.). *Stedman's Medical Dictionary.* The Williams & Wilkins Co., Baltimore, ND 1678 p. R121.S8 See Hensyl, W. R. 1990.

Stenesh, J. 1989 (2nd ed.). *Dictionary of Biochemistry and Molecular Biology.* Wiley Publishing Co., New York, NY 525 p. QP512.S73

Sterling, Keir B., Harmond, Richard P., Cevasco, George A., and Hammond, Lorne, F. 1997. *Biographical Dictionary of American and Canadian Naturalists and Environmentalists.* Greenwood Press, Westport, CT 937 p. QH26.B535

Stevenson, L. Harold and Wyman, Bruce C. 1991. *Dictionary of Environmental Science.* Facts On File, New York, NY 294 p. TD9.S74

Steward, Julian H. 1955. *Theory of Cultural Change: The Methodology of Multi-linear Evolution.* University of Illinois Press, Urbana, IL 244 p. PM101.S785

Stewart, Alfred W. 1919. *Recent Advances in Physical and Inorganic Chemistry.* Longmans, Green and Co., London, England 284 p. QD15.S8

Stewart, Ian. 1998. *Life's Other Secret.* Wiley Publishers, New York, NY 285 p. QH323.5S74

Stewart, Robert W. and Satterly, John. 1946. *Textbook of Heat.* University Tutorial Press, Ltd., London, England 410 p. QC255.S8

Streeter, Victor L. 1961. 1966 (4th ed.). *Handbook of Fluid Dynamics.* McGraw-Hill Book Co., New York, NY various pagings QA911.S83

Suh, Nam P. 1990. *The Principles of Design.* Oxford University Press, New York, NY 401 p. TA174.S89

Suppe, Frederick. 1974. *The Structure of Scientific Theories.* University of Illinois Press, Urbana, IL 682 p. Q174.S8

*Taber's Encyclopedic Medical Dictionary.* See Thomas, C. L. 1989.

Tabor, David. 1961. *The Hardness of Metals.* Clarendon Press, Oxford 175 p. TA460.T27

Tait, Peter G. 1899 (4th ed.). *Properties of Matter.* Adam and Charles Black, Publisher, London, England 340 p. QC171.T13

Talbott, John H. 1970. *A Biographical History of Medicine.* Grune & Stratton, New York, NY 1211. p. R134.T35

Tapley, B. D. 1990 (4th ed.). *Eshbach's Handbook of Engineering Fundamentals.* John Wiley & Sons, New York, NY various paging TA151.E8

Taylor, John G. 1974. 1975. *Black Holes: The End of the Universe.* Random House and Avon Publishers, New York, NY 174 p. QB843.B55T39

Taylor, N. B. (ed.). 1953. *Stedman's Medical Dictionary.* Williams and Wilkins Co, Baltimore, MD. Also Stedman, Thomas L. 1995 (26th ed.). various paging R121.S8 1995

Thewlis, James (ed.). 1961-1964. *Encyclopaedic Dictionary of Physics, etc.* Macmillan, New York, NY. In 9 volumes plus four supplementary volumes in 1966, 1967, 1969, 1971. QC5.E53

Thomas, Clayton L. (ed.). 1989 (16th ed.). *Taber's Cyclopedic Medical Dictionary.* F. A. Davis Co., Philadelphia, PA 2401 p. R121.T144. Previous editions before 1968 by Taber, Clarence W.

Thompson, J. J. 1967. *An Introduction to Chemical Energetics.* Houghton Mifflin Co., Boston, MA 106 p QD501.T64

Thomsen, Julius. 1908. (translated from Danish by Katharine Alice Burke). *Thermochemistry.* Longman's, Green and Co., London, England 495 p. QD511.T47

Thomson, William (Sir) and Tait, Peter G. 1867. 1890 (rev. ed.). *Treatise on Natural Philosophy.* University Press at Cambridge. In two volumes QA805.K31.T7

Thomson, William A. R. 1984 (34th ed.). *Black's Medical Dictionary.* Adams and Charles Black, London. 997 p. R121.B598

Thorne, J. O. and Collocott, T. C. (ed.). 1962. 1984. *Chambers Biographical Dictionary.* Chambers, Edinburgh, Scotland and Cambridge University Press, Cambridge and New York 1493 p. CT103.C4

Thorpe, Jocelyn and Whiteley, Martha, et al. 1940 (4th ed.). *Thorpe's Dictionary of Applied Chemistry.* Longmans, Green, New York and London. In 12 volumes TP9.T72

Thrush, Paul W. (ed.). 1968. *A Dictionary of Mining, Mineral, and Related Terms.* U. S. Bureau of Mines, Washington, DC 1269 p. TN9.T53

*(The) Timetables™ of Science.* See Hellemans, Alexander. 1988. See Bunch, Bryan and Hellemans, Alexander. 1993.

Toback, Ann B. (ed.). 1990. *Concise Dictionary of American Biography.* Charles Scribner's Sons, New York, NY 1536 p. E176.D564

Toffler, Alvin. 1980. 1989. *The Third Wave.* A Bantam Book, New York, NY 537 p. HN17.5T643

Tomkeieff, S. I. 1983. *Dictionary of Petrology.* John Wiley & Sons, New York, NY 680 p. QE423.T65

Toothill, Elizabeth. 1981. 1988. *Dictionary of Biology.* Facts On File, New York, NY 320 p. QH13.T66

Travers, Bridget E. (ed.). 1994. *World of Scientific Discovery.* Gale Research, Detroit, MI 776 p. Q126.W67

Travers, Bridget E. (ed.). 1996. *The Gale Encyclopedia of Science.* Gale Research, Detroit, MI 4136 p. In six volumes Q121.G35

Treybal, Robert E. 1963. *Liquid Extraction.* McGraw-Hill Book Co., New York, NY 666 p. TP155.T7

Trigg, George L. (ed.). 1991. 1996. *Encyclopedia of Applied Physics.* VCH Publishers, Inc., New York, NY. In 14 volumes QC5.E543

Truesdell, Clifford A. 1980. *The Tragicomical History of Thermodynamics, 1822-1854.* Springer-Verlag, New York, NY 372 p. QC311.T83

Truesdell, Clifford A. 1960. Revised by Flugge, Wilhelm. 1962.

Tuma, Jan J. and Walsh, Ronald A. 1970. 1998(4th ed.). *Engineering Mathematics Handbook.* McGraw-Hill Book Co., New York, NY 566 p. TA332.T85

Turkevich, John. 1963. *Soviet Men of Science.* Van Nostrand, Princeton, NJ 441 p. Q141.T83

Turner, Roland and Goulden, Steven. 1981. *Great Engineers and Pioneers in Technology.* St. Martin's Press, New York, NY 488 p. TA139.G7

Tver, David F. 1980. *The Petroleum Dictionary.* Van Nostrand Reinhold, New York, NY 374 p. TN865.T83

Tver, David F. 1981. *Dictionary of Dangerous Pollutants, Ecology, and Environment.* Industrial Press, New York, NY 347 p. TD173.T83

Ulicky, L. and Kemp, T. J. 1992. *Comprehensive Dictionary of Physical Chemistry.* Ellis Horwood and Prentice-Hall, New York, NY 472 p. QD5.C4555

Ushakov, Igor A. and Harrison, Robert A. 1994. *Handbook of Reliability Engineering.* John Wiley & Sons, New York, NY 663 p. TA169.S6713

Uvarov, E. B. and Chapman, D. R. 1964. *Dictionary of Science.* Rev. for 3rd edition by Isaacs, A. 1996. Penguin Books, Baltimore, MD 336 p. Q123.U8

Van Den Broek, J. A. 1942 (2nd ed.). *Elastic Energy Theory.* John Wiley and Sons, New York, NY 298 p. TG260.V22

*Van Nostrand Chemist's Dictionary.* See Honig, J. M. et al. 1953.

*Van Nostrand's Scientific Encyclopedia.* 1947 (2nd ed.). See Considine, D. M. 1976. 1995.

Van Valkenburg, Mac E. (ed.). 1993. 1995 (8th ed.). *Reference Data for Engineers.* SAMS, Prentice Hall Computer, Carmel, IN various paging ISBN 0-672-22753-3

Vecchio, Alfred (del). 1964. *Concise Dictionary of Atomics.* Philosophical Library, New York, NY 262 p. QC772.D4

Vennard, John K. and Street, Robert. 1947. 1954. 1961. 1982 (6th ed.). *Elementary Fluid Mechanics.* John Wiley & Sons, New York, NY 689 p. QA901.V4

Vernoff, Edward and Shore, Rima. 1987. *The International Dictionary of 20th Century Scientists.* New American Library, New York, NY 819 p. CT103.V8

Verplanck, William S. 1957. *A Glossary of Some Terms Used in the Objective Science of Behavior.* A Supplement to Psychological Review 64(6) pt. 2, 42 p., American Psychological Association.

Vinogradov, I. M. (ed.). 1987. *Soviet Mathematical Encyclopaedia.* Kluwer Academic Publishers, Dordrecht, Netherlands. In 10 volumes QA5.M3713 See Hazewinkel, M.

Walker, Peter M. B. (ed.). 1988. *Chambers Science and Technology Dictionary.* Chambers/Cambridge. 1008 p. Q123.C482

Walker, Peter M. B. (ed.). 1989. *Chambers Biology Dictionary.* Chambers/Cambridge. 324 p. QH307.2C33

Wallenquist, Ake. 1966. *Dictionary of Astronomical Terms.* National History Press, Garden City, New York, NY 265 p. QB14.W313

Washburn, Edward W. 1915. *An Introduction to the Principles of Physical Chemistry.* McGraw-Hill Book Co., New York, NY 445 p. QD453.W3

Wasson, Tyler. 1987. *Nobel Prize Winners: Biographical Directory.* H. W. Wilson Co., New York, NY 1165 p. AS911.N9N6

Watters, Carolyn, ed. 1992. *Dictionary of Information Science and Technology.* Academic Press, HBJ, San Diego, CA 300 p. Z1006.W35

Weast, Robert C. 1988 (69th ed.). See Lide, David R.

Weatherall, D. J., Ledingham, J. G. G. and Warrell, D. A.(ed.). 1987 (2nd ed.). 1996 (3rd ed.). *Oxford Textbook of Medicine.* Oxford University Press, Oxford 4376 p. In three volumes RC46.0995

Weaver, Jefferson Hane. 1987. *The World of Physics.* Simon and Schuster, New York, NY. In three volumes QC7.5.W6

*Webster's Biographical Dictionary.* 1959. G. & C. Merriam Co., Springfield, MA 1997 p.; *Webster's New Biographical Dictionary.* 1988. Merriam-Webster, Inc., Publishers, Springfield, MA 1130 p. CT103.W4; Also see *Merriam Webster's Biographical Dictionary. 1995.*

*Webster's Medical Dictionary.* See Merriam-Webster. 1993.

*Webster's New World Dictionary.* 1986. Merriam-Webster Co., Springfield, MA 2642 p. PE1625.W36

*Webster's New World Dictionary of Mathematics.* See Karush, William. 1989.

*Webster's New World/Stedman's Concise Medical Dictionary. 1987.* Williams & Wilkens, Baltimore, MD and Prentice-Hall, New York, NY 838 p. R121.S82 Also known as *Stedman's Concise Medical Dictionary.*

*Webster's Unabridged English Dictionary.* 1983. Senior editor, Jean L. McKechnie. Dorset & Baber, and Simon and Schuster, known as *Webster's New 20th Century Dictionary (2nd ed.).* 2129 p. PE1625.W4

Weik, Martin H. 1996 (3rd ed.). *Communication's Standard Dictionary.* Chapman & Hall, New York, NY 1192 p. TK5102.W437

Weld, LeRoy D., et al. 1937. *Glossary of Physics.* McGraw-Hill Book Co., New York, NY 255 p. QC5.W44

Wells, Dare E. and Slusher, Harold S. 1983. *Physics for Engineering and Science.* Schaum's Outline Series, McGraw-Hill Book Co., New York, NY 359 p. QC21.2.W44

West, Beverly H. et al. 1982. *The Prentice-Hall Encyclopedia of Mathematics.* Prentice-Hall, Englewood Cliffs, NJ 682 p. QA5.P7

Whitaker, Jerry C. 1996. *The Electronics Handbook.* CRC Press and IEEE Press, Boca Raton, FL 2575 p. TK7867.E4244

White, James T. and Company. Beginning in 1898. *The National Cyclopedia of American Biography.* Clifton, NJ. In numerous volumes E176.N27 E176.N283

Whitehead, Alfred N. 1949. 1954. *Science and the Modern World.* Macmillan, New York, NY 304 p. Q175.W65

*Who Was Who (1897-1993).* 1996. A & C Black, Ltd., London. In 11 volumes DA28.W65

*Who's Was Who in American History - Science and Technology.* 1976. *Marquis Who's Who,* Chicago, IL 688 p. Q149.U5W5

*Who's Who in Science, from Antiquity to the Present (Worldwide).* See Debus, A. G. 1968.

*Who's Who of Nobel Prize Winners.* See Schlessinger, B. S. and J. H. 1991 (2nd. ed.).

Wilder, Joseph F. 1967. *Stimulus and Response: Law of Initial Value.* Williams and Wilkins, Baltimore, MD 352 p. QP43.W67S8

Williams, E. T., et al. 1981 and other years. *Dictionary of National Biography.* DA28.D57

Williams, Peter L. 1989. *Gray's Anatomy.* C. Livingstone, Edinburgh, Scotland 1598 p. QM23.2G77

Williams, Roger J. and Lansford, Edwin M. 1967. *The Encyclopedia of Biochemistry.* Reinhold Publishing Co., New York, NY 876 p. QP512.W5

Williams, Trevor I. 1974 (2nd ed.). 1982(3rd ed.). 1994. *A Biographical Dictionary of Scientists.* A & C Black, London, England and HarperCollins Publishers, Glasgow and John Wiley, New York 674 p. Q141.B528

Wilson, John R. and the editors of Life. 1964. *The Mind.* Time, Inc., New York, NY 200 p. BF161.W49

Winfrid, Frances. 1961. *Coal.* Edward Arnold, London (not listed in NUC or LC)

Wolko, Howard S. 1981. *In Cause of Flight.* Smithsonian Inst., Washington, DC 121 p. TL539.W67

Wolman, Benjamin B. (ed.). 1973. 1989 (2nd ed.). *Dictionary of Behavioral Science.* Academic Press, New York, NY 370 p. BF31.W64; BF31.D48

Woodbury, Angus M. 1954. *Principles of General Ecology.* Blakiston Co., New York, NY 503 p. QH541.W6

World Book. 1989. *Men and Women of Science, vol. 8.* World Book, Inc. Chicago, IL 128 p. AE5.W55

*World Book Encyclopedia.* See Zeleny, R. O. 1990.

Young, Sidney. 1918. *Stoichiometry.* Longman's, Green, and Co., London, England 363 p.

Zeleny, Robert O. 1990. *The World Book Encyclopedia.* World Book, Inc., Chicago, IL. In 22 volumes AE5.W55

Znaniecki, Florian. 1967. *The Laws of Social Psychology.* Russell and Russell, New York, NY 320 p. HM251.Z6

Zusne, Leonard. 1987. *Eponyms in Psychology.* Greenwood Press, New York and London 339 p. BF31.Z87